CONTINUUM MECHANICS OF VISCOELASTIC LIQUIDS

CONTINUUM MECHANICS OF VISCOELASTIC LIQUIDS

R. R. HUILGOL

A HALSTED PRESS BOOK

HINDUSTAN PUBLISHING CORPN. (I)—DELHI

JOHN WILEY & SONS, INC.—NEW YORK

CONTINUUM MECHANICS OF
VISCOELASTIC LIQUIDS

Distributed in Western Hemisphere, Europe and Australia By Halsted Press, A Divison of John Wiley and Sons, Inc.,
New York

Library of Congress Cataloging in Publication Data

Huilgol, R R 1939-
Continuum mechanics of viscoelastic liquids.
Bibliography : p.
Includes index.
1. Viscous flow. 2. Continuum mechanics.
3. Liquids. I. Title.
QA929. H78 1975 532'. 0533 73-14413
ISBN 0-470-42043-X

Printed in India by
Hindustan Publishing Corporation (India), Delhi 110007 (India)

Dedicated to the memory
of
my grandfather
Budihal Ananthachar
(1896-1966)

PREFACE

Nonlinear continuum mechanics has seen a successful renascence since the second world war. Many fascinating theories have been put forward and much deeper understanding of diverse phenomena has been achieved. This book is a systematic outline of the issues involved, especially those relevant in the study of the flows of non-Newtonian fluids. It is hoped that the reader will gain, not only an overview of the subject, but also an entry to the vast terrain of unsolved problems.

The book is developed around the concept of the simple fluid, because the simple fluid and its parent, the simple material, offer a very fruitful approach to the study of nonlinear theories of mechanical behaviour. The book is written for people who use these materials—the engineers and rheologists—as well as for persons interested purely in academic enquiry.

When working with the simple fluid, it is first necessary to calculate the strain history of the material particle, and the steps by which this is done are explicated in the first two chapters. Examples illustrating the exact and approximate computations of the history are given. Other results concerning RIVLIN-ERICKSEN tensors, motions with constant stretch history, static continuation, jump history, and so on, are also included here.

The third chapter introduces observer transformations, two theorems on steady motions and the concept of symmetry, through the change of the local configuration. The fourth chapter is on balance laws, and some recent results of A. E. GREEN, M. E. GURTIN, W. NOLL and R. S. RIVLIN appear in it.

The fifth chapter is a full treatment of the constitutive equation of the simple fluid, and in the two appendices to this chapter, I have presented a summary of the methods of A. S. LODGE, J. G. OLDROYD, R. S. RIVLIN and compared them with that of W. NOLL. As a consequence of this comparison, a slightly different approach to the formulation of constitutive equations has been proposed here which, in the event, was found to have many similarities with the one followed by J. L. ERICKSEN.

The sixth chapter makes use of the objectivity restriction on the simple fluid. The enormous power of this restriction yields the form of the stress tensor with ease, e.g. in viscometric flows. This chapter includes some general solutions as well ; these are expanded further in the chapter seven and applied to viscometric flows, pipe flows, and other motions with constant stretch history. This chapter brings to an end the known solutions to problems solvable at the utmost level of generality.

Chapter eight develops approximate constitutive equations and includes integral models, order fluids, BKZ theory, additive functional theory, OLDROYD 6- and 8-constant models and so on. Chapter nine is on nearly viscometric flows and it is here that we begin to see the interplay between theory and experiment in deciding which of the available single integral models is a better approximation to the simple fluid. Chapter ten is on perturbation about the rest state and includes flows over and along slots, trough flow, COUETTE flow with a free surface, inertial corrections, initial value problems, and so on.

The last chapter is a detailed discussion of the experimental configurations (and their theoretical justifications) to measure viscosity, normal stress differences, extensional viscosity and to predict die swell.

In writing this book, I have chosen those topics which emphasize the peculiarities of viscoelastic fluids, their 'non-Newtonianness'. In addition, if a general solution is available, I have not included the numerous special models for which the same problem can be solved. I have thus sought generality, and depth and breadth at that level rather than particularity at the lower end of the spectrum. Only when the general solution is not available, have approximate solutions and equations been included. It is for this kind of reason that stability theory has been passed over for the subject lacks generality at the moment. However, it is hoped to include it in a second volume, along with a more extended treatment of initial value problems and finite element methods.

Although new discoveries have been made and published since this book went to press, some of these were communicated to me by their authors in the intervening period, and these materials have been incorporated wherever possible. Of course, not all errors and omissions could be rectified, but I hope only the minor ones remain.

For acting as my liaison with the publishers and accommodating all my wishes for changes, I must thank Mr. J. K. JAIN of HINDUSTAN PUBLISHING CORPORATION, New Delhi. On his initiative, this book has become a joint venture with the Halsted Press in New York and, I hope, will serve as a model for international cooperation in the technical book publishing field.

This book was begun at the United Engineering Conference at the Asilomar Grounds, Pacific Grove, California in February 1971 and finished in Chicago in June 1973. During the time, the facilities of the Mechanics, Mechanical and Aerospace Engineering Department at the Illinois Institute of Technology were available to me, and this book would not have been finished without the help of Mrs. ANGIE LUCAS, and the support of the chairman, Dr. S. KUMAR. In addition, my summers were left free for research and writing, because of a research grant from the National Science Foundation, held jointly with my colleague Professor B. BERNSTEIN. This help is deeply appreciated.

Over the years I have participated in many discussions and correspondence with a number of people about the subject matter covered here. The most fruitful of these have been with Professors B. BERNSTEIN, R. B. BIRD, J. L. ERICKSEN, M. W. JOHNSON, JR., A. S. LODGE, A. B. METZNER, A. C. PIPKIN,

R. S. RIVLIN, R. I. TANNER and C. TRUESDELL. In addition, I am particularly indebted to Professor R. I. TANNER for fulfilling, if somewhat haphazardly, the role of guru, and to GLYNN for fulfilling, in a similar manner, the role of wife and companion.

Finally, this book is dedicated to my mother's father. He brought me up in a small town in India, called Hospet, with love and affection. He was a lawyer, a politician, a Sanskrit scholar and above all, someone I miss very much.

School of Mathematical Sciences,
Flinders University of South Australia,
Bedford Park
June 30, 1975

R. R. HUILGOL

NOTATION

The notations of tensor calculus, of linear vector and matrix algebra, and of GIBBS' vector analysis will be employed. Physical components will be defined in orthogonal systems only and the symbol $\langle \cdot \rangle$ will be used for this purpose. Scalars and scalar fields are usually denoted by Greek letters and the first three letters of the English alphabet. Vectors are written as **x**, **X**, **u**, **v**, **w**, $\boldsymbol{\xi}$ or $\boldsymbol{\omega}$. Second-order tensors are designated by upper case letters, except for X. In a second-order tensor, the first index refers to the row and the second to the column. Bold face notation means that the quantity is either a vector or a second-order tensor. A function of a second-order tensor(s) is denoted by $\mathbf{f}(\cdot)$, or $\mathbf{g}(\cdot)$. The superscript T denotes the transpose and **1** the identity transformation or the unit tensor.

We now list an index of frequently used symbols.

Symbol	*Name*	*Place of definition or first occurrence*
$\mathbf{a}$, a^i	Acceleration field	(1.6)
$\mathbf{A}_n$, $(A_n)_{ij}$, $\overset{(n)}{A}_{ij}$	n-th RIVLIN-ERICKSEN tensor	(4.1)
$\mathbf{b}$, b^i	Body force field	(17.1)
$\mathbf{B}$, B^{ij}	Left CAUCHY-GREEN tensor	(1.11)
$\mathbf{C}$, $C^{\alpha\beta}$	Right CAUCHY-GREEN tensor	(1.12)
$\mathbf{C}_t(\tau)$	Right relative CAUCHY-GREEN tensor	(1.15)
$\mathbf{C}_t(t-s)$, $C_t(t-s)\langle ij\rangle$	Strain history	(1.16)
$°\mathbf{C}_t(t-s)$, $°\mathbf{C}$	Viscometric strain history	(5.13)
$\mathbf{D}$, $D\langle ij\rangle$	Rate of deformation tensor	Page 28
$\dfrac{d}{dt}$	Material derivative operator	Page 2
$\mathbf{E}$	Infinitesimal strain tensor	(5.8)
$\mathbf{E}_t(t-s)$	Perturbation on a viscometric strain history	(5.13)

$\mathbf{F}$	Deformation gradient	(1.9)
$\mathbf{F}_t(t-s)$	Relative deformation gradient	(1.13)
g_{ij}	Metric tensor in spatial coordinates	(1.12)
$G^{\alpha\beta}$	Metric tensor in material coordinates	(1.11)
$G(s)$	Relaxation modulus	(34.*b*5)
$\mathbf{G}(s)$	Difference history	(24.17)
G'	Storage modulus	(34.*b*7)
G''	Loss modulus	(34.*b*8)
$G'_{\parallel}, G''_{\parallel}$	Dynamic moduli in in-line oscillations	(39.8)
$G'_{\perp}, G''_{\perp}$	Dynamic moduli in transverse oscillations	(39.11)
$\mathbf{L}$	Velocity gradient	(1.8)
$\mathbf{L}_n$	*n*-th acceleration gradient	(4.12)
n_i	Normal stress coefficients, $i = 1, 2, 3$	(26.69)
$\mathbf{n}$	Unit normal	Sec. 17
N_1	First normal stress difference	(26.43)
N_2	Second normal stress difference	(26.44)
$N'_{1\parallel}, N''_{1\parallel}$	Dynamic first normal stress difference in in-line oscillations	(39.9)
$N'_{2\parallel}, N''_{2\parallel}$	Dynamic second normal stress difference in in-line oscillations	(39.10)
p	Pressure field	(24.30)
p_e	Pressure error	(41.40)
$\mathbf{R}$	Rotation tensor	(1.10)
$\mathbf{R}_t(t-s)$	Relative rotation tensor	(23.9)
$\overset{(i)}{\mathbf{S}}$	Slow flow stresses	(41.2)

$\mathbf{t}_{(\mathbf{n})}$	Stress vector	(17.1)
$\mathbf{T}$	CAUCHY stress tensor	(17.14)
$\mathbf{T}_E$	Extra stress tensor	(25.4)
$\mathbf{U}$	Right stretch tensor	(1.10)
$\mathbf{U}_t(t-s)$	Right relative stretch tensor	(23.9)
$\mathbf{V}$	Left stretch tensor	(1.10)
$\mathbf{v}$	Velocity field	(1.6)
$\mathbf{W}$	Spin tensor	(3.21)
$\mathbf{x}, x^i$	Spatial position of a particle	(1.1)
$\mathbf{X}, X$	Material position of a particle	(1.1)
$F, \bar{F}$	Constitutive operator of a simple material	Sec. 22
$G, H, \tilde{G}, \tilde{H}$	Constitutive operators of a simple fluid	Sec. 24
I, II, III	Invariants of a second-order tensor	(7.3)
$\mathcal{B}$	Body in current configuration	Page 1
$\mathcal{B}_{\mathcal{R}}$	Body in reference configuration	Page 1
$\mathcal{G}_{\mathcal{P}}$	$\mathcal{P}$-symmetry group	Page 71
$\mathcal{O}$	Orthogonal group	Page 92
$\mathcal{U}$	Unimodular group	Page 72
δS_{ijkl}	Nearly viscometric flow linear functionals	(38.6)
α_1, α_{-1}	Material functions in the additive functional theory	(36.20)
β, γ	Second-order fluid constants	(35.24)
$\dot{\varepsilon}$	Strain rate in extensional flow	Sec. 49
η	Viscosity in viscometric flows	(26.73)
η_0	Zero shear viscosity	(34.*c*5)

η_e	Extensional viscosity	(49.5)
η'	Dynamic viscosity	(34.*b*8)
$\eta'_{\parallel}$	Dynamic viscosity in in-line oscillations	(39.12)
$\eta'_{\perp}$	Dynamic viscosity in transverse oscillations	(39.13)
η^*	Complex viscosity	(34.*b*10)
$\varkappa$	Rate of shear in viscometric flows	Page 41
$\mu(s)$	Memory function in finite linear viscoelasticity	(34.10)
τ	Shear stress function in viscometric flows	(26.42)
ρ	Density field	Page 74
$\boldsymbol{\xi}$	Position of a particle at time $t - s, 0 \leqslant s < \infty$	(1.16)
$\boldsymbol{\omega}$	Vorticity vector	Page 11

In addition, some symbols, e. g., φ, ψ, are used in two different contexts. For instance, ψ represents the stream function and the modified pressure; but, this double usage does not occur at the same place and so no confusion should arise.

Finally, two works are cited by the letters CFT and NFTM. The full references to these are :

CFT : The Classical Field Theories, by C. TRUESDELL and R. TOUPIN, **In:** *Encycl. of Physics*, Vol. III/1, Ed. S. FLÜGGE, Springer-Verlag, New York, 1960.

NFTM : The Nonlinear Field Theories of Mechanics, by C. TRUESDELL and W. NOLL, **In** : *Encycl. of Physics*, Vol. III/3. Ed. S. FLÜGGE, Springer-Verlag, New York, 1965.

ACKNOWLEDGMENT

I wish to thank the following authors, editors, publishers and societies for permission to reproduce the following figures and a table in the text:

Figure	*Journal or Source*	*Authorization*
48.7, 49.2	*American Institute of Chemical Engineers Journal*, **18** (1972)	American Institute of Chemical Engineers
44.1, 50.1, 50.2	*Archive for Rational Mechanics and Analysis*, **49** (1973), **56** (1974)	Springer-Verlag
17.2	*Handbuch der Physik*, **VI** a/21	Springer-Verlag
2.2	*Journal of Fluid Mechanics* **43** (1970)	Cambridge University Press
39.4	*Journal of Physics, A : General Physics*, **4** (1971)	Institute of Physics and Professor K. Walters
2.1, 46.1	*Journal of Physics, D : Applied Physics*, **2** (1969), **4** (1971)	Institute of Physics, Dr. N. D. Waters and Professor K. Walters
50.3	*Journal of Polymer Science, Part A-2*, **8** (1970)	John Wiley and Sons, Inc.
48.6	*Mechanics Today*, **1** (1972)	Pergamon Press, Professors A. C. Pipkin and R. I. Tanner
47.2, 47.3	*Polymer Systems—Deformaiion and Flow*	Macmillan and Co., and Professor K. Walters
42.2	*Proceedings of the International Symposium on Second Order Effects, Haifa*	Pergamon Press and Professor R. S. Rivlin
31.1	*Quarterly of Applied Mathematics*, **29** (1971)	Professor W. Frieberger, the Editor
39.2, 39.3, 46.2, 49.1	*Rheologica Acta*, **4** (1965), **6** (1967), **7** (1968), **10** (1971)	Dr. Dietrich Steinkopff Verlag
41.1	*Society for Industrial and Applied Mathematics Journal of Applied Mathematics*, **24** (1973)	Society for Industrial and Applied Mathematics
39.1, 43.1, 47.1, 48.1, 48.2, 48.3, 48.5	*Transactions of the Society of Rheology*, **13** (1969), **14** (1970), **15** (1971), **18** (1974)	John Wiley and Sons, Inc. and Professor R. R. Myers, the Editor
48.4	. . .	Dr. H. van Es
49.1 (Table)	*American Institute of Chemical Engineers Journal*, **18** (1972)	American Institute of Chemical Engineers

CONTENTS

CHAPTER 1

KINEMATICS OF FLUID FLOW

The last twenty-five years of research, reinforcing much earlier work, indicates fairly conclusively that the most relevant measures of the kinematics of fluid flow are the strain history experienced by the fluid particle and its instantaneous derivatives. The calculation of these various measures is based, not on the absolute motion of a particle in space, but on the motion of the immediate vicinity of the particle relative to it. The methodology of calculating the strain history from the path lines derived from the velocity field, the instantaneous derivatives of this history and the effect of linearization, static continuation and jumps on the strain history therefore require our attention. This chapter treats these matters in a detailed manner.

In addition, this chapter provides a brief introduction to the methodology of attacking problems in mathematical physics. While exact solutions are always preferable and will therefore be sought eagerly, in those cases where exact solutions cannot be derived it is possible to linearize the problem to obtain approximate solutions either in terms of an important parameter or about the origin (rest state) or about an arbitrary ground state. Examples of such techniques are given.

To make the book self-contained and for ease of reference, useful results from tensor analysis are listed in the Appendix and can be consulted directly.

1. Kinematics

Let X be a particle of a body (a continuous medium) $\mathcal{B}$, and let X occupy a point in the three-dimensional Euclidean space E^3 at a fixed instant. We shall call the configuration occupied by the body $\mathcal{B}$, the *reference configuration* $\mathcal{B}_R$, and call the coordinates X^α, of the point where X is at that fixed instant, the *material coordinates* of the particle. The position vector of X will be denoted by $\mathbf{X}$.

Let X trace out a path in E^3. We shall denote this path by the curve

$$\mathbf{x} = \mathbf{M}(\mathbf{X}, t), \qquad x^i = M^i(X^\alpha, t) \tag{1.1}$$

in the space. Here t, the time coordinate, acts as a parameter and as t varies from $-\infty < t < \infty$, the function $\mathbf{M}(.\,, t)$ describes the path. Now, we are interested in the motion of the whole body $\mathcal{B}$ rather than a single particle. So, the domain of $\mathbf{M}$ is the Cartesian product $\mathcal{B}_R \times (-\infty, \infty)$. In other words, changing $\mathbf{X}$ in $\mathbf{M}(\mathbf{X}, .)$ gives us the path of another particle of $\mathcal{B}$, so that at time t, $\mathbf{M}(\mathbf{X}, t)$, $\mathbf{X} \in \mathcal{B}_R$, gives us the *spatial configuration* of $\mathcal{B}$.

We demand that **M** be continuously differentiable twice with respect to **X** and t, or $\mathbf{M} \in C^{2,2}$, though on the boundary **M** may be $C^{1,1}$. In this volume, the propagation of singularities of any kind is not treated and hence the motion **M** is required to have a continuous non-singular gradient with respect to **X**, or $\nabla_{\mathbf{X}}\mathbf{x}$ is nonsingular everywhere. One notes that this is equivalent to

$$\mathbf{X} = \mathbf{M}^{-1}(\mathbf{x}, t). \tag{1.2}$$

The velocity **v** and the acceleration **a** of a particle are defined through the functions:

$$\hat{\mathbf{v}} = \frac{\partial}{\partial t}\mathbf{M}(\mathbf{X}, t), \qquad \hat{v}^i = \left.\frac{\partial M^i}{\partial t}\right|_{\mathbf{X}}, \tag{1.3}$$ [1]

$$\hat{\mathbf{a}} = \frac{\partial^2}{\partial t^2}\mathbf{M}(\mathbf{X}, t), \qquad \hat{a}^i = \left.\frac{\partial^2 M^i}{\partial t^2}\right|_{\mathbf{X}}. \tag{1.4}$$

Hence, if the velocity **v** is defined over the body by the function $\hat{\mathbf{v}}(\mathbf{X}, t)$, the acceleration is given by

$$\mathbf{a} = \frac{\partial}{\partial t}\hat{\mathbf{v}}(\mathbf{X}, t). \tag{1.5}$$

However, we usually find it more convenient to express the velocity of X at time t in terms of the coordinates it occupies at time t or as a spatial field $\mathbf{v}(\mathbf{x}, t)$.

Let $$\bar{\mathbf{v}}(X, t) = \hat{v}^\alpha(X^\beta, t)\,\mathbf{G}_\alpha = v^i(x^j, t)\,\mathbf{g}_i, \tag{1.6}$$

where $\mathbf{G}_\alpha$ and $\mathbf{g}_i$ are the base vectors at X^α and x^i, respectively. Then (1.6) becomes

$$a^i = \frac{\partial v^i}{\partial t} + v^i_{\ ;j}\, v^j, \tag{1.7}$$

where ; denotes the covariant derivative. We call the derivative $(d/dt)(\cdot) = (\partial(\cdot)/\partial t) + v^i(\cdot)_{;i}$, the *material derivative* of $(\cdot)$; also, the notation $(d/dt)(\cdot) \equiv \dot{(\cdot)}$ will be employed wherever necessary.

A motion is said to be *steady* if the velocity field $\mathbf{v} = \mathbf{v}(\mathbf{x})$, or it is independent of t when expressed according to $(1.6)_2$. For nonsteady motions, the term $\partial\mathbf{v}/\partial t$ represents the *local acceleration*, while $\nabla\mathbf{v}\cdot\mathbf{v}$ represents the *convected terms.* We shall denote the velocity gradient $\nabla\mathbf{v}$ by **L**, or

$$L^i_j = v^i_{,j}, \qquad L_{ij} = v_{i;j}. \tag{1.8}$$

The gradient of **x** with respect to **X** is called the *deformation gradient*. We shall write:

$$\mathbf{F} = \nabla_{\mathbf{X}}\mathbf{x}\,; \qquad F^i_\alpha = \frac{\partial x^i}{\partial X^\alpha} = x^i_{,\alpha}. \tag{1.9}$$

1. BIERMANN has classified in [1971 : 5] the many types of velocity fields, e.g., irrotational, accelerationless, etc., by the use of Venn-Euler and Hasse diagrams.

By the fact that det **F** (det=determinant) is the Jacobian of the mapping, and hence a measure of the ratio of the volumes in the **X** and **x** spaces, we demand that $|\det \mathbf{F}| > 0$. This ensures that the mapping (1.1) is not degenerate, that is, the conservation of mass is assured, and that the inversion in (1.2) is possible everywhere.

Since **F** is nonsingular, it may be expressed as the product of a positive-definite and symmetric tensor and an isometric tensor. By the polar decomposition theorem[2]

$$\mathbf{F} = \mathbf{R}\mathbf{U} = \mathbf{V}\mathbf{R}, \tag{1.10}$$

where **U** and **V** are positive-definite and symmetric and **R** is the orthogonal tensor representing the isometry. Note that $\det \mathbf{R} = \pm 1$. The two Cauchy-Green tensors **B** and **C** are defined through

$$\mathbf{B} = \mathbf{V}^2 = \mathbf{F}\,\mathbf{F}^T, \qquad B^{ij} = G^{\alpha\beta}\, x^i_{,\alpha}\, x^j_{,\beta}, \tag{1.11}$$

$$\mathbf{C} = \mathbf{U}^2 = \mathbf{F}^T\,\mathbf{F}, \qquad C_{\alpha\beta} = g_{ij}\, x^i_{,\alpha}\, x^j_{,\beta}. \tag{1.12}$$

In (1.11), $G^{\alpha\beta}$ are the contravariant components of the metric tensor in the material coordinate system X^α, while in (1.12), g_{ij} are the covariant components of the metric tensor in the coordinate frame x^i.

If one assigns the position coordinates of the particle at time $t - s$ to be ξ^α, and recalls that x^i are the coordinates at time t, one can define the relative deformation gradient $\mathbf{F}_t\,(t-s)$ as

$$\mathbf{F}_t\,(t-s) = \nabla\, \boldsymbol{\xi}, \qquad [F_t(t-s)]^\alpha_i = \frac{\partial \xi^\alpha}{\partial x^i}. \tag{1.13}$$

By the chain rule,

$$\frac{\partial \xi^\alpha}{\partial x^i} = \frac{\partial \xi^\alpha}{\partial X^\beta}\,\frac{\partial X^\beta}{\partial x^i}, \qquad \mathbf{F}_t(t-s) = \mathbf{F}(t-s)\,\mathbf{F}(t)^{-1}, \tag{1.14}$$

where $\mathbf{F}(t)$ represents the deformation gradient at time t with respect to the fixed reference position **X**. The most useful strain measure in viscoelastic fluid flow is the right relative Cauchy-Green strain tensor $\mathbf{C}_t\,(\tau)$ defined through

$$\mathbf{C}_t\,(\tau) = \mathbf{F}_t(\tau)^T\,\mathbf{F}_t(\tau), \tag{1.15}$$

$$[C_t(t-s)]_{ij} = g_{\alpha\beta}\,\xi^\alpha_{,i}\,\xi^\beta_{,j}. \tag{1.16}$$

Note that $g_{\alpha\beta}$ are now measured in the ξ^α coordinate system at the point $\boldsymbol{\xi}$ in E^3, while the covariant components $[C_t(t-s)]_{ij}$ are calculated in the x^i frame.

2. Path Lines

If we consider $\boldsymbol{\xi}$ as a function of **X** and $\tau = t - s$, then

$$\frac{d}{d\tau}\,\boldsymbol{\xi}\,(\mathbf{X}, \tau) = \hat{\mathbf{v}}(\mathbf{X}, \tau) \tag{2.1}$$

2 For a proof, see MARTIN and MIZEL [1966 : 12], 291-293.

implies

$$\frac{d}{ds}\,\boldsymbol{\xi}\,(\mathbf{X},\,t-s) = -\,\hat{v}(\mathbf{X},\,t-s) = -\,\mathbf{v}[\,\boldsymbol{\xi}\,(t-s),\,t-s]. \tag{2.2}$$

In writting (2.2), we have expressed **X** as a function of $\boldsymbol{\xi}$ and $t-s$. Since, at $s = 0$,

$$\boldsymbol{\xi}\,(\mathbf{X},\,t-s)\,|_{s=0} = \mathbf{x}\,(\mathbf{X},\,t), \tag{2.3}$$

these may be used as the initial conditions to integrate the differential equation $(2.2)_2$. This integration yields the *path lines* (on suppressing the dependence on **X**) *in history format*:

$$\boldsymbol{\xi} = \mathbf{x} - \int_0^s \mathbf{v}[\,\boldsymbol{\xi}\,(v-\sigma),\,t-\sigma]\,d\sigma, \tag{2.4}$$ [3]

$$\xi^i = x^i - \int_0^s v^i[\xi^j(t-\sigma),\,t-\sigma]\,d\sigma. \tag{2.5}$$

As an example, let the velocity field at time t be steady and be given in component form in a curvilinear orthogonal coordinate system by

$$\dot{x}^1 = v^1 = 0, \qquad \dot{x}^2 = v^2 = v(x^1), \qquad \dot{x}^3 = v^3 = w(x^1), \tag{2.6}$$

where $v(.)$ and $w(.)$ are smooth functions of the coordinate x^1. In other words, at time $t-s$,

$$\dot{\xi}^1 = 0, \qquad \dot{\xi}^2 = v(\xi^1), \qquad \dot{\xi}^3 = w(\xi^1). \tag{2.7}$$

In (2.6) and (2.7) the superposed dots denote differentiation with respect to t. Expressing this differentiation with respect to s, one obtains

$$\frac{d}{ds}\,\xi^1 = 0, \qquad \frac{d}{ds}\,\xi^2 = -\,v(\xi^1), \qquad \frac{d}{ds}\,\xi^3 = -\,w(\xi^1). \tag{2.8}$$

Integration of (2.8) leads to the path lines:

$$\begin{aligned}\xi^1 &= x^1,\\ \xi^2 &= x^2 - \int_0^s v(x^1)\,d\sigma = x^2 - sv(x^1),\\ \xi^3 &= x^3 - \int_0^s w(x^1)\,d\sigma = x^3 - sw(x^1).\end{aligned} \tag{2.9}$$ [4]

Corresponding to these path lines the relative deformation gradient has the components $[F_t(t-s)]_i^\alpha$ given by

$$[F_t(t-s)]_i^\alpha = \begin{pmatrix} 1 & 0 & 0 \\ -sv' & 1 & 0 \\ -sw' & 0 & 0 \end{pmatrix}, \qquad v' = \frac{dv}{dx^1}, \qquad w' = \frac{dw}{dx^1}. \tag{2.10}$$

3. The crucial point to remember is that the integrand **v** is a function of $\boldsymbol{\xi}\,(\mathbf{X},\,t-\sigma)$ and $t-\sigma$, $0 \leqq \sigma \leqq s$. The argument $\boldsymbol{\xi}\,(\mathbf{X},\,t-\sigma)$ should not be confused with $\boldsymbol{\xi}\,(\mathbf{X},\,t-s)$, and in the integration, it is the dependence of **v** on the points in space occupied by X during $t-s$ to t, which creates analytical difficulty.

4. Since $\xi^1 \equiv x^1$, $v[\xi^1(t-\sigma)] \equiv v(x^1)$, etc.

As a second example, consider the velocity field in a Cartesian coordinate system defined according to

$$\begin{aligned} \dot{x} &= \varkappa y, \\ \dot{y} &= 0, \\ \dot{z} &= \frac{Ay\omega}{h} \cos\omega t, \end{aligned} \tag{2.11}$$

where $\varkappa$, A, ω and h are constants. By letting (ξ, η, ζ) be the coordinates of the particle at time $t - s$, one writes (2.11) as

$$\begin{aligned} \frac{d\xi}{ds} &= -\varkappa\eta, \\ \frac{d\eta}{ds} &= 0, \\ \frac{d\zeta}{ds} &= -\frac{A\eta\omega}{h} \cos\omega\,(t-s). \end{aligned} \tag{2.12}$$

The path lines corresponding to (2.12) are

$$\begin{aligned} \xi &= x - \varkappa s y, \\ \eta &= y, \\ \zeta &= z - \int_0^s \frac{Ay\omega}{h} \cos\omega(t-\sigma)\, d\sigma \\ &= z + \frac{Ay}{h}\,[\sin\omega(t-s) - \sin\omega t]. \end{aligned} \tag{2.13}$$

The calculation of path lines is not always as direct and straightforward as the above two examples may suggest. As a more difficult exercise, consider the velocity field

$$\dot{x}^1 = 0, \qquad \dot{x}^2 = -cx^2 + ex^3, \qquad \dot{x}^3 = fx^2 + cx^3, \tag{2.14}$$

where c, e and f are constants obeying the condition

$$c^2 + ef = 0. \tag{2.15}$$

Expressing (2.14) in terms of ξ^i, one obtains

$$\begin{aligned} \frac{d\xi^1}{ds} &= 0, \\ \frac{d\xi^2}{ds} &= c\xi^2 - e\xi^3, \\ \frac{d\xi^3}{ds} &= -f\xi^2 - c\xi^3. \end{aligned} \tag{2.16}$$

These are to be integrated and, on doing so, yield

$$\begin{aligned} \xi^1 &= x^1, \\ \xi^2 &= x^2 + c\int_0^s \xi^2 d\sigma - e\int_0^s \xi^3 d\sigma, \\ \xi^3 &= x^3 - f\int_0^s \xi^2 d\sigma - c\int_0^s \xi^3 d\sigma. \end{aligned} \tag{2.17}$$

Multiply $(2.17)_2$ by c and $(2.17)_3$ by e and subtract one from the other. We get, on using (2.15), the equation

$$c\xi^2 - e\xi^3 = cx^2 - ex^3. \tag{2.18}$$

Introducing this into $(2.17)_2$ leads to

$$\xi^2 = x^2 + \int_0^s (cx^2 - ex^3)\, d\sigma = x^2 + s(cx^2 - ex^3). \tag{2.19}$$

Alternatively, multiplication of $(2.17)_2$ by f and $(2.17)_3$ by c and adding leads to

$$f\xi^2 + c\xi^3 = fx^2 + cx^3. \tag{2.20}$$

On substitution of (2.20) into $(2.17)_3$, we arrive at

$$\xi^3 = x^3 - s(fx^2 + cx^3). \tag{2.21}$$

Hence, $(2.17)_1$, $(2.19)_2$ and (2.21) are the path lines for the velocity field defined by (2.14) and (2.15).

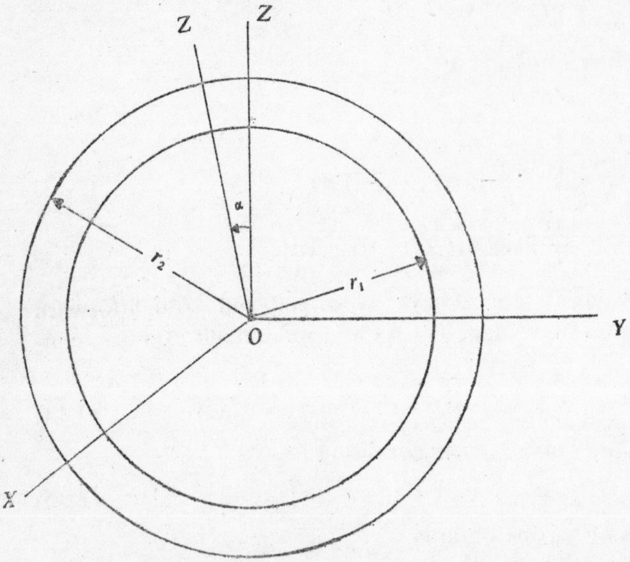

Fig. 2.1. **The KEPES apparatus**

So far, the path lines have been obtained by an exact integration procedure. However, there are instances when a solution is sought in terms of a certain parameter (say, α) and the solution is to be linear in this parameter α; in other words, it is assumed that α is so small that terms involving α^2 and higher powers of α are to be ignored. Such a problem occurs with the following velocity field[5] in spherical polar coordinates (see Figure 2.1):

$$v^r = 0, \qquad v^\theta = -\alpha\Omega\lambda\left(1 - \frac{r_1^3}{r^3}\right)\sin\varphi,$$

$$v^\varphi = \Omega - \alpha\Omega\lambda\left(1 - \frac{r_1^3}{r^3}\right)\cot\theta\cos\varphi. \tag{2.22}$$

5. JONES and WALTERS [1969 : 10].

In (2.22), (r, θ, φ) are the spherical coordinates with φ representing the "longitudinal" angle and θ the angle from the "north pole"; r_1 is the radius of the inner sphere which rotates about the axis $\theta = 0$ (or OZ) with an angular velocity Ω. The outer sphere of radius r_2 rotates with the same angular velocity about the axis $O\bar{Z}$, inclined to OZ at an angle α. The axis $O\bar{Z}$ lies in the OXZ plane. Finally, λ is the parameter defined through

$$\lambda = r_2^3 /(r_2^3 - r_1^3). \tag{2.23}$$

Letting (ξ, η, ζ) be the coordinates of a particle at $t - s$, the particle being at (r, θ, φ) at time t, we have

$$\begin{aligned} \xi &= r, \\ \eta &= \theta + \int_0^s \alpha\Omega\lambda\left(1-\frac{r_1^3}{\xi^3}\right) \sin \zeta \, d\sigma, \\ \zeta &= \varphi - \int_0^s \left\{\Omega-\alpha\Omega\lambda\left(1-\frac{r_1^3}{\xi^3}\right) \cot \eta \cos \zeta\right\} d\sigma. \end{aligned} \tag{2.24}$$

Here $(2.24)_2$ and $(2.24)_3$ are coupled together and in a complicated way. Consider $(2.24)_2$ first. Using $(2.24)_3$, $\sin \zeta$ at time $t - s$ can be expressed as

$$\sin \zeta = \sin [\varphi - \Omega s + O(\alpha)], \tag{2.25}$$

where $O(\alpha)$ is a term of order α. Hence at time $t - \sigma$,

$$\begin{aligned} \sin \zeta &= \sin (\varphi - \Omega\sigma) \cos [O(\alpha)] + \cos (\varphi-\Omega\sigma) \sin [O(\alpha)] \\ &\approx \sin (\varphi-\Omega\sigma) + O(\alpha). \end{aligned} \tag{2.26}$$

In writing (2.26), we have kept in mind that α occurs in $(2.24)_2$ already and is very small. Moreover, we also have

$$\cos [O(\alpha)] = 1 - O(\alpha^2), \qquad \sin [O(\alpha)] = O(\alpha). \tag{2.27}$$

On substituting (2.26), $(2.24)_2$ becomes

$$\begin{aligned} \eta &= \theta + \alpha\lambda \left(1-\frac{r_1^3}{r^3}\right) \cos (\varphi-\Omega\sigma) \Big|_0^s \\ &= \theta + \alpha\lambda \left(1-\frac{r_1^3}{r^3}\right) \left[\cos \varphi(\cos \Omega s-1)+\sin \varphi \sin \Omega s\right]. \end{aligned} \tag{2.28}$$

Next, in $(2.24)_3$ we have at time $t - \sigma$:

$$\cot \eta \approx \cot \theta + O(\alpha); \quad \cos \zeta \approx \cos (\varphi-\Omega\sigma) + O(\alpha). \tag{2.29}$$

From (2.29) follows

$$\begin{aligned} \zeta &= \varphi - \Omega s + \alpha\Omega\lambda \left(1-\frac{r_1^3}{r^3}\right) \int_0^s \cot \theta \cos (\varphi-\Omega\sigma) \, d\sigma = \varphi - \Omega s \\ &\quad - \alpha\lambda\left(1-\frac{r_1^3}{r^3}\right) \cot \theta \left[\sin \varphi (\cos \Omega s-1)-\cos \varphi \sin \Omega s\right]. \end{aligned} \tag{2.30}$$

Thus, to the first power of α, eqs. $(2.24)_1$, (2.28) and (2.30) represent the path lines[6] of the velocity field (2.22).

There is another, and of course, equivalent, way of computing the path lines. Since

$$\boldsymbol{\xi} = \boldsymbol{\xi}\,(\mathbf{X}, \tau), \tag{2.31}$$

we have

$$\frac{d}{dt}\;\boldsymbol{\xi}\,(\mathbf{X}, \tau)\Big|_{\mathbf{X}\text{ fixed}} = \mathbf{O}, \qquad t \neq \tau. \tag{2.32}$$

Now consider $\mathbf{X}$ as a funetion of $\mathbf{x}$ and t in (2.32). Then (2.32) may be written as

$$\frac{\partial \xi^i}{\partial t} + \frac{\partial \xi^i}{\partial x^j}\, v^j = 0, \qquad v^j = v^j(x^k, t). \tag{2.33}$$

This approach is due to OLDROYD[7] and we shall illustrate the use of (2.33) by an example.

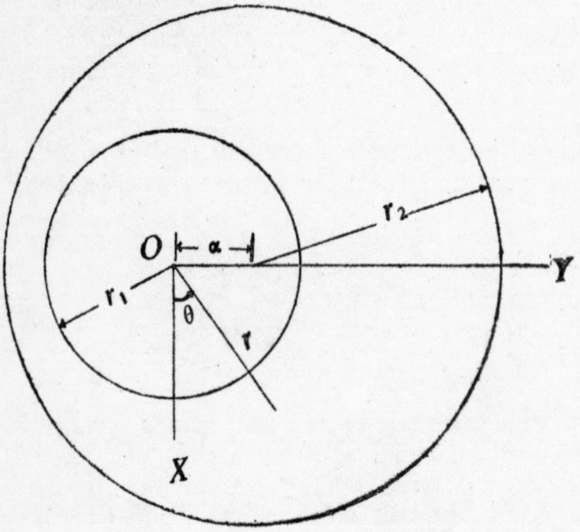

Fig. 2.2. Eccentrie cylinder rheometer

Consider the flow of a fluid between a pair of rotating eccentric cylinders[8] (see Figure 2.2). Both the cylinders rotate with an angular velocity Ω and the eccentricity is reflected by α, the distance between their centers. In cylindrical polar coordinates, the boundary conditions are

$$\begin{aligned} &v\langle r\rangle = 0, \quad v\langle\theta\rangle = \Omega r_1, \quad v\langle z\rangle = 0, \quad \text{on} \quad r = r_1; \\ &v\langle r\rangle = \Omega\alpha\cos\theta, \quad v\langle\theta\rangle = \Omega(r - \alpha\sin\theta), \\ &\qquad\qquad v\langle z\rangle = 0, \quad r = r_2 + \alpha\sin\theta. \end{aligned} \tag{2.34}$$

We shall assume that in the region between the cylinders,

$$\begin{aligned} &v\langle r\rangle = \Omega\alpha\, F(r)\, e^{i\theta}, \quad v\langle\theta\rangle = \Omega\left[r + i\alpha\,\frac{d}{dr}\,[rF(r)]\, e^{i\theta}\right], \\ &\qquad\qquad v\langle z\rangle = 0. \end{aligned} \tag{2.35}$$

Here $F(r)$ is unknown function, to be determined from the equations of motion[9]. However, (2.34) implies that

$$F(r_1) = F'(r_1) = F'(r_2) = 0, \qquad F(r_2) = 1. \tag{2.36}$$

6. JONES and WALTERS [1969 : 10]. See their eq. (20).
7 [1950 : 1], eq. (21).
8. ABBOTT and WALTERS [1970 : 2].
9. See Sec. 34 ter.

We also emphasize that only the real parts in (2.35) have any physical meaning.

Let (ξ, η, ζ) be the coordinates at time τ of the particle which is at (r, θ, z) at time t. Then (2.33) yields $\zeta(\tau) = z(t)$ and

$$\frac{\partial \xi}{\partial t} + \frac{\partial \xi}{\partial r} v\langle r\rangle + \frac{1}{r}\frac{\partial \xi}{\partial \theta} v\langle \theta\rangle = 0,$$

$$\frac{\partial \eta}{\partial t} + \frac{\partial \eta}{\partial r} v\langle r\rangle + \frac{1}{r}\frac{\partial \eta}{\partial \theta} v\langle \theta\rangle = 0. \tag{2.37}$$

We shall assume that there exist three functions, $f(r)$, $g(\theta)$, $h(t-\tau)$ such that

$$\xi = r + \alpha f(r)\, g(\theta)\, h(t-\tau), \qquad h(0) = 0. \tag{2.38}$$

The constant α appears in (2.38) because as $\alpha \to 0$ the motion is a rigid one. Neglecting all terms of order α^2, substitution of (2.38) into (2.37) yields

$$fg\dot{h} + \Omega\, F(r)\, e^{i\theta} + fg'h\Omega = 0. \tag{2.39}$$

Hence $g(\theta) = e^{i\theta}$, $f(r) = F(r)$. Then, with $\dot{h} = (dh(\sigma)/d\sigma)$, the differential equation

$$\dot{h} + \Omega + ih\Omega = 0 \tag{2.40}$$

leads to

$$h(\sigma) = i(1-e^{-i\Omega\sigma}). \tag{2.41}$$

Hence, to $O(\alpha)$ and putting $t - \tau = s$ we have

$$\xi = r + i\alpha\, F(r)\, e^{i\theta}\, [1-e^{-i\Omega s}]. \tag{2.42}$$

Similarly, by assuming $\eta = \theta - \Omega(t-\tau) + \alpha\, u(r)\, v(\theta)\, w(t-\tau)$, one obtains

$$\eta = \theta - \Omega s - \frac{\alpha}{r}\frac{d}{dr}[rF(r)]\, e^{i\theta}\,(1-e^{-i\Omega s}). \tag{2.43}$$

Thus (2.42), (2.43) and $\zeta(\tau) = z(t)$ are the path lines for the velocity field (2.35), correct to $O(\alpha)$.

3. Strain History

Suppose that the path lines have been determined. Then, using (1.17), we may calculate the *strain history* $\mathbf{C}_t(t-s)$, $0 \leqslant s < \infty$. For example, corresponding to the velocity field (2.6), the path lines are (2.9) and the strain history is given by

$$[C_t(t-s)]_{ij} = \begin{pmatrix} g_{11} + g_{22}s^2v'^2 + g_{33}s^2w'^2 & -g_{22}sv' & g_{33}sw' \\ \cdot & g_{22} & 0 \\ \cdot & \cdot & g_{33} \end{pmatrix} \tag{3.1}$$[10]

where, in checking the calculations, the following points need attention:

(*a*) since the coordinate system is orthogonal,

$$g_{\alpha\beta} = 0 \quad (\alpha\neq\beta;\ \alpha, \beta=1, 2, 3). \tag{3.2}$$

10. The dots indicate the symmetry of the matrix.

(*b*) g_{11}, g_{22} and g_{33} are functions of $\boldsymbol{\xi}$ at time $(t-s)$. If we demand that the motion (2.6) be such that the *components of the metric tensor do not change along the path lines of the particle,* then

$$g_{\underline{ii}}(\boldsymbol{\xi}) = g_{\underline{ii}}(\mathbf{x}), \qquad (i=1,\,2,\,3;\ \text{no sum}). \tag{3.3}$$

We define, in conformity with tensor analysis, the *physical components* of $\mathbf{C}_t(t-s)$ to be [cf. (A1.18)]:

$$C_t(t-s)\,\langle i\,j\rangle = [C_t(t-s)]_{\underline{ij}}/\sqrt{g_{\underline{ii}}\,g_{\underline{jj}}}, \qquad (\text{no sum}), \tag{3.4}$$

or the physical components are the *components of* $\mathbf{C}_t(t-s)$ *with respect to an orthonormal basis.* On using (3.3), (3.1) now yields

$$[C_t(t-s)\,\langle i\,j\rangle = \begin{pmatrix} 1+s^2\varkappa^2 & -s\sqrt{\dfrac{g_{22}}{g_{11}}}\,v' & -s\sqrt{\dfrac{g_{33}}{g_{11}}}\,w' \\ \cdot & 1 & 0 \\ \cdot & \cdot & 1 \end{pmatrix}, \tag{3.5}$$

where

$$\varkappa^2 = [g_{22}v'^2 + g_{33}w'^2]/g_{11}. \tag{3.6}$$

As a second example, consider the velocity field (2.22) and the path lines corresponding to that flow. The strain tensor is, again, to be computed to the first order in α only. We determine $g_{\alpha\beta}(\boldsymbol{\xi})$ in terms of $g_{ij}(\mathbf{x})$ first, as follows[11]:

$$\begin{aligned} g_{11}(\xi,\eta,\zeta) &= g_{11}(r,\theta,\varphi) = 1, \\ g_{22}(\xi,\eta,\zeta) &= \xi^2 = r^2, \\ g_{33}(\xi,\eta,\zeta) &= \xi^2\sin^2\eta \approx r^2\sin^2[\theta+O(\alpha)] \\ &\approx r^2\sin^2\theta[1+2\cot\theta\; O(\alpha)] \\ &\approx r^2\sin^2\theta\left[1+2\alpha\lambda\left(1-\frac{r_1^3}{r^3}\right)\right. \\ &\qquad \left.\times\cot\theta\left\{\cos\varphi(\cos\Omega s-1)+\sin\varphi\sin\Omega s\right\}\right]. \end{aligned} \tag{3.7}$$

In computing $(3.7)_3$, we have used (2.28). Using $(2.24)_1$, (2.28), (2.30) and (3.7) in (1.16) and (3.4) also, one obtains the physical components of $C_t(t-s)$, corret to $O(\alpha)$:

$$[C_t(t-s)\,\langle i\,j\rangle] = \begin{pmatrix} 1 & \dfrac{3\alpha\,\lambda\,r_1^3}{r^3}f(\varphi,\Omega,s) & \dfrac{3\alpha\lambda r_1^3}{r^3}\cos\theta h(\varphi,\Omega,s) \\ \cdot & 1 & 0 \\ \cdot & \cdot & 1 \end{pmatrix}, \tag{3.8}$$

where

$$\left.\begin{aligned} f(\varphi,\Omega,s) &= \sin\varphi\sin\Omega s+\cos\varphi(\cos\Omega s-1) \\ h(\varphi,\Omega,s) &= \cos\varphi\sin\Omega s+\sin\varphi(1-\cos\Omega s) \end{aligned}\right\}. \tag{3.9}$$

11. JONES and WALTERS [1969 : 10], eq. (21).

Again, we emphasize that in verifying these calculations, the reader should note that all terms are correct to $O(\alpha)$ only.

For the flow field (2.34), using the path lines (2.42), (2.43) and $\zeta(t-s) = z(t)$, the strain history $\mathbf{C}_t(t-s)$ is given in physical component form by

$$[C_t(t-s)\langle i\,j\rangle] = \begin{pmatrix} 1+2i\alpha F' e^{i\theta}(1-e^{-i\Omega s}) & -\alpha(rF'+r^2F'')\,e^{i\theta}(1-e^{-i\Omega s}) & 0 \\ \cdot & 1-2i\alpha r^2 F' e^{i\theta}(1-e^{-i\Omega s}) & 0 \\ \cdot & \cdot & 1 \end{pmatrix}, \tag{3.10}$$
[12]

where $F' = dF(r)/dr$ and we have used the following formulae for the metric tensor, derived from (2.42):

$$g_{\xi\xi} = 1, \qquad g_{\eta\eta} = \xi^2 = r^2 + 2i\alpha rF(r)\, e^{i\theta}(1-e^{-i\Omega s}), \qquad g_{\zeta\zeta} = 1. \tag{3.11}$$

These are again, computed to $O(\alpha)$.

An interesting method, depending on the motion in the vicinity of a *surface of rest* (which will be relaxed to one of rigid motion later on) $\mathcal{S}$ has been developed by CASWELL[13] [1967 : 6] to calculate the strain history about a point on the surface, which may be immersed in the fluid or be a bounding surface. We shall discuss this idea next.

Let P be a point on the surface $\mathcal{S}$. Then in the neighbourhood of P, say at Q in the fluid, the velocity $\mathbf{v}(Q)$ is given by

$$\mathbf{v}(Q) = \mathbf{L}(P)\,\mathbf{r} + O(r^2), \tag{3.12}$$
[14]

where $\mathbf{r}$ is the position vector from P to Q, and $\mathbf{L}(P)$ is the velocity gradient at P. Let $\mathbf{L}(P) + \mathbf{L}^T(P) = \mathbf{A}_1(P) \equiv \mathbf{A}_P^{(1)}$ and ω_P be the first RIVLIN-ERICKSEN tensor[15] and the vorticity at P respectively. Then (3.12) is equivalent to

$$\mathbf{v}(Q) = \frac{1}{2}\,\mathbf{A}_P^{(1)}\,\mathbf{r} - \frac{1}{2}\,\mathbf{r}\times\boldsymbol{\omega}_P. \tag{3.13}$$

Let $\mathbf{n}$ be the unit normal to $\mathcal{S}$ at P (into thc fluid if $\mathcal{S}$ is a bounding surface). The KELVIN's transformation applied to any reconcilable closed curve s lying in $\mathcal{S}$ reads[16]

$$\int_{\mathcal{S}} \varepsilon_{mrp}\, n_r v_{k,p}\, da = \oint_C t_m\, v_k\, ds, \tag{3.14}$$

12. ABBOTT and WALTERS [1970 : 2].
13. We will not use the dyadic notation of CASWELL.
14. If P is not at rest, obviously $\mathbf{v}(P)$ appears in (3.12).
15. Such tensors are discussed in detail in Sec. 4. Also it is understood that

$$\mathbf{A}_P^{(1)} = \lim_{Q\to P} \mathbf{A}_1(Q), \text{ etc.}$$

16. WEATHERBURN [1924 : 1], p. 25. Cartesian coordinates are used here.

where ε_{mrp} is the alternating tensor, and $\mathbf{t}$ is the unit tangent vector along $\mathcal{C}$. Since $v_k = 0$ on $\mathcal{C}$, (3.14) implies that

$$\varepsilon_{mrp}\, n_r\, v_{k,\,p} = 0 \text{ on } \mathcal{S}. \tag{3.15}$$

Contraction on k and m leads to

$$\varepsilon_{rpm}\, v_{m,\,p}\, n_r = \boldsymbol{\omega}\cdot\mathbf{n} = 0 \text{ on } \mathcal{S}, \tag{3.16}$$

implying the parallelality of $\boldsymbol{\omega}$ to the surface $\mathcal{S}$. Next, from (3.15) one derives

$$\varepsilon_{mks}\,\varepsilon_{mrp}\, n_r\, v_{k,\,p} = n_k\, v_{k,\,s} - v_{p,\,p}\, n_s = 0 \text{ on } \mathcal{S}, \tag{3.17}$$[17]

or

$$(\mathbf{n}\times\nabla)\times\mathbf{v} = \mathbf{Ln} - \boldsymbol{\omega}\times\mathbf{n} - \mathbf{n}(\nabla\cdot\mathbf{v}) = 0 \text{ on } \mathcal{S}. \tag{3.18}$$

From (3.17) and (3.18), it follows that

$$\mathbf{A}_1\mathbf{n} = \mathbf{Ln} + \mathbf{L}^T\mathbf{n} = \boldsymbol{\omega}\times\mathbf{n} + 2\mathbf{n}(\nabla\cdot\mathbf{v}) \text{ on } \mathcal{S}. \tag{3.19}$$

Moreover, (3.15) implies that

$$\varepsilon_{mrp}\, n_r(v_{k,\,p}+v_{p,\,k}) = -\,2\varepsilon_{mrp}\, n_r\, W_{kp}, \tag{3.20}$$

where the spin tensor $\mathbf{W}$ is related to the vorticity $\boldsymbol{\omega}$ through

$$W_{kp} = -\,\frac{1}{2}\,\varepsilon_{kps}\,\omega_s. \tag{3.21}$$

Hence (3.20) and (3.21) lead to

$$\varepsilon_{mrp}\, n_r\, A^{(1)}_{kp} = \omega_m\, n_k \text{ on } \mathcal{S}. \tag{3.22}$$

Since $\mathbf{n} \neq \mathbf{0}$, and $(\boldsymbol{\omega}\otimes\mathbf{n})^T\,\mathbf{n} = \mathbf{0}$, the unique solution for $\mathbf{A}_1$ obeying (3.19) and (3.22) is

$$\mathbf{A}_1 = 2(\nabla\cdot\mathbf{v})\,\mathbf{n}\otimes\mathbf{n} + (\boldsymbol{\omega}\times\mathbf{n})\otimes\mathbf{n} + [(\boldsymbol{\omega}\times\mathbf{n})\otimes\mathbf{n}]^T \text{ on } \mathcal{S}. \tag{3.23}$$[18]

Restricting the attention to isochoric motions, that is, *those obeying* $\nabla\cdot\mathbf{v} = 0$, *from now on* we have

$$\mathbf{A}_1 = (\boldsymbol{\omega}\times\mathbf{n})\otimes\mathbf{n} + \mathbf{n}\otimes(\boldsymbol{\omega}\times\mathbf{n}). \tag{3.24}$$

Let $\{x^i\}$ be a set of curvilinear orthogonal coordinates such that $\mathcal{S}$ is described by $x^1 = 0$. Then the deformation gradient at P becomes

$$\mathbf{F}_t(t-s) = \mathbf{1} - \int_0^s \mathbf{L}(x^2, x^3, t-\sigma)\, d\sigma, \tag{3.25}$$

where we have utilized the fact that $\{0, x^2, x^3\}$ are the coordinates of P, the definition of $\mathbf{F}_t(t-s)$ in (1.13) and the path lines in (2.4). From (3.16) and

17. This result is given by WEATHERBURN [1924 : 1], p. 126. The formula (3.18) which follows from it was rediscovered by BERKER [1951 : 1].

18. $\mathbf{a}\otimes\mathbf{b}$ is the tensor product of two vectors $\mathbf{a}$ and $\mathbf{b}$.

(3.24) it follows that

$$\mathbf{L} = (\boldsymbol{\omega} \times \mathbf{n}) \otimes \mathbf{n} \text{ on } S, \tag{3.26}$$

and

$$\mathbf{F}_t(t-s) = \mathbf{1} - \int_0^s (\boldsymbol{\omega}(P, t-\sigma) \times \mathbf{n}) \otimes \mathbf{n} \, d\sigma. \tag{3.27}$$

It is trivial to verify that the strain history is given by

$$\mathbf{C}_t(t-s)\,(P) = \mathbf{1} + [\boldsymbol{\Omega}^t(s)]^2 \, \mathbf{n} \otimes \mathbf{n} - \mathbf{n} \otimes (\boldsymbol{\Omega}^t(s) \times \mathbf{n}) - (\boldsymbol{\Omega}^t(s) \times \mathbf{n}) \otimes \mathbf{n}, \tag{3.28}$$

where $[\boldsymbol{\Omega}^t(s)]^2 = | \boldsymbol{\Omega}^t(s) |^2$, and

$$\boldsymbol{\Omega}(t-s) = \boldsymbol{\Omega}^t(s) = \int_0^s \boldsymbol{\omega}(P, t-\sigma) \, d\sigma. \tag{3.29}$$

Note that $\boldsymbol{\Omega}^t(s)$ has only two nonzero components with $\boldsymbol{\Omega}^t(s) \langle 1 \rangle = 0$, because $\boldsymbol{\omega} \cdot \mathbf{n} = 0$.

It is convenient to introduce a set of orthonormal base vectors $\mathbf{i}_1, \mathbf{i}_2, \mathbf{i}_3$ defined through

$$\mathbf{i}_1 = \mathbf{n}, \quad \mathbf{i}_2 = \boldsymbol{\omega}^* \times \mathbf{n}, \quad \mathbf{i}_3 = \boldsymbol{\omega}^*, \quad \boldsymbol{\omega}^* = \boldsymbol{\omega}(t)/\omega(t). \tag{3.30}$$ [19]

One can rewrite $\mathbf{C}_t(t-s)$ relative to this basis as

$$\mathbf{C}_t(t-s) = \mathbf{1} + [\boldsymbol{\Omega}^t(s)]^2 \, \mathbf{i}_1 \otimes \mathbf{i}_1 - U^t(s) \, (\mathbf{i}_1 \otimes \mathbf{i}_2 + \mathbf{i}_2 \otimes \mathbf{i}_1) + V^t(s)(\mathbf{i}_1 \otimes \mathbf{i}_3 + \mathbf{i}_3 \otimes \mathbf{i}_1), \tag{3.31}$$

where

$$V^t(s) = \boldsymbol{\Omega}^t(s) \cdot \mathbf{i}_2, \qquad U^t(s) = \boldsymbol{\Omega}^t(s) \cdot \mathbf{i}_3. \tag{3.32}$$

At a given instant, one can find an orthogonal, time dependent, linear mapping of the natural orthonormal basis along the $\{x^i\}$ onto the set $\{\mathbf{i}_j\}$. This fact will be useful later on in Sec. 26.

If the *vortex lines are steady*,[20] then

$$\frac{\partial \boldsymbol{\omega}}{\partial t} \times \boldsymbol{\omega} = \mathbf{0}, \qquad \text{or} \qquad \frac{\partial \boldsymbol{\omega}}{\partial t} = C(\mathbf{x}, t) \, \boldsymbol{\omega}. \tag{3.33}$$

Hence it follows that the magnitude of the vorticity is given by

$$\omega(t) = \omega(0) \exp \left(\int_0^t C(\mathbf{x}, s) \, ds \right), \tag{3.34}$$

whence

$$\frac{\partial \boldsymbol{\omega}^*}{\partial t} = \mathbf{0}. \tag{3.35}$$

So, for *steady vortex lines*, $V^t(s)$ in (3.32) is zero, because

$$\boldsymbol{\Omega}^t(s) = \boldsymbol{\omega}^* \int_0^s \omega(\mathbf{x}, t-\sigma) \, d\sigma, \quad \text{and} \quad (\boldsymbol{\omega}^* \times \mathbf{n}) \cdot \boldsymbol{\Omega}^t(s) = 0$$

19. If at a given instant $\boldsymbol{\omega}(t) = \mathbf{0}$, then one has to define $\mathbf{i}_2$ and $\mathbf{i}_3$ by continuity with respect to t.

20. The classic text on vorticity is due to TRUESDELL [1954 : 1].

here. Thus, for steady vortex lines, the strain history reduces to

$$\mathbf{C}_t(t-s) = \mathbf{1} + [\Omega^t(s)]^2 \, \mathbf{i}_1 \otimes \mathbf{i}_1 - U^t(s)(\mathbf{i}_1 \otimes \mathbf{i}_2 + \mathbf{i}_2 \otimes \mathbf{i}_1). \tag{3.36}$$

If the *vorticity itself is steady,* then

$$\Omega^t(s) = s\omega, \tag{3.37}$$

and the strain history in (3.36) becomes *viscometric*[21]. So steady flow near a smooth surface at rest is viscometric.

If we keep the requirement that div $\mathbf{v} = 0$, but that $\mathcal{S}$ is a material surface which moves as a rigid surface through the fluid so that all the points on $\mathcal{S}$ have the same velocity at a given time t, then the foregoing calculations for $\mathbf{C}_t(t-s)$ still remain valid, except that the vorticity $\boldsymbol{\omega} = \boldsymbol{\omega}_P$ will have to be replaced by the vorticity relative to P, that is,

$$\boldsymbol{\omega}(P, t) = \lim_{Q \to P} \text{curl} \, [\mathbf{v}(Q, t) - \mathbf{v}(P, t)]. \tag{3.38}$$

Here P may still be described by $(0, x^2, x^3)$ on $\mathcal{S}$, but in a fixed coordinate system its coordinates will be $\{y^i\}$. Unless the motion is such that the metric tensor can be computed along the path line of P in space, such a complicated motion does not lend itself to easy analysis. We shall revert to these points in Sec. 26.

4. Rivlin-Ericksen Tensors

Suppose that the path lines have been determined and that the strain history $\mathbf{C}_t(t-s)$, $0 \leq s < \infty$ has been calculated. If this is considered as a function of s, differentiable at $s = 0$, then one may define the n-th RIVLIN-ERICKSEN tensor[22] [1955: 4] $\mathbf{A}_n$ through

$$\frac{d^n}{ds^n} \mathbf{C}_t(t-s) \Big|_{s=0} = (-1)^n \, \mathbf{A}_n, \qquad n = 1, 2, 3, \ldots \; . \tag{4.1}$$

Note that

$$\mathbf{C}_t(t-s) \,|_{s=0} = \mathbf{1} \tag{4.2}$$

for as $s \to 0$, $\boldsymbol{\xi}(t-s) \to \mathbf{x}(t)$ and thus $\dfrac{\partial \xi^i}{\partial x^j} \to \delta^i_j$, and (4.2) follows. Hence, we may put $\mathbf{A}_0 = \mathbf{1}$ and write (4.1) as

$$\frac{d^n}{ds^n} \mathbf{C}_t(t-s) \Big|_{s=0} = (-1)^n \, \mathbf{A}_n, \qquad n = 0, 1, 2, \ldots \; . \tag{4.3}$$

From the path lines and the definition of $\mathbf{C}_t(t-s)$ in (1.16), it is obvious that

---

21. See Secs. 7 and 8.

22. These tensors occur for the first time in continuum mechanics, in the paper by YVONNE DUPONT [1931 : 1]. However, her work was ignored until mentioned in the paper [1955 : 4]. Independently of DUPONT'S work, OLDROYD [1950 : 1] had discovered these tensors in a different way. We believe that NOLL [1958 : 3] is responsible for these tensors being named as *Rivlin-Ericksen tensors*.

the n-th derivative $\mathbf{A}_n$ must be derivable from the velocity field. We may express this relation in the following manner. First of all, the relative deformation gradient has the properties listed below, viz.,

$$\mathbf{F}_t(t-s)\,|_{s=0} = \mathbf{1}, \tag{4.4}$$

$$\frac{d}{dt}\mathbf{F}(t) = \frac{d}{dt}\nabla_{\mathbf{X}}\mathbf{x} = \nabla_{\mathbf{X}}\frac{d}{dt}\mathbf{x} = \nabla_{\mathbf{X}}\mathbf{v}$$
$$= \nabla\mathbf{v}\,\nabla_{\mathbf{X}}\mathbf{x} = \mathbf{LF}, \tag{4.5}$$ [23]

or,
$$\frac{d}{dt}\left(\frac{\partial x^i}{\partial X^\alpha}\right) = \frac{\partial}{\partial X^\alpha}\left(\frac{d}{dt}x^i\right) = v^i_{;j}\,F^j_\alpha\,. \tag{4.6}$$

Hence, by using

$$\frac{d}{d\tau}\mathbf{F}_t(\tau) = \frac{d}{d\tau}[\mathbf{F}(\tau)\,\mathbf{F}(t)^{-1}] = \mathbf{L}(\tau)\,\mathbf{F}(\tau)\,\mathbf{F}(t)^{-1}$$
$$= \mathbf{L}(\tau)\,\mathbf{F}_t(\tau), \tag{4.7}$$

and replacing τ by $t-s$, we get

$$\frac{d}{ds}\mathbf{F}_t(t-s) = -\,\mathbf{L}(t-s)\,\mathbf{F}_t(t-s). \tag{4.8}$$

When this equation is evaluated at $s = 0$, we have through (4.4):

$$\frac{d}{ds}\mathbf{F}_t(t-s)\Big|_{s=0} = -\,\mathbf{L}(t). \tag{4.9}$$

We now turn to the definition of $\mathbf{C}_t(t-s)$ in (1.15) and derive

$$\frac{d}{ds}\mathbf{C}_t(t-s)\Big|_{s=0} = -\,[\mathbf{L}(t)+\mathbf{L}(t)^T], \tag{4.10}$$

and hence
$$\mathbf{A}_1(t) = \mathbf{L}(t) + \mathbf{L}(t)^T, \tag{4.11}$$

which establishes that the first RIVLIN-ERICKSEN tensor is twice the symmetric part of the velocity gradient. If we define $\mathbf{L}_n$ through

$$\mathbf{L}_n = (-1)^n \frac{d^n}{ds^n}\mathbf{F}_t(t-s)\Big|_{s=0}, \qquad n = 1, 2, 3, \ldots \tag{4.12}$$

then

$$\frac{d^n}{ds^n}\mathbf{C}_t(t-s)\Big|_{s=0} = (-1)^n \sum_{r=0}^{n}\binom{n}{r}\mathbf{L}^T_{n-r}\,\mathbf{L}_r, \tag{4.13}$$

where we have put

$$\binom{n}{r} = \frac{n!}{r!(n-r)!}, \qquad \mathbf{L}_1 = \mathbf{L}, \quad \mathbf{L}_0 = \mathbf{1}. \tag{4.14}$$

23. NOLL [1955 : **2**], eq, (2.10).

Thus, for example,

$$\mathbf{A}_2 = \mathbf{L}_2 + \mathbf{L}_2^T + 2\mathbf{L}^T\,\mathbf{L}, \tag{4.15}$$

$$\mathbf{L}_2 = \frac{d^2}{ds^2}\,\mathbf{F}_t(t-s)\bigg|_{s=0} = -\frac{d}{ds}\,[\mathbf{L}(t-s)\,\mathbf{F}_t(t-s)]_{s=0}$$

$$= \dot{\mathbf{L}}(t) + \mathbf{L}^2(t). \tag{4.16}$$

We note that

$$a^i_{;j} = \left[\frac{\partial v^i}{\partial t} + v^i_{;k}\,v^k\right]_{;j}$$

$$= \frac{d}{dt}\,(v^i_{;j}) + v^i_{;k}\,v^k_{;j}, \tag{4.17}$$

and hence we have that $\mathbf{L}_2 = \nabla\mathbf{a}$. Thus $\mathbf{L}_2$ is the gradient of the acceleration field. Similarly, if we consider the n-th acceleration field, then $\mathbf{L}_{n+1}$ is the gradient of that field. So, eq. (4.13) yields $\mathbf{A}_n$ in terms of the first n acceleration gradients. We now turn to OLDROYD's recursive formula [1950:1] for $\mathbf{A}_{n+1}$ in terms of $\mathbf{A}_n$ and $\mathbf{L}$.

Let dL^2 be the square of the distance between the points $\mathbf{X}$ and $\mathbf{X} + d\mathbf{X}$ in the material coordinate system. Let the square of the distance between these very same particles be dl^2 at time t. Then

$$dl^2(t) = g_{ij}\,dx^i\,dx^j = g_{ij}\,\frac{\partial x^i}{\partial X^\alpha}\,\frac{\partial x^j}{\partial X^\beta}\,dX^\alpha\,dX^\beta$$

$$= [C(t)]_{\alpha\beta}\,dX^\alpha\,dX^\beta. \tag{4.18}$$

Similarly,

$$dl^2(t-s) = [C(t-s)]_{\alpha\beta}\,dX^\alpha\,dX^\beta, \tag{4.19}$$

and since

$$\mathbf{C}_t(t-s) = \mathbf{F}_t(t-s)^T\,\mathbf{F}_t(t-s)$$

$$= [\mathbf{F}(t)^{-1}]^T\,\mathbf{F}(t-s)^T\,\mathbf{F}(t-s)\,\mathbf{F}(t)^{-1}$$

$$= [\mathbf{F}(t)^{-1}]^T\,\mathbf{C}(t-s)\,\mathbf{F}(t)^{-1}, \tag{4.20}$$

we have that

$$\frac{d^n}{ds^n}\,\mathbf{C}_t(t-s) = [\mathbf{F}(t)^{-1}]^T\left[\frac{d^n}{ds^n}\,\mathbf{C}(t-s)\right]\mathbf{F}(t)^{-1}. \tag{4.21}$$

Thus the n-th derivative of $dl^2(t-s)$ with respect to s at $s = 0$ is related to $\mathbf{A}_n$. From (4.21), we obtain

$$\mathbf{F}(t)^T\left[\frac{d^n}{ds^n}\,\mathbf{C}_t(t-s)\right]\mathbf{F}(t) = \frac{d^n}{ds^n}\,\mathbf{C}(t-s). \tag{4.22}$$

Multiply by $d\mathbf{X}^T$ from the left and $d\mathbf{X}$ from the right, getting

$$d\mathbf{X}^T\,\mathbf{F}(t)^T\left[\frac{d^n}{ds^n}\,\mathbf{C}_t(t-s)\right]\mathbf{F}(t)\,d\mathbf{X} = d\mathbf{X}^T\left[\frac{d^n}{ds^n}\,\mathbf{C}_t(t-s)\right]d\mathbf{X}. \tag{4.23}$$

Since $$d\mathbf{x}(t) = \mathbf{F}(t)\, d\mathbf{X}, \qquad dx^i = \frac{\partial x^i}{\partial X^\alpha}\, dX^\alpha, \tag{4.24}$$

(4.23) may be expressed, in indicial notation, as

$$(-1)^n (A_n)_{ij}\, dx^i\, dx^j = \frac{d^n}{ds^n}\, dl^2(t-s)\bigg|_{s=0} \tag{4.25}$$

$$= (-1)^n \frac{d^n}{dt^n}\, dl^2(t). \tag{4.26}$$

Thus, for example,

$$(A_1)_{ij}\, dx^i\, dx^j = \frac{d}{dt}\, dl^2(t). \tag{4.27}$$

Now, we derive a recursive formula for $\mathbf{A}_{n+1}$ in terms of $\mathbf{A}_n$ as follows. Since

$$(A_{n+1})_{ij}\, dx^i\, dx^j = \frac{d^{n+1}}{dt^{n+1}}\, [dl^2(t)] = \frac{d}{dt}\left(\frac{d^n}{dt^n} dl^2(t)\right) = \frac{d}{dt}\, [(A_n)_{ij}\, dx^i dx^j]$$

$$= \left[\frac{d}{dt}\, (A_n)_{ij} + (A_n)_{ik}\, v^k_{;j} + (A_n)_{kj}\, v^k_{;i}\right] dx^i\, dx^j,$$

$$n = 1, 2, \ldots, \tag{4.28}$$

we have $$\mathbf{A}_{n+1} = \frac{d}{dt}\, \mathbf{A}_n + \mathbf{A}_n\mathbf{L} + \mathbf{L}^T\mathbf{A}_n, \qquad n = 1, 2, \ldots, \tag{4.29}$$

or, $$(A_{n+1})_{ij} = \frac{\partial}{\partial t}\, (A_n)_{ij} + (A_n)_{ij;\,k}\, v^k + (A_n)_{ik}\, v^k_{;j}$$

$$+ (A_n)_{kj}\, v^k_{;i}, \qquad n = 1, 2, 3, \ldots. \tag{4.30}$$

In writting (4.28), we have used the fact that

$$\frac{d}{dt}\, dx^i = d\dot{x}^i = dv^i = v^i_{;j}\, dx^j. \tag{4.31}$$

As an example, for the velocity field

$$\dot{r} = 0, \qquad \dot{\theta} = \omega(r), \qquad \dot{z} = 0, \tag{4.32}$$

in cylindrical polar coordinates, we determine $\mathbf{A}_1$ and $\mathbf{A}_2$. Now [see (4.11)]:

$$(A_1)_{ij} = v_{i\,;\,j} + v_{j\,;\,i}\,, \tag{4.33}$$

and since the $v_{i\,;\,j}$ components of the velocity field are given in contravariant form, we find the covariant components:

$$v_1 = 0, \qquad v_2 = r^2\omega(r), \qquad v_3 = 0. \tag{4.34}$$

Next, we determine $v_{i\,;\,j}$ and thence compute the matrix of $L\langle ij\rangle$ to be:

$$[L\langle ij\rangle] = \begin{pmatrix} 0 & -\omega & 0 \\ \omega + r\omega' & 0 & 0 \\ 0 & 0 & 0 \end{pmatrix}, \qquad \omega' = \frac{d\omega}{dr}. \tag{4.35}$$

Thereupon,

$$[A_1\langle ij\rangle] = \begin{pmatrix} 0 & r\omega' & 0 \\ \cdot & 0 & 0 \\ \cdot & \cdot & 0 \end{pmatrix}, \tag{4.36}$$

and $\dot{\mathbf{A}}_1 = \dfrac{\partial \mathbf{A}_1}{\partial t} + \nabla\mathbf{A}_1\cdot\mathbf{v}$ has the components

$$[\dot{A}_1\langle ij\rangle] = \begin{pmatrix} -2\omega r\omega' & 0 & 0 \\ \cdot & 2\omega r\omega' & 0 \\ \cdot & \cdot & 0 \end{pmatrix}. \tag{4.37}$$

Hence the contravariant and physical components of $\mathbf{A}_2$ are equal, given by

$$[(A_2)_{ij}] = [A_2\langle ij\rangle] = \begin{pmatrix} 2r^2\omega'^2 & 0 & 0 \\ \cdot & 0 & 0 \\ \cdot & \cdot & 0 \end{pmatrix}, \tag{4.38}$$

We could have obtained $\mathbf{A}_1$ and $\mathbf{A}_2$ from $\mathbf{C}_t(t-s)$ as follows. Since

$$g_{11} = 1, \qquad g_{22} = r^2, \qquad g_{33} = 1, \tag{4.39}$$

along the path lines of the velocity field (4.32), that is, the components of the metric tensor at time $(t-s)$ are the same as those at time t for each particle, one sees that

$$[C_t(t-s)_{ij}] = \begin{pmatrix} 1 + s^2r^2\omega'^2 & -sr^2\omega' & 0 \\ \cdot & r^2 & 0 \\ \cdot & \cdot & 1 \end{pmatrix} \tag{4.40}$$

$$= [1]_{ij} - s\,[A_1]_{ij} + \frac{1}{2}\,s^2\,[A_2]_{ij}\,. \tag{4.41}$$

Thus, we recover (4.36) and (4.38).

The Taylor series expansion for $\mathbf{C}_t(t-s)$ is in terms of the tensors $\mathbf{A}_n(t)$, that is, in general, $\mathbf{A}_n$ depend on the present time t. But a trivial confusion that often arises is the following: in the example just cited, the components of $\mathbf{A}_1$ and $\mathbf{A}_2$ are independent of t. Why does this dependence on t disappear and yet $\dot{\mathbf{A}}_1 \neq \mathbf{0}$? Firstly, we note that the components are calculated with respect to the basis at the point $\mathbf{x}$ in space. This point is not occupied by the particle always. So the components, while they remain fixed with respect to the natural basis at the point where the particle is at any given time, do change if viewed from a fixed basis in space. Since the tensor $\mathbf{A}_1$ must be written as

$$\mathbf{A}_1(\mathbf{x}, t) = (A_1)_{ij}\,(\mathbf{x}, t)\,\mathbf{g}^i(\mathbf{x})\otimes\mathbf{g}^j(\mathbf{x}), \tag{4.43}$$

where the $\mathbf{g}^i$ is the vector basis at $\mathbf{x}$, we have that $(\partial/\partial t)\, A_{ij}^{(1)}(\mathbf{x}, t) = 0$ for the flow considered, but

$$\nabla \mathbf{A}_1(\mathbf{x}, t) \cdot \mathbf{v} \neq 0, \tag{4.44}$$

because $\nabla \mathbf{g}^i(\mathbf{x}) \neq \mathbf{0}$, in general, is a curvilinear coordinate system. In other words, the change in the magnitude and direction of the vector basis gives rise to a non-zero $\dot{\mathbf{A}}_1$. Another way of explaining this will be given in Sec. 8.

6. Approximations to the Strain History

5.1. Infinitesimal History. Clearly, if a Taylor series for $\mathbf{C}_t(t-s)$ is written about $s = 0$, then we can, in the vicinity of $s = 0$, approximate $\mathbf{C}_t(t-s)$ as closely as we desire by this series, or, as in Sec. 3, we may derive the approximation in terms of a parameter. We are not interested in such approximations here. What we are concerned with may be formulated thus: suppose that the motion is such that the fluid particle, at all times, is very "close" to its position in the reference configuration. Then, can we express $\mathbf{C}_t(t-s)$ in terms of the displacement gradient from the reference configuration?

Let $\mathbf{u}(\mathbf{X}, t)$ be the displacement at time t, that is

$$\mathbf{x}(\mathbf{X}, t) - \mathbf{X} = \mathbf{u}(\mathbf{X}, t). \tag{5.1}$$

Then, if $\mathbf{H} = \nabla_{\mathbf{X}} \mathbf{u}$, we have

$$\mathbf{F}(t) = \nabla_{\mathbf{X}} \mathbf{x} = \mathbf{1} + \mathbf{H}. \tag{5.2}$$

Let $\mathbf{H}$ be "small", that is, the norm $|\mathbf{H}|$ of $\mathbf{H}$ defined by

$$|\mathbf{H}| = \sqrt{\operatorname{tr} \mathbf{H}\mathbf{H}^T} \tag{5.3}$$

is such that its supremum ε:

$$\varepsilon = \sup_t |\mathbf{H}(t)| \tag{5.4}$$

is very small, or $\varepsilon \ll 1$, then we say that the deformation (5.1) is *infinitesimal.*[24] Thereupon,

$$\mathbf{F}(t)^{-1} = \mathbf{1} - \mathbf{H} + o(\varepsilon), \tag{5.5}$$

$$\mathbf{F}_t(\tau) = \mathbf{F}(\tau)\, \mathbf{F}(t)^{-1} = \mathbf{1} + \mathbf{H}(\tau) - \mathbf{H}(t) + o(\varepsilon), \tag{5.6}$$

$$\mathbf{C}_t(\tau) = \mathbf{F}_t(\tau)^T\, \mathbf{F}_t(\tau) = \mathbf{1} + \mathbf{H}(\tau)^T + \mathbf{H}(\tau) - \mathbf{H}(t)^T - \mathbf{H}(t) + o(\varepsilon). \tag{5.7}$$

Note that the infinitesimal strain $\mathbf{E}(t)$ is defined by

$$2\mathbf{E}(t) = \mathbf{H}(t) + \mathbf{H}(t)^T, \tag{5.8}$$

so that

$$\mathbf{C}_t(\tau) = \mathbf{1} + 2\mathbf{E}(\tau) - 2\mathbf{E}(t) + o(\varepsilon). \tag{5.9}$$

Since $\mathbf{U}(t)$ is the unique square root of $\mathbf{C}(t)$, we obtain

$$\mathbf{U}(t) = \mathbf{1} + \mathbf{E}(t) + o(\varepsilon), \tag{5.10}$$

24. Coleman and Noll [1961 : 1].

whence $$\mathbf{R}(t) = \mathbf{F}(t)\,\mathbf{U}(t)^{-1} = \mathbf{1} + \mathbf{W}(t) + o(\varepsilon), \tag{5.11}$$

where $$\mathbf{W}(t) = \frac{1}{2}\,[\mathbf{H}(t) - \mathbf{H}(t)^T] \tag{5.12}$$

is the *infinitesimal rotation tensor*.

We have thus formulated the relative strain in terms of the infinitesimal strain when $\mathbf{H}(t)$ is small for all t.

5.2. "Small on Large". There is another kind of approximation also. Suppose that a velocity field $\mathbf{v}^{\circ}(\mathbf{x}, t)$ is given and we calculate the relative strain history ${}^{\circ}\mathbf{C}_t(t-s)$, $0 \leq s < \infty$ corresponding to this. Then if a small perturbation is superposed on this velocity field, $\mathbf{C}_t(t-s)$ can be viewed as

$$\mathbf{C}_t(t-s) = {}^{\circ}\mathbf{C}_t(t-s) + \mathbf{E}_t(t-s), \tag{5.13}$$

where $\mathbf{E}_t(t-s)$ is the perturbation term. For example, corresponding to the velocity field (2.11), and the path lines (2.13), on assuming that A is small so that A^2 and higher powers can be neglected, we obtain

$$\mathbf{C}_t(t-s) = {}^{\circ}\mathbf{C}_t(t-s) + \mathbf{E}_t(t-s) + o(A), \tag{5.14}$$

where

$$[{}^{\circ}C_t(t-s)\,\langle i\,j\rangle] = \begin{pmatrix} 1 & -\varkappa s & 0 \\ \cdot & 1+\varkappa^2 s^2 & 0 \\ \cdot & \cdot & 1 \end{pmatrix}, \tag{5.15}$$

$$[E_t(t-s)\,\langle i\,j\rangle] = \begin{pmatrix} 0 & 0 & 0 \\ \cdot & 0 & f(A,t,s) \\ \cdot & \cdot & 0 \end{pmatrix}, \tag{5.16}$$

$$f(A, t, s) = \frac{A}{h}\,[\sin\omega(t-s) - \sin\omega t]. \tag{5.17}$$

We call such strain histories as those derived by "small on large", that is, by superposing a small disturbance, or displacement, on a given velocity or displacement field.

We now reformulate this concept in terms of small displacements superposed on large. Following PIPKIN[25] [1968 : 20], let $\mathbf{x}^0$ be the position occupied by the particle X at time t and let $\boldsymbol{\xi}^0$ be the position, at time $t-s$, of the particle which was at $\mathbf{x}^0$ at time t. Let us express this as

$$\boldsymbol{\xi}^0 = \boldsymbol{\xi}^0(\mathbf{x}^0, t, t-s). \tag{5.18}$$

Let $\mathbf{u}(\mathbf{x}^0, t)$ be the small displacement at time t, superposed on the particle which would have been at $\mathbf{x}^0$, but for this perturbation. Then, by definition,

$$\boldsymbol{\xi}(\mathbf{x}^0 + \mathbf{u}(\mathbf{x}^0, t), t, t-s) = \boldsymbol{\xi}^0(\mathbf{x}^0, t, t-s) + \mathbf{u}(\boldsymbol{\xi}^0(\mathbf{x}^0, t, t-s), t-s). \tag{5.19}$$

25. Though PIPKIN discusses viscometric flows as the base motions, his argument applies to all cases.

Stated in words, (5.19) reads: the position vector, at time $t-s$, of the particle which is at $\mathbf{x}^0+\mathbf{u}(\mathbf{x}^0, t)$ at time t is the same as the position vector, at time $t-s$, of the particle which is at $\mathbf{x}^0$ at time t plus the displacement vector of the particle which would have been at $\boldsymbol{\xi}^0$ at time $t-s$, had the disturbance not occurred. In (5.19), replace $\mathbf{x}^0$ by $\mathbf{x}^0-\mathbf{u}(\mathbf{x}^0, t)$. Then, one obtains

$$\boldsymbol{\xi}(\mathbf{x}^0, t, t-s) = \boldsymbol{\xi}^0(\mathbf{x}^0-\mathbf{u}(\mathbf{x}^0, t), t, t-s) + \mathbf{u}\{\boldsymbol{\xi}^0(\mathbf{x}^0-\mathbf{u}(\mathbf{x}^0, t), t, t-s), t-s\}. \tag{5.20}$$

Linearization with respect to $\mathbf{u}(\mathbf{x}^0, t)$ in (5.20) yields

$$\boldsymbol{\xi}(\mathbf{x}^0, t, t-s) = \boldsymbol{\xi}^0(\mathbf{x}^0, t, t-s) - \mathbf{u}(\mathbf{x}^0, t)\cdot\nabla\boldsymbol{\xi}^0(\mathbf{x}^0, t, t-s) + \mathbf{u}(\boldsymbol{\xi}^0(\mathbf{x}^0, t, t-s), t-s), \tag{5.21}$$

where we have written

$$\nabla\boldsymbol{\xi}^0 \equiv \nabla_{\mathbf{x}^0}\boldsymbol{\xi}^0 \equiv {}^\circ\mathbf{F}_t(t-s) = {}^\circ\mathbf{F}, \qquad {}^\circ F_{\alpha i} = \frac{\partial \xi^0_\alpha}{\partial x^0_i}. \tag{5.22}$$

Hence, by calculating $\nabla_{\mathbf{x}^0}\boldsymbol{\xi}$, one determines $\mathbf{C}_t(t-s)$ for the perturbed motion in Cartesian coordinates as

$$[\mathbf{C}_t(t-s)]_{ij} = {}^\circ C_{ij} - {}^\circ C_{ik}\, u_{k,j} - {}^\circ C_{jk}\, u_{k,i} + {}^\circ F_{\alpha i}{}^\circ F_{\beta j}\,[u_{\alpha,\beta}+u_{\beta,\alpha}] - u_m{}^\circ C_{ij,m}. \tag{5.23}$$

In (5.23), note that the derivatives $u_{i,j}$ and $u_{\alpha,\beta}$ are defined through

$$u_{i,j} = \frac{\partial u_i(\mathbf{x}^0, t)}{\partial x^0_j}, \tag{5.24}$$

$$u_{\alpha,\beta} = \frac{\partial u_\alpha(\boldsymbol{\xi}^0, t-s)}{\partial \xi^0_\beta}, \tag{5.25}$$

and

$$ {}^\circ\mathbf{C} \equiv {}^\circ\mathbf{C}_t(t-s). \tag{5.26}$$

For example if we superpose, in a Cartesian coordinate system, a displacement field in the z-direction:

$$\mathbf{u}(\mathbf{x}^0, t) = \frac{Ay^0}{h}\sin\omega t\,\mathbf{k}, \tag{5.27}$$

on a basic viscometric flow

$$\mathbf{v}(\mathbf{x}^0) = \varkappa y^\circ\,\mathbf{i}, \tag{5.28}$$

then, we have that

$$\boldsymbol{\xi}^0 = \mathbf{x}^0 - s\varkappa y^0\,\mathbf{i}, \tag{5.29}$$

$$ {}^\circ\mathbf{F} = \mathbf{1} - s\varkappa\, \mathbf{j}\otimes\mathbf{i}, \tag{5.30}$$

$$\nabla{}^\circ\mathbf{F} = \mathbf{0}. \tag{5.31}$$

Using (5.27)—(5.31) in (5.23) leads to

$$[C_t(t-s)\langle ij\rangle - {}^{\circ}C\langle ij\rangle]$$

$$= \begin{pmatrix} 0 & 0 & 0 \\ \cdot & 0 & \frac{A}{h}[\sin\omega(t-s) - \sin\omega t] \\ \cdot & \cdot & 0 \end{pmatrix}, \quad (5.32)$$

which is to be compared with (5.16)-(5.17).

6. Static Continuation and Jump Histories

In stress relaxation experiments, a material is held in a state of deformation without any additional motion for a long time and the decay of stress is recorded. Here we shall express the strain history, connected with a continuation of a state of deformation through a time interval, mathematically.

Let $\mathbf{x}(\mathbf{X}, t_0)$ be the position occupied by the particle at time t_0, and let this be also its position at time t, a time interval δ later. Thus,

$$\mathbf{x}(\mathbf{X}, t_0) = \mathbf{x}(\mathbf{X}, t), \qquad t - t_0 = \delta > 0. \qquad (6.1)$$

Then the deformation gradients $\mathbf{F}(t_0)$ and $\mathbf{F}(t)$ are identical for $t \geqq t_0$. We may write its history as:

$$\mathbf{F}(t-s) = \begin{cases} \mathbf{F}(t_0), & 0 \leqq s \leqq \delta, \\ \mathbf{F}(t_0 - s + \delta), & \delta \leqq s < \infty. \end{cases} \qquad (6.2)$$

We agree to call such a deformation gradient history a *static continuation of* $\mathbf{F}(t_0)$ *through* δ [1964 : 4, p. 34]. The relative deformation gradient is

$$\mathbf{F}_t(t-s) = \mathbf{F}(t-s)\,\mathbf{F}(t)^{-1} = \begin{cases} \mathbf{1}, & 0 \leqq s \leqq \delta, \\ \mathbf{F}(t_0 - s + \delta)\,\mathbf{F}(t_0)^{-1}, & \delta \leqq s < \infty: \end{cases} \qquad (6.3)$$

$$= \mathbf{F}(t_0 - \tau)\,\mathbf{F}(t_0)^{-1}, \qquad 0 \leqq \tau < \delta$$

$$= \mathbf{F}_{t_0}(t_0 - \tau), \qquad 0 \leqq \tau < \infty.$$

Hence the strain history is given by

$$\mathbf{C}_t(t-s) = \begin{cases} \mathbf{1}, & 0 \leqq s \leqq \delta, \\ \mathbf{C}_{t_0}(t_0 - \tau), & 0 \leqq \tau < \infty \end{cases} \qquad (6.4)$$

$$= \mathbf{C}_{t_0}(t_0 - s + \delta), \quad \delta \leqq s < \infty.$$

This form will be particularly useful in discussing stress relaxation later on.

A second important strain history is the following: suppose that at time t_0, the deformation gradient suffers a jump, that is, a discontinuity in $\mathbf{F}_{t_0}(t)$ exists, and that the material remains in this position after the jump. Then, we have, with $t - t_0 = \delta$,

$$\mathbf{F}_t(t-s) = \begin{cases} \mathbf{1}, & 0 \leqq s < \delta, \\ \mathbf{F}_t(t_0 - s + \delta), & \delta \leqq s < \infty, \end{cases}$$
$$= \mathbf{F}_{t_0}(t_0 - s + \delta)\, \mathbf{F}_t(t_0), \qquad \delta \leqq s < \infty, \tag{6.5}$$
$$= \mathbf{F}_{t_0}(t_0 - \tau)\, \mathbf{F}_t(t_0), \qquad 0 \leqq \tau < \infty.$$

[26]

Thus the strain history reads

$$\mathbf{C}_t(t-s) = \begin{cases} \mathbf{1}, & 0 \leqq s < \delta, \\ \mathbf{F}_t(t_0)^T\, \mathbf{C}_{t_0}(t_0 - s + \delta)\, \mathbf{F}_t(t_0), & \delta \leqq s < \infty \end{cases}$$
$$= \mathbf{F}_t(t_0)^T\, \mathbf{C}_{t_0}(t_0 - \tau)\, \mathbf{F}_t(t_0), \qquad 0 \leqq \tau < \infty. \tag{6.6}$$

This form $(6.6)_2$ will be particularly useful in discussing the elastic response of a viscoelastic fluid particle to a sudden change of its configuration (see Secs. 34 and 50).

Appendix: Some Basic Results from Tensor Analysis

This is a very brief outline of the various results from three-dimensional tensor analysis that one needs in the study of fluid flows.

Let $\{x^i\}$ be a curvilinear orthogonal coordinate system and let $\{X^K\}$ be a Cartesian coordinate system. The measures of the distance in these two systems are given by the covariant metric tensors g_{ij} end δ_{KL} respectively. These are defined, by using the summation convention, through the square of the length of an infinitesimal element, viz.

$$dl^2 = \delta_{KL}\, dX^K\, dX^L = g_{ij}\, dx^i\, dx^j, \tag{A1.1}$$

and considering $X^K = X^K(x^i)$, one has that

$$g_{ij} = \delta_{KL} \frac{\partial X^K}{\partial x^i} \frac{\partial X^L}{\partial x^j}, \tag{A1.2}$$

where δ_{KL} is the Kronecker delta. We reiterate that the summation convention will be employed from now on. Using

$$X^1 = x^1 \cos x^2, \qquad X^2 = x^1 \sin x^2, \qquad X^3 = x^3, \tag{A1.3}$$

where (x^1, x^2, x^3) stand for (r, θ, z) of the cylindrical polar coordinate system, one can show that

$$g_{11} = 1, \quad g_{22} = r^2, \quad g_{33} = 1, \quad g_{ij} = 0, \quad i \neq j. \tag{A1.4}$$

Similarly, for the spherical polar coordinate system with

$$(r, \theta, \varphi) \sim (x^1, x^2, x^3)$$

26. COLEMAN [1964 : 5].

and

$$X^1 = x^1 \sin x^2 \cos x^3, \quad X^2 = x^1 \sin x^2 \sin x^3, \quad X^3 = x^1 \cos x^2, \tag{A1.5}$$

we have that

$$g_{11} = 1, \quad g_{22} = r^2, \quad g_{33} = r^2 \sin^2\theta, \quad g_{ij} = 0, \quad i \neq j. \tag{A1.6}$$

The property that $g_{ij} = 0$, $i \neq j$, is valid in orthogonal coordinate systems only.

One defines the contravariant metric tensor g^{ij} in orthogonal coordinate systems as

$$g^{ij} = \begin{cases} 1/g_{ij}, & i = j, \\ 0, & i \neq j, \end{cases} \tag{A1.7}$$

so that g^{ij} is diagonal also.

Now let $\mathbf{v}(\mathbf{x})$ be a vector field. Then, with respect to the basis $\{g_i\}$ at the point $\mathbf{x}$ in a space which is related to the Cartesian basis $\{\mathbf{i}_K\}$ at $\mathbf{X}$ via

$$\mathbf{g}_i = \frac{\partial X^K}{\partial x^i}\mathbf{i}_K, \tag{A1.8}$$

the contravariant components $V^i(\mathbf{x})$ are defined through

$$\mathbf{v}(\mathbf{x}) = v^i(\mathbf{x})\, \mathbf{g}_i(\mathbf{x}). \tag{A1.9}$$

The reciprocal $\mathbf{g}^i(\mathbf{x})$ is defined by the usual dot product:

$$\mathbf{g}^i(\mathbf{x})\cdot\mathbf{g}_j(\mathbf{x}) = \delta^i_j, \tag{A1.10}$$

where δ^i_j is again the Kronecker delta. Then (A1.9) tells us that

$$v^j(\mathbf{x}) = \mathbf{v}(\mathbf{x})\cdot\mathbf{g}^j(\mathbf{x}). \tag{A1.11}$$

The covariant components V_i of $\mathrm{V}(\mathbf{x})$ are defined through

$$\mathbf{v}(\mathbf{x}) = v_i(\mathbf{x})\, \mathbf{g}^i(\mathbf{x}), \tag{A1.12}$$

and

$$v_j(\mathbf{x}) = \mathbf{v}(\mathbf{x})\cdot\mathbf{g}_j(\mathbf{x}). \tag{A1.13}$$

In terms of these bases,

$$g_{ij} = \mathbf{g}_i\cdot\mathbf{g}_j, \qquad g^{ij} = \mathbf{g}^i\cdot\mathbf{g}^j. \tag{A1.14}$$

The set $\{\mathbf{g}_i\}$ is called the *natural basis* at $\mathbf{x}$, and $\{\mathbf{g}^i\}$ the *reciprocal basis*. Since they are not necessarily of unit length, we shall introduce an orthonormal basis $\mathbf{e}_i(\mathbf{x})$ via

$$\mathbf{e}_i(\mathbf{x}) = \mathbf{g}_i/\sqrt{g_{\underline{ii}}} = \mathbf{g}^i\sqrt{g_{\underline{ii}}}, \tag{A1.15}$$

where the underscoring suspends summation. This convention will be adopted in this volume. The set $\{\mathbf{e}_i\}$ is called the *natural orthonormal basis.* We define the physical component $v\langle i\rangle$ of a vector $\mathbf{v}$ to be

$$\mathbf{v}\langle i\rangle = \mathbf{v}\cdot\mathbf{e}_i(\mathbf{x}) = v^i\sqrt{g_{\underline{ii}}} = v_i/\sqrt{g_{\underline{ii}}}. \tag{A1.16}$$

For a second order tensor $\mathbf{T}$, one defines the contravariant, covariant and mixed components via

$$\mathbf{T} = T^{ij}\, \mathbf{g}_i \otimes \mathbf{g}_j = T_{ij}\, \mathbf{g}^i \otimes \mathbf{g}^j$$
$$= T^i_{\ j}\, \mathbf{g}_i \otimes \mathbf{g}^j = T^{\ j}_{i}\, \mathbf{g}^i \otimes \mathbf{g}_j, \tag{A1.17}$$

where $\mathbf{g}_i \otimes \mathbf{g}_j$ is called the *outer product* of the vectors $\mathbf{g}_i$ and $\mathbf{g}_j$, etc. The set of nine such products forms a basis for the space of linear transformations. Analogous to (A1.15), we define the physical components $T\langle ij \rangle$ of the second order tensor $\mathbf{T}$ to be:

$$T\langle ij \rangle = T^{ij} \sqrt{g_{\underline{ii}}\, g_{\underline{jj}}} = T_{ij}/\sqrt{g_{\underline{ii}}\, g_{\underline{jj}}}$$
$$= T^i_{\ j} \sqrt{g_{\underline{ii}}/g_{\underline{jj}}} = T^{\ j}_{i} \sqrt{g_{\underline{jj}}/g_{\underline{ii}}}, \tag{A1.18}$$

where the summation convention is again suspended. For example.

$$T\langle 11 \rangle = T^{11} \sqrt{g_{11}^{2}}, \tag{A1.19}$$

etc. Note that in a Cartesian coordinate system, the physical and tensorial components of a vector (or a tensor) are identical.

The *covariant derivative* $v^i_{;j}$ of a contravariant vector field $v^i(\mathbf{x})$ is defined by

$$v^i_{;j} = \frac{\partial v^i}{\partial x^j} + \left(\begin{matrix} i \\ jk \end{matrix}\right) v^k, \tag{A1.20}$$

while the derivative $V_{i;j}$ of a contravariant vector field $v_i(\mathbf{x})$ is

$$v_{i;j} = \frac{\partial v_i}{\partial x^j} - \left(\begin{matrix} k \\ ij \end{matrix}\right) v_k. \tag{A1.21}$$

Here $\left(\begin{matrix} i \\ jk \end{matrix}\right) = \left(\begin{matrix} i \\ kj \end{matrix}\right)$ are called the *Christoffel symbols* of the second kind and the summation convention applies here too. We list the symbols in three coordinates:

(*i*) $\left(\begin{matrix} i \\ jk \end{matrix}\right) = 0$ in Cartesian coordinates, so that covariant differentiation reduces to partial differentiation.

(*ii*) $$\left(\begin{matrix} 1 \\ 22 \end{matrix}\right) = -r, \quad \left(\begin{matrix} 2 \\ 12 \end{matrix}\right) = \left(\begin{matrix} 2 \\ 21 \end{matrix}\right) = \frac{1}{r}, \tag{A1.22}$$

with the others zero in cylindrical polar coordinates

$$(x^1 = r,\ x^2 = \theta,\ x^3 = z).$$

(*iii*) $$\left(\begin{matrix} 1 \\ 22 \end{matrix}\right) = -r, \quad \left(\begin{matrix} 1 \\ 33 \end{matrix}\right) = -r \sin^2\theta, \quad \left(\begin{matrix} 2 \\ 33 \end{matrix}\right) = -\sin\theta\cos\theta,$$
$$\left(\begin{matrix} 2 \\ 12 \end{matrix}\right) = \left(\begin{matrix} 3 \\ 13 \end{matrix}\right) = \frac{1}{r}, \quad \left(\begin{matrix} 3 \\ 23 \end{matrix}\right) = \cot\theta, \tag{A1.23}$$

are the non-vanishing symbols in spherical polar coordinates

$$(x^1 = r, \quad x^2 = \theta, \quad x^3 = \varphi).$$

One can define the covariant derivative of a second order tensor T^{ij} through

$$T^{ij}_{;k} = \frac{\partial T^{ij}}{\partial x^k} + \begin{pmatrix} i \\ kl \end{pmatrix} T^{kj} + \begin{pmatrix} j \\ kl \end{pmatrix} T^{il}. \tag{A1.24}$$

Similarly, the covariant derivative of T_{ij} is given by

$$T_{ij;k} = \frac{\partial T_{ij}}{\partial x^k} - \begin{Bmatrix} l \\ ik \end{Bmatrix} T_{lj} - \begin{Bmatrix} l \\ jk \end{Bmatrix} T_{il}, \tag{A1.25}$$

and similarly derivatives of mixed tensors.

Applications. We now comment on certain applications.

(i) *Rising and lowering indices.* By this is meant the operation of converting a covariant index into a contravariant one and *vice versa.* For instance,

$$v^i = g^{ij} v_j, \qquad v_i = g_{ij} v^j, \tag{A1.26}$$

$$T^{ij} = g^{ik} g^{jl} T_{kl}, \text{ etc.}$$

(ii) *Divergence operator.* This is a differential operator which reduces the rank of the tensor by one. So a vector becomes a scalar on the application of this operator. Hence

$$\text{div } \mathbf{v} = v^i_{;i}, \tag{A1.27}$$

$$\text{div } \mathbf{T} = (T^{ij}_{;j})\, \mathbf{g}_i, \tag{A1.28}$$

where the fact that div **T** is a vector has been used.

(iii) *Curl operator.* We shall mention the curl of a vector field only here. We let curl $\mathbf{v} = \boldsymbol{\omega}$, where $\boldsymbol{\omega}$ is a vector such that

$$\omega^i = \varepsilon^{ijk} v_{k;j}, \tag{A1.29}$$

$$\varepsilon^{ijk} = e^{ijk}/\sqrt{g}, \qquad g = \det g_{ij}, \tag{A1.30}$$

$$e^{ijk} = \begin{cases} 1 & \text{if } i, j, k \text{ are an even permutation of } 1, 2, 3, \\ 0 & \text{if } i, j, k \text{ is not a permutation,} \\ -1 & \text{if } i, j, k \text{ is an odd permutation.} \end{cases} \tag{A1.31}$$

One may obtain the covariant component from $\omega_i = g_{ij}\, \omega^j$. Now (A1.31) means that

$$e^{123} = e^{231} = e^{312} = +1, \qquad e^{132} = e^{321} = e^{213} = -1, \tag{A1.32}$$

and that others are zero. Since e^{ijk} is a Cartesian tensor, $e^{ijk} \equiv e_{ijk}$.

(iv) *Physical components of derivatives.* The covariant differentiations listed above are all expressible in terms of the Christoffel symbols, the contravariant or covariant or mixed components, and their derivatives. In practice one usually needs the *physical components of the derivatives*, in terms of the *physical components of the vectors or tensors*. In such cases, it is preferable to perform the differentiations and then obtain the physical components of the derivatives.

As an example, consider the component $T^{1j}_{;j}$ in the cylindrical polar coordinate system. Let $T^{ij}_{;j} = A^i$, and thus

$$A^1 = \frac{\partial T^{1j}}{\partial x^j} + \begin{pmatrix} 1 \\ jk \end{pmatrix} T^{kj} + \begin{pmatrix} j \\ jk \end{pmatrix} T^{1k}. \qquad \text{(A1.34)}$$

Using the non-zero components of $\begin{pmatrix} i \\ jk \end{pmatrix}$ from (A1.22), one has

$$A^1 = \frac{\partial T^{rr}}{\partial r} + \frac{\partial T^{r\theta}}{\partial \theta} + \frac{\partial T^{rz}}{\partial z} - rT^{\theta\theta} + \frac{T^{rr}}{r}. \qquad \text{(A1.35)}$$

Since $A^1 = A^r = A\langle r \rangle$, $T^{rr} = T\langle rr \rangle$, $T^{rz} = T\langle rz \rangle$, $T^{r\theta} = T\langle r\theta \rangle / r$. $T^{\theta\theta} = T\langle \theta\theta \rangle / r^2$, one has that

$$A\langle r \rangle = \frac{\partial T\langle rr \rangle}{\partial r} + \frac{1}{r}\frac{\partial T\langle r\theta \rangle}{\partial \theta} + \frac{\partial T\langle rz \rangle}{\partial z} + \frac{1}{r}(T\langle rr \rangle - T\langle \theta\theta \rangle),$$

a result which will be useful in Chap. 6. (1.36)

(v) *Divergence and curl of vectors.* We list below the divergence and curl of vectors in the three coordinates systems.

(*a*) *div* **V**:

Cartesian: $$\frac{\partial v\langle x \rangle}{\partial x} + \frac{\partial v\langle y \rangle}{\partial y} + \frac{\partial v\langle z \rangle}{\partial z}. \qquad \text{(A1.37)}$$

Cyl. Polar: $$\frac{\partial v\langle r \rangle}{\partial r} + \frac{v\langle r \rangle}{r} + \frac{1}{r}\frac{\partial v\langle \theta \rangle}{\partial \theta} + \frac{\partial v\langle z \rangle}{\partial z}. \qquad \text{(A1.38)}$$

Spherical Polar: $$\frac{\partial v\langle r \rangle}{\partial r} + \frac{2v\langle r \rangle}{r} + \frac{1}{r}\frac{\partial v\langle \theta \rangle}{\partial \theta} + \frac{v\langle \theta \rangle}{r}\cot\theta + \frac{1}{r\sin\theta}\frac{\partial v\langle \varphi \rangle}{\partial \varphi}. \qquad \text{(A1.39)}$$

(*b*) $\boldsymbol{\omega}$ = curl **v**:

Cartesian: $$\omega\langle x \rangle = \frac{\partial v\langle z \rangle}{\partial y} - \frac{\partial v\langle z \rangle}{\partial z}, \text{ etc.} \qquad \text{(A1.40)}$$

Cyl. Polar:

$$\omega\langle r \rangle = \frac{1}{r}\frac{\partial v\langle z \rangle}{\partial \theta} - \frac{\partial v\langle \theta \rangle}{\partial z}, \qquad \text{(A1.41)}$$

$$\omega\langle \theta \rangle = \frac{\partial v\langle r \rangle}{\partial z} - \frac{\partial v\langle z \rangle}{\partial r},$$

$$\omega\langle z \rangle = \frac{\partial v\langle \theta \rangle}{\partial r} + \frac{v\langle \theta \rangle}{r} - \frac{1}{r}\frac{\partial v\langle r \rangle}{\partial \theta}.$$

Sph. Polar:

$$\omega\langle r\rangle = \frac{1}{r}\frac{\partial v\langle\varphi\rangle}{\partial\theta} + \frac{v\langle\varphi\rangle}{r}\cot\theta - \frac{1}{r\sin\theta}\frac{\partial v\langle\theta\rangle}{\partial\varphi},$$

$$\omega\langle\theta\rangle = \frac{1}{r\sin\theta}\frac{\partial v\langle r\rangle}{\partial\varphi} - \frac{v\langle\varphi\rangle}{r} - \frac{\partial v\langle\varphi\rangle}{\partial r}, \tag{A1.42}$$

$$\omega\langle\varphi\rangle = \frac{\partial v\langle\theta\rangle}{\partial r} + \frac{v\langle\theta\rangle}{r} - \frac{1}{r}\frac{\partial v\langle r\rangle}{\partial\theta}.$$

(vi) *Components of the rate of deformation tensor.* The physical components of the rate of deformation tensor $\mathbf{D} = \frac{1}{2}\mathbf{A}_1$ are listed in Cartesian, cylindrical and spherical polar coordinates below, in terms of the velocity field $\mathbf{v}$.

Cartesian:

$$D\langle ij\rangle = \frac{\partial v\langle i\rangle}{\partial x^j} + \frac{\partial v\langle j\rangle}{\partial x^i}. \tag{A1.43}$$

Cylindrical Polar:

$$D\langle rr\rangle = \frac{\partial v\langle r\rangle}{\partial r}, \qquad D\langle zz\rangle = \frac{\partial v\langle z\rangle}{\partial z},$$

$$D\langle\theta\theta\rangle = \frac{1}{r}\left(\frac{\partial v\langle\theta\rangle}{\partial\theta} + v\langle r\rangle\right), \tag{A1.44}$$

$$2D\langle r\theta\rangle = \frac{\partial v\langle\theta\rangle}{\partial r} - \frac{v\langle\theta\rangle}{r} + \frac{1}{r}\frac{\partial v\langle r\rangle}{\partial\theta},$$

$$2D\langle\theta z\rangle = \frac{1}{r}\frac{\partial v\langle z\rangle}{\partial\theta} + \frac{\partial v\langle\theta\rangle}{\partial z},$$

$$2D\langle rz\rangle = \frac{\partial v\langle r\rangle}{\partial z} + \frac{\partial v\langle z\rangle}{\partial r}.$$

Sph. Polar:

$$D\langle rr\rangle = \frac{\partial v\langle r\rangle}{\partial r},$$

$$D\langle\theta\theta\rangle = \frac{1}{r}\left(\frac{\partial v\langle\theta\rangle}{\partial\theta} + v\langle r\rangle\right),$$

$$\sin\theta\, D\langle\varphi\varphi\rangle = \frac{\partial v\langle\varphi\rangle}{\partial\varphi} + v\langle r\rangle\sin\theta + v\langle\theta\rangle\cos\theta, \tag{A1.45}$$

$$2D\langle r\theta\rangle = \frac{\partial v\langle\theta\rangle}{\partial r} - \frac{v\langle\theta\rangle}{\partial r} + \frac{1}{r}\frac{\partial v\langle r\rangle}{\partial\theta},$$

$$2r\sin\theta\, D\langle\theta\varphi\rangle = \sin\theta\frac{\partial v\langle\varphi\rangle}{\partial\theta} - v\langle\varphi\rangle\cos\theta + \frac{\partial v\langle\theta\rangle}{\partial\varphi},$$

$$2r\sin\theta\, D\langle r\varphi\rangle = \frac{\partial v\langle r\rangle}{\partial\varphi} + r\sin\theta\frac{\partial v\langle\varphi\rangle}{\partial r} - v\langle\varphi\rangle\sin\theta.$$

Finally, we note that the velocity field $\mathbf{v}$ gives rise to the vorticity $\boldsymbol{\omega}$

through curl $\mathbf{v} = \boldsymbol{\omega}$. Since we know the physical components of $\boldsymbol{\omega}$ in various coordinates, the physical components of the spin tensor

$$\mathbf{W} = \frac{1}{2}(\mathbf{L}-\mathbf{L}^T),$$

where **L** is the velocity gradient, can be found from:

$$[W\langle ij\rangle] = \frac{1}{2}\begin{pmatrix} 0 & -\omega\langle 3\rangle & \omega\langle 2\rangle \\ \omega\langle 3\rangle & 0 & -\omega\langle 1\rangle \\ -\omega\langle 2\rangle & \omega\langle 1\rangle & 0 \end{pmatrix}. \tag{A1.46}$$

CHAPTER 2

KINEMATICS OF MOTIONS WITH CONSTANT STRETCH HISTORY

The flows to be considered in this chapter have kinematical properties which are extremely fascinating, and their study has the added challenge due to their practical importance. For in the set under study, there exist many motions which can be shown to be dynamically possible, that is, they are possible in the laboratory under well-defined sets of forces. More importantly, the equations of motion can be solved without a full knowledge of the "nature of the memory" of the fluid, since the stresses are easily computed. We discuss this latter point briefly for the sake of motivation.

It will be shown later that the stress (T) constitutive equation for a viscoelastic fluid is

$$\mathbf{T}(t) = -p(\rho)\mathbf{1} + \underset{s=0}{\overset{\infty}{\mathscr{H}}} [\mathbf{C}_t(t-s); \rho], \qquad (*)$$

where the stress tensor at time t is a nonlinear functional of the entire strain history of the motion. [Eq. (*) implies that each component $T\langle ij\rangle$ depends *nonlinearly on the six components of the history of* $C_t(t-s)\langle ij\rangle$.] Different fluids react differently to the same strain history because of their own memory; especially how far back they can "remember", how fast is this "decay", and so on. If, for example, the strain history obeys (7.2), then

$$T\langle ij\rangle(\tau) = T\langle ij\rangle(t), \quad \text{if } \rho = \text{const}; \qquad (**)$$

or the time lapse between τ and t has no effect, or we have a *translation invariance* of the stresses along the path line of the particle. So the stress tensor at time t must depend only on the present set of tensors determining $\mathbf{C}_t(t-s)$, e.g. the velocity gradient $\mathbf{L}$ in Sec. 7. Similarly, in Sec. 8 we find flows obeying (8.2), and there one can show that (see Chap. 6)

$$T\langle ij\rangle(\tau) = T\langle ij\rangle(t), \quad (\rho = \text{const.}) \qquad (***)$$

calculated with respect to the rotating basis $\mathbf{b}_k(t)$, defined in (8.3), or one has invariance of the stress tensor with respect to the rotating basis, and is thus easily computed. Once the stress field is known, the dynamics of the motion can be discussed. It is these aspects which make the exponential histories (Sec. 7) and motions with constant strech history (hereafter abbreviated MWCSH) kinematically and dynamically a basic subject in viscoelasticity.

In Sec. 7, we shall find examples of motions obeying (7.2) and in Sec. 8 study them from a more general viewpoint. Additional properties are explored in Secs. 9–11.

7. Exponential Histories

As we pointed out in the opening remarks here, we seek those flows which have *translation invariance properties.* So our aim is to verify if there exist velocity fields such that the corresponding strain histories have the form

$$\mathbf{C}_t(t-s) = e^{-s\mathbf{L}^T} e^{-s\mathbf{L}}, \qquad 0 \leqq s < \infty, \tag{7.1}$$

where $\mathbf{L}$ is a *constant tensor* and $e^{-s\mathbf{L}}$ is the exponential function. Note that if $\mathbf{L}$ is a constant tensor, then

$$\mathbf{C}_\tau(\tau-s) = \mathbf{C}_t(t-s) \tag{7.2}$$

for any other instant τ. We emphasize again that any quantity which may depend in an extremely complicated manner on the history $\mathbf{C}_t(t-s)$ is, mathematically speaking, known once $\mathbf{L}$ is known: for $\mathbf{L}$ determines $\mathbf{C}_t(t-s)$, and any scalar, or vector, or tensor-valued functional of $\mathbf{C}_t(t-s)$ reduces to a function of $\mathbf{L}$.

Let us examine the distinct types of second-order tensors in E^3. A linear transformation $\mathbf{L}$ in E^3 is either *trivially zero*, or

(i) nilpotent of order two, that is, $\mathbf{L}^2 = \mathbf{0}$; or
(ii) nilpotent of order three and not two, that is, $\mathbf{L}^2 \neq \mathbf{0}$, $\mathbf{L}^3 = \mathbf{0}$; or
(iii) not nilpotent, that is, $\mathbf{L}^n \neq \mathbf{0}$ for $n = 1, 2, 3, \ldots$

Excluding the trivial case, for then $\mathbf{C}_t(t-s) \equiv \mathbf{1}$, that is, the particle is in a state of rest for ever, we are left with the three possibilities listed above.[1] We shall study them in detail next.

First of all, we derive an interesting property of strain histories of type (i) and (ii). In E^3, the Hamilton-Cayley Theorem yields

$$\mathbf{L}^3 - I\,\mathbf{L}^2 + II\,\mathbf{L} - III\,\mathbf{1} = \mathbf{0}, \tag{7.3}$$

where I, II and III are the invariants of $\mathbf{L}$ (tr=trace of, det=determinant of):

$$I = \text{tr}\,\mathbf{L}, \qquad II = \frac{1}{2}\{I^2 - \text{tr}\,\mathbf{L}^2\}, \qquad III = \det \mathbf{L}. \tag{7.4}$$

Since $\mathbf{L}^3 = \mathbf{0}$ for the case (ii), we have

$$I\,\mathbf{L}^2 - II\,\mathbf{L} + III\,\mathbf{1} = \mathbf{0}. \tag{7.5}$$

Multiply by $\mathbf{L}^2$ and noting that $\mathbf{L}^2 \neq 0$, derive the following:

$$III = 0. \tag{7.6}$$

Similarly, observe that multiplication of (7.5) by $\mathbf{L}$ yields that $II = 0$. Thus from (7.5) it follows that if $\mathbf{L}^3 = 0$, then

$$\text{tr}\,\mathbf{L} = 0, \qquad \text{tr}\,\mathbf{L}^2 = 0, \qquad \det \mathbf{L} = 0. \tag{7.7}$$

Similarly, if $\mathbf{L}^2 = \mathbf{0}$, then (7.3) reads

$$II\,\mathbf{L} - III\,\mathbf{1} = \mathbf{0}; \tag{7.8}$$

1. Noll [1962 : 7].

from which derive that if $\mathbf{L}^2 = \mathbf{0}$, then

$$II = 0, \qquad III = 0. \tag{7.9}$$

But (7.4)$_2$ implies for $\mathbf{L}^2 = \mathbf{0}$, that

$$II = \frac{1}{2} I^2 = \frac{1}{2} (\text{tr } \mathbf{L})^2. \tag{7.10}$$

Thus, again, when $\mathbf{L}^2 = \mathbf{0}$, we obtain that

$$\text{tr } \mathbf{L} = 0 \tag{7.11}$$

also. Thus for histories of the type (7.1), falling under cases (i) and (ii), we have the important results:

$$\text{tr } \mathbf{L} = 0, \qquad \text{tr } \mathbf{L}^2 = 0, \qquad \det \mathbf{L} = 0. \tag{7.12}$$

From (4.11) we know that

$$\mathbf{A}_1 = (-1) \frac{d}{ds} \mathbf{C}_t(t-s) \bigg|_{s=0} = \mathbf{L} + \mathbf{L}^T. \tag{7.13}$$

Hence, for motions of exponential type (i) and (ii),

$$\text{tr } \mathbf{A}_1 = 0. \tag{7.14}$$

We call exponential histories of these types *isochoric motions* for reasons that will become explicit later in Chap. 4.

Exponential histories of the type (iii) may or may not be isochoric. However, $\mathbf{L}^n \neq \mathbf{0}$ is obeyed by both symmetric $\mathbf{L}$ (always) and some asymmetric $\mathbf{L}$ and thus (iii) has actually two sub-cases.

We now list the examples for each of the three cases in Cartesian coordinates.

(i) $\mathbf{L}^2 = \mathbf{0}$. Let the velocity field be defined through

$$\dot{x} = \varkappa y, \qquad \dot{y} = \dot{z} = 0. \tag{7.15}$$

Denoting the coordinates of a particle by (ξ, η, ζ) and (x, y, z) at time $(t-s)$ and t respectively, the path lines are

$$\xi = x - s\varkappa y, \qquad \eta = y, \qquad \zeta = z. \tag{7.16}$$

The strain history has the physical components

$$[C_t(t-s).\langle ij \rangle] = \begin{pmatrix} 1 & -s\varkappa & 0 \\ \cdot & 1 + s^2\varkappa^2 & 0 \\ \cdot & & 1 \end{pmatrix}. \tag{7.17}$$

Hence

$$\mathbf{C}_t(t-s) = (\mathbf{1} - s\mathbf{L}^T)(\mathbf{1} - s\mathbf{L}) = e^{-s\mathbf{L}^T} e^{-s\mathbf{L}}, \tag{7.18}$$

where the matrix of $\mathbf{L}$ has the form

$$[L\langle ij \rangle] = \begin{bmatrix} 0 & \varkappa & 0 \\ 0 & 0 & 0 \\ 0 & 0 & 0 \end{bmatrix}, \qquad \mathbf{L}^2 = \mathbf{0}. \tag{7.19}$$

Note that $\mathbf{L}$ is also the gradient of the velocity field (7.15).

(ii) $\mathbf{L}^2 \neq \mathbf{0}$, $\mathbf{L}^3 = \mathbf{0}$. Let the velocity field be given by[2]

$$\dot{x} = 0, \qquad \dot{y} = lx, \qquad \dot{z} = mx + ny. \tag{7.20}$$

The path lines, with the usual notation, are

$$\xi = x, \qquad \eta = y - slx, \tag{7.21}$$

$$\zeta = z - \int_0^s (m\xi + n\eta)\, d\sigma = z - \int_0^s (mx + ny - n\sigma lx)\, d\sigma$$
$$= z - s(mx+ny) + \frac{1}{2} s^2 lnx. \tag{7.22}$$

The velocity gradient $\mathbf{L}$ associated with (7.20) is given by

$$[L\langle ij\rangle] = \begin{pmatrix} 0 & 0 & 0 \\ l & 0 & 0 \\ m & n & 0 \end{pmatrix}, \qquad [L^2\langle ij\rangle] = \begin{pmatrix} 0 & 0 & 0 \\ 0 & 0 & 0 \\ ln & 0 & 0 \end{pmatrix}. \tag{7.23}$$

The strain history may readily be computed and found to be given by

$$\mathbf{C}_t(t-s) = \left(\mathbf{1} - s\mathbf{L}^T + \frac{1}{2} s^2 \mathbf{L}^{2T}\right)\left(\mathbf{1} - s\mathbf{L} + \frac{1}{2} s^2 \mathbf{L}^2\right) = e^{-s\mathbf{L}^T} e^{-s\mathbf{L}}, \tag{7.24}$$

with $\mathbf{L}$ defined in (7.23).

(iii) $\mathbf{L}^n \neq \mathbf{0}$ for all $n = 1, 2, 3, \ldots$.

(*a*) $\mathbf{L} = \mathbf{L}^T$ *or* $\mathbf{L}$ *is symmetric.* Let the velocity field be that of *simple extension*[3]:

$$\dot{x} = a_1 x, \quad \dot{y} = a_2 y, \quad \dot{z} = a_3 z; \qquad a_i = \text{const.}, \quad i = 1, 2, 3. \tag{7.25}$$

The path lines are

$$\xi = xe^{-sa_1}, \qquad \eta = ye^{-sa_2}, \qquad \zeta = ze^{-sa_3}. \tag{7.26}$$

With the gradient of the velocity field (7.25) having the form

$$[L\langle ij\rangle] = \text{diag}\,[a_1, a_2, a_3], \tag{7.27}$$

the strain history reads:

$$[C_t(t-s)\langle ij\rangle] = \begin{pmatrix} e^{-2sa_1} & 0 & 0 \\ \cdot & e^{-2sa_2} & 0 \\ \cdot & \cdot & e^{-2sa_3} \end{pmatrix}. \tag{7.28}$$

Since $\mathbf{L} = \mathbf{L}^T$, or $\mathbf{L}$ and $\mathbf{L}^T$ commute, we must have

$$e^{-s\mathbf{L}^T} e^{-s\mathbf{L}} = e^{-s(\mathbf{L}^T + \mathbf{L})} = e^{-s\mathbf{A}_1}. \tag{7.29}$$ [4]

2. Noll [1962 : 7], Eq. (3.14). See also: Astarita [1967: 2], Eq. (24), who discussed

$$\dot{x} = \beta yz, \qquad \dot{y} = \gamma z, \qquad \dot{z} = 0.$$

3. Coleman and Noll [1962 : 3].

4. Hence every steady homogeneous velocity field with a symmetric velocity gradient is equivalent to simple extension.

It is seen that $\mathbf{C}_t(t-s)$ in (7.28) is $\exp(-s\mathbf{A}_1)$.

It is trivial to note that *a motion is a simple extension*, that is, *it obeys* (7.29) *if and only if*

$$\mathbf{A}_1^n(t) \equiv \mathbf{A}_n(t), \qquad n = 1, 2, 3, \ldots$$

for all t.

(*b*) $\mathbf{L} \neq \mathbf{L}^T$. Let the velocity field now be that occurring in the Maxwell

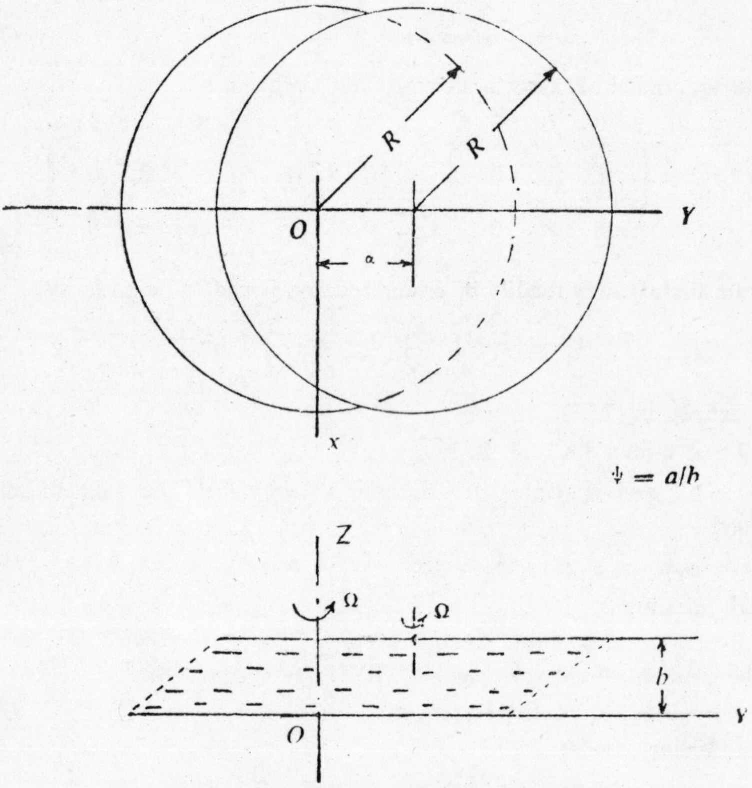

Fig. 7.1. Maxwell orthogonal rheometer

orthogonal rheometer[5] (see Fig. 7.1):

$$\dot{x} = -\Omega y + \Omega\psi z, \qquad \dot{y} = \Omega x, \qquad \dot{z} = 0, \tag{7.30}$$

where Ω and ψ are constants.

Since (7.30) is equivalent to

$$\frac{d\xi}{ds} = \Omega\eta - \Omega\psi\zeta, \quad \frac{d\eta}{ds} = -\Omega\xi, \qquad \frac{d\zeta}{ds} = 0, \tag{7.31}$$

5. BIRD and HARRIS [1968 : 2]; HUILGOL [1969 : 8]; MAXWELL and CHARTOFF [1965 : 11].

one has $\zeta = z$, and

$$\frac{d^2\xi}{ds^2} = -\Omega^2\xi,$$
$$\frac{d^2\eta}{ds^2} = -\Omega^2\eta + \Omega^2\psi\zeta. \tag{7.32}$$

The solutions to (7.32) are thus given by:

$$\begin{aligned} \xi &= A(x, y, z)\cos\Omega s + B(x, y, z)\sin\Omega s, \\ \eta &= C(x, y, z)\cos\Omega s + D(x, y, z)\sin\Omega s + \psi z, \\ \zeta &= z. \end{aligned} \tag{7.33}$$

In (7.33), $A(x, y, z)$ etc., are functions of (x, y, z) and are to be determined from the initial conditions and (7.31). The initial conditions that $\xi = x$, $\eta = y$ at $s = 0$ yield

$$A(x, y, z) = x; \quad C(x, y, z) = y - \psi z. \tag{7.34}$$

Using these functions in (7.33), computing $d\xi/ds$ and $d\eta/ds$ and comparing the results with (7.31) yield

$$D(x, y, z) = -x, \qquad B(x, y, z) = y - \psi z. \tag{7.35}$$

Hence the path lines are

$$\begin{aligned} \xi &= x\cos\Omega s + (y-\psi z)\sin\Omega s, \\ \eta &= (y-\psi z)\cos\Omega s - x\sin\Omega s + \psi z, \\ \zeta &= z, \end{aligned} \tag{7.36}$$

and from these the strain history may be computed. There is a more direct way of calculating this history, however. We shall outline this approach next.

The velocity gradient $\mathbf{L}$ for the field (7.30) is given by

$$[L\langle ij\rangle] = \begin{pmatrix} 0 & -\Omega & \Omega\psi \\ \Omega & 0 & 0 \\ 0 & 0 & 0 \end{pmatrix}, \tag{7.37}$$

and hence

$$\frac{d}{dt}\mathbf{L} = \frac{\partial}{\partial t}\mathbf{L} + \nabla\mathbf{L}\cdot\mathbf{v} = \mathbf{0}. \tag{7.38}$$

But, from (4.5), we know that

$$\frac{d}{dt}\mathbf{F}(t) = \mathbf{L}(t)\,\mathbf{F}(t), \tag{7.39}$$

and if $\mathbf{L}$ is a *constant*, we may integrate (7.39) to obtain

$$\mathbf{F}(t) = e^{t\mathbf{L}}\,\mathbf{F}(0). \tag{7.40}$$

Next, the eauation (7.40) yields

$$\mathbf{F}_t(t-s) = \mathbf{F}(t-s)\,\mathbf{F}(t)^{-1} = e^{(t-s)\,\mathbf{L}}\,e^{-t\mathbf{L}} = e^{-s\mathbf{L}}, \tag{7.41}$$

and we immediately read off the strain history:

$$\mathbf{C}_t(t-s) = e^{-s\mathbf{L}^T} e^{-s\mathbf{L}}, \tag{7.42}$$

where $\mathbf{L}$ is given by (7.37). When one notes that

$$\mathbf{L}^{2n+1} = (-1)^n \Omega^{2n} \mathbf{L}, \qquad n = 1, 2, 3, \ldots, \tag{7.43}$$

$$\mathbf{L}^{2n} = (-1)^{n+1} \Omega^{2(n-1)} \mathbf{L}, \qquad n = 1, 2, 3, \ldots, \tag{7.44}$$

$\mathbf{C}_t(t-s)$ has the form

$$[C_t(t-s) \langle ij \rangle] = \begin{pmatrix} 1 & 0 & -\psi \sin \Omega s \\ \cdot & 1 & -\psi(1-\cos \Omega s) \\ \cdot & \cdot & 1 + 2\psi^2(1-\cos \Omega s) \end{pmatrix}. \tag{7.45}$$

In fact, in the three earlier examples, we may have performed the integration given by (7.39) and obtained (7.42). We have proved[6]

Let the material derivative $\dot{\mathbf{L}} = \mathbf{0}$ *for all* $-\infty < t < \infty$. *Then the strain history has the form* (7.1).

In the next section, we generalize the equations (7.1) and (7.2) and obtain motions in curvilinear coordinate systems.

8. Motions with Constant Stretch History

From linear algebra[7] we know that if $\mathbf{T}$ is a linear transformation in E^3, and $\mathbf{b}_i (i=1, 2, 3)$ is an *orthonormal basis* for E^3, then the components of $\mathbf{T}$ relative to $\mathbf{b}_i$, which are also the physical components $T\langle ij \rangle$ of $\mathbf{T}$, are given by

$$T\langle ij \rangle = \mathbf{b}_i \cdot \mathbf{T}\mathbf{b}_j \,. \tag{8.1}$$

Let $\mathbf{C}_t(t-s)$ and $\mathbf{C}_0(0-s)$ be the strain histories of the motion of a particle computed with respect to t and a fixed instant 0 respectively. Suppose that

$$\mathbf{C}_t(t-s) = \mathbf{Q}(t)\, \mathbf{C}_0(0-s)\, \mathbf{Q}(t)^T, \tag{8.2}$$

where $\mathbf{Q}(t)$ is an orthogonal tensor function of t. Further, let $\mathbf{b}_i(0)$ $(i=1, 2, 3)$, be an orthonormal basis at time 0 and let us choose a basis $\mathbf{b}_i(t)$ at time t through

$$\mathbf{b}_i(t) = \mathbf{Q}(t)\, \mathbf{b}_i(0), \qquad \mathbf{b}_i(0) = \mathbf{Q}(t)^T \mathbf{b}_i(t), \qquad i = 1, 2, 3. \tag{8.3}$$

Then

$$C_t(t-s)\langle ij \rangle = \mathbf{b}_i(t) \cdot \mathbf{Q}(t)\, \mathbf{C}_0(0-s)\, \mathbf{Q}(t)^T \mathbf{b}_j(t) \tag{8.4}$$

$$= \mathbf{b}_i(t) \cdot \mathbf{Q}(t)\, \mathbf{C}_0(0-s)\, \mathbf{b}_j(0), \tag{8.5}$$

because of $(8.3)_2$. Also, we can rewrite (8.5) as

$$C_t(t-s)\langle ij \rangle = \mathbf{Q}^T(t)\, \mathbf{b}_i(t) \cdot \mathbf{C}_0(0-s)\, \mathbf{b}_j(0) \tag{8.6}$$

$$= C_0(0-s)\langle ij \rangle. \tag{8.7}$$

6. HUILGOL [1971 . 12].
7. See, for example, MARTIN and MIZEL [1966 : 12], Chap. 4.

From this we conclude that if the strain history obeys (8.2) then we can choose a time-dependent vector basis (8.3) such that the components, with respect to this basis, of $\mathbf{C}_t(t-s)$ and $\mathbf{C}_0(0-s)$ are identical. Motions obeying (8.2) were called *substantially stagnant motions* by COLEMAN [1962 : 2]. The term *motions with constant stretch history* (abbreviated MWCSH) was conceived by NOLL in the same year [1962: 7] for such flows because of the equality of

$$C_t(t-s)\,\langle ij\rangle \qquad \text{and} \qquad C_0(0-s)\,\langle ij\rangle.$$

For, by definition[8], the stretch is given by

$$\lambda_t^i\,(t-s)/1, \quad \text{where} \quad \lambda_t^i\,(t-s)\ (i=1,2,3)$$

are the eigenvalues of $\mathbf{U}_t(t-s)$, the unique square root of $\mathbf{C}_t(t-s)$. For motions obeying (8.2), because of (8.7),

$$\lambda_t^i\,(t-s) = \lambda_0^i\,(0-s), \qquad i = 1, 2, 3, \qquad 0 \leqq s < \infty. \tag{8.8}$$

or the stretch has the same value at $(t-s)$ or at $(0-s)$. We conclude therefore that its *history* is the same and hence follows NOLL's title.

WANG [1965 : 20] defined an *invariant motion* to be one which satisfies [*cf.* (7.40)] :

$$\mathbf{F}(t-s) = e^{(t-s)\mathbf{A}}\,\mathbf{F}(0), \qquad \mathbf{A} = \text{constant}, \tag{8.9}$$

and showed that this is equivalent to (8.2).

So far we have not discussed what form the strain history $\mathbf{C}_t(t-s)$ should take if it obeys (8.2). We discuss this below, though we may anticipate the result to a large extent because of (7.1) and (7.2).

THEOREM 8.1 (NOLL)[9]. *A motion is a motion with constant stretch history if and only if the deformation gradient $\mathbf{F}_0(\tau)$ relative to some fixed time 0 has the representation*

$$\mathbf{F}_0(\tau) = \mathbf{Q}(\tau)\,e^{\tau\mathbf{M}}, \qquad \mathbf{Q}(0) = \mathbf{1}, \tag{8.10}$$

where $\mathbf{M}$ *is a constant tensor and* $\mathbf{Q}(\tau)$ *is an orthogonal tensor function.* The proof is presented in the Appendix.

We now give a list of formulae which are derivable from (8.10). Let

$$\mathbf{L}_1(t) = \mathbf{Q}(t)\,\mathbf{M}\mathbf{Q}(t)^T, \qquad \mathbf{Z}(t) = \dot{\mathbf{Q}}(t)\,\mathbf{Q}(t)^T. \tag{8.11}$$

Clearly, $\mathbf{Z}(t)$ is skew, because it follows from $\mathbf{Q}(t)\,\mathbf{Q}(t)^T = \mathbf{1}$ that

$$\dot{\mathbf{Q}}(t)\,\mathbf{Q}(t)^T + \mathbf{Q}(t)\,\dot{\overline{\mathbf{Q}(t)^T}} = \mathbf{0}; \tag{8.12}$$

or,

$$\begin{aligned}\mathbf{Z}(t) = \dot{\mathbf{Q}}(t)\,\mathbf{Q}(t)^T &= -\,\mathbf{Q}(t)\,\dot{\overline{\mathbf{Q}(t)^T}} = -\,\mathbf{Q}(t)\,\dot{\mathbf{Q}}(t)^T \\ &= -\,[\dot{\mathbf{Q}}(t)\,\mathbf{Q}(t)^T]^T = -\,\mathbf{Z}(t)^T.\end{aligned} \tag{8.13}$$

8. CFT, Secs. 25—26.

9. [1967 : 7] Thm. 1. The tensor function $\mathbf{Q}(\tau)$ which appears in (8.10) yields the tensor $\mathbf{Q}(t)$ of (8.3), when evaluated at $\tau = t$.

Note that $\mathbf{Z}(t)$ is the spin tensor with components equal to those of the angular velocity of the base vector frame $\mathbf{b}_i(t)$. Next,

$$\mathbf{F}_t(\tau) = \mathbf{F}_0(\tau)\,\mathbf{F}_0(t)^{-1} = \mathbf{Q}(\tau)e^{(\tau-t)\,\mathbf{M}}\,\mathbf{Q}(t)^T$$

$$= \mathbf{Q}(\tau)\,\mathbf{Q}(t)^T\;e^{(\tau-t)\,\mathbf{L}_1(t)}\,, \tag{8.14}$$

$$\mathbf{C}_t(\tau) = \mathbf{Q}(t)\,e^{(\tau-t)\,\mathbf{M}^T}\,e^{(\tau-t)\,\mathbf{M}}\,\mathbf{Q}(t)^T \tag{8.15}$$

$$= \mathbf{Q}(t)\,e^{(\tau-t)\,\mathbf{M}^T}\,\mathbf{Q}(t)^T\,\mathbf{Q}(t)\,e^{(\tau-t)\,\mathbf{M}}\,\mathbf{Q}(t)^T$$

$$= e^{(\tau-t)\,\mathbf{Q}(t)\,\mathbf{M}^T\,\mathbf{Q}(t)^T}\,e^{(\tau-t)\,\mathbf{Q}(t)\mathbf{M}\,\mathbf{Q}(t)^T} \tag{8.16}$$

$$= e^{(\tau-t)\,\mathbf{L}_1^T}\,e^{(\tau-t)\,\mathbf{L}_1}, \tag{8.17}$$

because, for any $\mathbf{A}$ such that $\mathbf{A}^{-1}$ exists,

$$\mathbf{A}e^{\mathbf{M}}\,\mathbf{A}^{-1} = e^{\mathbf{AMA}^{-1}}. \tag{8.18}$$

Moreover,

$$\frac{d}{d\tau}\,\mathbf{F}_t(\tau)\Big|_{\tau=t} = \mathbf{L}(t) = \left\{\dot{\mathbf{Q}}(\tau)\,e^{(\tau-t)\,\mathbf{M}}\,\mathbf{Q}(t)^T + \mathbf{Q}(\tau)\,e^{(\tau-t)\,\mathbf{M}}\,\mathbf{M}\mathbf{Q}(t)^T\right\}_{\tau=t}$$

$$= \mathbf{Z}(t) + \mathbf{L}_1(t). \tag{8.19}$$

Hence,

$$\dot{\mathbf{L}}(t) = \dot{\mathbf{Z}}(t) + \dot{\mathbf{L}}_1(t)$$

$$= \dot{\mathbf{Z}}(t) + \mathbf{Z}(t)\,\mathbf{L}_1(t) - \mathbf{L}_1(t)\,\mathbf{Z}(t). \tag{8.20}$$

Finally,

$$\mathbf{A}_n(t) = \mathbf{Q}(t)\,\mathbf{A}_n(0)\,\mathbf{Q}(t)^T, \qquad n = 1, 2, 3, \ldots, \tag{8.21}$$

$$\dot{\mathbf{A}}_n(t) = \mathbf{Z}(t)\,\mathbf{A}_n(t) - \mathbf{A}_n(t)\,\mathbf{Z}(t), \quad n = 1, 2, 3, \ldots \tag{8.22}$$

so that

$$\mathbf{A}_1(t) = \mathbf{L}_1(t) + \mathbf{L}_1(t)^T = \mathbf{L}(t) + \mathbf{L}(t)^T, \tag{8.23}$$

$$\mathbf{A}_2(t) = \mathbf{A}_1(t)\,\mathbf{L}_1(t) + \mathbf{L}_1(t)^T\,\mathbf{A}_1(t), \tag{8.24}$$

and, more generally,

$$\mathbf{A}_{n+1}(t) = \mathbf{A}_n(t)\,\mathbf{L}_1(t) + \mathbf{L}_1(t)^T\,\mathbf{A}_n(t), \qquad n = 1, 2, 3, \ldots. \tag{8.25}$$

The formula (8.25), *applicable only in* **motions with constant stretch history** (MWCSH), is to be compared with (4.29), valid in all flows.

From (8.22) we see immediately that if the basis attached to the particle rotates, that is, $\mathbf{Z}(t) \neq \mathbf{0}$, then $\dot{\mathbf{A}}_n$ may not be zero. This explans, why even in MWCSH, one may obtain a non-zero value for $\dot{\mathbf{A}}_n$ (*cf.* p. 19).

Retracing our steps, we see that the tensor $\mathbf{L}_1(t)$ is related to the velocity gradient $\mathbf{L}(t)$ through (8.19) and is the velocity gradient in the rotating frame of the base vectors $\mathbf{b}_i(t)$; and if $\mathbf{L}_1(t)$ is known, then $\mathbf{C}_t(t-s)$ is completely determined through (8.17), on replacing τ by $(t-s)$.

Strain histories obeying (7.1) are MWCSH, because they are special cases of (8.2) with $\mathbf{Q}(t) \equiv \mathbf{1}$ for all t, or $\mathbf{L}_1 \equiv \mathbf{L}$. When $\mathbf{Q}(t) \equiv \mathbf{1}$, the vector basis $\mathbf{b}_i(t)$ is a fixed one, at least for each particle, and thus such examples are best given in a Cartesian coordinate system, though this is not always the case. Returning to the general case, we classify MWCSH as the three possibilities apply here also. Thus we classify MWCSH as

(i) MWCSH — I : $\mathbf{L}_1^2 = \mathbf{0}$. Also called as *viscometric flows*[10].

(ii) MWCSH — II : $\mathbf{L}_1^2 \neq \mathbf{0}$; $\mathbf{L}_1^3 = \mathbf{0}$.

(iii) MWCSH — III: $\mathbf{L}_1^n \neq \mathbf{0}$ for all $n = 1, 2, 3, \ldots$

We now list some kinematical examples of such motions.

(i) **Viscometric Flows or MWCSH—I.** In such flows $\mathbf{L}_1^2 = \mathbf{0}$ and hence there exists an orthonormal basis such that the matrix of $\mathbf{L}_1(t)$ and that of $\mathbf{C}_t(t-s)$ are:

$$[L_1(t) \langle ij \rangle] = \begin{pmatrix} 0 & 0 & 0 \\ \varkappa & 0 & 0 \\ 0 & 0 & 0 \end{pmatrix}, \quad [C_t(t-s) \langle ij \rangle] = \begin{pmatrix} 1+\varkappa^2 s^2 & -\varkappa s & 0 \\ \cdot & 1 & 0 \\ \cdot & \cdot & 1 \end{pmatrix}. \tag{8.26}$$

Moreover, $\mathbf{A}_n(t) \equiv \mathbf{0}$, $n \geqq 3$. This implies that in viscometric flows

$$\mathbf{C}_t(t-s) = \mathbf{1} - s\mathbf{A}_1(t) + \frac{1}{2} s^2 \mathbf{A}_2(t), \qquad 0 \leqq s < \infty. \tag{8.27}$$

We shall now give an example.

Consider a *curvilinear orthogonal coordinate* system (x^1, x^2, x^3), and the following steady velocity field (called *steady curvilineal flow*) in it:

$$\dot{x}^1 = 0, \qquad \dot{x}^2 = v(x^1), \qquad \dot{x}^3 = w(x^1). \tag{8.28}$$

We shall show that this is viscometric provided the components of the metric tensor do not change along the path line of the particle[11]; for, the path lines, given by (2.9), yield the strain history (3.1).

Let $\mathbf{e}_i(\mathbf{x})$ be the orthonormal basis of the coordinate system, along the coordinate lines. Such a basis is called the *natural orthonormal basis*. Since, in general, this natural basis at the point occupied by the particle at time 0, say at $\mathbf{x}(0)$, is different from that occupied by the particle at time t, say at $\mathbf{x}(t)$, we can attach a rotating vector basis $\tilde{\mathbf{b}}_i(\tau)$ to the particle such that

$$\tilde{\mathbf{b}}_i(\tau) = \mathbf{e}_i[\mathbf{x}(\tau)], \qquad i = 1, 2, 3, \tag{8.29}$$

10. COLEMAN [1962 : 2] introduced this term.
11. NOLL [1962 : 7]. Thm. 4.

or this rotating basis always coincides with the natural basis. Then with respect to $\tilde{\mathbf{b}}_i(t)$, the physical components of $\mathbf{C}_t(t-s)$ for the flow (8.28) are

$$[C_t(t-s)\langle ij\rangle] = \begin{pmatrix} 1+\varkappa^2 s^2 & -\alpha\varkappa s & -\beta\varkappa s \\ \cdot & 1 & 0 \\ \cdot & \cdot & 1 \end{pmatrix}, \tag{8.30}$$

where

$$\varkappa^2 = (g_{22}v'^2+g_{33}w'^2)/g_{11}, \qquad \alpha^2\varkappa^2 = g_{22}v'^2/g_{11}, \qquad \beta^2\varkappa^2 = g_{33}w'^2/g_{11}. \tag{8.31}$$

Now, choose the basis $\mathbf{b}_i(t)$ such that

$$\begin{aligned} \mathbf{b}_1(t) &= \tilde{\mathbf{b}}_1(t), \\ \mathbf{b}_2(t) &= \alpha\tilde{\mathbf{b}}_2(t) + \beta\tilde{\mathbf{b}}_3(t), \\ \mathbf{b}_3(t) &= -\beta\tilde{\mathbf{b}}_2(t) + \alpha\tilde{\mathbf{b}}_3(t). \end{aligned} \tag{8.32}$$

Then with respect to this new basis by using (8.1), we can show that

$$[C_t(t-s)\langle ij\rangle] = \begin{bmatrix} 1+\varkappa^2 s^2 & -\varkappa s & 0 \\ \cdot & 1 & 0 \\ \cdot & \cdot & 1 \end{bmatrix}, \tag{8.33}$$

and hence (8.28) is indeed viscometric.

NOLL's steady curvilineal flows do not exhaust all the kinematical possibilities though they include those usually found in practical rheometers and the like. In all curvilineal flows, the surfaces x^1 = const. are material surfaces, and they slide over one another without stretching. We list a few in Table 1 to emphasize this point.

TABLE 1

Coordinate system	*Velocity field*			*Name of flow*
Cartesian (x, y, z)	$\dot{x} = \kappa y$,	$\dot{y} = \dot{z} = 0$		Simple shear
	$\dot{x} = f(y)$,	$\dot{y} = \dot{z} = 0$		Plane Poiseuille
Cyl. polar (r, θ, z)	$\dot{r} = 0$,	$\dot{\theta} = \omega(r)$,	$\dot{z} = 0$	Couette
	$\dot{r} = \dot{\theta} = 0$,	$\dot{z} = u(r)$		Poiseuille
	$\dot{r} = 0$,	$\dot{\theta} = \omega(r)$,	$\dot{z} = u(r)$	Helical
	$\dot{r} = 0$,	$\dot{\theta} = \omega(z)$,	$\dot{z} = 0$	Torsional
	$\dot{r} = \dot{\theta} = 0$,	$\dot{z} = c\theta$		Fanned plane motion[12]
Sph. polar (r, θ, ϕ)	$\dot{r} = \dot{\theta} = 0$,	$\dot{\phi} = c\theta$		Cone-and-plate

12. PIPKIN [1968 : 21], Eq. (7.20).

The above list naturally suggests the following questions: are all viscometric flows characterized by rigid material surfaces sliding over one another? Can these bend? Also, does the definition of MWCSH demand that the motion be steady with respect to some frame of reference?

In a detailed investigation, YIN and PIPKIN [1970: 31][13] have attempted to answer the above questions by applying results from differential geometry to the streamline, the lines of "velocity gradient" and the mutual normal. In the next few paragraphs we shall summarize their conclusions, refering the reader to the original memoir for details.

Let $\mathbf{X}$ be the position occupied by a material particle at time $t = 0$, this fixed time being the one mentioned in Theorem 8.1. Let $\mathbf{a}(\mathbf{X}, t)$, $\mathbf{b}(\mathbf{X}, t)$ and $\mathbf{c}(\mathbf{X}, t)$ be mutually orthogonal fields of unit vectors. The notation $\mathbf{a}^1 = \mathbf{a}$, $\mathbf{a}^2 = \mathbf{b}$, $\mathbf{a}^3 = \mathbf{c}$ will be used in the sequel whenever necessary, along with $\mathbf{a}^i_0 = \mathbf{a}^i(\mathbf{X}, 0)$, $i = 1, 2, 3$, for the vector fields at time $t = 0$. The components of $\mathbf{a}^i(\mathbf{X}, t)$ are a^i_k, while those of $\mathbf{a}^i_0$ are a^i_α. Now according to Theorem 8.1 and the ensuing discussion, a deformation gradient history is viscometric, if

$$\left.\begin{aligned} \mathbf{F}_0(t) &= \mathbf{F}(\mathbf{X}, t) = \mathbf{Q}(t)\,[\mathbf{1}+t\mathbf{M}(\mathbf{X})] \\ \mathbf{Q}(0) &= \mathbf{1}, \quad \mathbf{M}^2 = \mathbf{0} \end{aligned}\right\}. \tag{8.34}$$

Since $\mathbf{M}(\mathbf{X})$ is a constant tensor and $\mathbf{M}^2 = \mathbf{0}$, let it have the component form

$$M_{\alpha\beta}(\mathbf{X}) = \varkappa(\mathbf{X})\, a_\alpha(\mathbf{X})\, b_\beta(\mathbf{X}). \tag{8.35}$$

Define $Q_{k\alpha}\, a^i_\alpha = a^i_k$. Then (8.34) becomes

$$F_{k\alpha}(\mathbf{X}, t) = a^i_k\, a^i_\alpha + t\varkappa(\mathbf{X})\, a_k b_\alpha, \tag{8.36}$$

with summation on $i = 1, 2, 3$. Since at every instant t one can draw the trajectories of the fields $\mathbf{a}^i$, one can obtain the $\mathbf{a}^i$-lines. Some properties of these vector fields are now listed below:

(*a*) **a**-lines and **c**-lines are material fibers which move without stretching. For if at $t = 0$, $d\mathbf{X} = \mathbf{a}_0\, dl$, then at time t, this material element is

$$d\mathbf{x} = \mathbf{F}\, d\mathbf{X} = \mathbf{a}\, dl, \tag{8.37}$$

since $\mathbf{F}\mathbf{a}_0 = \mathbf{a}$ from (8.36). So a material element at time $t = 0$ lying in the **a**-direction lies in the **a**-direction at time t, and *its length is not changed.*[14] The same argument holds for the **c**-lines.

(*b*) The **b**-lines are not inextensible. For, if $d\mathbf{X} = \mathbf{b}_0\, dl$, then

$$d\mathbf{x} = \mathbf{F}\, d\mathbf{X} = (\mathbf{b}+t\varkappa\, \mathbf{a})\, dl, \tag{8.38}$$

so that the element in the **b**-direction is sheared toward the **a**-direction at a

13. It is precisely the properties to be listed below which make it impossible for the superposition of two viscometric flows to lead to a viscometric flow always.

14. This explains the need for demanding that "the components of the metric tensor do not change along the path line of the particle."

constant rate $\varkappa$. So the flow is locally a simple shearing flow, with **a** being the direction of motion, **b** the direction of "velocity gradient" and **c** mutually orthogonal to these vectors.

(*c*) The **a**- and **c**-lines always mesh to form material surfaces which will be called *slip surfaces*. The **b**-lines are orthogonal to these slip surfaces.

(*d*) The slip surfaces move without stretching; however, they may move *rigidly* or *bend*.

(*e*) Rigid slip surfaces need not be co-axial, though the list in Table 1 may suggest that this is always the case. It is even possible to generate intrinsically unsteady (that is, not steady with respect to any reference frame whatsoever[15]) viscometric flows with rigid slip surfaces, provided these are not co-axial or congruent at $t = 0$.

(*f*) Slip surfaces that bend will be called *flexible slip surfaces*. We give below an example of such a motion with a constant shear rate $\varkappa$.

Let $\{\mathbf{i}, \mathbf{j}, \mathbf{k}\}$ be a constant orthonormal basis, and let

$$\begin{aligned} \mathbf{a} &\equiv \mathbf{a}(\alpha) = \mathbf{i} \cos \alpha + \mathbf{j} \sin \alpha, \\ \mathbf{b} &\equiv \mathbf{b}(\alpha) = -\mathbf{i} \sin \alpha + \mathbf{j} \cos \alpha, \\ \mathbf{c} &\equiv \mathbf{k}. \end{aligned} \tag{8.39}$$

Consider a material cylinder, initially of radius r_0, and the motion

$$\begin{aligned} \mathbf{x}(\mathbf{X}, t) &= r_0(1+t^2\varkappa^2)^{-1} [\mathbf{a}(\alpha) - t\varkappa\, \mathbf{b}(\alpha)] + z_0\mathbf{k}, \\ \alpha &= \theta_0 + t\varkappa \ln r_0 + t^2\varkappa^2\theta_0, \end{aligned} \tag{8.40}$$

where θ_0 is a constant. Take $z_0 = 0$ for example, and consider the radial line along $\theta_0 = 0$ at $t = 0$. This is an **a**-line at $t = 0$, or at $t = 0$, the radial lines are **a**-lines and the azimuthal circles are **b**-lines. Eq. (8.40) tells us that the length of a given **a**-line does not change as t varies, but the **a**-line does not remain straight, rather it bends. Indeed, any segment of an azimuthal circle grows in length by a factor of $(1+t^2\varkappa^2)^{1/2}$ according to (8.38) and will thus overlap itself, as the motion proceeds. So the motion is kinematically possible for a finite period of time only.

Considering the **a**-line along $\theta = 0$ as a plane, we have here an example of a flexible slip surface.

The investigation by YIN and PIPKIN is not complete because their kinematical study does not exhaust all viscometric flows with flexible slip surfaces. However, their study is dynamically complete in one sense for they have determined all "partially controllable flows", which we shall discuss in Chap. 7. Since the industrial viscometers rely on such flows, it could be said that from a practical point, we now know all viscometric flows.

The reader will notice that intrinsic unsteadiness was mentioned in the above summary. This point will be discussed further in Sec. 13.

(ii) **MWCSH II.** In such flows, $\mathbf{L}_1^2 \neq \mathbf{0}$, but $\mathbf{L}_1^3 = \mathbf{0}$ and hence there

15. See Sec. 13 for additional discussion.

exists an orthonormal basis such that the matrix of $\mathbf{L}_1(t)$ has the form

$$[L_1(t)\langle ij\rangle] = \varkappa \begin{pmatrix} 0 & 0 & 0 \\ l & 0 & 0 \\ m & n & 0 \end{pmatrix}, \qquad \varkappa > 0, \tag{8.41}$$

whence $\mathbf{A}_n(t) \equiv \mathbf{0}$ for $n \geqq 5$, and the strain history has the form[16]

$$\mathbf{C}_t(t-s) = \mathbf{1} - s\mathbf{A}_1 + \frac{1}{2!}\, s^2\mathbf{A}_2 - \frac{1}{3!}\, s^3\mathbf{A}_3 + \frac{1}{4!}\, s^4\mathbf{A}_4. \tag{8.42}$$

Since $\mathbf{L}_1$ in (8.41) can be obtained by adding three matrices of the type in $(8.26)_1$, it is intuitively clear that the present type of motion may be generated by superposing two or three viscometric flows. As an example, refer to the motion of NOLL[17] in Sec. 7. Here we generalize this idea of superposition to *flows in curvilinear orthogonal coordinate systems when the components of the metric tensor do not change along the path line of each particle.*

The simplest example is[18] :

$$\dot{x}^1 = 0, \quad \dot{x}^2 = v(x^1), \quad \dot{x}^3 = cx^2, \quad c = \text{const.}, \tag{8.43}$$

that is, the velocity in x^2 direction is an arbitrary function of x^1, while that in x^3 direction is linear in x^2. A generalization of (8.43) is[18]

$$\dot{x}^1 = 0, \quad \dot{x}^2 = v(x^1), \quad \dot{x}^3 = cx^2 + w(x^1). \tag{8.44}$$

In examples 2 and 3 of Table 2 below, we superpose the torsional flow (viscometric) on Poiseuille and helical flows (viscometric) respectively, while in example 4, we superpose the axial motion of fanned planes[19] on a helical flow.

TABLE 2

Co-ordinate system	*Velocity field*			*Name of flow*
Cartesian : x, y, z	$\dot{x} = 0.$	$\dot{y} = \kappa l x,$	$\dot{z} = \kappa(mx+ny)$	
Cylindrical polar : r, θ, z	$\dot{r} = 0,$	$\dot{\theta} = cz,$	$\dot{z} = u(r)$	Poiseuille-torsional[20] flow
,,	$\dot{r} = 0,$	$\dot{\theta} = \omega(r) + cz,$	$\dot{z} = u(r)$	Helical-torsional[21] flow
,,	$\dot{r} = 0,$	$\dot{\theta} = \omega(r),$	$\dot{z} = c\theta + u(r)$	Helical-axial motion[21] of fanned planes

16. [1962 : 7], Thm. 3. In view of (8.42), ASTARITA [1967 : 2] says that OLDROYD [1965 : 13] calls such flows as *fourth order flows.*
17. [1962 : 7], Eq. (3.14).
18. HUILGOL [1971 : 12].
19. PIPKIN [1968 : 21], Eq. (7.20).
20. OLDROYD [1965 : 13], Eq. (75).
21. HUILGOL [1971 : 12].

In fact, the flow (8.43) may be generalized[21] further to

$$\left.\begin{array}{l} x^1 = 0, \quad \dot{x}^2 = v(x^1) - cx^2 + ex^3, \quad \dot{x}^3 = w(x^1) + fx^2 + cx^3, \\ \qquad c^2 + ef = 0, \qquad (c, e, f \text{ being constants}). \end{array}\right\} \tag{8.45}$$

In addition to the above motion, one also has

$$\dot{x}^1 = 0, \quad \dot{x}^2 = v(x^1), \quad \dot{x}^3 = x^2 w(x^1) \tag{8.46}$$[22]

as another example of MWCSH-*II*, with the usual restriction on the metric tensor.

(iii) **MWCSH III.** At present there do not exist many examples of such motions in curvilinear coordinates. We may, however, recast the simple extensional flow (7.25) into a MWCSH in cylindrical polar coordinates (r, θ, z) when $a_1 = a_2$; that is, if

$$\dot{x} = a_1 x, \quad \dot{y} = a_1 y, \quad \dot{z} = a_3 z, \tag{8.47}$$

then in cylindrical polar coordinates, we have

$$\dot{r} = a_1 r, \quad \dot{\theta} = 0, \quad \dot{z} = a_3 z. \tag{8.48}$$

The path lines are

$$\begin{aligned} \xi &= re^{-a_1 s}, \\ \eta &= \theta, \\ \zeta &= ze^{-a_3 s}, \end{aligned} \tag{8.49}$$

and the strain history has the physical components

$$[C_t(t-s)\langle ij\rangle] = \begin{pmatrix} e^{-2a_1 s} & 0 & 0 \\ \cdot & e^{-2a_1 s} & 0 \\ \cdot & \cdot & e^{-2a_3 s} \end{pmatrix}. \tag{8.50}$$

This finishes the list of known examples of MWCSH.

REMARK. A final word about MWCSH will be in order to indicate the usefulness of the formula (8.22)[23]. For example, for the Couette flow

$$v^r = 0, \quad v^\theta = \omega(r), \quad v^z = 0, \tag{8.51}$$

the calculation of $\dot{\mathbf{A}}_1$ through

$$(\dot{A}_1)_{ij} = \frac{\partial}{\partial t} A^{(1)}_{ij} + A^{(1)}_{ij;k} v^k \tag{8.52}$$

is very tedious. However, one notes through (8.22) that if we know the angular velocity of the rotating vector basis $\mathbf{b}_k(t)$, then $\mathbf{Z}(t)$ is easily computed. Clearly, in the present case, the basis attached to the particle rotates

22. This generalizes the example of ASTARITA [1967 : 2].
23. Due to COLEMAN [1962 : 2].

with an angular velocity of $\omega(r)$ around the z-axis. Hence

$$[Z\langle ij\rangle] = \begin{pmatrix} 0 & -\omega(r) & 0 \\ \omega(r) & 0 & 0 \\ 0 & 0 & 0 \end{pmatrix}. \tag{8.53}$$

With this known, we may use $\mathbf{A}_1$ from (4.35) and substitute it into (8.22) to calculate

$$[\dot{A}_1\langle ij\rangle] = \begin{pmatrix} -2r\omega\omega' & 0 & 0 \\ \cdot & 2r\omega\omega' & 0 \\ \cdot & \cdot & 0 \end{pmatrix}, \qquad \omega' = \frac{d\omega}{dr}, \tag{8.54}$$

which is (4.37). Also, we can use $\mathbf{A}_2$ from (4.38) and use (8.53) to show that

$$[\dot{A}_2\langle ij\rangle] = \begin{pmatrix} 0 & 2r^2\omega\omega'^2 & 0 \\ \cdot & 0 & 0 \\ \cdot & \cdot & 0 \end{pmatrix}. \tag{8.55}$$

9. Importance of the First Three Rivlin-Ericksen Tensors

Given a velocity field $\mathbf{v}(\mathbf{x})$ in a MWCSH, one can compute the path lines and determine the strain history. However, this calculation is not always easy to perform, and so one should seek easier methods of determining the strain history. Now, this computation becomes algebraic if one can find the tensor $\mathbf{L}_1(t)$, which determines $\mathbf{C}_t(t-s)$, $0 \leqq s < \infty$. One way to find $\mathbf{L}_1(t)$ is to calculate $\mathbf{L}(t)$ for a particle and use (8.19); but the skew-tensor $\mathbf{Z}(t)$ is not known in advance. So the determination of $\mathbf{L}_1(t)$ does not seem to be straightforward.

However, one may indeed ask: is $\mathbf{L}_1(t)$ necessary at all? For, one is motivated to calculate the strain history to compute the stress tensor (assumed, for the present, to be determined by this history)—and since $\mathbf{L}_1(t)$ determines $\mathbf{C}_t(t-s)$, $0 \leqq s < \infty$, the stress tensor is a function of $\mathbf{L}_1(t)$. Now suppose one can deduce that the strain history [or $\mathbf{L}_1(t)$] is uniquely determined in a MWCSH by some other kinematical tensors which can be found in a straightforward way—then the stress tensor is a function of these kinematical tensors only. To be specific, it will be demonstrated that in general three kinematical tensors are needed and that these are the tensors $\mathbf{A}_1$, $\mathbf{A}_2$, $\mathbf{A}_3$ in a MWCSH.

The result may be anticipated to a certain extent since in:

MWCSH — *I* : $\mathbf{A}_1$ and $\mathbf{A}_2$ determine $\mathbf{C}_t(t-s)$ uniquely;
MWCSH — *II* : $\mathbf{A}_1$, $\mathbf{A}_2$, $\mathbf{A}_3$, $\mathbf{A}_4$ suffice;
MWCSH — *III* : (*a*) $\mathbf{A}_1$ suffices in simple extension.

However, as always, the proof of a statement is to be given, and hence the upcoming theorem[24]:

THEOREM 9.1. *In a MWCSH, the first three Rivlin-Ericksen tensors* $\mathbf{A}_1(t)$, $\mathbf{A}_2(t)$, $\mathbf{A}_3(t)$ *determine* $\mathbf{C}_t(t-s)$ *uniquely for they determine* $\mathbf{L}_1(t)$ *so as to yield a unique strain history.*

We shall prove this after disposing of the next lemma of basic character.

LEMMA 9.1. *Let* $[S\langle ij\rangle]$ *be a* 3×3 *diagonal matrix and* $[W\langle ij\rangle]$ *be a* 3×3 *skew matrix*

$$[S\langle ij\rangle] = \begin{pmatrix} a & 0 & 0 \\ 0 & b & 0 \\ 0 & 0 & c \end{pmatrix},$$

$$[W\langle ij\rangle] = \begin{pmatrix} 0 & x & y \\ -x & 0 & z \\ -y & -z & 0 \end{pmatrix}. \tag{9.1}$$

Let $[SW]$ *be the matrix* $[S\langle ik\rangle\ W\langle kj\rangle]$ *and* $[WS]$ *be the matrix* $[W\langle ik\rangle\ S\langle kj\rangle]$. *Then*

(i) *if* $a \neq b \neq c \neq a$, $[SW] = [WS]$ *if and only if* $x = y = z = 0$;

(ii) *if* $a = b \neq c$, $[SW] = [WS]$ *if and only if* $y = z = 0$.

(iii) *if* $a = b = c$, $[SW] = [WS]$ *for all* x, y, z.

PROOF. By multiplication

$$[SW] = \begin{bmatrix} 0 & ax & ay \\ -bx & 0 & bz \\ -cy & -cz & 0 \end{bmatrix},$$

$$[WS] = \begin{bmatrix} 0 & bx & cy \\ -ax & 0 & cz \\ -ay & -bz & 0 \end{bmatrix}. \tag{9.2}$$

Therefore $[SW] = [WS]$ if and only if

$$(a-b)\,x = 0, \quad (a-c)\,y = 0, \quad (b-c)\,z = 0, \tag{9.3}$$

implying the lemma.

We now use this to prove the main theorem of WANG.

PROOF OF THEOREM 9.1. **WI.** *Suppose* $\mathbf{A}_1$ is diagonalized and *has three distinct eigenvalues.* Then, we claim that $\mathbf{A}_1$ and $\mathbf{A}_2$ determine $\mathbf{L}_1$ *uniquely* and hence $\mathbf{C}_t(t-s)$ *uniquely.*

24. WANG [1965 : 20], Thm. in Sec. 2. Also, Lemma is in Sec. 2.

To cstablish this, assume the contrary. Then, let

$$\mathbf{A}_1 = \mathbf{L}_1 + \mathbf{L}_1^T = \bar{\mathbf{L}}_1 + \bar{\mathbf{L}}_1^T, \tag{9.4}$$

$$\mathbf{A}_2 = \mathbf{A}_1\mathbf{L}_1 + \mathbf{L}_1^T\mathbf{A}_1 = \mathbf{A}_1\bar{\mathbf{L}}_1 + \bar{\mathbf{L}}_1^T\mathbf{A}_1, \tag{9.5}$$

that is, we assume that $\mathbf{L}_1$ and $\bar{\mathbf{L}}_1$ are two tensors yielding the same $\mathbf{A}_1$ and $\mathbf{A}_2$. Using (9.5), we have

$$\mathbf{A}_1(\mathbf{L}_1 - \bar{\mathbf{L}}_1) = (\bar{\mathbf{L}}_1^T - \mathbf{L}_1^T)\,\mathbf{A}_1. \tag{9.6}$$

But $(9.4)_3$ yields that

$$\mathbf{L}_1 - \bar{\mathbf{L}}_1 = \bar{\mathbf{L}}_1^T - \mathbf{L}_1^T = -\,(\mathbf{L}_1 - \bar{\mathbf{L}}_1)^T, \tag{9.7}$$

and hence $\mathbf{L}_1 - \bar{\mathbf{L}}_1$ is skew. Now, (9.6) becomes

$$\mathbf{A}_1(\mathbf{L}_1 - \bar{\mathbf{L}}_1) = (\mathbf{L}_1 - \bar{\mathbf{L}}_1)\,\mathbf{A}_1, \tag{9.8}$$

or $\mathbf{A}_1$ commutes with the skew-tensor $\mathbf{L}_1 - \bar{\mathbf{L}}_1$. But the distinct eigenvalues of $\mathbf{A}_1$ exclude this possibility unless $\mathbf{L}_1 - \bar{\mathbf{L}}_1 = 0$. Thus $\mathbf{L}_1$ is uniquely determined by $\mathbf{A}_1$ and $\mathbf{A}_2$ and hence $\mathbf{C}_t(t-s)$ is uniquely determined by $\mathbf{A}_1$ and $\mathbf{A}_2$.

W2. Let $\mathbf{A}_1$ have two distinct eigenvalues so that relative to an orthonormal basis, the matrix of $\mathbf{A}_1$ is

$$[A_1\langle ij\rangle] = \begin{pmatrix} a & 0 & 0 \\ 0 & a & 0 \\ 0 & 0 & c \end{pmatrix}, \quad a \neq c. \tag{9.9}$$

We actually have two subcases:

(*a*) Relative to the basis of $\mathbf{A}_1$, let the matrix of $\mathbf{A}_2$ be

$$[A_2\langle ij\rangle] = \begin{pmatrix} u & 0 & 0 \\ 0 & u & 0 \\ 0 & 0 & v \end{pmatrix}. \tag{9.10}$$

We claim that

$$u = a^2, \qquad v = c^2, \tag{9.11}$$

and that $\mathbf{C}_t(t-s)$ is unique and given by

$$\mathbf{C}_t(t-s) = e^{-s\mathbf{A}_1}, \tag{9.12}$$

but $\mathbf{L}_1$ is not unique, and has the form

$$[L_1\langle ij\rangle] = \begin{bmatrix} \frac{a}{2} & x & 0 \\ -x & \frac{a}{2} & 0 \\ 0 & 0 & \frac{c}{2} \end{bmatrix}, \quad x \text{ arbitrary.} \tag{9.13}$$

To prove this, the most general form of $[\mathbf{L}_1]$ consistent with $[\mathbf{A}_1]$ in (9.9) is

$$[L_1\langle ij\rangle] = \begin{bmatrix} \frac{a}{2} & x & y \\ -x & \frac{a}{2} & z \\ -y & -z & \frac{c}{2} \end{bmatrix}. \tag{9.14}$$

Then, from (8.24) we have that

$$[A_2\langle ij\rangle] = \begin{bmatrix} a^2 & 0 & (a-c)\,y \\ \cdot & a^2 & (a-c)\,z \\ \cdot & \cdot & c^2 \end{bmatrix}. \tag{9.15}$$

Comparing (9.15) with (9.10), we obtain (9.11) and (9.13), for $y = z = 0$.

Now, if $\mathbf{L}_1$ has the form (9.13), then

$$\mathbf{L}_1\,\mathbf{L}_1^T = \mathbf{L}_1^T\,\mathbf{L}_1. \tag{9.16}$$

Thus this commutativity yields

$$\mathbf{C}_t(t-s) = e^{-s\mathbf{L}_1^T}\,e^{-s\mathbf{L}_1} = e^{-s\mathbf{A}_1}, \tag{9.17}$$

which is (9.12). Thus $\mathbf{C}_t(t-s)$ is uniquely determined by $\mathbf{A}_1$, though $\mathbf{L}_1$ is not. Incidentally, we have shown that in a MWCSH, $\mathbf{A}_1^2 = \mathbf{A}_2$ implies that the motion is a simple extension, because in such flows (9.17) holds [*cf.* (7.29)].

(*b*) Suppose that $[A_2\langle ij\rangle]$ does not have the form obeying (9.10) and (9.11). Then we claim that $\mathbf{A}_1$, $\mathbf{A}_2$ and $\mathbf{A}_3$ determine $\mathbf{L}_1$, and thus $\mathbf{C}_t(t-s)$, uniquely.

To prove this, assume the contrary and let $\mathbf{L}_1$ and $\bar{\mathbf{L}}_1$ be the two solutions so that

$$\mathbf{A}_1 = \mathbf{L}_1 + \mathbf{L}_1^T = \bar{\mathbf{L}}_1 + \bar{\mathbf{L}}_1^T, \tag{9.18}$$

$$\mathbf{A}_2 = \mathbf{A}_1\mathbf{L}_1 + \mathbf{L}_1^T\,\mathbf{A}_1 = \mathbf{A}_1\bar{\mathbf{L}}_1 + \bar{\mathbf{L}}_1^T\,\mathbf{A}_1, \tag{9.19}$$

$$\mathbf{A}_3 = \mathbf{A}_2\mathbf{L}_1 + \mathbf{L}_1^T\,\mathbf{A}_2 = \mathbf{A}_2\bar{\mathbf{L}}_1 + \bar{\mathbf{L}}_1^T\,\mathbf{A}_2. \tag{9.20}$$

From (9.18), we have that $\mathbf{L}_1 - \bar{\mathbf{L}}_1$ is skew. By (9.19), $\mathbf{A}_1$ commutes with $\mathbf{L}_1 - \bar{\mathbf{L}}_1$, just as in Wi above. Since $[A_1\langle ij\rangle]$ is given by (9.9), $(\mathbf{L}_1 - \bar{\mathbf{L}}_1)$ must be of the type

$$[(L_1-\bar{L}_1)\,\langle ij\rangle] = \begin{pmatrix} 0 & x & 0 \\ -x & 0 & 0 \\ 0 & 0 & 0 \end{pmatrix}. \tag{9.21}$$

In obtaining (9.21), we have used (ii) of Lemma 9.1. Now, by (9.20), $\mathbf{A}_2$ commutes with $\mathbf{L}_1 - \bar{\mathbf{L}}_1$, but $\mathbf{A}_2$ is not of the form (9.10), obeying (9.11). So, we retrace our steps and note that the most general forms of $\mathbf{L}_1$ and $\bar{\mathbf{L}}_1$, consistent with (9.9) and (9.21) are:

$$\left.\begin{aligned} [L_1\langle ij\rangle] &= \begin{bmatrix} \frac{a}{2} & x & y \\ -x & \frac{a}{2} & z \\ y & -z & \frac{c}{2} \end{bmatrix} \\ [\bar{L}_1\langle ij\rangle] &= \begin{bmatrix} \frac{a}{2} & 0 & y \\ 0 & \frac{a}{2} & z \\ -y & -z & \frac{c}{2} \end{bmatrix} \end{aligned}\right\}. \tag{9.22}$$

Now, from (9.9) and $(9.22)_{1 \text{ and } 2}$, we have

$$[A_2\langle ij\rangle] = \begin{bmatrix} a^2 & 0 & (a-c)y \\ \cdot & a^2 & (a-c)z \\ \cdot & \cdot & c^2 \end{bmatrix} \tag{9.23}$$

This $\mathbf{A}_2$ does not commute with $(\mathbf{L}_1 - \bar{\mathbf{L}}_1)$ in (9.21) unless $x = 0$, that is, $\mathbf{L}_1 - \bar{\mathbf{L}}_1 = 0$. Thus, $\mathbf{A}_1$, $\mathbf{A}_2$ and $\mathbf{A}_3$ determine $\mathbf{L}_1$ and hence $\mathbf{C}_t(t-s)$ uniquely.

(*c*) All eigenvalues of $\mathbf{A}_1$ are equal to a. Then $\mathbf{A}_1$ commutes with all tensors, because $\mathbf{A}_1 = a\mathbf{1}$. Hence

$$\mathbf{L}_1 = \frac{a}{2}\mathbf{1} + \mathbf{W}, \tag{9.24}$$

$\mathbf{W}$ is skew and completely arbitrary. But with this form, $\mathbf{L}_1\mathbf{L}_1^T = \mathbf{L}_1^T\mathbf{L}_1$ and hence

$$\mathbf{C}_t(t-s) = e^{-sa\mathbf{1}}, \tag{9.25}$$

or $\mathbf{C}_t(t-s)$ is uniquely determined by $\mathbf{A}_1$, though $\mathbf{L}_1$ is not.

To conclude, $\mathbf{A}_1$, $\mathbf{A}_2$ and $\mathbf{A}_3$ determine $\mathbf{C}_t(t-s)$ uniquely in all MWCSH.

The reader will note that in Theorem 9.1, the eigenvalues of $\mathbf{A}_1$ played the central role. So we examine this matter closely next.

10. Eigenvalues of the First Rivlin-Ericksen Tensor

Let the matrix of $\mathbf{A}_1$ relative to an orthonormal basis be

$$[A_1\langle ij\rangle] = \varkappa \begin{bmatrix} a_1 & l & m \\ l & a_2 & n \\ m & n & a_3 \end{bmatrix}, \qquad \varkappa > 0, \tag{10.1}$$

$$a_1^2 + a_2^2 + a_3^2 + l^2 + m^2 + n^2 = 1, \tag{10.2}$$

$$a_1 + a_2 + a_3 = 0. \tag{10.3}$$

The condition (10.2) is introduced to "normalize" $\mathbf{A}_1$ and (10.3) is very important because it expresses the volume-preserving nature of the motion, as will be seen later; in incompressible materials, (10.3) is always satisfied.

Let λ_1, λ_2 and λ_3 be the three eigenvalues of $\varkappa^{-1}\mathbf{A}_1$. Then (10.3) prohibits that $\lambda_1 = \lambda_2 = \lambda_3 = \lambda$. For this would imply that $\lambda = 0$, which in turn implies that $\mathbf{A}_1$ is zero, or the motion is rigid. Leaving this aside, we look at the characteristic equation of $\varkappa^{-1} A_1$. This is given by

$$\lambda^3 - (1 + a_1 a_2 - a_3^2)\lambda + a_3 l^2 + a_2 m^2 + a_1 n^2 - a_1 a_2 a_3 - 2lmn = 0. \tag{10.4}$$

If the three roots of (10.4) are λ_1, λ_2 and λ_3, then by well-known results in the theory of equations,

$$\lambda_1 + \lambda_2 + \lambda_3 = 0, \tag{10.5}$$

$$\lambda_1\lambda_2 + \lambda_2\lambda_3 + \lambda_3\lambda_1 = a_3^2 - a_1 a_2 - 1, \tag{10.6}$$

$$\lambda_1\lambda_2\lambda_3 = 2lmn + a_1 a_2 a_3 - a_3 l^2 - a_2 m^2 - a_1 n^2. \tag{10.7}$$

Without loss of generality, take $\lambda_1 = \lambda_2 \neq \lambda_3$. Then (10.5) – (10.7) become

$$\lambda_3 = -2\lambda_1, \qquad 3\lambda_1^2 = 1 + a_1 a_2 - a_3^2, \tag{10.8}$$

$$2\lambda_1^3 = a_3 l^2 + a_2 m^2 + a_1 n^2 - a_1 a_2 a_3 - 2lmn. \tag{10.9}$$

Hence

$$\lambda_1 = \frac{3(a_3 l^2 + a_2 m^2 + a_1 n^2 - a_1 a_2 a_3 - 2lmn)}{2(1 + a_1 a_2 - a_3^2)}. \tag{10.10}$$

Thence $\lambda_1 = \lambda_2$ whenever λ_1^6 given by (10.9) is equal to that given by $(10.8)_2$; or, whenever

$$27(a_3 l^2 + a_2 m^2 + a_1 n^2 - a_1 a_2 a_3 - 2lmn)^2 = 4(1 + a_1 a_2 - a_3^2)^3. \tag{10.11}$$

Thus, if (10.11) is not satisfied, the eigenvalues of $\mathbf{A}_1$, obeying (10.1) – (10.3), are all distinct.

Now in MWCSH, of the types I and II in NOLL's classification, $\mathbf{L}_1$ is nilpotent. Thus, either $\mathbf{L}_1^2 = \mathbf{0}$ or $\mathbf{L}_1^3 = \mathbf{0}$. In either case, the matrix of $\mathbf{L}_1$

with respect to an orthonormal basis, must have the form

$$[L_1 \langle ij \rangle] = \varkappa \begin{bmatrix} 0 & 0 & 0 \\ l & 0 & 0 \\ m & n & 0 \end{bmatrix}, \quad \varkappa > 0, \quad l^2 + m^2 + n^2 = 1. \tag{10.12}$$

If $\mathbf{L}_1^2 = 0$, then $m = n = 0$ in (10.12); if $\mathbf{L}_1^2 \neq 0$, $\mathbf{L}_1^3 = 0$, then l, m, n obey $(10.12)_3$ only. In any case, the matrix of $\mathbf{A}_1$ is of the form

$$[A_1 \langle ij \rangle] = \varkappa \begin{bmatrix} 0 & l & m \\ l & 0 & n \\ m & n & 0 \end{bmatrix}, \quad \varkappa > 0, \quad l^2 + m^2 + n^2 = 1, \tag{10.13}$$

because in MWCSH, $\mathbf{A}_1 = \mathbf{L}_1 + \mathbf{L}_1^T$.

Now, given the form (10.13), $\mathbf{A}_1$ has two eigenvalues if and only if

$$27\, l^2 m^2 n^2 = 1, \tag{10.14}$$

which is obtained from (10.11) by inserting there $a_1 = a_2 = a_3 = 0$. Thus $10.12)_3$ and (10.14) together imply that

$$\frac{l^2+m^2+n^2}{3} = \frac{1}{3}, \quad \sqrt[3]{l^2m^2n^2} = \frac{1}{3}, \tag{10.15}$$

or, the arithmetic mean and geometric mean are equal. Now it is well known[25] that this equality is possible if and only if

$$l^2 = m^2 = n^2 = 1/3. \tag{10.16}$$

Thus we have proved the following

THEOREM. *Let the matrix of* $\mathbf{A}_1$ *relative to an orthonormal basis be given by* (10.13). *Then the eigenvalues of* $\mathbf{A}_1$ *are distinct if and only if* (10.16) *is not satisfied.*[26]

Returning to Sec. 8, we note that in MWCSH;

(i) If $\mathbf{L}_1^2 = 0$, then $\mathbf{A}_1$ is of the form (10.13), but $m = n = 0$.

Hence (10.16) is not satisfied, or $\mathbf{A}_1$ has three distinct eigenvalues. By WANG's theorem, $\mathbf{A}_1$ and $\mathbf{A}_2$ determine $\mathbf{L}_1$ uniquely, and thus $\mathbf{C}_t(t-s)$ also.

(ii) If $\mathbf{L}_1^2 \neq 0$, but $\mathbf{L}_1^3 = 0$, then $\mathbf{A}_1$ is of the form (10.13). Hence:

(*a*) if $l^2 = m^2 = n^2 = 1/3$, then $\mathbf{A}_1$ has *two distinct eigenvalues.* Now

25. HARDY, LITTLEWOOD and POLYA [1964 : 10], p.17.
26. HUILGOL [1971 : 12].

with $\mathbf{L}_1$ having the form (10.12), $\mathbf{C}_t(t-s)$ is a polynomial of degree 4 in s, that is,

$$\mathbf{C}_t(t-s) = \sum_{n=0}^{4} (-1)^n \frac{s^n}{n!} \mathbf{A}_n(t), \qquad \mathbf{A}_0 \equiv \mathbf{1}. \tag{10.17}$$

In other words, this MWCSH-*II* falls under W2(b) in WANG's theorem. Thus, only when $l^2 = m^2 = n^2 = 1/3$ do we require $\mathbf{A}_1$, $\mathbf{A}_2$ and $\mathbf{A}_3$ to determine $\mathbf{L}_1$ uniquely.

(b) if $l^2 \neq m^2 \neq n^2 \neq 1/3$, then $\mathbf{A}_1$ has three distinct eigenvalues. Thereupon $\mathbf{A}_1$ and $\mathbf{A}_2$ determine $\mathbf{L}_1$ uniquely and thus $\mathbf{C}_t(t-s)$ uniquely.

Thus we have placed MWCSH-*II* under WANG's classification W1 or W2(b) by discovering when $\mathbf{A}_1$ has three distinct eigenvalues or has two distinct eigenvalues. Also, MWCSH-*I* always falls under WANG's classification W1 because $\mathbf{A}_1$ has always got three distinct eigenvalues.

(iii) Regarding MWCSH-*III* we note that the strain history is an infinite series and hence we have the following:

(a) If $\operatorname{tr} \mathbf{A}_1 = 0$ and (10.16) is not true, then $\mathbf{A}_1$ has three distinct eigenvalues. So this MWCSH falls under W1 of WANG's theorem. Our example of the flow in the MAXWELL rheometer (Sec. 7) belongs to this class.

(b) If $\operatorname{tr} \mathbf{A}_1 = 0$, and $\mathbf{A}_1 = a\mathbf{1}$, then $a = 0$. Hence WANG's case W3 does not apply.

(c) If $[\mathbf{A}_1]$ has the form (9.9) and $[A_2]$ obeys (9.10) and (9.11), then $\mathbf{C}_t(t-s)$ is given by (9.12) and it falls under W2(a), Also we have[27]

$$\mathbf{A}_1^k = \mathbf{A}_k, \qquad k = 1, 2, 3, \ldots . \tag{10.18}$$

Such flows are called *simple extensional flows*[28] [*cf.* (7.29)]. We close by deriving another criterion for MWCSH-*I*. For the velocity field (7.15), we have

$$[A_1\langle ij\rangle] = \begin{bmatrix} 0 & \varkappa & 0 \\ \varkappa & 0 & 0 \\ 0 & 0 & 0 \end{bmatrix}, \quad [A_2\langle ij\rangle] = \begin{bmatrix} 0 & 0 & 0 \\ 0 & 2\varkappa^2 & 0 \\ 0 & 0 & 0 \end{bmatrix}. \tag{10.19}$$

The eigenvalues of A_1 are:

$$\lambda_1 = \varkappa, \qquad \lambda_2 = -\varkappa, \qquad \lambda_3 = 0. \tag{10.20}$$

Hence the diagonal form of $\mathbf{A}_1$ is $\mathbf{A}_1^* = \mathbf{Q}\mathbf{A}_1\mathbf{Q}^T$, and

27. WANG [1965 : 21], Cor. 2.

28. COLEMAN and NOLL [1962 : 3]. These authors used $\operatorname{tr} \mathbf{A}_1 = 0$ also, but (10.18) does not need this restriction.

$$[A_1^* \langle ij \rangle] = \begin{bmatrix} \varkappa & 0 & 0 \\ 0 & -\varkappa & 0 \\ 0 & 0 & 0 \end{bmatrix} \tag{10.21}$$

with the orthogonal tensor Q given by

$$[Q \langle ij \rangle] = \frac{1}{\sqrt{2}} \begin{bmatrix} 1 & 1 & 0 \\ -1 & 1 & 0 \\ 0 & 0 & \sqrt{2} \end{bmatrix} \tag{10.22}$$

Hence $\mathbf{A}_2^* = \mathbf{Q}\mathbf{A}_2\mathbf{Q}^T$ has the form

$$[A_2^* \ \langle ij \rangle] = 2 \begin{bmatrix} \varkappa^2 & \varkappa^2 & 0 \\ \varkappa^2 & \varkappa^2 & 0 \\ 0 & 0 & 0 \end{bmatrix} \tag{10.23}$$

Hence a MWCSH is of Type *I* if and only if (10.21) and (10.23) are respectively satisfied by the $\mathbf{A}_1$ and $\mathbf{A}_2$ belonging to that motion[29].

Note that the condition

$$\operatorname{tr} \mathbf{A}_2 = \operatorname{tr} \mathbf{A}_1^2 , \tag{10.24}$$

is not sufficient to characterize viscometric flows, because if tr $\mathbf{A}_1 = 0$ or $v^i_{;i} = 0$, then using Cartesian coordinates, we get

$$\overset{(2)}{A}_{ii} = \frac{d}{dt} \overset{(1)}{A}_{ii} + \overset{(1)}{A}_{ik} \, v_{k,\,i} + \overset{(1)}{A}_{ki} \, v_{k,\,i}$$

$$= 2\overset{(1)}{A}_{ik} \, v_{k,\,i} = 2\overset{(1)}{A}_{ik} \, (D_{ki} + W_{k,i})$$

$$= \overset{(1)}{A}_{ik} \, (2D_{ki}) = \overset{(1)}{A}{}^2_{ii}, \qquad \overset{(1)}{\mathbf{A}} \equiv \mathbf{A}_1, \qquad \overset{(2)}{\mathbf{A}} \equiv \mathbf{A}_2, \tag{10.25}$$

or tr $\mathbf{A}_2 \equiv$ tr $\mathbf{A}_1^2$ if tr $\mathbf{A}_1 = 0$. In deriving (10.25), we have used the fact that the symmetric part of the velocity gradient $v_{k,i}$ is

$$D_{ki} = \frac{1}{2} \overset{(1)}{A}_{ki} = \frac{1}{2} \, (v_{k,i} + v_{i,k}), \tag{10.26}$$

and that if W_{ki} is the skew part of $v_{k,i}$, then

$$W_{ki} \, \overset{(1)}{A}_{ki} = 0. \tag{10.27}$$

---

29. WANG [1965 : 20], Cor. 1. Note that if we take $x = \dot{z} = 0$, $\dot{y} = \kappa x$, we would get WANG's Eqs. (3.10) and (3.11). Thus our results do not contradict the corollary

11. Compatibility Conditions

What we have done so far is to assume that a motion is a MWCSH and then to find the various relations between the RIVLIN-ERICKSEN tensors and the tensor $\mathbf{L}_1$ as well as the strain history. Can one look at a velocity field and decide whether it represents a MWCSH (on a global scale)? Suppose we are given the *three Rivlin-Ericksen tensors* $\mathbf{A}_1$, $\mathbf{A}_2$ and $\mathbf{A}_3$ pertaining to a motion for all time t; is the motion one with CSH? This can be answered in the affirmative quite easily[30] by the following procedure:

(i) Decide whether $\mathbf{A}_1$ has three distinct, or two distinct, or three equal eigenvalues.

(ii) If $\mathbf{A}_1$ has three distinct eigenvalues a, b, c and it is diagonalized, then with respect to this basis, does $[A_2 \langle ij \rangle]$ have the form

$$[A_2 \langle ij \rangle] = \begin{bmatrix} a^2 & d & e \\ \cdot & b^2 & f \\ \cdot & \cdot & c^2 \end{bmatrix}, \tag{11.1}$$

where d, e and f are arbitrary? If the answer is affirmative then the given motion is a MWCSH.

(iii) If $\mathbf{A}_1$ has two distinct eigenvalues a and c, and $\mathbf{A}_1$ is diagonalized to read $[A_1 \langle ij \rangle] = \text{diag}\,[a, a, c]$, then with respect to this basis, does $\mathbf{A}_2$ have the form

$$[A_2 \langle ij \rangle] = \begin{bmatrix} a^2 & 0 & 0 \\ \cdot & a^2 & 0 \\ \cdot & \cdot & c^2 \end{bmatrix} ? \tag{11.2}$$

if so, the motion is a MWCSH, and is the *simple extensional flow*.

(iv) Again, let $\mathbf{A}_1$ have two distinct eigenvalues a and c, and let it be diagonalized so that $[A_1 \langle ij \rangle] = \text{diag}\,[a, a, c]$. Then, with respect to this basis, if

$$[A_2 \langle ij \rangle] = \begin{bmatrix} a^2 & 0 & d \\ 0 & a^2 & e \\ d & e & c^2 \end{bmatrix}, \tag{11.3}$$

and

$$[A_3 \langle ij \rangle] = \begin{bmatrix} a^3 - \dfrac{2d^2}{a-c} & -\dfrac{2de}{a-c} & -df + \dfrac{3\,(a+c)d}{2} \\ & a^3 - \dfrac{2e^2}{a-c} & -ef + \dfrac{3\,(a+b)d}{2} \\ & & c^3 + \dfrac{2\,(d^2+e^2)}{a-c} \end{bmatrix} \tag{11.4}$$

30. WANG [1965 : 20], Sec. 3.

hold, then the motion is a MWCSH.

(v) Finally, let $\mathbf{A}_1 = a\mathbf{1}$. Then, if

$$\mathbf{A}_n = a^n\mathbf{1}, \qquad n = 1, 2, 3, \ldots, \tag{11.5}$$

the given motion is a MWCSH.

The above results are the *compatibility relations* between $\mathbf{A}_1$, $\mathbf{A}_2$ and $\mathbf{A}_3$ in a MWCSH, and are obtained quite easily by appealing to WANG's theorem in Sec. 9.

The results cited here are somewhat unconventional because we are claiming that if we know the first two or three derivatives of a function $[\mathbf{C}_t(t-s)]$ at $s = 0$, we know its entire history. As every student of mathematics knows this is not true in general. For example, knowing $f(0)$, $f'(0)$, $f''(0)$, and $f'''(0)$ is not equivalent to knowing $f(x)$ for all x. But the reason the above claims go through for $\mathbf{C}_t(t-s)$ is because we demand that these three derivatives of this history obey some restrictions for all t. We shall prove the correctness of our reasoning by a detailed study of one case.

According to WANG[31], if for a particle

$$\mathbf{A}_2 = \mathbf{A}_1^2 \text{ for all } t, \tag{11.6}$$

then the motion is a simple extension. We shall now prove this. Let the matrix of $\mathbf{A}_1$ relative to an orthonormal basis be

$$[A_1 \langle ij \rangle] = \text{diag}\,[a, b, c]. \tag{11.7}$$

The most general form of velocity gradient $\mathbf{L}$ compatible with (11.7) must have the matrix form

$$[L \langle ij \rangle] = \begin{bmatrix} a/2 & x & y \\ -x & b/2 & z \\ -y & -z & c/2 \end{bmatrix}. \tag{11.8}$$

Thus

$$[A_2 \langle ij \rangle] = \left[\frac{d}{dt}(A_1 \langle ij \rangle)\right] + \begin{bmatrix} a^2 & 0 & 0 \\ 0 & b^2 & 0 \\ 0 & 0 & c^2 \end{bmatrix}$$

$$+ [A_1 \langle ik \rangle\, W \langle kj \rangle] - [W \langle ik \rangle\, A_1 \langle kj \rangle], \tag{11.9}$$

where $\mathbf{W}$ is the skew matrix defined by

$$\mathbf{W} = \frac{1}{2}(\mathbf{L} - \mathbf{L}^T). \tag{11.10}$$

31. [1965 : 20], p. 337.

Thus, the condition (11.6) implies that

$$\frac{d\mathbf{A}_1}{dt} = \mathbf{W}(t)\,\mathbf{A}_1(t) - \mathbf{A}_1(t)\,\mathbf{W}(t). \tag{11.11}$$

Now, define an orthogonal tensor function $\mathbf{Q}(t)$ such that

$$\dot{\mathbf{Q}}(t)\,\mathbf{Q}(t)^T = \mathbf{W}(t). \tag{11.12}$$

The unique solution to this differential equation is:

$$\mathbf{Q}(t) = e^{\int_0^t \mathbf{W}(\tau)\,d\tau} \tag{11.13}$$

where the value of $\mathbf{Q}(t)$ at time $t = 0$ has been taken to be $\mathbf{1}$. This completes the construction of $\mathbf{Q}(t)$. Moreover, the solution to (11.11) is

$$\mathbf{A}_1(t) = \mathbf{Q}(t)\,\mathbf{A}_1(0)\,\mathbf{Q}(t)^T, \tag{11.14}$$

where $\mathbf{A}_1(0)$ is the value of $\mathbf{A}_1(t)$ at $t = 0$ [*cf.* (8.21)]. Next, $\mathbf{A}_1^2(t) = \mathbf{A}_2(t)$ says that

$$\mathbf{A}_2(t) = \mathbf{Q}(t)\,\mathbf{A}_1^2(0)\,\mathbf{Q}(t)^T, \tag{11.15}$$

implying

$$\frac{d}{dt}\mathbf{A}_2(t) = \mathbf{W}(t)\,\mathbf{A}_2(t) - \mathbf{A}_2(t)\,\mathbf{W}(t). \tag{11.16}$$

Hence

$$\mathbf{A}_3(t) = \frac{d}{dt}\mathbf{A}_2(t) + \mathbf{A}_2(t)\,\mathbf{L}(t) + \mathbf{L}^T(t)\,\mathbf{A}_2(t) = \mathbf{A}_1^3(t). \tag{11.17}$$

Similarly,

$$\mathbf{A}_n(t) = \mathbf{A}_1^n(t), \qquad n = 1, 2, 3, \ldots \tag{11.18}$$

Hence the given motion is a simple extension because we have established that for the given motion, (11.6) implies (11.18). Thus,

$$\mathbf{C}_t(t-s) = \sum_{n=0}^{\infty} (-1)^n \frac{s^n}{n!}\mathbf{A}_n(t), \qquad \mathbf{A}_0 = \mathbf{1}, \tag{11.19}$$

for all t.

Stated in words,[32] we have shown that *a motion of a particle is a simple extension if and only if the Zaremba time flux*[33] *of* $\mathbf{A}_1$ *is zero for all* t.

From (11.14), we see that in simple extensional flow there is a basis such that the flow at a particle is irrotational, this being intrinsic in the construc-

32. HUILGOL [1973 : 3].

33. CFT, Sec. 148; This flux of $\mathbf{A}_1$ is $\mathring{\mathbf{A}}_1 = \dot{\mathbf{A}}_1 + \mathbf{A}_1\mathbf{W}_1 - \mathbf{W}\mathbf{A}_1$.

tion of $\mathbf{Q}(t)$ through (11.12).

Finally, if $\dot{\mathbf{A}}_1 = \mathbf{0}$, then the motion of a particle is a simple extension if and only if $\mathbf{A}_1\mathbf{W} = \mathbf{W}\mathbf{A}_1$. If the vorticity $\boldsymbol{\omega} = \{w_1, w_2, w_3\}$, then $\mathbf{A}_1\mathbf{W} = \mathbf{W}\mathbf{A}_1$ if and only if one of the following is true [see (11.7)]:

(i) if $a = b = c$, then w_1, w_2 and w_3 are arbitrary;
(ii) if $a = b \neq c$, then $w_1 = w_2 = 0$, w_3 arbitrary;
(iii) if $a \neq b \neq c \neq a$, then $w_1 = w_2 = w_3 = 0$.

In materials for which tr $\mathbf{A}_1 \equiv 0$, (i) is not a simple extension, this motion is rigid. Hence *in such materials, a motion such that* $\dot{\mathbf{A}}_1 = \mathbf{0}$ *is a simple extension if and only if* (ii) *or* (iii) *are satisfied.* For example,

$$\dot{x} = a_1 x - \Omega(t) y, \qquad \dot{y} = a_1 y + \Omega(t) x, \qquad \dot{z} = 0$$

obey (ii).

Concluding Remarks. We end this chapter by pointing out that there do exist motions which yield a *finite number of terms for the strain history*, but these are not MWCSH. Such examples will be given in Sec. 13.

Also, what is physically interesting is a motion which is globally a MWCSH[34]. In other words, in field theories, we seek motions which yield CSH for all particles. The flows in Secs. 7–8 are of this class, and such velocity fields can be subjected to global dynamic analysis, as will become apparent in Chap. 7.

34. COLEMAN [1962 : 2] introduced the term globally viscometric flows.

Appendix: Proof of Noll's Theorem

Here we present NOLL's proof[1] of his theorem on MWCSH enunciated in Sec. 8 above.

PROOF. Clearly, if the deformation gradient relative to a fixed reference configuration at time 0 has the form

$$\mathbf{F}_0(\tau) = \mathbf{Q}(\tau)\, e^{\tau\mathbf{M}}, \qquad \mathbf{Q}(0) = \mathbf{1}, \tag{A2.1}$$

where $\mathbf{M}$ is a constant tensor and $\mathbf{Q}(\tau)$ is an orthogonal tensor function, then

$$\mathbf{C}_t(t-s) = \mathbf{Q}(t)\, \mathbf{C}_0(0-s)\, \mathbf{Q}(t)^T, \qquad 0 \leqq s < \infty, \tag{A2.2}$$

trivially. So we prove the converse by assuming that the motion obeys (A2.2). Choose the orthogonal tensor function $\mathbf{Q}(t)$ in (A2.2) such that $\mathbf{Q}(0) = \mathbf{1}$ and define

$$\mathbf{E}(t) = \mathbf{Q}(t)^T\, \mathbf{F}_0(t), \qquad \mathbf{H}(\tau) = \mathbf{C}_0(0-\tau). \tag{A2.3}$$

Then, $\mathbf{E}(0) = \mathbf{H}(0) = \mathbf{1}$, and

$$\begin{aligned} \mathbf{H}(s-t) = \mathbf{C}_0(t-s) &= \mathbf{F}_0(t)^T\, \mathbf{C}_t(t-s)\, \mathbf{F}_0(t) \\ &= \mathbf{E}(t)^T\, \mathbf{C}_0(0-s)\, \mathbf{E}(t), \qquad s \geqq 0, \end{aligned} \tag{A2.4}$$

because of (A2.2). Differentiate (A2.4) with respect to t and put $t = 0$, obtaining

$$-\dot{\mathbf{H}}(s) = \mathbf{M}^T\, \mathbf{H}(s) + \mathbf{H}(s)\, \mathbf{M}, \qquad s \geqq 0, \tag{A2.5}$$

where $\mathbf{M} = \dot{\mathbf{E}}(0)$. The solution to this differential equation is

$$\mathbf{H}(s) = e^{-s\mathbf{M}^T}\, e^{-s\mathbf{M}}, \qquad s \geqq 0. \tag{A2.6}$$

Substitute (A2.6) back into (A2.4) and insert $s - t = \tau$, yielding

$$\mathbf{H}(\tau) = \mathbf{E}(t)^T\, e^{-t\mathbf{M}^T}\, e^{-\tau\mathbf{M}^T}\, e^{-t\mathbf{M}}\, e^{-\tau\mathbf{M}}\, \mathbf{E}(t), \quad \tau \geqq -t, \tag{A2.7}$$

or, $\mathbf{H}(\tau)$ depends analytically on τ for all $\tau \geqq -t$. Since t is arbitrary, $\mathbf{H}(\tau)$ is analytic in τ for all τ. From (A2.6) and the principle of analytic continuation,

$$\mathbf{H}(-\tau) = e^{\tau\mathbf{M}^T}\, e^{\tau\mathbf{M}} = \mathbf{C}_0(\tau) = \mathbf{F}_0(\tau)^T\, \mathbf{F}_0(\tau) \tag{A2.8}$$

for all τ. But (A2.8) is equivalent to

$$\left(\mathbf{F}_0(\tau)\, e^{-\tau\mathbf{M}}\right)^T \left(\mathbf{F}_0(\tau)\, e^{-\tau\mathbf{M}}\right) = \mathbf{1}, \tag{A2.9}$$

or that the function $\mathbf{Q}(\tau)$ defined by

$$\mathbf{Q}(\tau) = \mathbf{F}_0(\tau)\, e^{-\tau\mathbf{M}} \qquad \text{(A2.9}$$

is orthogonal, implying that (A2.1) holds.

1. [1962: 7], Thm. 1.

CHAPTER 3

OBJECTIVITY AND CHANGES OF LOCAL CONFIGURATION

So far we have assigned a fixed reference vector to a particle in a reference configuration and observed its motion. By using certain well-known rules, e.g. chain rule, we have calculated the strain with respect to the present configuration at time $t—s$. The scheme of this chapter is as follows :

(i) Firstly, we take up the components of various kinematical and strain measures as they are seen by a second observer, in relative motion to the first. The second observer is so chosen that he assigns the same reference configuration to the whole body as the first and from then on is in a state of translation and rotation with respect to the first.

(ii) Secondly, the first observer will be called upon to go back to the reference configuration and alter the neighbourhood of the particle X, leaving its position $\mathbf{X}$ invariant. This alteration of the neighbourhood will be done so as to preserve the local density, i.e., the mapping of the neighbourhood from one local shape around X to another will be unimodular. After this is done, it will be assumed that the motion is such that if X and X' are two particles at $\mathbf{X}$ and $\mathbf{X} + d\mathbf{X}$ in the first shape of the neighbourhood about X, and are at $\mathbf{X}$ and $\mathbf{X} + d\mathbf{X}'$ in the second shape of the neighbourhood of X, then X and X' will be at $\mathbf{x}$ and $\mathbf{x} + d\mathbf{x}$ at time t in each case. This leads to changes in the strain measures as we shall see below.

These two operations are *totally independent*. This must be understood by the student of continuum mechanics, for the two operations are used widely and yield us a number of restrictions on the constitutive equations as we shall see later.

In addition, this chapter shows why the RIVLIN-ERICKSEN tensors (Sec. 4) arise in a natural way as the measures of gradients of velocity and acceleration fields if an observer moves himself in a "co-rotational" manner with the particle. Also, we make transparent the meaning of steadiness and intrinsic unsteadiness (*cf.* Sec. 8).

12. Change of Observer : Objective Quantities

In Sec. 1, we called the reference configuration $\mathcal{B}_R$ of the body $\mathcal{B}$ as that occupied by it in the space E^3 at some fixed time t_0. This is not necessary.

All that is really needed is a triplet of numbers to describe a particle unambiguously. So the reference configuration of a body may never be occupied by a body at all. However, it is computationally more tractable if one were to choose the initial configuration during a motion (at time t_0) as the reference configuration—or for an elastic solid, one could choose the unstressed state as the reference state.

Having chosen the reference state of the body as that occupied by the body at time t_0, let $\mathcal{O}$ be an observer with a clock and an orthonormal triad of rigid base vectors, and let $\mathcal{O}$ *measure* the coordinates of the particle X and call them X_α. Let $\mathcal{O}^*$ be another observer similarly equipped, in relative motion to $\mathcal{O}$. At time $t = t_0$, he assigns the reference coordinates of X to be X_α, whatever may be the *measured* coordinates of X at time $t = t_0$ in the $\mathcal{O}^*$ frame. This has certain interesting consequences as we shall see later.

Now, let $M_i(\mathbf{X}, t)$, $\mathbf{X} \in \mathcal{B}_R$, $-\infty < t < \infty$, be the motion of $\mathcal{B}$ so that

$$x_i(t) = M_i(X_\alpha, t) \tag{12.1}$$

are the components of the position vector occupied by X at time t, as measured by $\mathcal{O}$. If $\mathcal{O}^*$ is in relative motion to $\mathcal{O}$ then at time t, the *measured* coordinates of X (see, Fig. 12.1) are

$$x_i^*(t) = c_i(t) + \hat{x}_i(t), \tag{12.2}$$

where $\hat{x}_i(t)$ are the components of $\mathbf{x}(X, t)$ with respect to the base vectors of $\mathcal{O}^*$. However, since the two sets of base vectors are orthonormal, at any

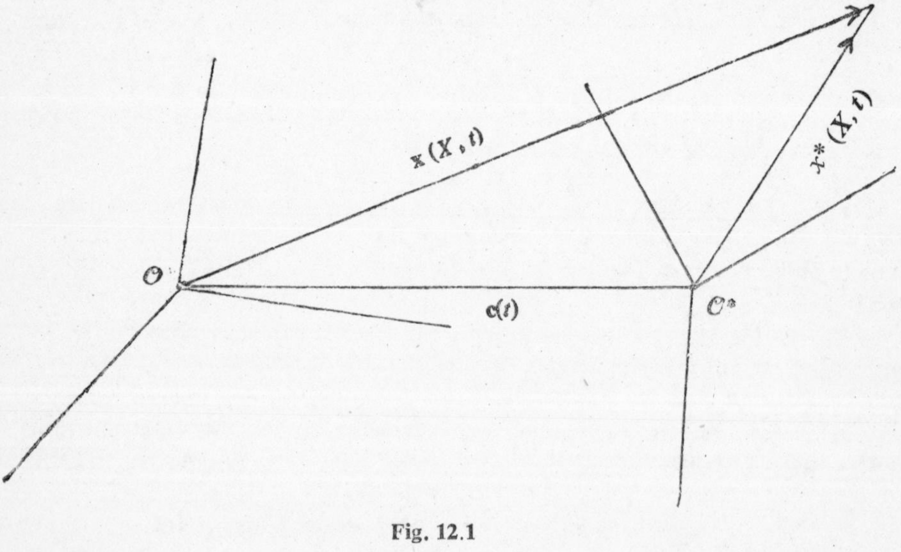

Fig. 12.1

time t, there exists a unique orthogonal tensor function $\mathbf{Q}(t)$ of t such that

$$\hat{x}_i(t) = Q_{ij}(t)\, x_j(t), \tag{12.3}$$

where $x_j(t)$ are measured by the observer $\mathcal{O}$. So (12.1) reads

$$x_i^*(i) = c_i\,(t) + Q_{ij}\,(t)\,x_j\,(t). \tag{12.4}$$

We shall write (12.4) in direct notation as

$$\mathbf{x}^*(t) = \mathbf{c}(t) + \mathbf{Q}(t)\,\mathbf{x}(t), \tag{12.5}$$

though (12.5) is not a tensor relation. We call (12.5), with the assumption about reference coordinates, an *objective motion*. Since

$$\mathbf{M}^*(X_\alpha, t) = \mathbf{c}(t) + \mathbf{Q}(t)\,\mathbf{M}(X_\alpha, t) \tag{12.6}$$

for all $X_\alpha \in \mathcal{B}_R$, $-\infty < t < \infty$, we may calculate the following :

$$F_{i\alpha}^*\,(\mathbf{X}, t) = \frac{\partial x_i^*}{\partial x_\alpha} = Q_{ij}(t)\;F_{j\alpha}(\mathbf{X}, t) \tag{12.7}$$

or,

$$\mathbf{F}^* = \mathbf{QF}. \tag{12.8}$$

Similarly, suppressing the dependence of $\mathbf{Q}$ on t,

$$\begin{aligned} B_{ij}^*\,(\mathbf{X}, t) = F_{i\alpha}^* F_{j\alpha}^* &= Q_{ik}\,F_{k\alpha}\,Q_{jm}\,F_{m\alpha} \\ &= Q_{ik}\,B_{km}\,Q_{jm}, \end{aligned} \tag{12.9}$$

or,

$$\mathbf{B}^* = \mathbf{QBQ}^T\,. \tag{12.10}$$

Next,

$$\mathbf{C}^* = \mathbf{F}^{*T}\,\mathbf{F}^* = \mathbf{C}. \tag{12.11}$$

However,

$$\mathbf{F}_t^*\,(t-s) = \mathbf{F}^*(t-s)\,\mathbf{F}^*(t)^{-1} = \mathbf{Q}(t-s)\,\mathbf{F}_t(t-s)\,\mathbf{Q}(t)^T\,, \tag{12.12}$$

and hence

$$\mathbf{C}_t^*\,(t-s) = \mathbf{Q}(t)\,\mathbf{C}_t(t-s)\,\mathbf{Q}(t)^T. \tag{12.13}$$

So a pattern begins to emerge. Some strain measures, e.g. $\mathbf{B}$, $\mathbf{C}_t(t-s)$ transform so that their components change with the observer, while others such as $\mathbf{C}^*$ do not. In other words, certain measures transform as "tensors", while other do not. We call a scalar φ, a vector $\mathbf{u}$ or a tensor of second order $\mathbf{A}$ *objective* if

$$\varphi^* = \varphi, \quad \mathbf{u}^* = \mathbf{Qu}, \quad \mathbf{A}^* = \mathbf{QAQ}^T. \tag{12.14}$$

Here $\mathbf{Q}$ is a function of time t; but once t is chosen, $\mathbf{Q}(t)$ is a *constant*, for it has only one value at one time.

As can be anticipated from (12.13), the RIVLIN-ERICKSEN tensors are objective, that is,

$$\mathbf{A}_n^*(t) = \mathbf{Q}(t)\,\mathbf{A}_n(t)\,\mathbf{Q}(t)^T, \qquad n = 1, 2, 3, \ldots, \tag{12.15}$$

but the velocity gradient $\mathbf{L}$ is not. For the velocity $\hat{\mathbf{v}}^*(\mathbf{X}, t)$ is given by

$$\hat{\mathbf{v}}^*(\mathbf{X}, t) = \dot{\mathbf{c}}(t) + \mathbf{Q}(t)\,\hat{\mathbf{v}}(\mathbf{X}, t) + \dot{\mathbf{Q}}(t)\,\mathbf{x}(t). \tag{12.16}$$

Now express the velocity in a spatial field form and note that

$$\mathbf{L}^* = \nabla_{\mathbf{x}^*}\mathbf{v}^* = \nabla_{\mathbf{x}}\mathbf{v}^*\,\nabla_{\mathbf{x}^*}\mathbf{x} = [\,\mathbf{Q}(t)\mathbf{L} + \dot{\mathbf{Q}}(t)\,]\,\mathbf{Q}(t)^T \tag{12.17}$$

$$= \mathbf{Q}(t)\,\mathbf{L}\mathbf{Q}(t)^T + \dot{\mathbf{Q}}(t)\,\mathbf{Q}(t)^T. \tag{12.18}$$

Hence $\mathbf{L}^*$ is not objective, but is altered by an amount equal to the spin (or skew) tensor associated with the rotation of $\mathcal{O}^*$ relative to $\mathcal{O}$. Because of (12.18), we have that when $\mathbf{L} = \mathbf{D} + \mathbf{W}$,

$$\mathbf{D}^* = \frac{1}{2}\,\mathbf{A}_1^* = \mathbf{Q}\,\mathbf{D}\,\mathbf{Q}^T, \tag{12.19}$$

and

$$\mathbf{W}^* = \mathbf{Q}\,\mathbf{W}\,\mathbf{Q}^T + \dot{\mathbf{Q}}\,\mathbf{Q}^T. \tag{12.20}$$

All these results are traced to NOLL[1]. Note that (12.19) implies that $\mathbf{A}_1$ is objective, that is,

$$\mathbf{A}_1^* = \mathbf{Q}\mathbf{A}_1\mathbf{Q}^T. \tag{12.21}$$

We now show that all $\mathbf{A}_n$, $n = 1, 2, 3, \ldots$ are objective, by induction on n.

Let all the RIVLIN-ERICKSEN tensors $\mathbf{A}_n$ upto a positive integer n be objective. Then

$$\begin{aligned}\mathbf{A}_{n+1}^* &= \frac{d}{dt}\,\mathbf{A}_n^* + \mathbf{A}_n^*\,\mathbf{L}^* + \mathbf{L}^{*T}\,\mathbf{A}_n^* \\ &= \frac{d}{dt}\,(\mathbf{Q}\mathbf{A}_n\mathbf{Q}^T) + \mathbf{Q}\mathbf{A}_n\mathbf{Q}^T(\mathbf{Q}\mathbf{L}\mathbf{Q}^T + \dot{\mathbf{Q}}\mathbf{Q}^T) + (\mathbf{Q}\mathbf{L}^T\mathbf{Q}^T + \mathbf{Q}\dot{\mathbf{Q}}^T)\,\mathbf{Q}\mathbf{A}_n\mathbf{Q}^T \\ &= \mathbf{Q}(\dot{\mathbf{A}}_n + \mathbf{A}_n\mathbf{L} + \mathbf{L}^T\mathbf{A}_n)\,\mathbf{Q}^T + \\ &\quad + \dot{\mathbf{Q}}\mathbf{A}_n\mathbf{Q}^T + \mathbf{Q}\mathbf{A}_n\dot{\mathbf{Q}}^T + \mathbf{Q}\mathbf{A}_n\mathbf{Q}^T\dot{\mathbf{Q}}\mathbf{Q}^T + \mathbf{Q}\dot{\mathbf{Q}}^T\mathbf{Q}\mathbf{A}_n\mathbf{Q}^T.\end{aligned} \tag{12.22}$$

Since $\dot{\mathbf{Q}}^T = \dot{\overline{\mathbf{Q}^T}}$, and $\mathbf{Q}\mathbf{Q}^T = \mathbf{1}$ implies [*cf.* (8.12)] that

$$\dot{\mathbf{Q}}\mathbf{Q}^T + \mathbf{Q}\dot{\mathbf{Q}}^T = \mathbf{0}, \tag{12.23}$$

we obtain

$$\dot{\mathbf{Q}} = -\,\mathbf{Q}\,\dot{\mathbf{Q}}^T\mathbf{Q}, \qquad \text{or} \qquad \dot{\mathbf{Q}}^T = -\,\mathbf{Q}^T\dot{\mathbf{Q}}\,\mathbf{Q}^T. \tag{12.24}$$

1. [1955 : 2].

Thus the last four terms in (12.22) cancel, yielding the result:

$$\mathbf{A}^*_{n+1} = \mathbf{Q}\,\mathbf{A}_{n+1}\,\mathbf{Q}^T\,, \tag{12.25}$$

and hence if all the $\mathbf{A}_n$ upto n are objective, $\mathbf{A}_{n+1}$ is also objective. Since this result is true for $n = 1$, it must be true for $n = 2, 3, \ldots$ etc.

An interesting explanation to the way the $\mathbf{A}_n$ appear as the non-vanishing kinematical quantities can be given as follows.[2] Eq. (12.20) says that given $\mathbf{L}$, one can choose an observer $\mathcal{O}^*_1$ such that $\mathbf{W}^* = \mathbf{0}$ in his frame for a particle X, that is, $\mathcal{O}^*_1$ can rotate himself such that $\mathbf{L}^* = \mathbf{D}^*$ in his frame of reference. Or, he chooses $\dot{\mathbf{Q}}(t)$ such that

$$\dot{\mathbf{Q}}(t) = \mathbf{Q}(t)\,\overline{\mathbf{W}}(X, t) \tag{12.26}$$

for a single particle X, and thus $\mathbf{L}^* - \mathbf{L}^{*T} = \mathbf{0}$ in his frame of reference. So, $\mathcal{O}^*_1$ sees that $\mathbf{A}^*_1$ is the non-zero kinematical measure of the motion, since $\mathbf{A}^*_1 = 2\mathbf{D}^*$.

Now suppose that the observer $\mathcal{O}$ measures the acceleration gradient $\mathbf{L}_2$ in his frame. Can $\mathcal{O}^*_1$ choose $\ddot{\mathbf{Q}}(t)$ such that $\mathbf{L}^*_2 - \mathbf{L}^{*T}_2 = \mathbf{0}$ in his frame ? We now show that this can be done. Since

$$\mathbf{L}^*_2 = \mathbf{Q}\mathbf{L}_2\mathbf{Q}^T + 2\dot{\mathbf{Q}}\mathbf{L}\mathbf{Q}^T + \ddot{\mathbf{Q}}\mathbf{Q}^T\,, \tag{12.27}$$

demanding that $\mathbf{L}^*_2$ be symmetric yields

$$2\ddot{\mathbf{Q}} = \mathbf{Q}(\mathbf{L}^T_2 - \mathbf{L}_2) + 2\mathbf{Q}\mathbf{L}^T\dot{\mathbf{Q}}^T\mathbf{Q} - 2\dot{\mathbf{Q}}\mathbf{L} - 2\dot{\mathbf{Q}}\dot{\mathbf{Q}}^T\mathbf{Q}, \tag{12.28}$$

where we have used

$$\ddot{\mathbf{Q}}\mathbf{Q}^T + 2\dot{\mathbf{Q}}\dot{\mathbf{Q}}^T + \mathbf{Q}\ddot{\mathbf{Q}}^T = \mathbf{0}, \tag{12.29}$$

derived from (12.23). Replace $\dot{\mathbf{Q}}$ in (12.28) by $\mathbf{Q}(t)\mathbf{W}$ from (12.26). Then,

$$2\ddot{\mathbf{Q}} = \mathbf{Q}(\mathbf{L}^T_2 - \mathbf{L}_2) + \mathbf{Q}\mathbf{L}^T\mathbf{W} + \mathbf{Q}\mathbf{W}\mathbf{L} + \frac{1}{2}\,\mathbf{Q}\mathbf{W}^2. \tag{12.30}$$

With this choice of $\ddot{\mathbf{Q}}$, one can show that

$$\mathbf{L}^*_2 + \mathbf{L}^{*T}_2 = \mathbf{Q}\left(\mathbf{L}_2 + \mathbf{L}^T_2 + 2\,\mathbf{L}^T\mathbf{L} - \frac{1}{2}\,\mathbf{A}^2_1\right)\mathbf{Q}^T. \tag{12.31}$$

Now, since $\mathbf{L}^* = \mathbf{D}^* = \mathbf{L}^{*T}$,

$$\mathbf{Q}\,\mathbf{A}^2_1\,\mathbf{Q}^T = \mathbf{A}^{*2}_1 = 4\mathbf{L}^{*2} = 4\mathbf{L}^{*T}\,\mathbf{L}. \tag{12.32}$$

2. RIVLIN and ERICKSEN [1955 : 4].

Hence (12.31) becomes

$$\mathbf{L}_2^* + \mathbf{L}_2^{*T} + 2\mathbf{L}^{*T}\,\mathbf{L}^* = \mathbf{Q}(\mathbf{L}_2 + \mathbf{L}_2^T + 2\mathbf{L}^T\,\mathbf{L})\,\mathbf{Q}^T\,, \tag{12.33}$$

which is nothing but a statement that

$$\mathbf{A}_2^* = \mathbf{Q}\,\mathbf{A}_2\,\mathbf{Q}^T\,. \tag{12.34}$$

Indeed, the paper of RIVLIN and ERICKSEN[3] was devoted to proving that one can choose $\dot{\mathbf{Q}}, \ddot{\mathbf{Q}}, \ldots, \overset{(n)}{\mathbf{Q}}$ such that $\mathbf{L}_n^* = \mathbf{L}_n^{*T}$, and that in doing so the tensors $\mathbf{A}_n$ appear as objective quantities and that $\mathbf{L}_n^*$ is a combination of the powers of $\mathbf{A}_n, \ldots, \mathbf{A}_1$. We have, of course, demonstrated this by a detailed examination of $\mathbf{L}^*$ and $\mathbf{L}_2^*$ only.

REMARK. We now come to the consequence of both $\mathcal{O}$ and $\mathcal{O}^*$ assigning X_α to X at time t_0. By (12.8) it follows that

$$\lim_{t \to t_0} \mathbf{F}^*(t) = \mathbf{Q}(t_0). \tag{12.35}$$

In other words, $\mathcal{O}^*$ may find that the deformation gradient is not **1** in the reference state. It is $\mathbf{Q}(t_0)$, which may be any orthogonal tensor, including **1**. If, for example, $\mathcal{O}$ is right-handed and $\mathcal{O}^*$ left-handed, $\mathbf{Q}(t)$ will be improper, and det $Q(t) = -1$.

By this definition of the reference coordinates of a particle, we have thus introduced both proper and improper deformation gradients into continuum mechanics. We shall see at the end of Chap. 5, how this viewpoint (due to NOLL) is different from that adopted by other workers in continuum mechanics and how it affects the formulation of constitutive equations.

13. Rigid Motions and Velocity Fields

Given a motion of a body

$$\mathbf{x}(t) = \mathbf{M}(\mathbf{X}, t), \quad \mathbf{X} \in \mathcal{B}_R\,, -\infty < t < \infty, \tag{13.1}$$

to discover whether the velocity field associated with (13.1) is steady or not, we

(i) differentiate (13.1) with respect to t, holding $\mathbf{X}$ fixed, that is, calculate $\hat{\mathbf{v}}(\mathbf{X}, t)$ for all $\mathbf{X}$;

(ii) and then use (13.1) to write $\hat{\mathbf{v}}(\mathbf{X}, t)$ as a field

$$\hat{\mathbf{v}}(\mathbf{X}, t) = \mathbf{v}(\mathbf{x}, t). \tag{13.2}$$

If, in (13.2), $\mathbf{v}(\mathbf{x}, t)$ does not depend on t, the velocity field is steady. Instead of (13.1), if we are given

$$\mathbf{x}^*(t) = \mathbf{c}(t) + \mathbf{Q}(t)\,\mathbf{M}(\mathbf{X}, t) \tag{13.3}$$

3. [1955 : 4].

we determine $\hat{\mathbf{v}}^*(\mathbf{X}, t)$ getting $\mathbf{v}^*(\mathbf{x}^*, t)$ and then see if $\mathbf{v}^*(\mathbf{x}^*, t)$ is independent of t, so that the velocity field is steady in another reference frame.

Two questions arise here: (i) is there a different way of determining whether a given motion leads to a steady velocity field or not? (ii) Suppose a velocity field is not steady in one reference frame, under what conditions will it be steady in another reference frame, that is, can the second observer rotate and translate himself such that the velocity field is steady in his frame of reference?

Before we proceed to answer these questions, it is desirable to dispel the notion that a time-dependent velocity field $\mathbf{c}(t)$ added to an existing velocity field $\mathbf{v}(\mathbf{x}, t)$ is equivalent to superposing a rigid motion on the body.[4]

For, let $\mathbf{x}(\mathbf{X}, t) = \mathbf{M}(\mathbf{X}, t)$ be the motion associated with $\mathbf{v}(\mathbf{x}, t)$. Then $\mathbf{c}(t) + \mathbf{v}(\mathbf{x}, t)$ may not be associated with $\mathbf{c}(t) + \mathbf{M}(\mathbf{X}, t)$. We shall now present an example to make this obvious. Let the following velocity field in Cartesian coordinates be given:

$$\dot{x} = u_0, \quad \dot{y} = lx^2, \quad \dot{z} = ny, \tag{13.4}$$ [5]

where u_0, l, n are constants. The path lines corresponding to (13.4) are

$$\begin{aligned} \xi &= x - su_0, \\ \eta &= y - l(x^2s - xu_0s^2 + \tfrac{1}{3}u_0s^3), \\ \zeta &= z - n\left(ys - \frac{lx^2}{2}s^2 + \frac{lxu_0}{3}s^3 - \frac{lu_0}{12}s^4\right). \end{aligned} \tag{13.5}$$

We can visualize (13.4) as the addition of $u_0\mathbf{i}$ to $\mathbf{v}(\mathbf{x}) = lx^2\mathbf{i} + ny\mathbf{k}$. While $u_0\mathbf{i}$ by itself leads to a rigid motion, its addition to $\mathbf{v}(\mathbf{x})$ alters the strain history enormously.

(i) if $u_0 = 0$, then (13.4) gives rise to a MWCSH-II;

(ii) if $u_0 \neq 0$, then

$$\mathbf{C}_1(t-s) = \sum_{m=0}^{6} (-1)^m \frac{s^m}{m!} \mathbf{A}_m, \quad \mathbf{A}_0 \equiv \mathbf{1}, \tag{13.6}$$

or the strain history has a finite number of terms in it. It is not a MWCSH, and since (13.6) is not a MWCSH-II, the velocity field (13.4) with $u_0 = 0$, is not equivalent, modulo a rigid motion, to the velocity field with $u_0 \neq 0$. The reader should keep this in mind when we come to the discussion of adding velocity fields to global viscometric flows in Sec 29.

4. The flow in the MAXWELL rheometer is not equivalent to a rigid rotation superposed on simple shearing, as we have already seen in Sec. 7.

5. HUILGOL [1971 : 12].

Returning to the two questions posed earlier, we have two theorems concerning the first question.

THEOREM 13.1 (GREENBERG[6]). *In one dimension, a motion M is steady if and only if*

$$x(t) = M(X, t) = f(\alpha), \qquad \alpha = \frac{X}{V_0} + t, \tag{13.7}$$

where V_0 is a constant and $df/d\alpha = f'(\alpha) > 0$ almost everywhere.

The proof will be omitted, being straightforward, and also because a three-dimensional version will be presented below. Before doing this, since $f'(\alpha) > 0$, one can invert (13.7) and write

$$f^{-1}[M(X, t)] = \alpha = \frac{X}{V_0} + t. \tag{13.8}$$

Let $w(\cdot)$ be the inverse of $f(\cdot)$. Then (13.8) implies that

$$w[x(t)] - w[\xi(\tau)] = t - \tau, \tag{13.9}$$

where $x(t)$ and $\xi(\tau)$ are the positions of the same particle at times t and τ respectively.

We now outline the proof of the three-dimensional version of the above theorem next.

THEOREM 13.2 (BERNSTEIN[7]). *A motion* $\mathbf{x}(t)$ *is steady if and only if*

$$\mathbf{x}(t) = \overline{\mathbf{M}}((X, t) = \mathbf{f}(\boldsymbol{\xi}, t-\tau), \tag{13.10}$$

where $\mathbf{x}(t)$ *and* $\xi(\tau)$ *are the positions ocuupied by the particle X at times t and τ respectively.*

PROOF. *Sufficiency.* Let (13.10) hold. Define

$$\mathbf{f}'(\boldsymbol{\xi}, \mu) = \frac{\partial}{\partial \mu}\mathbf{f}(\boldsymbol{\xi}, \mu). \tag{13.11}$$

Then

$$\begin{aligned}\overline{\mathbf{V}}(X, t) &= \frac{\partial}{\partial t}\overline{\mathbf{M}}(X, t) = \left.\frac{\partial \mathbf{f}}{\partial t}\right|_{X \text{ fixed},\ t=\tau} \\ &= \mathbf{f}'\Big|_{\mu=0} = \mathbf{f}'(\overline{\mathbf{M}}(X, t), 0) = \mathbf{f}'[\mathbf{x}(t)],\end{aligned} \tag{13.12}$$

or $\overline{\mathbf{v}}(X, t)$ is a function of $\mathbf{x}(t)$, the spatial position occupied by X. So the velcocity field is steady.

6. [1967 : 12].
7. [1971 : 3].

Necessity. Let $\mathbf{v}(\mathbf{x})$ be such that $v_i v_i \neq 0$ in $[\mathbf{x}_1, \mathbf{x}_2]$. Define a function $w^1(\mathbf{x})$ such that

$$w^1_{,i}\, v_i = 1, \tag{13.13}$$

whence we find two other functionally independent functions $w^2(\mathbf{x})$ and $w^3(\mathbf{x})$ such that

$$w^A_{,i}\, v_i = 0, \qquad A = 2, 3. \tag{13.14}$$

Combining (13.13) and (13.14), we have

$$w^A_{,i}\, v_i = \delta^A_1 , \qquad A = 1, 2, 3. \tag{13.15}$$

Or, in terms of material coordinates,

$$\frac{\partial \hat{w}^A}{\partial t}(\mathbf{X}, t) = \delta^A_1 \cdot \tag{13.16}$$

Thus the solution to (13.16) is:

$$\hat{w}^A(\mathbf{X}, t) = w^A[\mathbf{x}(t)] = \delta^A_1\, t + F^A(\mathbf{X}), \qquad A = 1, 2, 3, \tag{13.17}$$

where the $F^A(\mathbf{X})$ are functions of $\mathbf{X}$ alone. Hence, the analogue of (13.9) is:

$$w^A[\mathbf{x}(t)] - w^A[\boldsymbol{\xi}(\tau)] = (t-\tau)\, \delta^A_1 \cdot \tag{13.18}$$

Since the w^A are functionally independent, we have

$$x_i(t) = W_i[(t-\tau)\, \delta^A_1 + w^A(\boldsymbol{\xi}(\tau))], \qquad i = 1, 2, 3, \tag{13.19}$$

where W_i are the inverses to w^A, that is,

$$W_i[w^A(\mathbf{x})] = x_i. \tag{13.20}$$

Now (13.19) has the desired form (13.10).

Another interesting result stems from (13.10). Since

$$\begin{aligned} 0 = \frac{\partial x_i}{\partial \tau}\bigg|_X &= \frac{\partial f_i}{\partial \xi_\alpha}\, \frac{\partial \xi_\alpha}{\partial \tau}\bigg|_X - f'_i \\ &= \frac{\partial x_i}{\partial \xi_\alpha}\, \hat{v}_\alpha - \hat{v}_i, \end{aligned} \tag{13.21}$$

we get

$$\hat{\mathbf{v}}(\mathbf{X}, t) = \mathbf{F}_\tau(t)\, \hat{\mathbf{v}}(\mathbf{X}, \tau), \tag{13.22}$$

or,

$$\mathbf{F}(\mathbf{X}, t)^{-1}\, \hat{\mathbf{v}}(\mathbf{X}, t) = \mathbf{F}(\mathbf{X}, \tau)^{-1}\, \hat{\mathbf{v}}(\mathbf{X}, \tau). \tag{13.23}$$

We have now assembled the apparatus to decide whether a given motion yields a steady velocity field or not. Suppose a given motion does not yield a steady velocity field. Is it steady in another frame of reference? The necessary and sufficient condition governing this has been known for some time, but no answer is at hand. We shall derive this condition next.

Rewrite (13.3) as

$$x_k = Q_{i'k}\, x^*_{i'} + b_k = \tilde{Q}_{ki'}\, x^*_{i'} + b_k, \tag{13.24}$$

where we have put $b_k = -Q_{i'k}\, c_{i'}$, and defined a new orthogonal function $\tilde{Q}_{ki'}(t) = Q_{i'k}(t)$. Now for any function $\mathbf{f} = \mathbf{f}(\mathbf{x}, t)$, put

$$\mathbf{f}^*(\mathbf{x}^*, t) = \mathbf{f}(\mathbf{x}(\mathbf{x}^*, t), t). \tag{13.25}$$

Then keeping $\mathbf{x}^*$ constant, we get

$$\frac{\partial f_i^*}{\partial t} = \frac{\partial f_i}{\partial t} + \frac{\partial f_i}{\partial x_k}\frac{\partial x_k}{\partial t} \tag{13.26}$$

$$= \frac{\partial f_i}{\partial t} + \frac{\partial f_i}{\partial x_k}[\dot{\tilde{Q}}_{ki'}\, x^*_{i'} + \dot{b}_k]$$

$$= \frac{\partial f_i}{\partial t} + \frac{\partial f_i}{\partial x_k}[\omega_{kl}x_l + a_k], \tag{13.27}$$

inserting

$$a_k = \dot{b}_k - \omega_{kl}b_l,$$

$$x^*_{i'} = \tilde{Q}_{li'}(x_l - b_l) \tag{13.28}$$

$$\omega_{kl} = \dot{\tilde{Q}}_{ki'}\, \tilde{Q}_{li'}.$$

Identify $\mathbf{f}(\mathbf{x}, t)$ with $\mathbf{v}(\mathbf{x}, t)$. Then, keeping $\mathbf{x}^*$ fixed, (13.27) yields

$$\frac{\partial v_k}{\partial t} + (\omega_{ln}x_n + a_l)\, v_{k,l} = \frac{\partial}{\partial t}(\dot{b}_k + \dot{\tilde{Q}}_{ki'}\, x^*_{i'} + \tilde{Q}_{ki'}\, \dot{x}^*_{i'})$$

$$= \ddot{b}_k + \ddot{\tilde{Q}}_{ki'}\, x^*_{i'} + 2\dot{\tilde{Q}}_{ki'}\, \dot{x}^*_{i'} + \tilde{Q}_{ki'}\frac{\partial \dot{x}^*_{i'}}{\partial t}. \tag{13.29}$$

Replacing $x^*_{i'}$ and $\dot{x}^*_{i'}$ by their expressions in terms of x_k and t, one gets that $\partial \dot{x}^*_{i'}/\partial t = 0$ if and only if[8]

$$\frac{\partial v_k}{\partial t}(\omega_{mn}x_n + a_m)\, v_{k,m} - \dot{a}_k - \dot{\omega}_{km}x_m - \omega_{km}v_m = 0, \tag{13.30}$$

which is expressed entirely in terms of $\mathbf{v}(\mathbf{x}, t)$. If $\mathbf{a}$ and $\boldsymbol{\omega}$ exist, then $\mathbf{v}^*(\mathbf{x}^*, t)$ will be steady but not otherwise. At present, the existence of solutions to (13.30) has not been resolved.

8. CFT, Sec. 146.

Suppose **a** and **ω** do not exist. Then $\mathbf{v}(\mathbf{x}, t)$ is *intrinsically unsteady*, in other words $\mathbf{v}(\mathbf{x}, t)$ is not steady in any frame of reference. It is in this context that the intrinsic unsteadiness of the motion (8.40) has to be interpreted, and a laborious calculation will convince the reader of this fact.

14. Change of Local Configuration[9]

As stipulated earlier, let X and X' be two adjacent particles and let $\mathbf{X}$ and $\mathbf{X} + d\mathbf{X}$ be their positions in the fixed reference configuration. Let the motion be given by

$$\mathbf{x}(X, t) = \mathbf{M}(\mathbf{X}, t), \quad \forall\, \mathbf{X} \in \mathcal{B}_R, \quad -\infty < t < \infty, \tag{14.1}$$

or X goes to $\mathbf{x}$ and X' goes to $\mathbf{x} + d\mathbf{x}$ at time t, so that $\mathbf{F}, \mathbf{B}, \mathbf{C}, \ldots$ are all

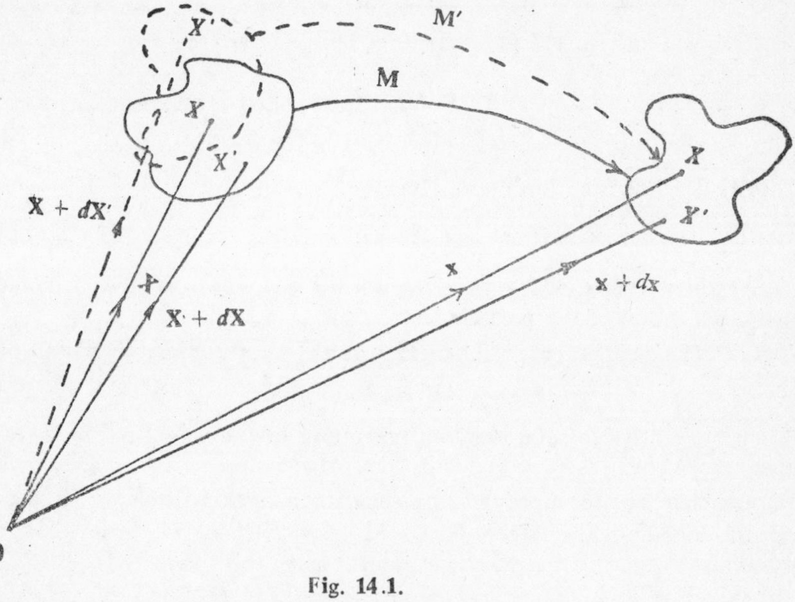

Fig. 14.1.

well defined. Now let the neighbourhood (see, Fig. 14.1) be changed in the reference configuration so that

X remains at $\mathbf{X}$,

X' goes to $\mathbf{X} + d\mathbf{X}'$.

This mapping $\mathbf{H}$:

$$d\mathbf{X}' = H d\mathbf{X} \tag{14.2}$$

is linear, independent of time, and is unimodular, in symbols $|\det \mathbf{H}| = 1$.

Now if the motion is such that X again goes to $\mathbf{x}$, and X' goes to $\mathbf{x} + d\mathbf{x}$,

9. Based on NOLL [1955 : 2 ; 1958 : 3].

then we have a new motion $\mathbf{M}'$ compared to the old $\mathbf{M}$:

$$\mathbf{M} : (X \text{ or}) \ \mathbf{X} \rightarrow \mathbf{x}; \quad \mathbf{M} : (X' \text{ or}) \ \mathbf{X} + d\mathbf{X} \rightarrow \mathbf{x} + d\mathbf{x}. \tag{14.3}$$

$$\mathbf{M}' : (X \text{ or}) \ \mathbf{X} \rightarrow \mathbf{x}; \quad \mathbf{M}' : (X' \text{ or}) \ \mathbf{X} + d\mathbf{X}' \rightarrow \mathbf{x} + d\mathbf{x}. \tag{14.4}$$

Obviously, $\nabla_{\mathbf{X}}\mathbf{M}$ and $\nabla_{\mathbf{X}}\mathbf{M}'$ will be different. They will be different, because two shapes about X are being mapped into the same shape about X in the position $\mathbf{x}$ in space. We calculate $\mathbf{F}' = \nabla_{\mathbf{X}}\mathbf{M}'$ from the following. Since

$$d\mathbf{x} = \mathbf{F}d\mathbf{X} = \mathbf{F}' \, d\mathbf{X}' = \mathbf{F} \, \mathbf{H}dX, \tag{14.5}$$

we have

$$\mathbf{F}' \, \mathbf{H} = \mathbf{F} \quad \text{or} \quad \mathbf{F}' = \mathbf{F}\mathbf{H}^{-1}, \tag{14.6}$$

implying

$$\mathbf{B}' = \mathbf{F}' \, \mathbf{F}'^{T} = \mathbf{F}\mathbf{H}^{-1} \, (\mathbf{H}^{-1})^{T} \, \mathbf{F}^{T}, \tag{14.7}$$

$$\mathbf{C}' = \mathbf{F}'^{T} \, \mathbf{F}' = (\mathbf{H}^{-1})^{T} \, \mathbf{C}\mathbf{H}^{-1}, \tag{14.8}$$

$$\begin{aligned} \mathbf{F}'_{t} \, (t-s) &= \mathbf{F}'(t-s) \, \mathbf{F}'(t)^{-1} \\ &= \mathbf{F}(t-s) \, \mathbf{F}(t)^{-1} = \mathbf{F}_{t}(t-s), \end{aligned} \tag{14.9}$$

because $\mathbf{H}$ is not a function of t. Hence

$$\mathbf{C}'_{t} \, (t-s) = \mathbf{C}_{t}(t-s), \tag{14.10}$$

or the strain measure $\mathbf{C}_{t}(t-s)$ is not sensitive to non-singular changes of the neighbourhood of the particle.

Now, let $\mathbf{H}$ be *orthogonal*. Then $\mathbf{H}^{-1} = \mathbf{H}^{T}$, and under this condition,

$$\mathbf{B}' = \mathbf{B}, \tag{14.11}$$

or $\mathbf{B}$ is not sensitive to rotations and reflections of the neighbourhood of the particle.

Suppose that we are interested in measuring a certain property $\mathcal{P}$ depending on the kinematical variable $\mathbf{F}(\tau)$. Moreover, let it so happen that some unimodular changes of local shape about X leave this quantity $\mathcal{P}$ unchanged, though clearly $\mathbf{F}(\tau)$ is altered. This set $\mathcal{G}_{\mathcal{P}} = \{H_1, H_2, \ldots\}$ of unimodular transformations (finite or infinite in number) then *preserves* $\mathcal{P}$, or $\mathcal{P}$ is unaffected by such alterations of the neighbourhood. Expressing the functional relationship between $\mathcal{P}$ and $\mathbf{F}(\tau)$ as

$$\mathcal{P} = \mathfrak{p}[\mathbf{F} \, (\tau)], \tag{14.12}$$

we have that for all $\mathbf{F}(\tau)$

$$\mathfrak{p}[\mathbf{F}(\tau)] = \mathfrak{p}[\mathbf{F}(\tau)\mathbf{H}_1], \quad \mathfrak{p}[\mathbf{F}(\tau)] = \mathfrak{p}[\mathbf{F}(\tau)\mathbf{H}_2], \tag{14.13}$$

etc. Let us replace $\mathbf{F}(\tau)$ in $(14.13)_2$ by $\mathbf{F}(\tau)\mathbf{H}_1$. This implies that

$$\mathfrak{p}(\mathbf{F}(\tau)\mathbf{H}_1) = \mathfrak{p}(\mathbf{F}(\tau)\mathbf{H}_1 \, \mathbf{H}_2), \tag{14.14}$$

or

$$H_1 \, H_2 \in \mathcal{G}_{\mathcal{P}}, \quad \text{if} \quad H_1, \, H_2 \in \mathcal{G}_{\mathcal{P}}. \tag{14.15}$$

Next, replace $\mathbf{F}(\tau)$ by $\mathbf{F}(\tau)\,\mathbf{H}_1^{-1}$. Then $(14.13)_1$ demands that

$$\mathfrak{p}(\mathbf{F}(\tau)\mathbf{H}_1^{-1}) = \mathfrak{p}(\mathbf{F}(\tau)\mathbf{H}_1^{-1}\,\mathbf{H}_1) = \mathfrak{p}[\mathbf{F}(\tau)], \tag{14.16}$$

or

$$\mathbf{H}_1^{-1} \in \mathcal{G}_{\mathcal{P}}, \quad \text{if} \quad \mathbf{H}_1 \in \mathcal{G}_{\mathcal{P}}. \tag{14.17}$$

Hence the set $\mathcal{G}_{\mathcal{P}}$ forms a group under the binary operation of composition:

$$\mathbf{H}_1\,\mathbf{H}_2(\mathbf{u}) = \mathbf{H}_1[\mathbf{H}_2(\mathbf{u})] \tag{14.18}$$

where $\mathbf{u}$ is a vector.

So we say that the set $\mathcal{G}_{\mathcal{P}}$ is a *$\mathcal{P}$-symmetry group*, that is, the group of unimodular changes of the neighbourhood of X which leave $\mathcal{P}$ invariant.

Suppose we wish to find the group $\mathcal{G}_{\mathcal{P}}$ with respect to another position of X. In the other words, let $\mathbf{X}$ and $\mathbf{X}_1$ be two positions of X and let X' be the neighbourhood particle of X. With respect to $\mathbf{X}$, X' will be assumed to be at $\mathbf{X} + d\mathbf{X}$, and with respect to $\mathbf{X}_1$, X' will be at $\mathbf{X}_1 + d\mathbf{X}_1$. Again, let X and X' go to $\mathbf{x}$ and $\mathbf{x} + d\mathbf{x}$ respectively. Then

$$d\mathbf{x} = \mathbf{F}d\mathbf{X} = \mathbf{F}_1\,d\mathbf{X}_1. \tag{14.19}$$

If the motion from $\mathbf{X}$ to $\mathbf{X}_1$ is such that

$$d\mathbf{X}_1 = \mathbf{N}\,d\mathbf{X}, \tag{14.20}$$

or $\mathbf{N}$ is the gradient of the mapping from $\mathbf{X}$ to $\mathbf{X}_1$ at a fixed time τ_0, then

$$\mathbf{F}_1 = \mathbf{F}\mathbf{N}(\tau_0)^{-1}. \tag{14.21}$$

Hence the property $\mathcal{P}$ is given by

$$\mathcal{P} = \mathfrak{p}[\mathbf{F}(\tau)] = \mathfrak{p}_1[\mathbf{F}_1(\tau)] = \mathfrak{p}_1[\mathbf{F}(\tau)\,\mathbf{N}(\tau_0)^{-1}], \tag{14.22}$$

where we emphasize that the measurement of $\mathbf{F}$ (or $\mathbf{F}_1$) depends on $\mathbf{X}$ (or $\mathbf{X}_1$). Now, arguing as previously,

$$\begin{aligned}\mathfrak{p}[\mathbf{F}(\tau)] &= \mathfrak{p}[\mathbf{F}(\tau)\,\mathbf{H}_1] \\ &= \mathfrak{p}_1[\mathbf{F}(\tau)\,\mathbf{N}(\tau_0)^{-1}] = \mathfrak{p}_1[\mathbf{F}(\tau)\,\mathbf{H}_1\,\mathbf{N}(\tau_0)^{-1}].\end{aligned} \tag{14.23}$$

Now, replace $\mathbf{F}(\tau)$ by $\mathbf{F}(\tau)\,\mathbf{N}(\tau_0)$ in (14.23). Then

$$\begin{aligned}\mathfrak{p}\,[\mathbf{F}(\tau)\,\mathbf{N}(\tau_0)] &= \mathfrak{p}[\mathbf{F}(\tau)\,\mathbf{N}(\tau_0)\,\mathbf{H}_1] \\ &= \mathfrak{p}_1[\mathbf{F}(\tau)\,] = \mathfrak{p}_1(\mathbf{F}(\tau)\,\mathbf{N}(\tau_0)\,\mathbf{H}_1\,\mathbf{N}(\tau_0)^{-1}).\end{aligned} \tag{14.24}$$

Hence, the last two yield that if $\mathbf{H}_1 \in \mathcal{G}_{\mathcal{P}}$, then

$$\mathbf{N}(\tau_0)\,\mathbf{H}_1\,\mathbf{N}(\tau_0)^{-1} \in \mathcal{G}_1^{\mathcal{P}}, \tag{14.25}$$

where $\mathcal{G}_1^{\mathcal{P}}$ is the new $\mathcal{P}$-symmetry group. Hence

$$N(\tau_0)\, \mathcal{G}_{\mathcal{P}}\, N(\tau_0)^{-1} = \mathcal{G}_1^{\mathcal{P}}, \tag{14.26}$$

or the change in the reference configuration leads to a change in the $\mathcal{P}$-symmetry group to within a *conjugation.*

Now, if $\mathcal{G}_{\mathcal{P}} = \mathcal{U}$, the full unimodular group, then $\mathcal{G}_1^{\mathcal{P}} = \mathcal{U}$ also for all N in (14.26). In words, a unimodular group is mapped onto itself by conjugation.

This statement leads to the following important fact. *Any and all unimodular changes of reference configuration leave* $C_t(t-s)$ *invariant*, though objective motions do not do so. Hence, it follows that $C_t(t-s)$ appears in the constitutive equation of a simple material (which we shall study in Chap. 5), whatever its symmetry group is.

CHAPTER 4

BALANCE EQUATIONS

The purpose of this chapter is to derive certain balance equations for physical quantities which arise in the study of continuous media. Specifically, we are interested in the form the equations take for conservation of mass, balance of linear and angular momenta and balance of energy. These relations will be derived for situations wherein the motion $\mathbf{x}(\mathbf{X}, t)$ is continuous and has continuous partial derivatives of order two at least.

15. Reynolds' Transport Theorem[1]

Let $\Psi(\mathbf{x}, t)$ be a field defined over the volume $\mathcal{V}$ occupied by the body $\mathcal{B}$ at time t. Then, REYNOLDS' theorem asserts that

$$\frac{d}{dt}\int_{\mathcal{V}} \Psi \, dv = \int_{\mathcal{V}} (\dot{\Psi} + \Psi \operatorname{div} \mathbf{v}) \, dv, \tag{15.1}$$

where

$$\dot{\Psi} = \frac{\partial \Psi}{\partial t} + \Psi_{;k} \, v^k. \tag{15.2}$$

To prove this theorem, let $\mathcal{V}_0$ be the material volume occupied by $\mathcal{B}$ in its reference configuration $\mathcal{B}_R$. Assuming that the motion $\mathbf{x}(\mathbf{X}, t)$ is continuous and twice continuously differentiable, write

$$\Psi(\mathbf{x}, t) = \overline{\Psi}(\mathbf{X}, t). \tag{15.3}$$

Then

$$\frac{d}{dt}\int_{\mathcal{V}} \Psi \, dv = \frac{d}{dt}\int_{\mathcal{V}_0} \overline{\Psi} \, J \, dV, \tag{15.4}$$

where dV represents the volume in the fixed configuration $\mathcal{B}_R$, and J is the Jacobian of the transformation, that is, $J = |\det \mathbf{F}(\mathbf{X}, t)|$. Now, because $\mathcal{V}_0$ is independent of t, we may differentiate the integrand with respect to t and obtain

$$\frac{d}{dt}\int_{\mathcal{V}_0} \overline{\Psi} \, J \, dV = \int_{\mathcal{V}_0} \frac{d}{dt}(\overline{\Psi} J) \, dV. \tag{15.5}$$

Next,

$$\begin{aligned} \frac{d}{dt}\overline{\Psi}(\mathbf{X}, t)\Big|_{\mathbf{X} \text{ fixed}} &= \frac{d}{dt}\Psi(\mathbf{x}(\mathbf{X}, t), t)\Big|_{\mathbf{X} \text{ fixed}} \\ &= \frac{\partial \Psi}{\partial t}\Big|_{\mathbf{x} \text{ fixed}} + \nabla\Psi \cdot \mathbf{v} \end{aligned} \tag{15.6}$$

1. Proved by REYNOLDS in 1903. For a modern derivation, see CFT, Sec. 81.

Also

$$\frac{dJ}{dt} = \frac{d}{dt} (\,|\det \mathbf{F}\,|) = I_d\, J, \tag{15.7}$$

where

$$I_d = \operatorname{div} \mathbf{v}. \tag{15.8}$$

Note that in proving (15.7), the rule for differentiating a determinant ought to be kept in view.

Collecting all the above results, and reverting to the current material volume $\mathcal{V}$, we obtain

$$\begin{aligned} \frac{d}{dt}\int_{\mathcal{V}} \Psi\, dv &= \int_{\mathcal{V}} (\dot{\Psi} + \operatorname{div} \mathbf{v})\, dv \\ &= \int_{\mathcal{V}} \left[\frac{\partial \Psi}{dt} + (\Psi\, v^k)_{;k}\right] dv \\ &= \int_{\mathcal{V}} \frac{\partial \Psi}{\partial t}\, dv + \int_{\mathcal{S}} \Psi\, \mathbf{v}\,.\,\mathbf{n}\, da, \end{aligned} \tag{15.9}$$

where the divergence theorem has been employed. In (15.9), $\mathcal{S}$ denotes the surface of $\mathcal{V}$, $\mathbf{n}$ the unit external normal to it and da the elemental area. Eq. (15.9) is called the *transport theorem of Reynolds.*

16. Conservation of Mass

In continuum mechanics, the conservation of mass is assumed to be a postulate. In other words, for all material volumes $\mathcal{V}_0$ (that is, those containing the same particles for all times) with a mass density field $\rho = \rho(\mathbf{x}, t)$:

$$\frac{d}{dt}\int_{\mathcal{V}_0} \rho\, dv = 0. \tag{16.1}$$

By the transport theorem, when the current volume $\mathcal{V}$ of $\mathcal{V}_0$ is assigned,

$$0 = \int_{\mathcal{V}} \left[\frac{\partial \rho}{\partial t} + (\rho v^k)_{;k}\right] dv. \tag{16.2}$$

Since $\mathcal{V}$ is arbitrary, a necessary and sufficient condition for the *conservation of mass* is the *continuity equation*

$$\frac{\partial \rho}{\partial t} + (\rho v^k)_{;k} = \dot{\rho} + \rho \operatorname{div} \mathbf{v} = 0. \tag{16.3}$$

In incompressible materials, only *isochoric* (volume-preserving) motions are possible and since ρ is a costant everywhere, conservation of mass implies and is guaranteed by

$$\operatorname{div} \mathbf{v} = v^k_{\ ;k} = 0. \tag{16.4}$$

But

$$v^i_{\ ;i} = \operatorname{tr} \mathbf{L} = \tfrac{1}{2} \operatorname{tr} \mathbf{A}_1, \tag{16.5}$$

and thus *an equivalent assertion is that* tr $\mathbf{A}_1 = 0$. This conclusion motivates the discussions in the earlier chapters of velocity fields obeying tr $\mathbf{A}_1 = 0$.

Also, we note that since $d\mathbf{x} = \mathbf{F}\, d\mathbf{X}$, the material volume dv is altered on deformation to $dV = J\, dV$, where $J = |\det \mathbf{F}|$. Hence, if ρ_R is the density in the reference configuration and ρ the present density, conservation of mass implies

$$\rho\, dv = \rho_R\, dV, \tag{16.6}$$

which in turn yields

$$\rho_R = \rho J. \tag{16.7}$$

In view of this, another necessary and sufficient condition for isochoric motions is :

$$|\det \mathbf{F}| = 1, \quad \det \mathbf{C}_t(t-s) = 1, \quad s \in [0, \infty). \tag{16.8}$$

We wish to reemphasise that *in all isochoric motions*

$$\text{tr}\, \mathbf{A}_1^2 \equiv \text{tr}\, \mathbf{A}_2, \tag{16.9}$$

which was proved in Sec. 11.

17. Balance of Linear Momentum

The forces that act on the body or its part will be divided into two categories : those that act by application to the surface, called *contact forces*, and those that act at a distance, called *body surfaces*. Let $\mathcal{V}$ denote a part of the material volume occupied by the body $\mathcal{B}$ at time t, $\mathbf{t}_{(\mathbf{n})}$ the contact force per unit area on its surface $\mathcal{S}$, exerted by the part of the body on the side of $\mathcal{S}$ towards which the unit normal vector $\mathbf{n}(\mathbf{x})$ is directed, and $\mathbf{b}$ the body force per unit mass. Then the equation for the balance of linear momentum in an inertial frame of reference reads

$$\frac{d}{dt}\int_{\mathcal{V}} \rho\, \mathbf{v}\, dv = \int_{\mathcal{S}} \mathbf{t}_{(\mathbf{n})}\, da + \int_{\mathcal{V}} \rho\, \mathbf{b}\, dv = \int_{\mathcal{V}} \rho\, \mathbf{a}\, dv, \tag{17.1}$$

where we have used the transport theorem and assumed that *mass is conserved* so that $\mathbf{a}$ is the acceleration field. We wish to convert the surface integral in (17.1) to a volume integral and apply the divergence theorem to obtain a differential equation for the balance of linear momentum. To accomplish this, we shall follow CAUCHY, and introduce the stress tensor in the following manner.

Consider a tetrahedron with three sides which are mutually orthogonal, the fourth having the outward unit normal $\mathbf{n}$ (see Fig. 17.1). Let the altitude of the tetrahedron be h; the area of the inclined surface A. If the projections of $\mathbf{n}$ onto the three axes are n_1, n_2, n_3, the area A_i of the face normal to the x_i direction is $n_i A = A_i$, $i = 1, 2, 3$. Assume that the fields $\mathbf{a}$ and $\mathbf{b}$ are bounded and that $\mathbf{t}_{(\mathbf{n})}$ and $\mathbf{t}_i$ are continuous functions of both $\mathbf{x}$ and t, where $\mathbf{t}_{(\mathbf{n})}$ is the stress vector field on the face with the unit normal $\mathbf{n}$, and $\mathbf{t}_i$ ($i=1, 2, 3$) are the stress vector fields on the faces with the unit normals in the $-x_i$ directions. The volume integrals may now be estimated

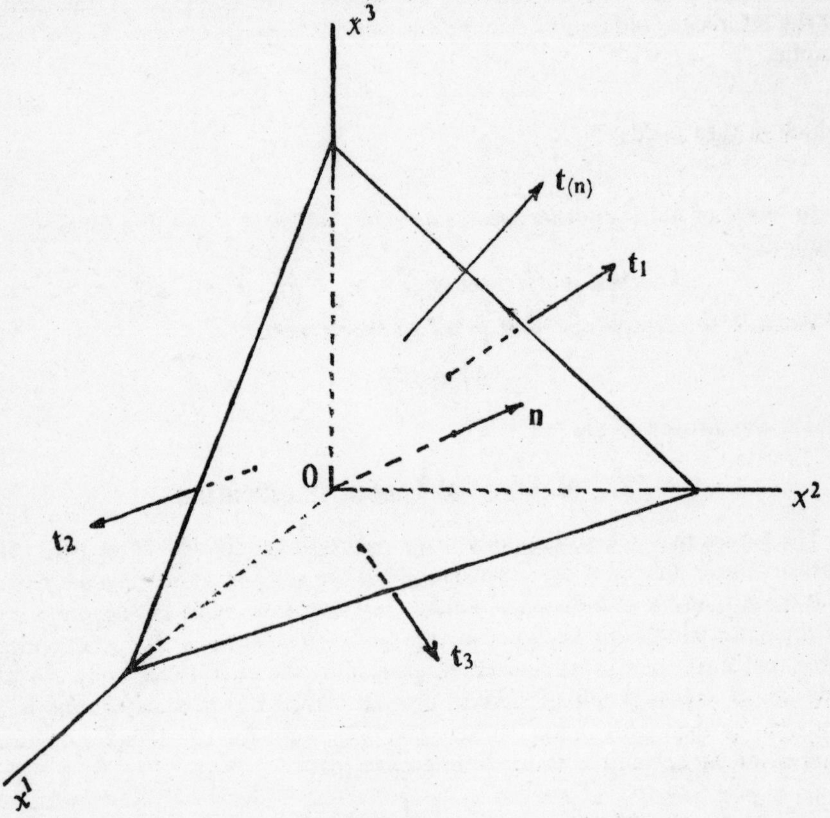

Fig. 17.1

and the mean-value theorem for surface integrals applied to (17.1) to yield

$$A\,[n_1\,\mathbf{t}_1^* + n_2\,\mathbf{t}_2^* + n_3\,\mathbf{t}_3^* + \mathbf{t}_{(\mathbf{n})}^*] + hAM = 0, \tag{17.2}$$

where M is a bound, and $\mathbf{t}_{(\mathbf{n})}^*$ and $\mathbf{t}_i^*$ ($i=1, 2, 3$) are the stress vectors at certain points on the surfaces. Cancel A and let $h \to 0$ and obtain the stress vector $\mathbf{t}_{(\mathbf{n})}$ at 0 on a plane with the unit normal $\mathbf{n}$, in terms of the three stress vectors $\mathbf{t}_i$ ($i=1, 2, 3$) at $\mathbf{0}$:

$$\mathbf{t}_{(\mathbf{n})} = -\,(\mathbf{t}_1 n_1 + \mathbf{t}_2 n_2 + \mathbf{t}_3 n_3). \tag{17.3}$$

By definition, $\mathbf{t}_i$ do not depend upon $\mathbf{n}$. Replacing $\mathbf{n}$ by $-\mathbf{n}$ in (17.3) yields

$$\mathbf{t}_{(\mathbf{n})} = -\,\mathbf{t}_{(-\mathbf{n})}, \tag{17.4}$$

or the stress vector $\mathbf{t}_{(\mathbf{n})}$ at 0 on a plane with a normal $\mathbf{n}$ is equal and opposite to $\mathbf{t}_{(-\mathbf{n})}$ on the same plane with the unit normal $-\mathbf{n}$. However, the above proof fails when $n_1 = 1, n_2 = n_3 = 0$, that is, when the construction of the tetrahedron is no longer valid.

So we present the following elegant proof due to GURTIN[2] to make (17.4) exact for all $\mathbf{n}$. Let $\mathbf{x}_0$ be the center of the parallelopiped P_ε (Fig. 17.2) such

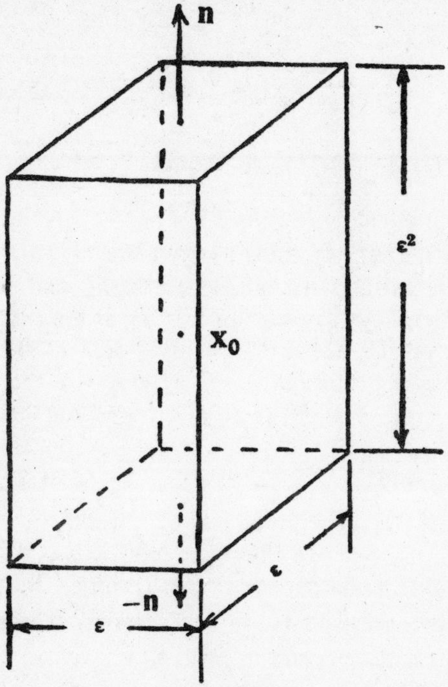

Fig. 17.2.

that its top and bottom surfaces $\mathcal{S}_\varepsilon^+$ and $\mathcal{S}_\varepsilon^-$ are squares of length ε and unit normals $\mathbf{n}$ and $-\mathbf{n}$, respectively, Let the height of P_ε be ε^2. Let $\mathcal{S}_\varepsilon$ be the union of the four lateral surfaces. Then the surface $\mathcal{S}(P_\varepsilon)$ of P_ε is the union:

$$\mathcal{S}(P_\varepsilon) = \mathcal{S}_\varepsilon^+ \cup \mathcal{S}_\varepsilon^- \cup \bar{\mathcal{S}}_\varepsilon, \tag{17.5}$$

and the volume and areas are

$$v(P_\varepsilon) = \varepsilon^4, \qquad A(\mathcal{S}_\varepsilon^\pm) = \varepsilon^2, \qquad A(\bar{\mathcal{S}}_\varepsilon) = 4\varepsilon^3. \tag{17.6}$$

Letting

$$k(t) = \sup_{\mathbf{x} \in P_\varepsilon} |\rho\mathbf{b} - \rho\mathbf{a}| < \infty, \tag{17.7}$$

we have that

$$\left| \int_{\mathcal{S}(P_\varepsilon)} \mathbf{t}_{(\mathbf{n})}\, da \right| \leq kv(P_\varepsilon). \tag{17.8}$$

2. [1972 : 5] : Sec. 16.

Hence

$$\frac{1}{\varepsilon^2}\int_{\mathcal{S}(\tilde{P}_\varepsilon)} \mathbf{t}_{(\mathbf{n})}\, da \to 0 \quad \text{as} \quad \varepsilon \to 0. \tag{17.9}$$

Moreover, from the assumed continuity of $\mathbf{t}_{(\mathbf{m})}$ for each unit fixed unit vector $\mathbf{m}$ in the body, it follows that

$$\frac{1}{\varepsilon^2}\int_{\mathcal{S}_\varepsilon^\pm} \mathbf{t}_{(\pm\mathbf{n})}\, da \to \mathbf{t}_{(\pm\mathbf{n})}(\mathbf{x}_0) \quad \text{as} \quad \varepsilon \to 0; \tag{17.10}$$

$$\frac{1}{\varepsilon^2}\int_{\mathcal{S}_\varepsilon^-} \mathbf{t}_{(\mathbf{n})}\, da \to 0 \qquad \text{as} \quad \varepsilon \to 0. \tag{17.11}$$

Hence, by (17.5), (17.9)—(17.11), we obtain

$$\mathbf{t}_{(\mathbf{n})}(\mathbf{x}_0) + \mathbf{t}_{(-\mathbf{n})}(\mathbf{x}_0) = \mathbf{0}, \tag{17.12}$$

proving completely that (17.4) holds everywhere.

Now we return to CAUCHY's original treatment and choose a Cartesian coordinate system, with the convention that T_{ij} is the i-th component of the stress vector acting on the outward (positive) side of the plane x_j = const. Consider the plane $x = 0$ through 0. Choose $\mathbf{n} = \mathbf{i}$, the unit vector in the x-direction. Then $\mathbf{t}_{(-\mathbf{i})} = \mathbf{t}_i$ in the notation used for the tetrahedron and by (17.4) we obtain

$$-t_{1x} = T_{xx}, \quad -t_{1y} = T_{yx}, \quad -t_{1z} = T_{zx}. \tag{17.13}$$

Similarly, we can obtain the components of $\mathbf{t}_2$ and $\mathbf{t}_3$.

Returning to (17.3), we derive that $\mathbf{t}_{(\mathbf{n})k} = T_{km}\, n_m$, where the quantities T_{km} are independent of $\mathbf{n}$ and depend on $(\mathbf{x}, t)$ only. By the quotient law of tensors, T_{km} are the components of a Cartesian tensor of second order, called CAUCHY's stress tensor, and in general notation,

$$t^k_{(\mathbf{n})} = T^{km}\, n_m, \quad \mathbf{t}_{(\mathbf{n})} = \mathbf{T}\mathbf{n}. \tag{17.14}$$

GURTIN[3] has also presented NOLL's elegant proof of the above result; this proof is straightforward and does not in any way depend upon the use of coordinates. It rests upon showing that

$$\mathbf{t}_{(\mathbf{n})}(\mathbf{x}, t) \overset{\text{def}}{=\!=} \mathbf{t}(\mathbf{x}, \mathbf{n}) \tag{17.15}$$

is additive in $\mathbf{n}$, that is,

$$\mathbf{t}(\mathbf{x}, \mathbf{u}_1 + \mathbf{u}_2) = \mathbf{t}(\mathbf{x}, \mathbf{u}_1) + \mathbf{t}(\mathbf{x}, \mathbf{u}_2), \tag{17.16}$$

for all linearly independent vectors $\mathbf{u}_1$ and $\mathbf{u}_2$; and that

$$\mathbf{t}(\mathbf{x}, \alpha\mathbf{n}) = \alpha\mathbf{t}(\mathbf{x}, \mathbf{n}), \tag{17.17}$$

or that $\mathbf{t}(\mathbf{x}, \cdot)$ is *homogeneous*. From Eqs. (17.16)—(17.17), one concludes that the mapping $\mathbf{n} \to \mathbf{t}_{(\mathbf{n})}(\mathbf{x}, t)$ is the restriction to the set of unit vectors

3. [1972 : 5; Sec. 16].

of a linear function on the vector space. By the linear functional theorem of finite-dimensional vector spaces, one has that

$$\mathbf{t}_{(\mathbf{n})} = \mathbf{T}\mathbf{n}, \tag{17.18}$$

where $\mathbf{T}$ is independent of $\mathbf{n}$, and depends on $\mathbf{x}$ and t only.

Eq. (17.1) now becomes

$$\int_{\mathcal{CV}} \rho\, \mathbf{a}\, dv = \int_{\mathcal{S}} \mathbf{T}\mathbf{n}\, da + \int_{\mathcal{CV}} \rho\, \mathbf{b}\, dv. \tag{17.19}$$

By the divergence theorem, this may by written as

$$\int_{\mathcal{CV}} (\operatorname{div} \mathbf{T} + \rho\mathbf{b} - \rho\mathbf{a})\, dv = 0, \tag{17.20}$$

and since $\mathcal{CV}$ is arbitrary, the necessary and sufficient condition for the balance of linear momentum is the following differential equation of balance:

$$\operatorname{div} \mathbf{T} + \rho\mathbf{b} = \rho\mathbf{a},$$
$$T^{ij}_{;j} + \rho b^i = \rho a^i. \tag{17.21}$$

The above form is *Cauchy's first law of motion.*

REMARK. The proof of the existence of the stress tensor is a consequence of the vector-valued version of the following. Let $f(\mathbf{x}, \mathbf{n})$ be a scalar field defined over the surface $\mathcal{S}$ of an arbitrary regular region D such that

$$\int_{\mathcal{S}} f(\mathbf{x}, \mathbf{n})\, da = \int_D \mathbf{b}(\mathbf{x})\, dv, \tag{17.22}$$

where $\mathbf{n}$ is the unit exterior normal to $\mathcal{S}$. This permits one to prove the next theorem.

THEOREM[4]. *Let* $f(\cdot, \mathbf{n})$ *(for each* $\mathbf{n}$*) and* $\mathbf{b}(\cdot)$ *be integrable over a region* R *and let* (17.22) *hold for every tetrahedron* D *in* R *for which the left-hand integral exists. Then there is a vector field* $\mathbf{f}$ *on* R *such that for each* $\mathbf{n}$,

$$f(\mathbf{x}, \mathbf{n}) = \mathbf{f}(\mathbf{x}) \cdot \mathbf{n}, \tag{17.23}$$

for almost every $\mathbf{x}$ *in* R.

Now consider each component of $\mathbf{t}_{(\mathbf{n})}$ as a scalar field. Thereupon the vector-valued analogue of the above theorem implies that provided (17.1) is integrable over R and (17.1) holds for every tetrahedron D in R, for which $\int_{\mathcal{S}} \mathbf{t}_{(\mathbf{n})}\, da$ exists, then

$$\mathbf{t}_{(\mathbf{n})}(\mathbf{x}, t) = \mathbf{T}(\mathbf{x}, t)\, \mathbf{n}, \tag{17.24}$$

where $\mathbf{T}(\mathbf{x}, t)$ is a linear transformation.

The reason the above theorem has a wider applicability is that it does not demand that $\mathbf{t}_{(\mathbf{n})}(\mathbf{x}, t)$ be continuous in R, and hence one could speak of the existence of the stress tensor in regions with shock waves, for example.

4. GURTIN, *et. al.* [1968 : 10]. More recently STIPPES [1971 : 26] has provided another proof of the above theorem by using variational methods.

18. Balance of Angular Momentum

Assuming that there do not exist any internal angular momentum, body couples and couple stresses, the balance of angular momentum equation with respect to an inertial frame for a material volume v reads

$$\frac{d}{dt}\int_{\mathcal{V}} \rho(\mathbf{x} \times \mathbf{v})\, dv = \int_{\mathcal{S}} \mathbf{x} \times \mathbf{t}_{(\mathbf{n})}\, da + \int_{\mathcal{V}} \rho(\mathbf{x} \times \mathbf{b})\, dv, \tag{18.1}$$

where $\mathbf{x}$ is the position vector from the fixed origin of the coordinate frame, and $\mathcal{S}$ is the surface of the volume $\mathcal{V}$. Using the transport theorem,

$$\frac{d}{dt}\int_{\mathcal{V}} \rho(\mathbf{x} \times \mathbf{v})\, dv = \int_{\mathcal{V}} (\dot{\rho} + \rho \operatorname{div} \mathbf{v})\, (\mathbf{x} \times \mathbf{v})\, dv + \int_{\mathcal{V}} \rho(\mathbf{x} \times \mathbf{a})\, dv, \tag{18.2}$$

where we have used that $\mathbf{v} \times \mathbf{v} = \mathbf{0}$. Using the continuity Eq. (16.3), (18.2) becomes

$$\frac{d}{dt}\int_{\mathcal{V}} \rho\, (\mathbf{x} \times \mathbf{v})\, dv = \int_{\mathcal{V}} \rho\, (\mathbf{x} \times \mathbf{a})\, dv. \tag{18.3}$$

Moreover, by the existence of the stress tensor $\mathbf{T}$,

$$\int \mathbf{x} \times \mathbf{t}_{(\mathbf{n})}\, da = \int (\mathbf{x} \times \mathbf{T}\mathbf{n})\, da. \tag{18.4}$$

Introducing Cartesian coordinates for convenience, the right side of (18.4) is

$$\int_{\mathcal{S}} e_{ijk}\, x_j\, T_{km}\, n_m\, da, \tag{18.5}$$

where e_{ijk} is the alternating tensor (A 1.31) – (A 1.32).

By the divergence theorem,

$$\begin{aligned}\int_{\mathcal{S}} e_{ijk}\, x_j\, T_{km}\, n_m\, da &= \int_{\mathcal{V}} (e_{ijk}\, x_j\, T_{km})_{,m}\, dv \\ &= \int_{\mathcal{V}} (e_{ijk}\, \delta_{jm}\, T_{km} + e_{ijk}\, x_j\, T_{km,m})\, dv \\ &= \int_{\mathcal{V}} e_{ijk}\, T_{kj} + e_{ijk}\, x_j\, T_{km,m})\, dv.\end{aligned} \tag{18.6}$$

In direct notation

$$\int_{\mathcal{S}} (\mathbf{x} \times \mathbf{T}\mathbf{n})\, da = \int_{\mathcal{V}} (\mathbf{t}_A + \mathbf{x} \times \operatorname{div} \mathbf{T})\, dv, \tag{18.7}$$

where $\mathbf{t}_A$ is the axial vector corresponding to $\mathbf{T}$ and is defined by

$$(t_A)_i = e_{ijk}\, T_{kj}. \tag{18.8}$$

Collecting together all of the above, the balance of angular momentum now reads

$$\int_{\mathcal{V}} \rho(\mathbf{x} \times \mathbf{a})\, dv = \int_{\mathcal{V}} (\mathbf{t}_A + \mathbf{x} \times \operatorname{div} \mathbf{T})\, dv + \int_{\mathcal{V}} \rho\, \mathbf{x} \times \mathbf{b}\, dv. \tag{18.9}$$

Assuming that the balance of linear momentum holds, that is,

$$\operatorname{div} \mathbf{T} + \rho\mathbf{b} - \rho\mathbf{a} = \mathbf{0} \text{ in } \mathcal{V}, \tag{18.10}$$

(18.9) reduces to

$$\int_{\mathcal{CV}} \mathbf{t}_A \, dv = 0. \tag{18.11}$$

Since $\mathcal{CV}$ is arbitrary, it follows that $\mathbf{t}_A = \mathbf{0}$ or that the stress tensor $\mathbf{T}$ is symmetric :

$$\mathbf{T} = \mathbf{T}^T, \qquad T^{ij} = T^{ji}, \qquad T\langle ij\rangle = T\langle ji\rangle. \tag{18.12}$$

Hence, in the absence of body couples and couple stresses, the necessary and sufficient condition for the balance of angular momentum, given the conservation of mass and balance of linear momentum, is that (18.12) holds. This is *Cauchy's second law af motion.*

In this book it will always be assumed that *the stress tensor is symmetric*, and thus the second law of motion is automatically satisfied.

19. Energy Balance

Let K be the kinetic energy of the body and E its internal energy so that for any material volume $\mathcal{CV}$,

$$K(\mathcal{CV}) = \int_{\mathcal{CV}} \frac{1}{2} \rho \, |\mathbf{v}|^2 \, dv, \quad E(\mathcal{CV}) = \int_{\mathcal{CV}} \rho \, \varepsilon \, dv, \tag{19.1}$$

where ε is the specific energy. The mechanical power exerted by the outside world on $\mathcal{CV}$ is given by

$$\int_{\mathcal{S}} \mathbf{t}_{(\mathbf{n})} \cdot \mathbf{v} \, da + \int_{\mathcal{CV}} \rho \mathbf{b} \cdot \mathbf{v} \, dv. \tag{19.2}$$

Let the efflux of energy out of $\mathcal{S}$ be $\mathbf{q}$ per unit area and the supply be r per unit mass. Then one postulates the balance of energy equation as

$$\frac{d}{dt}\left\{\int_{\mathcal{CV}}\left(\frac{1}{2}\rho\mathbf{v}\cdot\mathbf{v} + \rho\varepsilon\right) dv\right\}$$
$$= \int_{\mathcal{CV}} (\mathbf{t}_{(\mathbf{n})} \cdot \mathbf{v} - \mathbf{q} \cdot \mathbf{n}) \, da + \int_{\mathcal{CV}} (\rho r + \rho \mathbf{b} \cdot \mathbf{v}) \, dv. \tag{19.3}$$

Using the existence of the stress tensor, we obtain

$$\frac{d}{dt}\left\{\int_{\mathcal{CV}}\left(\frac{1}{2}\rho\mathbf{v}\cdot\mathbf{v} + \rho\varepsilon\right)\right\}$$
$$= \int_{\mathcal{S}} (\mathbf{v} \cdot \mathbf{Tn} - \mathbf{q} \cdot \mathbf{n}) \, da + \int_{\mathcal{CV}} \rho \, (\mathbf{b} \cdot \mathbf{v} + r) \, dv. \tag{19.4}$$

Assuming that mass is conserved and using the transport theorem, the left side can be written in tensorial notation as

$$\int_{\mathcal{CV}} \rho \, (a^k v_k + \dot{\varepsilon}) \, dv. \tag{19.5}$$

The divergence theorem applied to the surface integral on the right yields

$$\int_{\mathcal{S}} (v_i T^{ij} n_j + q^i n_i) \, da = \int_{\mathcal{CV}}\left(T^{ij}_{\ ;j} v_i + T^{ij} v_{i;j} - q^i_{\ ;i}\right) dv. \tag{19.6}$$

Using the balance of linear momentum equation and the symmetry of the stress tensor implied by the balance of angular momentum, we get

$$\int_{\mathcal{V}} \rho \, \dot{\varepsilon} \, dv = \int_{\mathcal{V}} T^{ij} \, D_{ij}, - v^{i}_{;i} + \rho r \,) \, dv, \qquad (19.7)$$

implying the differential equation of energy balance:

$$\rho \dot{\varepsilon} = T^{ij} \, D_{ij} - q^{i}_{;i} + \rho r, \qquad (19.8)$$

where D_{ij} is the symmetric part of the velocity gradient. Eq. (19.8) can be rewritten as

$$\rho \dot{\varepsilon} = \mathrm{tr} \;\; (\mathbf{TL}) - \mathrm{div} \, \mathbf{q} + \rho r, \qquad (19.9)$$

which is the coordinate free equation.

The reader may notice that the conservation of mass equation is basic. This is needed in obtaining CAUCHY's first law of motion, and these two together are needed for CAUCHY's second law of motion. All these three are necessary to obtain the energy equation.

Can one postulate an energy balance an recover the conservation of mass and the three balance laws ? GREEN and RIVLIN[5] postulated an energy balance of the form (in Cartesian coordinates) :

$$\int_{\mathcal{V}} [\rho \, v_i \, \dot{v} + \rho \, \dot{\varepsilon} + (\varepsilon + \frac{1}{2} \, v_i \, v_i) \, (\dot{\rho} + \rho v_{j,j})] \, dv$$

$$= \int_{\mathcal{V}} \rho \, (r + b_i \, v_i) \, dv - \int_{\mathcal{S}} (h - t_i \, v_i) \, da, \qquad (19.10)$$

where h is the heat flux across the surface $\mathcal{S}$, and the other terms have the usual meaning. Assuming that the energy balance is translation invariant, that ρ, $\dot{\rho}$, ε, $\dot{\varepsilon}$, t_i, b_i, h and r are unaffected by this translation, and that the body occupies the same position in space at time t, we can replace v_i by $(v_i + c_i)$ in (19.10), and obtain ($c_i = c_i \, (t)$ only) :

$$\left[\int_{\mathcal{V}} \{ \rho \, (a_i - b_i) + v_i \, (\dot{\rho} + \rho \, v_{j,j}) \} \, dv - \int_{\mathcal{S}} t_i \, da \right] c_i$$

$$+ \frac{1}{2} \, c_i \, c_i \int_{\mathcal{V}} (\dot{\rho} + \rho \, v_{j,j}) \, dv = 0. \qquad (19.11)$$

Since c_i is arbitrary, it follows that

$$\dot{\rho} + \rho \, v_{j,j} = 0, \qquad (19.12)$$

which is the conservation of mass equation, and

$$\int_{\mathcal{V}} \rho \, (a_i - b_i) \, dv = \int_{\mathcal{S}} t_i \, da, \qquad (19.13)$$

which is the balance of linear momentum equation. Eq. (19.13) implies the existence of the stress tensor as in Sec. 17, yielding $t_i = T_{ij} \, n_j$, and the

5. [1964 : 9]

first law of motion (17.21). Using this and (19.12) in (19.10), we have

$$\int_{\mathcal{V}} \rho \, \dot{\varepsilon} \, dv = \int_{\mathcal{V}} (\rho r + T_{ij} \, v_{i,j}) \, dv - \int_{\mathcal{S}} h \, da. \tag{19.14}$$

Now (19.14) holds for all v_i, so let us replace v_i by $v_i + \Omega_{ik} \, x_k$ where Ω_{ik} is a constant skew-symmetric tensor. Assuming that this rigid rotation leaves ρ, $\dot{\varepsilon}$, r, T_{ij} and h invariant and that the body occupies the same position at time t, one derives

$$\Omega_{ij} \int_{\mathcal{V}} T_{ij} \, dv = 0. \tag{19.15}$$

Since Ω_{ij} is arbitrary, $T_{ij} = T_{ji}$, which is CAUCHY's second law (18.12).

Consider (19.14) again. From the theorem of GURTIN, *et al.*,[6] it follows that $h = \mathbf{q}(\mathbf{x}) \cdot \mathbf{n}$, where $\mathbf{q}(\mathbf{x})$ is a vector field. Using this information in (19.14), one recovers the energy equation (19.9).

It is shown by BEATTY[7] that the balance laws of mechanics (excluding the energy balance) can be derived by postulating the frame indifference of mechanical power, which he defines as the rate of working of all forces on the body, including inertia. He claims his method to be more fundamental than that employed by GREEN and RIVLIN, since the conservation of mass, balance of linear and angular momentum equations can be derived by a false energy postulate. For example, in (19.10), one could neglect ε, retain $\dot{\varepsilon}$, let $h \equiv 0$, $r \neq 0$ and still obtain these three laws. However, to obtain the energy Eq. (19.9), BEATTY is forced to introduce non-mechanical power and in this case, his method becomes equivalent to that of GREEN and RIVLIN.

20. Other Forms of Equations of Motion[8]

The stress tensor $\mathbf{T}$ is a spatial field. We may introduce other stress tensors related to it as follows. We define the PIOLA-KIRCHHOFF stress tensor $\mathbf{T}_R$ through

$$\mathbf{T} = J^{-1} \, \mathbf{T}_R \, \mathbf{F}^T; \qquad T^{km} = J^{-1} \, T_R^{k\alpha} \, x^m_{,\alpha}. \tag{20.1}$$

The usefulness of this tensor lies in the following: if da and $\mathbf{n}$ are the surface element and the outward unit normal in the spatial configuration respectively and dA and $\mathbf{N}$ the corresponding ones in the reference configuration, then the contact force $\mathbf{f}_c$ is given by

$$\mathbf{f}_c = \int_{\mathcal{S}} \mathbf{T}\mathbf{n} \, da = \int_{\mathcal{S}_R} \mathbf{T}_R \, \mathbf{N} \, dA, \tag{20.2}$$

where $\mathcal{S}_R$ is the surface of the body in the reference configuration; the stress

---

6. [1968 : 10].
7. [1967 : 4].
8. CFT, Sec. 210; NFTM, Sec. 43.

vector $\mathbf{t}_{(N)R}$ in the reference configuration is, by definition:

$$\mathbf{t}_{(N)R} = \mathbf{T}_R\mathbf{N}. \tag{20.3}$$

The equations of motion for $\mathbf{T}_R$ become

$$\text{Div } \mathbf{T}_R + \rho_R \mathbf{b} = \rho_R \ddot{\mathbf{x}}, \tag{20.4}$$

where 'Div' is the divergence operator in the reference configuration. Eq. (20.4) may be expressed as

$$T^{\alpha}_{Rk\,;\,\alpha} + \rho_R\, b_k = \rho \ddot{x}_k. \tag{20.5}$$

In terms of $\mathbf{T}_R$ the balance of angular momentum reads

$$\mathbf{T}_R\,\mathbf{F}^T = \mathbf{F}\,\mathbf{T}^T_R; \quad \mathbf{T}^{k\alpha}_R\, x^{\;m}_{,\alpha} = x^{\;k}_{,\alpha}\,\mathbf{T}^{m\alpha}_R. \tag{20.6}$$

One may construct other forms of stress tensors such as (see also Appendix *B* to Chap. 5):

$$\tilde{\mathbf{T}} = \mathbf{F}^{-1}\,\mathbf{T}_R; \qquad \tilde{\mathbf{T}}^{\alpha\beta} = X^{\alpha}_{,m}\, T^{m\beta}_{,R}; \tag{20.7}$$

$$J\overline{\mathbf{T}} = \mathbf{C}\,\overline{\mathbf{T}}\,\mathbf{C}. \tag{20.8}$$

Although the last two are not very useful, the PIOLA-KIRCHHOFF tensor $\mathbf{T}_R$ is extremely fruitful in discussing wave propagation in viscoelastic materials.

CHAPTER 5

FORMULATION OF CONSTITUTIVE EQUATIONS. SIMPLE FLUIDS

21. Introduction

In Secs. 16-17, we assembled the continuity equation and the equations of motion. If we now wish to solve boundary value problems, which in fluid flow means the determination of the velocity field in the body due to the application of surface tractions or the prescription of the velocity field on the boundary, we note that the continuity equation, which determines a relation between ρ and $\mathbf{v}$, and in the absence of the body forces the equations of motion, which define ρ and $\mathbf{v}$ in terms of $\mathbf{T}$ (inside the body), *are sufficient* to ensure the existence of a 'stress-density function' in Euclidean space-time from which one could derive ρ, $\mathbf{v}$ and $\mathbf{T}$. An elegant exposition of this is given in CFT, Secs. 224-230, and we quote these results due to B. FINZI[1] below:

$$-\rho = \varepsilon^{smn}\,\varepsilon^{pqr}\,A_{mnqr\,;\,sp}, \tag{21.1}$$

$$-\rho v^k = \varepsilon^{ksm}\,\varepsilon^{pqr}\,(A'_{smqr\,;\,p} + 2A_{4mqr\,;\,sp}), \tag{21.2}$$

$$T^{km} - \rho v^k v^m = 4\,\varepsilon^{kpq}\,\varepsilon^{msn}\,A_{4p4s\,;\,qn} + 2\,(\varepsilon^{kpq}\,\varepsilon^{msn} + \varepsilon^{ksn}\,\varepsilon^{mpq})\,A'_{4npq\,;\,s} + \varepsilon^{kpq}\,\varepsilon^{msn}\,A''_{pqsn}, \tag{21.3}$$

where ε^{ijk} is the alternating tensor:

$$\varepsilon^{ijk} = e^{ijk}/\sqrt{g}, \qquad g \equiv \det g_{ij}, \tag{21.4}$$

and e^{ijk} is the Cartesian tensor defined by (A1.31). Also in (21.1) – (21.3), A_{pqrs} is the four-dimensional '*stress-density function*', the primes denote $\partial/\partial t$, and it is assumed that the body force does not exist. The 'four dimensions' are supposed to be (x^1, x^2, x^3, ρ), so that the fourth dimension concerns the continuity equation. Suppose then, given some boundary and initial data, we can find the field $\mathbf{A}$ everywhere. Then (21.2) above yields the velocity field completely. However, the equations of motion and the surface tractions are not in themselves sufficient to find $\mathbf{A}$. For example, in linear elasticity, the two dimensional elastostatic problem leads to the Airy stress function from the equation of equilibrium. But this function is not determinable from these equations. However, by choosing a particular constitutive equation (e.g. linear, isotropic elastic solid), inverting it so as to express strain in terms of the stresses and hence the Airy stress function, and then using these formulae for the strain in the compatibility conditions leads to

---

1. [1934 : 1].

the equation governing the Airy stress function. Finding this function with the requisite boundary conditions yields the complete picture of the deformed body.

As this example suggests, the laws of mechanics are insufficient, in themselves, to yield a total description of the motion in the interior of the body, and we thus need a constitutive relation adjoining the motion to the stress tensor. In fluid flow, we anticipate that this relation and the balance of linear momentum yield an equation from which one may compute the velocity field. We now study this matter of postulating a constitutive relation in detail.

For viscoelastic liquids, the subject of the present volume, no 'linear relation' between the stress tensor and velocity field suffices to predict the properties of viscoelastic liquids given below:

(i) In the so-called viscometric flows, the shear stress is a nonlinear function of the shear rate;
(ii) The normal stresses in these flows are not equal to one another;
(iii) If a ball of the viscoelastic fluid is thrown against the wall, it bounces like rubber (e.g. silly putty);
(iv) The fluid cannot withstand shear stress; it flows, though it may do so very slowly (e.g. silly putty stuck to a ceiling);
(v) If a sudden strain is applied, the stresses in the fluid shoot up and if the material is held in this strained configuration, the stresses relax gradually to zero;
(vi) The fluid exhibits creep under a steady loaded condition;
(vii) On exit from a die, an appreciable die swell is noted[2];
(viii) If a rod is rotated in a container full of some viscoelastic liquid, the liquid climbs up the rod.

One could list many such examples of striking behavior. Essentially what one notes is this:

(*a*) The previous history of the motion determines the present stress;
(*b*) Elastic recoil, creep and stress relaxation occur;
(*c*) The relation between stress and the velocity field is highly nonlinear even in situations where the history of the strain is highly repetitive.

The problem of proposing a constitutive equation was taken up by ERICKSEN, GREEN, LODGE, NOLL, OLDROYD, PIPKIN and RIVLIN (though not in that order). In addition, the works of BERNSTEIN, KEARSLEY and ZAPAS, as well as that of BIRD and his coworkers also deserve mention. All of these attempts lead to theories which fall under the category of the simple material of NOLL, which shall be studied here. However, we shall start with OLDROYD's concepts and tie these up with those of NOLL to show the close ties between these ideas.

2. It is *erroneous* to conclude that die swell does not occur with Newtonian fluids at low REYNOLDS numbers. See Sec. 50.

22. Basic Principles of Formulation. Simple Materials

In 1950, OLDROYD[3] stated that constitutive equations (equations of state) must be based on

(i) the history of the motion;
(ii) the motion of the neighbourhood of a particle relative to the motion of the particle as a whole in space;
(iii) convected coordinates imbedded in the material and deforming with it.

Then the convected components of a stress flux (see below) are obtained as a set of integro-differential equations in the strain history, temperature history and on certain tensors which define the symmetry of the material.

We shall now follow NOLL[4] and reformulate the above four statements.

P. A.: Let the stress tensor at the particle X at time t depend on the motion of the entire body $\mathcal{B}$; Or, let there exist a functional relationship between $\mathbf{T}(X, t)$ and $\overline{\mathbf{M}}(\mathcal{B}, t)$ such that;

$$\mathbf{T}(X, t) = \bar{F}\{\overline{\mathbf{M}}(\mathcal{B}, \tau)\}, \quad -\infty < \tau < \infty. \tag{22.1}$$

P. B.: Postulate that the stress vector $\mathbf{t}_{(\mathbf{n})}$ is objective[5]. Then

$$\mathbf{t}^*_{(\mathbf{n}^*)} = \mathbf{Q}(t)\,\mathbf{t}_{(\mathbf{n})}. \tag{22.2}$$

Moreover, since $\mathbf{n}$ is the vector from a point $\mathbf{y}$ to $\mathbf{x}$ in the current configuration of $\mathcal{B}$, the assumed objective motion implies that [*cf.* (12.5)]:

$$\mathbf{n}^* = \mathbf{y}^* - \mathbf{x}^* = \mathbf{Q}(t)(\mathbf{y} - \mathbf{x}) = \mathbf{Q}(t)\,\mathbf{n}. \tag{22.3}$$

Hence $\mathbf{n}$ is objective. Now the stress vectors $\mathbf{t}_{(\mathbf{n})}$ and $\mathbf{t}^*_{(\mathbf{n}^*)}$ are related by:

$$\mathbf{T}^*\mathbf{Q}\mathbf{n} = \mathbf{T}^*\mathbf{n}^* = \mathbf{t}^*_{(\mathbf{n}^*)} = \mathbf{Q}\mathbf{t}_{(\mathbf{n})} = \mathbf{Q}\mathbf{T}\mathbf{n}, \tag{22.4}$$

which implies that $\mathbf{T}$ is objective, that is, $\mathbf{T}^* = \mathbf{Q}\,\mathbf{T}\,\mathbf{Q}^T$.

P. C.: If $\mathbf{T}$ is objective, then any *stress flux* such as $\tilde{\mathbf{T}}$

$$\tilde{\mathbf{T}} = \mathbf{T} + \mathbf{T}\,\mathbf{L} + \mathbf{L}^T\mathbf{T} \tag{22.5}$$

is objective [*cf.* (12.14)], that is,

$$\tilde{\mathbf{T}}^* = \mathbf{Q}\,\tilde{\mathbf{T}}\,\mathbf{Q}^T. \tag{22.6}$$

Since $\mathbf{T}$ and $\mathbf{A}_1$ are objective, we may form another quantity:

$$\bar{\mathbf{T}} = \dot{\mathbf{T}} - \mathbf{W}\mathbf{T} + \mathbf{T}\mathbf{W}, \tag{22.7}$$

which is also objective[6]. Actually there exist many stress fluxes. One

---

3. [1950 : 1].
4. [1958 : 3].
5. This is also called the ***principle of frame indifference***. See NFTM, Sec. 19.
6. See Eq. (6) of OLDROYD [1958 : 4]. In Eq. (5) of this paper, ω is defined to be given by $-\mathbf{W}$. Hence, our (22.7) agrees with Eq. (6) to within a sign.

can add any objective and symmetric quantity to the right side of (22.5) or (22.7) at pleasure. In this sense, OLDROYD's choice of a particular stress flux[7] (given by (22.5) above) is not unique. But the CAUCHY stress tensor is unique and hence a constitutive equation in terms of the stress, such as (22.1), is postulated.

P. D. If (22.1) defines a constitutive equation then in an objective motion,

$$\mathbf{T}^* (X, t) = \bar{F} [(\overline{\mathbf{M}}^* (\mathcal{B}, \tau)], \qquad -\infty < \tau < \infty, \tag{22.8}$$

that is, the *functional operator* $\bar{F}$ *is unaltered.*

P. E. By the derived objectivity of T, we may prove a part of the statement (ii) of OLDROYD as follows : In [*cf.* (12.5)]

$$\overline{\mathbf{M}}^* = \mathbf{Q}(t) \overline{\mathbf{M}} + \mathbf{c}(t), \tag{22.9}$$

let $\mathbf{Q}(t) \equiv \mathbf{1}$, and $\mathbf{c}(t) = -\overline{\mathbf{M}}(X, t)$, where X is the particle under consideration. Since X is fixed, $\overline{\mathbf{M}}(X, t)$ is a function of t only. Then

$$\overline{\mathbf{M}}^* (\mathcal{B}, t) = \overline{\mathbf{M}} (\mathcal{B}, t) - \overline{\mathbf{M}} (X, t). \tag{22.10}$$

Substitution of (22.10) into (22.8) yields that $\mathbf{T}^* = \mathbf{T}$ and.

$$\mathbf{T} (X, t) = \bar{F} [\overline{\mathbf{M}} (\mathcal{B}, \tau) - \overline{\mathbf{M}} (X, \tau)], \qquad -\infty < \tau < \infty, \tag{22.11}$$

or that the stress depends on the motion of the body relative to the particle.

P. F. If we now postulate that if there exist two motions $\tilde{\mathbf{M}}$ and $\overline{\mathbf{M}}$ such that

$$\tilde{\mathbf{M}} (\mathcal{B}, \tau) - \tilde{\mathbf{M}} (X, \tau) = \overline{\mathbf{M}} (\mathcal{B}, \tau) - \overline{\mathbf{M}} (X, \tau), \qquad \tau \leqq t, \tag{22.12}$$

and if under this condition the following holds :

$$\mathbf{T} (X, t) = \bar{F} [\tilde{\mathbf{M}} (\mathcal{B}, \tau) - \mathbf{M} (X, \tau)] = \bar{F} [(\overline{\mathbf{M}} (\mathcal{B}, \tau) - \overline{\mathbf{M}} (\mathrm{X}, \tau)], \tag{22.13}$$

then in all frames :

$$\mathbf{T} (X, t) = \bar{\mathbf{F}} [\overline{\mathbf{M}} (\mathcal{B}, \tau) - \mathbf{M} (X, \tau)], \qquad -\infty < \tau \leqq t, \tag{22.14}$$

or the stress is determined by *the history of the motion upto t only, and not on future events.*

To prove this, one has to go back to the definition of objective motions and relax the condition (tacitly assumed there) that time value τ^* assigned by $\mathcal{O}^*$ is the same as that assigned ($= \tau$) by $\mathcal{O}$. We now remove this restriction and let $\tau^* = \tau - b$, where b is a constant, and put

$$\overline{\mathbf{M}}^* (X, \tau^*) = \mathbf{c} (\tau) + \mathbf{Q} (\tau) \overline{\mathbf{M}} (X, \tau). \tag{22.15}$$

7. See Eq. (60) of [1950 : 1].

Now, assign $\mathbf{c}(\tau) = \mathbf{0}$ and $\mathbf{Q}(\tau) = \mathbf{1}$, and $b = t$ in (22.15). Then $\tau^* = \tau - t$, and

$$\overline{\mathbf{M}}^*(X, \tau^*) = \overline{\mathbf{M}}^*(X, \tau - t) = \overline{\mathbf{M}}(X, \tau). \tag{22.16}$$

Also, at time t, $t^* = t - t = 0$, and

$$\overline{\mathbf{M}}(X, \tau) = \overline{\mathbf{M}}(x, \tau^* + t). \tag{22.17}$$

Hence the stress tensor at X at time t^* as measured by $\mathcal{O}^*$ is

$$\mathbf{T}^*(X, t^*) = \mathbf{T}^*(X, 0). \tag{22.18}$$

But from (22.8), and (22.11)

$$\mathbf{T}^*(X, 0) = \mathcal{F}[\overline{\mathbf{M}}^*(\mathcal{B}, \tau^*) - \overline{\mathbf{M}}^*(X, \tau^*)], \quad -\infty < \tau^* < \infty, \tag{22.19}$$

and from (22.16)

$$\mathcal{F}[\overline{\mathbf{M}}^*(\mathcal{B}, \tau^*) - \overline{\mathbf{M}}^*(X, \tau^*)] = \mathcal{F}[\overline{\mathbf{M}}(\mathcal{B}, \tau) - \overline{\mathbf{M}}(X, \tau)],$$
$$-\infty < \tau, \tau^* < \infty. \tag{22.20}$$

Comparing (22.11) and (22.19) — (22.20), we obtain

$$\mathbf{T}^*(X, 0) = \mathbf{T}(X, t). \tag{22.21}$$

Now, let $\tilde{\mathbf{M}}$ and $\overline{\mathbf{M}}$ be two motions obeying (22.12) for all $\tau \leqq t$. Then from (22.13), and (22.20) — (22.21),

$$\mathbf{T}^*(X, 0) = \mathcal{F}[\tilde{\mathbf{M}}^*(\mathcal{B}, \tau^*) - \tilde{\mathbf{M}}^*(\mathbf{X}, \tau^*)]$$
$$= \mathcal{F}[\overline{\mathbf{M}}^*(\mathcal{B}, \tau^*) - \overline{\mathbf{M}}^*(X, \tau^*)], \quad -\infty < \tau^* \leqq 0. \tag{22.22}$$

Now (22.22) states that if we assume that in one frame the constitutive equation obeys (22.12) — (22.13), then in all frames, it obeys (22.14). Or the stress is determined by the history of the relative motion upto time t.

P. G .Now, we examine the assumption that the constitutive equation must be based on the relative motion of the neighbourhood of X. Note that if the neighbourhood is 'very small', then

$$\overline{\mathbf{M}}(\mathbf{X}', t) - \overline{\mathbf{M}}(\mathbf{X}, t) = \mathbf{F}(\mathbf{X}, t)\, d\mathbf{X} + o(|d\mathbf{X}|), \tag{22.23}$$

where $\mathbf{X}$ and $\mathbf{X}'$ are the position vectors in the reference configuration of X and X' respectively and $d\mathbf{X} = \mathbf{X}' - \mathbf{X}$. Equation (22.23) shows us why the deformation gradient is important in continuum mechanics.

P. H. Bearing in mind the history of the relative motion, we postulate the *constitutive equation of a homogeneous simple material* as (we restrict attention to homogeneous materials in this volume),

$$\mathbf{T}(X, t) = \mathop{\mathcal{F}}_{s=0}^{\infty} [\mathbf{F}(t - s)], \tag{22.24}$$

where $\mathop{\mathcal{F}}\limits_{s=0}^{\infty}(\cdot)$ is a nonlinear tensor-valued functional of the history of the deformation gradient, computed with respect to a fixed reference configuration, upto time t. Note that (22.24) says that each component of the stress tensor is a nonlinear functional of the histories of all the nine components of $\mathbf{F}(t-s)$.

Now, a word about the meaning of the word functional. A functional, as used in mathematics, is a scalar-valued mapping of a vector space into the set of real numbers. The elements of the vector space may actually denote an equivalence class, rather than a single entity. (These two points will be elaborated, with respect to the constitutive equation of the simple fluid below in Chap. 8). As far as each component $T\langle ij\rangle$ of the stress tensor is concerned, it is a scalar-valued functional of the nine components of $\mathbf{F}(t-s)$. But the components $T\langle ij\rangle$ together are components of a tensor of second order. Hence $\underset{s=0}{\overset{\infty}{F}}[\mathbf{F}(t-s)]$ is a tensor-valued functional.

REMARKS. The crucial ideas in the development of any constitutive equation are :

(i) the motion of the neighbourhood of a particle relative to the particle is the significant variable;

(ii) the (past) history of this relative motion determines the stress;

(iii) the stress tensor is objective;

(iv) the form of the constitutive operator in one frame of reference is the same in another frame, though the stress tensor and the history of the relative motion [that is, $\mathbf{F}(t-s)$] vary from one frame to another, the former objectively and the latter not.

23. Objectivity Restriction on the Constitutive Equation

By objectivity requirements, $\mathbf{F}^*(\tau)=\mathbf{Q}(\tau)\,\mathbf{F}(\tau)$ and hence the constitutive equation (22.24) is subject to the restriction :

$$\mathbf{T}^*(X,t)=\mathbf{Q}(t)\,\mathbf{T}(X,t)\,\mathbf{Q}(t)^T. \tag{23.1}$$

Hence for all orthogonal tensors $\mathbf{Q}(t)$,

$$\underset{s=0}{\overset{\infty}{F}}[\mathbf{Q}(t-s)\,\mathbf{F}(t-s)]=\mathbf{Q}(t)\underset{s=0}{\overset{\infty}{F}}[\mathbf{F}(t-s)]\,\mathbf{Q}(t)^T \tag{23.2}$$

Thus (23.2) is the restriction on any constitutive equation. Let us follow NOLL[8] and find the form $F(\cdot)$ must have because of (23.2). By the polar decomposition theorem,

$$\mathbf{F}(t-s)=\mathbf{R}(t-s)\,\mathbf{U}(t-s),\qquad s\in[0,\infty), \tag{23.3}$$

where $\mathbf{R}(\cdot)$ and $\mathbf{U}(\cdot)$ are respectively orthogonal, and positive definite and symmetric. In (23.2), let

$$\mathbf{Q}(\tau)=\mathbf{R}(\tau)^T\qquad\text{for all } \tau. \tag{23.4}$$

8. [1958 : 3].

Then, we obtain,

$$\mathbf{R}(t)^T \mathbf{T}(t) \mathbf{R}(t) = \overset{\infty}{\underset{s=0}{F}} [\mathbf{U}(t - s)], \tag{23.5}$$

or,

$$\mathbf{T}(t) = \mathbf{R}(t) \overset{\infty}{\underset{s=0}{F}} [\mathbf{U}(t - s)] \mathbf{R}(t)^T, \tag{23.6}$$

so that the stress is independent of the past rotation. We now perform some manipulations towards a more definite constitutive equation. We note that

$$\begin{aligned} \mathbf{F}(\tau) &= \mathbf{F}_t(\tau) \mathbf{F}(t) = \mathbf{R}_t(\tau) \mathbf{U}_t(\tau) \mathbf{R}(t) \mathbf{U}(t) \\ &= \mathbf{R}_t(\tau) \mathbf{R}(t) \mathbf{R}(t)^T \mathbf{U}_t(\tau) \mathbf{R}(t) \mathbf{U}(t) \\ &= \mathbf{R}_t(\tau) \mathbf{R}(t) \mathbf{U}_t^*(\tau) \mathbf{U}(t), \end{aligned} \tag{23.7}$$

where we have put

$$\mathbf{U}_t^*(\tau) = \mathbf{R}(t)^T \mathbf{U}_t(\tau) \mathbf{R}(t). \tag{23.8}$$

Now, put $\tau = t - s$ in (23.7). Then

$$\mathbf{F}(t - s) = \mathbf{R}_t(t - s) \mathbf{R}(t) \mathbf{U}_t^*(t - s) \mathbf{U}(t). \tag{23.9}$$

Next, choose the orthogonal tensor $\mathbf{Q}(t - s)$, given by

$$\mathbf{Q}(t - s)^T = \mathbf{R}_t(t - s) \mathbf{R}(t), \tag{23.10}$$

and hence

$$\mathbf{Q}(t - s) \mathbf{F}(t - s) = \mathbf{U}_t^*(t - s) \mathbf{U}(t). \tag{23.11}$$

Then (23.2) implies that

$$\mathbf{T}(t) = \mathbf{R}(t) \overset{\infty}{\underset{s=0}{F}} [\mathbf{U}_t^*(t - s) \mathbf{U}(t)] \mathbf{R}(t)^T. \tag{23.12}$$

Define a new functional $\overset{\infty}{\underset{s=0}{G}} [\mathbf{C}_t^*(t - s); \mathbf{C}(t)]$ through

$$\overset{\infty}{\underset{s=0}{G}} [\mathbf{C}_t^*(t - s); \mathbf{C}(t)] = \overset{\infty}{\underset{s=0}{F}} [\mathbf{U}_t^*(t - s) \mathbf{U}(t)]. \tag{23.13}$$

Combining (23.12) and (23.13) we obtain

$$\mathbf{T}(t) = \mathbf{R}(t) \overset{\infty}{\underset{s=0}{G}} [\mathbf{C}_t^*(t - s); \mathbf{C}(t)] \mathbf{R}(t)^T, \tag{23.14}$$

or

$$\mathbf{R}(t)^T \mathbf{T}(t) \mathbf{R}(t) = \overset{\infty}{\underset{s=0}{G}} [\mathbf{C}_t^*(t - s); \mathbf{C}(t)], \tag{23.15}$$

where we emphasize that [*cf.* (23.8)]

$$\mathbf{C}_t^* (t - s) = [\mathbf{U}_t^* (t - s)]^2 = \mathbf{R}(t)^T \mathbf{C}_t (t - s) \mathbf{R}(t). \qquad (23.16)$$

Now, Eq. (23.15) is the solution we have sought for the objectivity restriction (23.2). It applies to *all simple materials.*

Other reductions are possible. However, they offer no help in writing constitutive equations for isotropic materials and hence will not be discussed here.

24. Restriction Due to Symmetry. Isotropic Solids and Simple Fluids[9]

Let us assume that the material is an isotropic solid, that is, we shall postulate that in the reference configuration $\mathcal{G}_T = \mathcal{O}$, the full orthogonal group. Then

$$\mathop{\mathcal{F}}_{s=0}^{\infty}\left(\mathbf{F}(t - s)\right) = \mathop{\mathcal{F}}_{s=0}^{\infty}\left(\mathbf{F}(t - s)\mathbf{H}\right), \qquad \forall\, \mathbf{H} \in \mathcal{O}. \qquad (24.1)$$

From Sec. 14, we now note that under symmetry operations with **H** *orthogonal*, that is, $\mathbf{H}^T = \mathbf{H}^{-1}$,

$$\mathbf{F}(\tau) \text{ is replaced by } \mathbf{F}(\tau)\mathbf{H}. \qquad (24.2)$$

$$\mathbf{C}(\tau) \text{ ,, ,, ,, } \mathbf{H}^T\mathbf{C}(\tau)\mathbf{H}. \qquad (24.3)$$

$$\left\{\begin{matrix} \mathbf{V}(\tau) \text{ ,, ,, ,, } \mathbf{V}(\tau) \\ \mathbf{B}(\tau) \text{ ,, ,, ,, } \mathbf{B}(\tau) \end{matrix}\right\}. \qquad (24.4)$$

Hence

$$\mathbf{F}(\tau) = \mathbf{V}(\tau)\mathbf{R}(\tau) \text{ becomes } \mathbf{F}(\tau)\mathbf{H} = \mathbf{V}(\tau)\mathbf{R}(\tau)\mathbf{H}, \qquad (24.5)$$

or the new rotation tensor is $\hat{\mathbf{R}}(\tau) = \mathbf{R}(\tau)\mathbf{H}$. Then

$$\mathbf{F}(\tau) = \mathbf{R}(\tau)\mathbf{U}(\tau) \text{ becomes } \hat{\mathbf{R}}(\tau)\hat{\mathbf{U}}(\tau) = \mathbf{F}(\tau)\mathbf{H}, \qquad (24.6)$$

and

$$\hat{\mathbf{U}}(\tau) = \mathbf{H}^T\mathbf{U}(\tau)\mathbf{H}. \qquad (24.7)$$

Now [*cf.* (23.16)] :

$$\mathbf{C}_t^* (t - s) \text{ is changed to } \mathbf{H}^T \mathbf{C}_t^* (t - s)\mathbf{H}, \qquad (24.8)$$

because $\mathbf{C}_t (t - s)$ is unaltered.

Thus for a solid, isotropic in the reference state, the above operations preserving the symmetry of the solid particle should not yield any change in the

9. Based on NOLL [1958 : 3].

stress tensor. So (23.15) gives us that :

$$\mathbf{H}^T \mathbf{R}(t)^T \mathbf{T}(t) \mathbf{R}(t) \mathbf{H} = \overset{\infty}{\underset{s=0}{G}} (\mathbf{H}^T \mathbf{C}_t^*(t-s)\mathbf{H}\,;\, \mathbf{H}^T \mathbf{C}(t)\mathbf{H}). \quad (24.9)$$

Given t, $\mathbf{R}(t)$ has unique value and this is a constant. So we may choose our mapping $\mathbf{H}$ in Sec. 14 to be

$$\mathbf{H}^T = \mathbf{R}(\tau)\,|_{\tau=t}\,. \quad (24.10)$$

Then (24.9) and (23.16) yield

$$\mathbf{T}(t) = \overset{\infty}{\underset{s=0}{G}} [\mathbf{C}_t(t-s)\,;\,\mathbf{B}(t)], \quad (24.11)$$

where we have also used the fact that (Sec. 1)

$$\mathbf{B}(t) = \mathbf{R}(t)\,\mathbf{C}(t)\,\mathbf{R}(t)^T\,. \quad (24.12)$$

Equation (24.11) represents *the constitutive equation of an isotropic solid.*

We say that a particle is a fluid particle if $\mathcal{G}_T = \mathcal{U}$, the full unimodular group. As noted in Sec. 14, if $\mathbf{H} \in \mathcal{U}$, $C_t(t-s)$ is not altered but

$$\mathbf{B}(t) \text{ is altered to } \mathbf{F}(t)\,\mathbf{H}\,\mathbf{H}^T\,\mathbf{F}(t)^T. \quad (24.13)$$

Now, let us choose $\mathbf{H}$ so that

$$\mathbf{H} = \{\,|\det \mathbf{F}(\tau)|\,\mathbf{F}(\tau)^{-1}\}\,|_{\tau=t}\,. \quad (24.14)$$

Hence for a simple fluid, (24.11) becomes

$$\mathbf{T}(t) = \overset{\infty}{\underset{s=0}{G}} \left(\mathbf{C}_t(t-s)\,;\,\frac{\rho_R^2}{\rho^2(t)}\mathbf{1} \right) = \overset{\infty}{\underset{s=0}{\bar{G}}} [\,\mathbf{C}_t(t-s);\,\rho(t)\,], \quad (24.15)$$

where we have used (16.7), and ρ_R and $\rho(t)$ are the reference and current densities, respectively. We may rewrite (24.15) as

$$\mathbf{T}(t) = -p[\rho(t)]\,\mathbf{1} + \overset{\infty}{\underset{s=0}{H}} [\mathbf{C}_t(t-s);\,\rho(t)], \quad \overset{\infty}{\underset{s=0}{H}} [\mathbf{1};\,\rho(t)] = 0. \quad (24.16)$$

or, equivalently as

$$\mathbf{T}(t) = -p[\rho(t)]\,\mathbf{1} + \overset{\infty}{\underset{s=0}{\tilde{H}}} [\mathbf{G}(s);\,\rho(t)], \quad \mathbf{G}(s) = \mathbf{C}_t(t-s) - \mathbf{1}, \quad \overset{\infty}{\underset{s=0}{\tilde{H}}} [\mathbf{0};\,\rho(t)] = 0. \quad (24.17)$$

Both types of equations will be used in this volume. The constitutive equation (24.16) or (24.17) defines a simple fluid. In (24.17), $p(\rho)$ is the hydrostatic pressure term.

We now derive (24.16) in a more direct manner. Recall (22.24) and consider (24.1) with it. Choose $\mathbf{H}$ to be given by (24.14). Then for a simple fluid

$$\mathbf{T}(t) = \overset{\infty}{\underset{s=0}{F}} [\mathbf{F}(t-s)] = \overset{\infty}{\underset{s=0}{F}} (\mathbf{F}_t(t-s)\,|\det \mathbf{F}(t)|\,). \quad (24.18)$$

This equation is subject to the objectivity condition

$$\mathbf{Q}(t)\,\mathbf{T}(t)\,\mathbf{Q}(t)^T = \mathop{F}_{s=0}^{\infty}(\mathbf{Q}(t-s)\,\mathbf{F}_t(t-s)\,\mathbf{Q}(t)^T \mid \det \mathbf{F}(t) \mid). \qquad (24.19)$$

Now choose

$$\mathbf{Q}(t-s) = \mathbf{R}_t(t-s)^T. \qquad (24.20)$$

Then we obtain

$$\mathbf{Q}(t)\,\mathbf{T}(t)\,\mathbf{Q}(t)^T = \mathop{F}_{s=0}^{\infty}(\mathbf{Q}(t)\,\mathbf{U}_t(t-s)\,\mathbf{Q}(t)^T \mid \det \mathbf{F}(t) \mid), \qquad (24.21)$$

valid for all orthogonal $\mathbf{Q}(t)$. Hence, by letting $\mathbf{Q}(t) = \mathbf{1}$, (24.21) yields

$$\mathbf{T}(t) = \mathop{F}_{s=0}^{\infty}(\mathbf{U}_t(t-s) \mid \det \mathbf{F}(t) \mid). \qquad (24.22)$$

Now define a new functional $\mathop{\tilde{G}}_{s=0}^{\infty}[\mathbf{C}_t(t-s); \rho(t)]$ to be given by

$$\mathop{\tilde{G}}_{s=0}^{\infty}[\mathbf{C}_t(t-s); \rho(t)] = \mathop{F}_{s=0}^{\infty}(\mathbf{U}_t(t-s) \mid \det \mathbf{F}(t) \mid). \qquad (24.23)$$

Then

$$\mathbf{T}(t) = \mathop{\tilde{G}}_{s=0}^{\infty}[\mathbf{C}_t(t-s)\,;\,\rho(t)], \qquad (24.24)$$

which is of the form (24.15).

Clearly, the discussions of Sec. 22 imply that for all orthogonal tensors $\mathbf{Q}(t)$,

$$\mathbf{Q}(t)\,\mathbf{T}(t)\,\mathbf{Q}(t)^T = \mathop{\tilde{G}}_{s=0}^{\infty}[\mathbf{Q}(t)\,\mathbf{C}_t(t-s)\,\mathbf{Q}(t)^T; \rho(t)]. \qquad (24.25)$$

We now indicate the reason for the appearance of $p(\rho)$ in (24.16) or (24.17). Suppose the fluid particle has been held in a state of rest forever so that

$$\mathbf{C}_t(t-s) = \mathbf{1}, \quad 0 \leqq s < \infty. \qquad (24.26)$$

Then (24.24) — (24.25) imply that

$$\mathbf{T}(t) = \mathop{\tilde{G}}_{s=0}^{\infty}[\mathbf{1}\,;\,\rho(t)], \qquad (24.27)$$

$$\mathbf{Q}(t)\,\mathbf{T}(t)\,\mathbf{Q}(t)^T = \mathop{\tilde{G}}_{s=0}^{\infty}[\mathbf{Q}(t)\,\mathbf{1}\,\mathbf{Q}(t)^T; \rho(t)] = \mathop{\tilde{G}}_{s=0}^{\infty}[\mathbf{1}\,;\,\rho(t)] = \mathbf{T}(t). \qquad (24.28)$$

Or,

$$\mathbf{QT} = \mathbf{TQ}, \qquad \forall\, \mathbf{Q} \in \mathcal{O}. \qquad (24.29)$$

Thus the state of stress due to the particle being in a state of rest commutes with all orthogonal tensors, which is possible if and only if

$$\mathbf{T}(t) = -\,p(\rho)\,\mathbf{1}, \qquad (24.30)$$

a scalar multiple of $\mathbf{1}$. Now, whether $p(\rho) > 0$ or < 0 depends on thermostatics. A mechanical constitutive equation cannot tell us that. However, physical experience tells us that $p(\rho) > 0$, or the stress in a state of rest is a pressure ($\equiv$ negative tension).

Because conjugation leaves the symmetry group of a simple fluid unaltered (Sec. 14), a simple fluid *is always isotropic*, that is, *its symmetry group always contains* $\mathcal{O}$.

Thus, the constitutive equation of a simple fluid is always given by

$$\mathbf{T}(t) = -p\,[\rho\,(t)]\,\mathbf{1} + \underset{s=0}{\overset{\infty}{\tilde{\boldsymbol{H}}}}\,[\mathbf{G}(s);\,\rho(t)], \tag{24.31}$$

whatever the current configuration may be.

At this point, some comments are in order to emphasize what may be implicit in the development of constitutive equations:

(i) The formulation of constitutive equations is in terms of Cartesian components, since the objectivity restriction is in Cartesian coordinates, and so is the symmetry group, when it contains orthogonal tensors only.

(ii) Since $\mathbf{C}_t(t-s)$ is insensitive to unimodular changes of the neighborhood of a particle, *no additional restriction can be put on* (24.24) *due to symmetry.*

(iii) The restriction (24.25) is a *consequence of objectivity and not the isotropy of the fluid.* For the general simple material (23.15), a restriction along the lines of (24.25) arises if and only if $\mathbf{Q} \in \mathcal{G}_T$ of that material. It is for this reason that (24.25) is said to be due to isotropy, but this is misleading.

Despite this remark, we shall say, for example, that a function $\mathbf{f}(\mathbf{A}_1)$ is *an isotropic function of* $\mathbf{A}_1$ if and only if

$$\mathbf{Q}\,\mathbf{f}\,(\mathbf{A}_1)\,\mathbf{Q}^T = \mathbf{f}(\mathbf{Q}\mathbf{A}_1\,\mathbf{Q}^T), \quad \forall\, \mathbf{Q} \in \mathcal{O}. \tag{24.32}$$

Or loosely speaking, we say that $\mathbf{f}(\cdot)$ *obeys the isotropy condition.* Sometimes we shall be more precise and call (24.32) *the objectivity condition* on $\mathbf{f}(\cdot)$. Similar language will be used for functionals also.

25. Incompressible Simple Fluids

In incompressible materials, all velocity fields $\mathbf{v}(\mathbf{x}, t)$ obey the condition $\text{div}\,\mathbf{v} = \text{tr}\,\mathbf{L} = 0$, and the density ρ is a constant everywhere. These constraints introduce an indeterminacy in the constitutive equation as we shall see below.

Under static conditions, $\mathbf{C}_t(t-s) = \mathbf{1}$ and hence, from Sec. 24 we expect that $\mathbf{T}$ should be a pressure. But this pressure cannot depend on ρ, because ρ is no longer a variable. A situation analogous to this arises when we look at the isothermal equation of state for an ideal gas:

$$p = c\rho, \tag{25.1}$$

where c is a constant. If the gas were incompressible, (25.1) cannot hold, or the pressure is not determined by an equation of state. Similarly in incompressible simple fluids, as long as the pressure is finite, no change of volume (or density) can take place and so p cannot be determined by ρ. So p is an unknown quantity, not derivable from the constitutive equation. However, as in hydrostatics, we may determine p for the simple fluid from the equations of equilibrium, to within a constant.

In dynamic situations, our aim is to determine the velocity field and the temperature field through the balance of linear momentum equation (17.21) and the balance of energy equation (19.9), respectively. But we note that in the latter equation, for incompressible materials $\text{tr}\,\mathbf{L} = 0$ and hence $\text{tr}\,\mathbf{TL} = 0$ whenever $\mathbf{T}$ is a scalar multiple of $\mathbf{1}$. Thus, as far as the energy equation is concerned the stress may differ by a scalar multiple of $\mathbf{1}$ without affecting the outcome. Now, if $\mathbf{T}$ is symmetric, then $\text{tr}\,\mathbf{TL} = \text{tr}\,\mathbf{TD}$, where $2\mathbf{D} = \mathbf{L} + \mathbf{L}^T$. The question may be asked: if $\text{tr}\,\mathbf{L} = \text{tr}\,\mathbf{D} = 0$, $\text{tr}\,\mathbf{TD} = 0$ and $\mathbf{T} = \mathbf{T}^T$, then what is the most general form $\mathbf{T}$ can have?

The answer is simple: Let $\mathbf{D}$ be diagonalised so that

$$[\mathbf{D}] = \text{diag}\,[d_1, d_2, -(d_1 + d_2)],$$

where the eigenvalues d_1 and d_2 are arbitrary. Then

$$0 = \text{tr}\,\mathbf{TD} = T_{11}\,d_1 + T_{22}\,d_2 - T_{33}\,(d_1 + d_2), \tag{25.2}$$

where T_{11}, T_{22}, T_{33} are the diagonal elements of $\mathbf{T}$. The arbitrariness of d_1 and d_2 forces us to conclude from (25.2) that $T_{11} = T_{22} = T_{33} = \lambda$ (say).

Then, one has that

$$\text{tr}\left\{\begin{bmatrix} \lambda & T_{12} & T_{13} \\ \cdot & \lambda & T_{23} \\ \cdot & \cdot & \lambda \end{bmatrix}\begin{bmatrix} d_1 & 0 & 0 \\ \cdot & d_2 & 0 \\ \cdot & \cdot & -(d_1+d_2) \end{bmatrix}\right\} = 0. \tag{25.3}$$

However, a suitable choice of rotations will yield the result that if d_1 and d_2 are arbitrary, then (25.3) will demand that $T_{12} = T_{13} = T_{23} = 0$ and hence the only non-zero symmetric tensor $\mathbf{T}$ that satisfies $\text{tr}\,\mathbf{TD} = 0$ for all symmetric $\mathbf{D}$ obeying $\text{tr}\,\mathbf{D} = 0$ is $\mathbf{T} = \lambda\mathbf{1}$[10]. So, in dynamic motions, the stress may be indeterminate to within a pressure as far as the balance of energy equation is concerned.

But such an addition, though constitutive equationally indeterminate both in the static and dynamic cases, cannot be totally arbitrary.

Even if we postulate that the constitutive equation defines the stress upto a

10. A similar elementary proof was given by LANGLOIS [1971 : 16].

pressure term, we are not at liberty to assign to this field any value we like, for the total stress field enters into the equations of equilibrium or motion and these must be used to determine p. Hence, as far as the constitutive equation is concerned, we postulate that

$$\mathbf{T} + p\,\mathbf{1} = \mathbf{T}_E = \overset{\infty}{\underset{s=0}{\mathscr{H}}} [\mathbf{C}_t(t-s)] = \overset{\infty}{\underset{s=0}{\tilde{\mathscr{H}}}} [\mathbf{G}(s)], \tag{25.4}$$

where $\mathbf{T}_E$ is called the exrta stress, and

$$\overset{\infty}{\underset{s=0}{\mathscr{H}}} (\mathbf{1}) = \overset{\infty}{\underset{s=0}{\tilde{\mathscr{H}}}} (\mathbf{0}) = \mathbf{0}, \tag{25.5}$$

and use the laws of motion or equilibrium to find $p(\mathbf{x}, t)$. In conformity with statics, we may, if we so desire, choose

$$p = -\frac{1}{3}\,\mathrm{tr}\,\mathbf{T}, \tag{25.6}$$

and hence

$$\mathrm{tr}\,\mathbf{T}_E = 0. \tag{25.7}$$

The equations of motion now become

$$-\nabla p + \mathrm{div}\,\mathbf{T}_E + \rho\mathbf{b} = \rho\mathbf{a}, \tag{25.8}$$

and this together with

$$\nabla \cdot \mathbf{v} = 0 \tag{25.9}$$

yields four equations for the four unknown functions p and $\mathbf{v}$.

For future reference, we shall recall that the condition (24.25) implies :

$$\mathbf{Q}\overset{\infty}{\underset{s=0}{\mathscr{H}}}(\cdot)\,\mathbf{Q}^T = \overset{\infty}{\underset{s=0}{\mathscr{H}}}(\mathbf{Q}\,\mathbf{C}_t(t-s)\,\mathbf{Q}^T); \quad \mathbf{Q}\overset{\infty}{\underset{s=0}{\tilde{\mathscr{H}}}}(\cdot)\,\mathbf{Q}^T = \mathbf{Q}\overset{\infty}{\underset{s=0}{\tilde{\mathscr{H}}}}(\mathbf{Q}\,\mathbf{G}(s)\,\mathbf{Q}^T),$$
$$\forall\, \mathbf{Q} \in \mathcal{O}. \tag{25.10}$$

In addition, we shall place another restriction on the constitutive functional in any given flow, equivalent to the irreversibility of the process. This is the condition, called *non-negativity of the stress power,* that is, that in all flows

$$\mathrm{tr}\,(\mathbf{T}_E \mathbf{A}_1) \geqq 0. \tag{25.11}$$

For example, this will mean that in viscometric flows, the viscosity function is non-negative, and so on. Note that (25.11) can be viewed as a restriction on the constitutive functional, or in determining whether a dynamically feasible flow (obeying all the laws of mechanics) can really occur in a world obeying the second law of thermodynamics.

One final property of the fluid that we shall assume is that it obeys the *adherence condition,* that is, that the fluid particle in contact with a solid boundary has the same velocity as the solid boundary.

Appendix A : Oldroyd's Method of Formulating Constitutive Equations[1]

The purpose of this section is to present the convected coordinate methodology of OLDROYD for forming constitutive equations. We present it here for reference and defer until Sec. 37 a comparison between simple fluids, RIVLIN-ERICKSEN fluids and the eight-constant model of OLDROYD (A5 22).

Let us recapitulate the basic ideas in formulating constitutive equations. These are that the absolute motion of a particle does not influence the stress, but the motion of the particle relative to its neighboring particles does. Secondly, as far as fluids are concerned, what matters is the strain history of the motion as experienced by the particle when this history is computed by using the present configuration for reference. As we have stated earlier, these ideas were formulated by OLDROYD in 1950. But the basic difference in the approach we have outlined and OLDROYD's method lies in the fact that the earlier technique relies on a fixed coordinate system to describe the motion, the strain history, and the stress, while OLDROYD used imbedded material surfaces. At time t, these material surfaces which form a coordinate system are described by $\xi^j \equiv$ constant, $j = 1, 2, 3$. The coordinates of a material particle X at time t are denoted by $\boldsymbol{\xi} = \{\xi^j\}$, and these are called *convected coordinates*. These do not change during the motion, being defined once and for all.

Let $\boldsymbol{\gamma}(\boldsymbol{\xi}, t)$† be the metric tensor at X at time t, with $\boldsymbol{\gamma}(\boldsymbol{\xi}, t')$ being the metric tensor at X at time t. If a certain time t_0 has a permanent significance, then we call

$$2\,\mathbf{E}(\boldsymbol{\xi}, t, t_0) = \boldsymbol{\gamma}(\boldsymbol{\xi}, t) - \boldsymbol{\gamma}(\boldsymbol{\xi}, t_0)$$

the *strain tensor* at time t. Next, the time differentiation of any quantity, keeping the coordinates $\boldsymbol{\xi} = \{\xi^j\}$ fixed, will be called *convected differentiation,* and denoted by D/Dt. Thus

$$\frac{D}{Dt}\mathbf{E}(\boldsymbol{\xi}, t, t_0) = \frac{1}{2}\frac{D}{Dt}\boldsymbol{\gamma}(\boldsymbol{\xi}, t) = \boldsymbol{\gamma}^{(1)}(\boldsymbol{\xi}, t) \qquad \text{(A5.2)}$$

is called the *rate of strain tensor* at time t. We now paraphrase OLDROYD : *A mechanical constitutive equation (or an equation of state), for a homogeneous and isotropic continuum is an invariant set of integro-differential equations relating* $\boldsymbol{\gamma}(\boldsymbol{\xi}, t)$ and $\pi(\boldsymbol{\xi}, t')$, for *all* $t' \leq t$, *where* $\pi(\boldsymbol{\xi}, t')$ *is the convected stress tensor at time* t', *at* $\boldsymbol{\xi}$ [2].

The difficulty with using such a stress tensor is that equations of motion in a moving coordinate frame are very complicated[3]. So a transformation to fixed coordinates must be made to solve the boundary value problems.

1. [1950 : 1].
2. [1950 : 1 ; p. 526].
3. LODGE [1951 : 2].

†Since the convected coordinates are fixed, and since the distance between particles may change in space during a motion, $\boldsymbol{\gamma}$ depends on t.

If the convected axes $\{\xi^i\}$ are chosen (at time t) so that they coincide with a fixed set of axes $\{x^i\}$ at time t, the stress tensors $\mathbf{T}(\mathbf{x}, t)$ and $\pi(\boldsymbol{\xi}, t)$ have the same components because this tensor is objective. But the equations of motion cannot be the same in these two coordinates[4] even if the axes coincide at time t. Also, if the two axes coincide at time t, they will not coincide in general at time t'. So we must find a way of converting $\pi(\boldsymbol{\xi}, t')$ to $\mathbf{T}(\mathbf{x}', t')$ where $\boldsymbol{\xi}$ and $\mathbf{x}'$ are respectively the *convected and the spatial coordinates* of a particle X, using the configuration at time t as the reference configuration. This conversion is essential because the constitutive equation may contain $\pi(\boldsymbol{\xi}, t')$ for all $t' \leq t$.

Suppose we let x^i and ξ^j be related at time t through

$$x^i = x^i(\xi^j, t), \qquad \xi^j = \xi^j(x^i, t). \tag{A5.3}$$

Then at time t', the position occupied by $\boldsymbol{\xi}$ is

$$x'^i = x'^i(\boldsymbol{\xi}, t') = x'^i(\mathbf{x}, t, t'). \tag{A5.4}$$ [5]

We have already used (A5.4) in Secs. 2—3 to derive the path lines connected with a velocity field $\mathbf{v}(\mathbf{x}, t)$, and we recall that (A5.4) is equivalent to

$$0 = \frac{\partial x'^i}{\partial t} + \frac{\partial x'^i}{\partial x^k} v^k. \tag{A5.5}$$

So, given $v^k(x^i, t)$, we can compute x'^i at time t, and express it as $\mathbf{x}' = \mathbf{x}'(\mathbf{x}, t, t')$. Thus a constitutive equation relating $\gamma(\boldsymbol{\xi}, t')$ and $\pi(\boldsymbol{\xi}, t')$ can be reduced to one involving $\partial x'^i/\partial x^j$, $\mathbf{g}(\mathbf{x}')$ and $\mathbf{T}(\mathbf{x}', t^j)$, where $\mathbf{g}(\mathbf{x}')$ is the metric tensor, in the fixed coordinate system at the position $\mathbf{x}'$ in space†.

At this juncture, let us consider the relation between $(D/Dt)\beta(\boldsymbol{\xi}, t)$ of any tensor β at $(\boldsymbol{\xi}, t)$ and the time derivative of this tensor in the fixed coordinate system $\mathbf{x}$ at time t in case such quantities arise in constitutive equations. Let

$$\beta^l_j(\boldsymbol{\xi}, t) = \left\| \frac{\partial \mathbf{x}}{\partial \boldsymbol{\xi}} \right\|^W \frac{\partial x^i}{\partial \xi^j} \frac{\partial \xi^l}{\partial x^k} b^k_i(\mathbf{x}, t), \tag{A5.6}$$

where W is the weight of the tensor and $\| \partial \mathbf{x}/\partial \boldsymbol{\xi} \|$ is the Jacobian of the transformation. We have

$$\frac{D}{Dt} \frac{\partial x^i}{\partial \xi^j} = \frac{\partial}{\partial \xi^j} \frac{Dx^i}{Dt} = \frac{\partial v^i}{\partial \xi^j}; \tag{A5.7}$$

$$\frac{D}{Dt} \left\| \frac{\partial \mathbf{x}}{\partial \boldsymbol{\xi}} \right\| = \left\| \frac{\partial \mathbf{x}}{\partial \boldsymbol{\xi}} \right\| \frac{\partial v^i}{\partial x^i}. \tag{A5.8}$$

On rewriting (A5.6) as

$$\frac{\partial x^k}{\partial \xi^l} \beta^l_j = \left\| \frac{\partial \mathbf{x}}{\partial \boldsymbol{\xi}} \right\|^W \frac{\partial x^i}{\partial \xi^j} b^k_i, \tag{A5.9}$$

4. As Tadjbaksh and Toupin [1964 : 24] have shown, errors occur if one ignores this distinction.

5. Oldroyd calls $(A5.4)_2$ *the displacement functions.*

†For examplel if $\xi^i \equiv x^i$,

$$\gamma_{jl}(\boldsymbol{\xi}, t') = g_{mn}(\mathbf{x}') \frac{\partial x'^m}{\partial x^j} \frac{\partial x'^n}{\partial x^l}.$$

with b_i^k being components of the tensor at $(\mathbf{x}, t)$, we obtain

$$\frac{\partial v^k}{\partial \xi^l}\beta_j^l + \frac{\partial x^k}{\partial \xi^l}\frac{D}{Dt}\beta_j^l = \left\| \frac{\partial \mathbf{x}}{\partial \boldsymbol{\xi}} \right\|^W \left\{ W \frac{\partial v^l}{\partial x^l}\frac{\partial x^i}{\partial \xi^j} b_i^k + \frac{\partial v^i}{\partial \xi^j} b_i^k + \frac{\partial x^i}{\partial \xi^j}\left[\frac{\partial b_i^k}{\partial t} + \frac{\partial b_i^k}{\partial x^r} v^r\right]\right\}. \tag{A5.10}$$

This is valid for $\boldsymbol{\xi} = \boldsymbol{\xi}\,(\mathbf{x}, t)$ and $\mathbf{x} = \mathbf{x}\,(\boldsymbol{\xi}, t)$. Choose $\boldsymbol{\xi} \equiv \mathbf{x}$. Then $\boldsymbol{\beta}\,(\boldsymbol{\xi}, t) = \mathbf{b}\,(\mathbf{x}, t)$. Thus, we have, on replacing D/Dt by $\delta/\delta t$,

$$\frac{\delta}{\delta t} b_j^k = \frac{\partial b_j^k}{\partial t} + \frac{\partial b_j^k}{\partial x^r} v^r + \frac{\partial v^i}{\partial x^j} b_i^k - \frac{\partial v^k}{\partial x^l} b_j^l + W \frac{\partial v^l}{\partial x^l} b_j^k . \tag{A5.11}$$

We call the operator $\delta/\delta t$ the *convected differentiation operator.* So any quantity with the convected components $(D/Dt)\,\Psi\,(\xi, t)$ has components $(\delta/\delta t)\,\Psi(\mathbf{x}, t)$ in the fixed coordinate system. Thus the rate of the strain tensor $\boldsymbol{\gamma}^{(1)}(\boldsymbol{\xi}, t)$ is denoted by $\mathbf{D}\,(\mathbf{x}, t)$ with

$$D_{ik}(\mathbf{x}, t) = \frac{1}{2}\frac{\delta}{\delta t} g_{ik}(\mathbf{x}, t) = \frac{1}{2}(v_{k;i} + v_{i;k}). \tag{A5.12}$$

However,

$$\frac{\delta g^{ik}}{\delta t} = -2\,D^{ik} . \tag{A5.13}$$

So the operations of raising or lowering a suffix and of convected differentiation do not commute. Motivated by this, in 1958, OLDROYD[6] introduced a new operator $\mathscr{D}/\mathscr{D}t$ through

$$\frac{\mathscr{D}b_i^k}{\mathscr{D}t} = \frac{\partial b_i^k}{\partial t} + \frac{\partial b_i^k}{\partial x^r} v^r - W_i^m\, b_m^k + W_m^k\, b_i^m , \tag{A5.14}$$

$$W_{ik} = \frac{1}{2}(v_{k;i} - v_{k;i}). \tag{A5.15}$$

In (A5.14), the coordinate system following the particle is *not the convected system*, but one which has the same velocity as the particle along with its spin **W**. This operator $\mathscr{D}/\mathscr{D}t$ is such that

$$\frac{\mathscr{D}g_{ik}}{\mathscr{D}t} = 0. \tag{A5.16}$$

We shall call this the *irrotational operator,* and emphasize that because of (A5.16) raising and lowering of indices commutes with $\mathscr{D}/\mathscr{D}t$.

6. [1958 : 4].

So far we have discussed some of the concepts needed in the OLDROYD theory, as formulated in his papers. Since $\mathscr{D}\mathbf{T}/\mathscr{D}t$ and $\mathscr{D}\mathbf{D}/\mathscr{D}t$ are objective, as are $\delta\mathbf{T}/\delta t$, $\delta\mathbf{D}/\delta t$, and $D\mathbf{D}/Dt$, these can occur in the constitutive equations. Of course **T** and **D** are objective; thus their products can also occur in the constitutive relations. Viewed in this context, the OLDROYD constitutive relations listed below are derived very easily, whether in convected coordinates or fixed coordinates. We now summarize some of OLDROYD's constitutive relations:

(i) A linear combination of the convected derivatives of π and $\mathbf{r}^{(1)}$ (or **T** and **D**) and their current values at time t (for an incompressible fluid) is as follows:

$$\left(1+\lambda_1\frac{D}{Dt}\right)\pi^E_{ik}=2\eta_0\left(1+\lambda_2\frac{D}{Dt}\right)\gamma^{(1)}_{ik};$$

$$\left(1+\lambda_1\frac{\delta}{\delta t}\right)T^E_{ik}=2\eta_0\left(1+\lambda_2\frac{\delta}{\delta t}\right)D_{ik}, \tag{A5.17}$$

where $\pi_{ik}=\pi^E_{ik}-\pi''\gamma_{ik}$ with π^E_{ik} being the extra stress and π'' the hydrostatic pressure; (A5.17) is called liquid A, and using contravariant components, that is, π^{ik}, etc., leads to liquid B;

(ii) or, in convected coordinates an integro-differential equation of state is given by:

$$\frac{D}{Dt}\gamma^{(1)j}_{j}=0,$$

$$\pi^E_{jl}(\xi,t)+\lambda\frac{D}{Dt}\pi^E_{jl}(\xi,t)-\int_{-\infty}^{t}\psi(t-t')\,\pi^E_{jl}(\xi,t')\,dt'$$

$$=\mu\left\{\frac{D\gamma_{jl}}{Dt}(\xi,t)-\int_{-\infty}^{t}\psi(t-t')\frac{D\gamma_{jl}}{Dt}(\xi,t')\,dt'\right\}, \tag{A5.18}$$

while in fixed coordinates:

$$T^E_{ik}(\mathbf{x},t)+\lambda\frac{\delta}{\delta t}T^E_{ik}(\mathbf{x},t)-\int_{-\infty}^{t}\psi(t-t')\,T^E_{mr}(\mathbf{x}',t')\frac{\partial x'^m}{\partial x^i}\frac{\partial x'^r}{\partial x^k}\,dt'$$

$$=2\mu\left\{D_{ik}(\mathbf{x},t)-\int_{-\infty}^{t}\psi(t-t')\,D_{mr}(\mathbf{x}',t')\frac{\partial x'^m}{\partial x^i}\frac{\partial x'^r}{\partial x^k}\,dt'\right\}, \tag{A5.19}$$

with $D^i_i=0$.

The liquid called A' by WALTERS[7] is obtained from (A5.19) by letting $\lambda=0$, omitting the integral involving $T^E_{ik}(\mathbf{x}',t')$ and on the right side by omitting $D_{ik}(\mathbf{x},t)$, and absorbing $-\mu$ into the memory function. It is given by

$$T_{ik}(\mathbf{x},t)=2\int_{-\infty}^{t}\psi(t-t')\,D_{mr}(\mathbf{x}',t')\frac{\partial x'^m}{\partial x^i}\frac{\partial x'^r}{\partial x^k}\,dt', \tag{A5.20}$$

7. [1962 : 12].

while the liquid B' is its contravariant counterpart :

$$T_E^{ik}(\mathbf{x}, t) = 2 \int_{-\infty}^{t} \psi(t - t')\, D^{mr}(\mathbf{x}', t') \frac{\partial x^i}{\partial x'^m} \frac{\partial x^k}{\partial x'^r}\, dt'. \qquad \text{(A5.21)}$$

(iii) In 1958, OLDROYD proposed that instead of (A5.17), one should use a constitutive equation of the type (in Cartesian coordinates) :

$$T^E_{ik} + \lambda_1 \frac{\mathcal{D} T^E_{ik}}{\mathcal{D} t} + \mu_0 T^E_{jj} D_{ik} - \mu_1 \left(T^E_{ij} D_{jk} + D_{ij} T^E_{jk} \right) + \nu_1 T^E_{jl} D_{jl} \delta_{ik}$$

$$= 2\, \eta_0 \left(D_{ik} + \lambda_2 \frac{\mathcal{D} D_{ik}}{\mathcal{D} t} - 2\mu_2 D_{ij} D_{jk} + \nu_2 D_{jl} D_{jl} \delta_{ik} \right). \qquad \text{(A5.22)}$$

We shall discuss the properties of (A5.22) in Sec. 37 in detail.

Appendix B : Formulation of Constitutive Equations (contd.). Comments on the Body Stress Tensor, Objectivity Principle and Symmetry Groups

In Appendix A, we showed how one could derive constitutive equations in convected coordinates. Since these are fixed, once and for all, one could think of them as material coordinates, and recently this idea has been explored further by LODGE[1], who calls them body coordinates. The purpose of this section is to explain this idea as well as to contrast the methodology of NOLL and RIVLIN. We begin by listing these three different approaches :

(i) writing the constitutive equation in the fixed reference configuration (that is, body manifold) or the current configuration (that is, space manifold);

(ii) in using the current configuration, one method uses the objectivity principle (in Sec. 12) for proper orthogonal tensors (det $\mathbf{Q}(t) = +1$) only, while another accepts the use of improper orthogonal (det $\mathbf{Q}(t) = -1$) as well;

(iii) the way material symmetry is introduced, and the way isotropy of a material is defined.

We shall examine these questions as they relate to the simple material in the order listed.

(i) While a reference configuration (body manifold), in principle, need never coincide with the spatial position occupied by the body in the course of its motion, nor does it have to be a Euclidean manifold, this argument is hard to apply to the formulation of constitutive equations, which will be used in E^3. So one is forced to consider the body manifold[2] as the configuration $\mathcal{B}_R$ occupied by the body in the space E^3 at some time t_0. If during the motion, the body occupies $\mathcal{B}_S$ in space at time t, we can introduce

1. [1951 : 2], [1964 : 14], [1972 : *A*] and [1972 : *B*].
2. The terminology and ideas are due to LODGE [1951 : 2]

the following transformation between the CAUCHY stress tensor $\mathbf{T}(t)$ and the body stress tensor $\boldsymbol{\pi}(t)$:

$$\boldsymbol{\pi}(t) = \mathbf{M}(t)\,\mathbf{T}(t)\,\mathbf{M}(t)^T, \qquad \mathbf{M}(t) = \mathbf{F}(t)^{-1}. \tag{B5.1}$$

Here $\mathbf{F}(t)$ is the deformation gradient from $\mathcal{B}_R$ to $\mathcal{B}_S$, evaluated for a particle P at time t. Eq. (B5.1) is identical to the rules of transformation proposed by LODGE[3].

Next, whether or not $\mathbf{Q}(t)$ in Sec. 12 is proper, the rule of transformation (12.8)

$$\mathbf{F}^* = \mathbf{QF} \tag{B5.2}$$

still holds. In an objective motion, using the objectivity of $\mathbf{T}$, one concludes that

$$\boldsymbol{\pi}^* = \boldsymbol{\pi}, \tag{B5.3}$$

or π is invariant. So the constitutive equation for a simple material of the form (22.24)

$$\mathbf{T}(t) = \mathop{\boldsymbol{F}}_{s=0}^{\infty}\,[\,\mathbf{F}(t-s)\,] \tag{B5.4}$$

is equivalent to

$$\pi(t) = \mathbf{M}(t)\,\mathop{\boldsymbol{F}}_{s=0}^{\infty}\,[\,\mathbf{F}(t-s)\,]\,\mathbf{M}(t)^T \tag{B5.5}$$

Since the functional $\mathop{\boldsymbol{F}}\limits_{s=0}^{\infty}(\cdot)$ obeys

$$\mathbf{Q}(t)\,\mathop{\boldsymbol{F}}_{s=0}^{\infty}\,[\,\mathbf{F}(t-s)\,]\,\mathbf{Q}(t)^T = \mathop{\boldsymbol{F}}_{s=0}^{\infty}\,[\,\mathbf{Q}(t-s)\,\mathbf{F}(t-s)\,], \tag{B5.6}$$

one can rewrite (B5.5) as

$$\pi(t) = \mathbf{M}(t)\,\mathop{\boldsymbol{F}}_{s=0}^{\infty}\,[\mathbf{U}(t-s)]\,\mathbf{M}(t)^T, \tag{B5.7}$$

where $\mathbf{U}(t-s)$ is the positive definite part of $\mathbf{F}(t-s)$ in its polar decomposition. Let us define a new functional:

$$\mathop{\boldsymbol{F}}_{s=0}^{\infty}\,[\mathbf{U}(t-s)] = \mathop{\bar{\boldsymbol{F}}}_{s=0}^{\infty}\,[\mathbf{C}(t-s)], \tag{B5.8}$$

where $\mathbf{C} = \mathbf{U}^2$. Here the components $C_{\alpha\beta}$ of $\mathbf{C}$ are given by

$$C_{\alpha\beta} = g_{ij}(\mathbf{x})\,x^i_{,\alpha}\,x^j_{,\beta}, \tag{B5.9}$$

and these are measured in the reference configuration (body manifold). $C_{\alpha\beta}$ are the covariant components of the body metric tensor $\boldsymbol{\gamma}(P,\ t-s)$ of

3. [1964 : 14], Eqs. (12.19) and (12.21).
4. [1964 : 14], Eq. (12.15).

LODGE[4]. One can rewrite (B5.7) as

$$\boldsymbol{\pi}(t) = \mathbf{M}(t) \underset{s=0}{\overset{\infty}{\bar{\boldsymbol{F}}}} [\mathbf{C}(t-s)]\, \mathbf{M}(t) \tag{B5.10}$$

Let us redefine (B5.10) via a new functional $\underset{s=0}{\overset{\infty}{\boldsymbol{L}}} (\cdot\, ; \cdot)$ as:

$$\boldsymbol{\pi}(t) = \underset{s=0}{\overset{\infty}{\boldsymbol{L}}} [\mathbf{C}(t-s); \mathbf{M}(t)]. \tag{B5.11}$$

Since $\mathbf{M}(t)$ is altered in an objective motion to $\mathbf{M}(t)\, \mathbf{Q}(t)^T$, we can write (B5.11) as:

$$\boldsymbol{\pi}(t) = \underset{s=0}{\overset{\infty}{\bar{\boldsymbol{L}}}} [\mathbf{C}(t-s); \mathbf{C}(t)]. \tag{B5.12}$$

No additional simplifications are possible here, except those due to symmetry. These may be interpreted in terms of coordinate transformations within the reference configuration.

In the preceding few paragraphs we have shown how to derive the constitutive equation for the body stress tensor from the spatial constitutive equation. The reason why we have followed the spatial description in this volume is because the CAUCHY stress tensor is unique, while the body stress tensor depends on the reference configuration chosen. This configuration is not unique as in the case of the 'natural state' for a perfectly elastic solid, and for a fluid, no such state exists. If one were to define the natural state as a stress free state (or a state of hydrostatic pressure for incompressible materials), one does not obtain a unique state for a viscoelastic material. In addition to this difficulty, while equations of motion may be set up and solved in body coordinates, as TADJBAKSH and TOUPIN[5] showed, such an approach could lead one into very subtle errors caused by overlooking the laws of transformation for mechanics from an inertial to a non-inertial system. These two reasons have caused us to use the traditional approach in this volume.

(ii) in recent years, RIVLIN[6] has called attention to the fact that NOLL's objectivity principle (Sec. 12), which permits $\mathbf{Q}(t)$ to be proper or improper yields incorrect results sometimes. For, as RIVLIN showed, errors arise in the formulation of electromagnetic constitutive equations if one uses NOLL's objectivity principle. It seems to us that this criticism is valid. Indeed, the objectivity principle of Sec. 12 seems to rule out the existence of a material which shows optical activity. So, in electromagnetic constitutive equations at least, the objectivity principle used in this volume could lead one into errors.

---

5. [1964 ; 26].
6. [1970 : 21]

Should we then abandon the principle as we have presented here in Sec. 12 and used in Secs. 21—25?

(iii) The matter is not as straightforward as may appear. For, the principle of objectivity and the concept of symmetry groups, while independent, are used in an entwined manner to obtain constitutive equations. Here again, NOLL[7] and RIVLIN[8] adopt different procedures. We shall compare their methods through the stored energy function of finite elasticity. Let $\sigma = \sigma(\mathbf{F})$ be this function.

	NOLL	RIVLIN
(*a*) Deformation gradient	F may be proper or improper.	F must always be proper.
(*b*) Objectivity	det $\mathbf{Q} = \pm 1$. $\sigma(\mathbf{F}) = \sigma(\mathbf{QF}) = \sigma(\mathbf{U}) = \bar{\sigma}(\mathbf{C})$ for both proper and improper **F** by choosing **Q** correctly.	det $\mathbf{Q} = +1$. $\sigma(\mathbf{F}) = \sigma(\mathbf{QF}) = \sigma(\mathbf{U}) = \bar{\sigma}(\mathbf{C})$, since **F** is always proper.
(*c*) Symmetry	$\sigma(\mathbf{F}) = \sigma(\mathbf{FH})$, where **H** is a unimodular tensor (det $\mathbf{H} = \pm 1$).	$\bar{\sigma}(\mathbf{C}) = \bar{\sigma}(\mathbf{QCQ}^T)$, where **Q** may be proper or improper, and orthogonal.
(*d*) Types of symmetry groups	Orthogonal and non-orthogonal subgroups of $\mathcal{U}$ are permitted.	Only orthogonal subgroups of $\mathcal{O}$ are permitted[9].

Suppose we use $\sigma(\mathbf{U}) = \sigma(\mathbf{FH}) = \sigma(\mathbf{RUH})$, $\mathbf{F} = \mathbf{RU}$, from the left column. Then, if $\mathbf{H} \in \mathcal{G}_\sigma$ (the symmetry group of σ) and $\mathcal{G}_\sigma \subset \mathcal{O}$, we get $\sigma(\mathbf{U}) = \sigma(\mathbf{QUQ}^T)$, by choosing $\mathbf{Q} = \mathbf{R}$, for $\forall\, \mathbf{Q} \in \mathcal{G}_\sigma$. So (*b*) and (*c*) in the left column imply (*b*) and (*c*) in the right, but not conversely.

Now, while RIVLIN permits proper and improper orthogonal tensors to be present in symmetric groups, there is no place for improper deformation gradients. This is why (*b*) and (*c*) take the form they do in his approach.

In 1955, NOLL[10] also restricted attention to proper $\mathbf{Q}(t)$ but defined the symmetry group as in (*c*) above. Since **H**, for a solid, can be an improper orthogonal transformation, one cannot claim that deformation gradients must always be proper **F**, **FH** can become improper. To remove this inconsistency, we believe that in 1958, NOLL[11] accepted the use of both proper and improper $\mathbf{Q}(t)$ and adopted the definition in Sec. 12 to turn **F** into an improper one for another observer.

7. [1958 : 3].
8. [1966 : 16].
9. See. e.g. SMITH and RIVLIN [1958 : 5], RIVLIN [1966 : 16].
10. [1955 : 2].
11. [1958 : 3].

More than this, in NOLL's approach, non-orthogonal symmetry groups are permitted. This facilitates the transition from an isotropic solid to a simple fluid very easily, as we have shown in Sec. 24. Also, liquid crystal type of behavior can be described[12]. Because RIVLIN does not accept non-orthogonal symmetry groups, it is difficult to derive that $\sigma(\mathbf{F}) = \tilde{\sigma}(\rho)$ for an elastic fluid, where ρ is the current density, by his technique. Indeed the transition from an isotropic elastic solid (24.11) to a simple fluid (24.15), that is, that in a fluid the dependence on $\mathbf{B}(t)$ must reduce to a dependence on $\rho(t)$, is not possible either, unless one postulates it[13]. It seems that in anticipating this difficulty, RIVLIN and SAWYERS[14] state that the 'concept (of fluidity) expressed by OLDROYD and used by RIVLIN and ERICKSEN[15] is embodied in NOLL's definition of the simple fluid.'

Moreover, RIVLIN and SAWYERS[16] also accept the use of proper and improper orthogonal tensors as restrictions on constitutive equations, as in (25.10). They claim that these restrictions are a consequence of the csotorphy of the fluid in its initial state, which is at variance with the concept of symmetry adopted in this volume.

Thus, whichever procedure one adopts, it will be found that both the methods of NOLL and RIVLIN have drawbacks — the former's ideas leading to the artificial improper deformation gradients and the non-existence of optically sensitive materials and the latter's ideas to a restrictive definition of symmetry.

It will now be shown that the following procedure avoids these two conceptual difficulties:

1°. Assume that $\det \mathbf{F} > 0$ always and hence that in objective motions $\mathbf{Q}(t)$ is always proper;

2°. Let the symmetry group be a subgroup of the *proper* unimodular group or equal to it.

The above two steps yield all the results in Chaps. 3 and 5, except that $\mathbf{Q}(t)$ is always proper and thus the restriction (25.10) is valid for $\forall\, \mathbf{Q} \in \mathcal{O}^+$, the subgroup of $\mathcal{O}$ containing proper tensors only. Next,

3°. In Eq. (25.10), let $\mathbf{Q} \in \mathcal{O}^+$. Replacing $\mathbf{Q}$ by $-\mathbf{Q}$ (which is improper), yields no new restriction because $-\mathbf{1}$ commutes with every tensor.

12. See COLEMAN [1965 : 4]. GREEN [1964 : 7], [1964 : 8] has derived such equations by letting the constitutive equation depend on the rotation tensor. This approach is not the same as that of using a non-orthogonal symmetry group, while the results look alike. See also WANG [1965 : 18].
13. In [1950 : 1], OLDROYD has explicitly stated that for a fluid, the constitutive equation cannot depend on a configuration which has permanent significance. Equivalently, it must depend on $\mathbf{C}_t(t-s)$ and absolute scalars, e.g. ρ.
14. [1971 : 21].
15. [1955 : 4].
16. [1971 : 21], Eq. (4.4).

Thus, *accept* the extension of the restriction (25.10) to $\forall\, \mathbf{Q} \in \mathcal{O}$, because replacing a proper orthogonal tensor by an improper orthogonal tensor implies *no additional* restriction[17]. It is our contention that this extension is in the nature of extending the domain of the functional and should not be considered to be derived from physical concepts such as objectivity and symmetry.

The implications of the above point of view to constitutive equations for non-simple materials have been discussed elsewhere[18]. However, for simple materials, the difference between NOLL's ideas and the one suggested here, while existing, is not vast and thus we have adopted NOLL's method in this treatise.

With these remarks, we turn to the remaining chapters.

17. I acknowledge some useful discussions with Professor J. L. ERICKSEN in formulating these ideas.

18. HUILGOL [1974, A].

CHAPTER 6

ISOTROPY AND THE EQUATIONS OF MOTION

Let us assume a velocity field, and let it be desired to examine whether it is dynamically possible in an imcompressible simple fluid. To keep the discussion at the utmost generality, let it be desired further to use only the constitutive equation (25.4) along with the restriction (25.10) due to objectivity.

In what follows, it will be shown that an application of the condition (25.10) will yield a substantial information regarding the stress field in a given flow, provided the strain history is known. For example, for the flow (26.1), it will be shown that the shear stresses $T\langle xz\rangle$ and $T\langle yz\rangle$ are zero. Similar reductions are possible for the steady curvilineal flow (26.50). Additional examples of the condition (25.10) appear throughout the book, especially in Chaps. 7 and 8.

Now, let it be supposed that the stress field is known, as a function of a coordinate and time (say), then to test whether the assumed velocity field is dynamically possible, one has to solve the equations of motion for the stresses (17.21). Examples of this technique are presented in Sec. 27; more will be found in Chap. 7. The reader accustomed to solving NAVIER-STOKES equations will find the methodology of this section totally new and should study it in detail. He will find that the task of deciding the dynamical feasibility of a motion is made easy by the assumption of incompressibility, which leaves the pressure field constitutive equationally indeterminate but determinable by the equations of motion. Examples are given where the pressure field is to be found; other instances where the emphasis is on the computation of the total stress field are also given.

26. Consequences of the Isotropy Condition

It is trivial to see that if the physical components of $\mathbf{C}_t\,(t-s)$ or $\mathbf{G}\,(s)$ depend on a coordinate x^1 and time t only, then the physical components of $\mathbf{T}_E$ (for incompressible fluids) and those of $\mathbf{T}$ (for compressible fluids) depend on x^1 and t only.

Suppose that the velocity field is given in a Cartesian coordinate system by

$$\dot{x} = v\,(y, t), \quad \dot{y} = 0, \quad \dot{z} = 0. \tag{26.1}$$

The path lines are given by

$$\xi = x - \int_0^s v\,(y, t-\sigma)\,d\sigma, \quad \eta = y, \quad \zeta = z. \tag{26.2}$$

The physical components of $\mathbf{C}_t\,(t - s)$ are :

$$[C_t\,(t - s)\,\langle ij \rangle] = \begin{bmatrix} 1 & -\lambda^t\,(s) & 0 \\ \cdot & 1 + [\lambda^t\,(s)]^2 & 0 \\ \cdot & \cdot & 1 \end{bmatrix}, \tag{26.3}$$

where

$$\lambda^t_{(s)} \equiv \lambda\,(t - s) = \frac{\partial}{\partial y} \int_0^s v\,(y,\, t - \sigma)\, d\sigma \tag{26.4}$$

is called the *amount of shear.*

It is straightforward then to note that for an incompressible fluid, the stresses are all functionals of y and t only. Moreover, the condition (25.7) on the trace of $\mathbf{T}_E$, viz.

$$T_E\,\langle 11 \rangle + T_E\,\langle 22 \rangle + T_E\,\langle 33 \rangle = 0, \tag{26.5}$$

implies that only two of the above three functionals are independent. So, we put

$$\underset{s=0}{\overset{\infty}{\tilde{N}_1}}\;[\lambda^t\,(s)] = T_E\langle 11 \rangle - T_E\langle 22 \rangle = T\langle 11 \rangle - T\,\langle 22 \rangle, \tag{26.6}$$

$$\underset{s=0}{\overset{\infty}{\tilde{N}_2}}\;[\lambda^t\,(s)] = T_E\,\langle 22 \rangle - T_E\,\langle 33 \rangle = T\,\langle 22 \rangle - T\,\langle 33 \rangle, \tag{26.7}$$

and call $\tilde{N}_1$ and $\tilde{N}_2$ the first and second normal stress difference functionals. Because of (26.5), one has that

$$T_E\,\langle 11 \rangle = \frac{1}{3}(2\,\tilde{N}_1 + \tilde{N}_2), \tag{26.8}$$

$$T_E\,\langle 22 \rangle = \frac{1}{3}\,(\tilde{N}_2 - \tilde{N}_1), \tag{26.9}$$

$$T_E\,\langle 33 \rangle = -\,\frac{1}{3}\,(\tilde{N}_1 + 2\,\tilde{N}_2). \tag{26.10}$$

Now, let us turn to the isotropy condition :

$$\mathbf{Q}\,\mathbf{T}_E\,\mathbf{Q}^T = \underset{s=0}{\overset{\infty}{H}}\;(\mathbf{Q}\,\mathbf{C}_t\,(t - s)\,\mathbf{Q}^T), \quad \forall\;\mathbf{Q} \in \mathcal{O}. \tag{26.11}$$

Let us choose Q to be the diagonal matrix with the components

$$[Q\,\langle ij \rangle] = \mathrm{diag}\,\{1,\, 1,\, -1\}. \tag{26.12}$$

Then, using (26.3) for $\mathbf{C}_t\,(t - s)$ and (26.12) for $\mathbf{Q}$, one notes that

$$\mathbf{Q}\,\mathbf{C}_t\,(t - s)\,\mathbf{Q}^T = \mathbf{C}_t\,(t - s), \tag{26.13}$$

or the components of the strain tensor are unaltered by replacing z by $-z$. Hence replacing z by $-z$ must leave the stress tensor $\mathbf{T}_E$ invariant. In other words,

$$\mathbf{T}_E = \underset{s=0}{\overset{\infty}{H}} \; (\mathbf{Q}\,\mathbf{C}_t\,(t-s)\mathbf{Q}^T). \tag{26.14}$$

However, from (26.11), we have that

$$[(Q\, T_E\, Q^T)\,\langle ij\rangle] = \begin{bmatrix} T_E\langle 11\rangle & T\langle 12\rangle & -T\langle 13\rangle \\ \cdot & T_E\langle 22\rangle & -T\langle 23\rangle \\ \cdot & \cdot & T_E\langle 33\rangle \end{bmatrix}, \tag{26.15}$$

or changing z to $-z$ alters the shear stress components $T\langle 13\rangle$ and $T\langle 23\rangle$. So we have the following:

(i) According to (26.14), the stresses are not altered from their original values;

(ii) According to (26.15), the shear stresses $T\langle 13\rangle$ and $T\langle 23\rangle$ become $-T\langle 13\rangle$ and $-T\langle 23\rangle$ respectively. Hence

$$\begin{bmatrix} T_E\langle 11\rangle & T\langle 12\rangle & T\langle 13\rangle \\ \cdot & T_E\langle 22\rangle & T\langle 23\rangle \\ \cdot & \cdot & T_E\langle 33\rangle \end{bmatrix} = \begin{bmatrix} T_E\langle 11\rangle & T\langle 12\rangle & -T\langle 13\rangle \\ \cdot & T_E\langle 22\rangle & -T\langle 23\rangle \\ \cdot & \cdot & T_E\langle 33\rangle \end{bmatrix}, \tag{26.16}$$

which is true if and only if

$$T\langle 13\rangle = T\langle 23\rangle = 0. \tag{26.17}$$

Or, we have proved that *if the velocity field has the form* (26.1), *then* (26.17) *holds*[1].

Not only this, it will now be proved that the normal stress differences $\tilde{N}_1$ and $\tilde{N}_2$ are even functionals of $\lambda^t(s)$ and that the shear stress functional defined by

$$T\langle 12\rangle = \underset{s=0}{\overset{\infty}{\tilde{\tau}}} \; [\lambda^t(s)] \tag{26.18}$$

is an odd functional of $\lambda^t(s)$.

For, if we choose Q to have the diagonal components

$$[Q\,\langle ij\rangle] = \text{diag}\,\{1, -1, 1\} \text{ or } \text{diag}\,\{-1, 1, 1\}, \tag{26.19}$$

then, through (26.3) one has that

$$[\,(Q\,C_t(t-s)\,Q^T)\,\langle ij\rangle\,] = \begin{bmatrix} 1 & \lambda^t(s) & 0 \\ \cdot & 1+(\lambda^t(s))^2 & 0 \\ \cdot & \cdot & 1 \end{bmatrix}, \tag{26.20}$$

1. COLEMAN and NOLL [1961 : 2]. These authors actually investigated the velocity field $\dot{x} = 0$, $\dot{y} = v\,(x, t)$, $\dot{z} = 0$. However, the conclusions (26.17) and (26.23), (26.26) and (26.36) hold there also, as do (26.22) and (26.29).

that is, changing the y axis to $-y$ axis (or x axis to $-x$ axis) is equivalent to changing the direction of flow, and replacing $\lambda^t(s)$ by $-\lambda^t(s)$. The stresses corresponding to (26.20) are given by

$$[(Q\, T_E\, Q_T)\langle ij\rangle] = \begin{bmatrix} T_E\langle 11\rangle & -T\langle 12\rangle & 0 \\ \cdot & T_E\langle 22\rangle & 0 \\ \cdot & \cdot & T_E\langle 33\rangle \end{bmatrix}, \tag{26.21}$$

whence

$$\tilde{N}_i\,[\lambda^t(s)] = \tilde{N}_i\,[-\lambda^t(s)], \quad i = 1, 2, \tag{26.22}$$

$$\tilde{\tau}\,[\lambda^t(s)] = -\,\tilde{\tau}[-\lambda^t(s)]. \tag{26.23}$$

Turning to motions that are periodic with a period $T > 0$, that is, for velocity fields of the type (26.1) with

$$v\,(y, t) = v\,(y, t + T), \tag{26.24}$$

we have that

$$\begin{aligned} \lambda\,(t + T - s) &= \frac{\partial}{\partial y}\int_0^s v\,(y, t + T - \sigma)\, d\sigma, \\ &= \frac{\partial}{\partial y}\int_0^s v\,(y, t - \sigma)\, d\sigma = \lambda(t - s), \end{aligned} \tag{26.25}$$

and hence

$$T_E\langle ij\rangle\,(t+T) = T_E\langle ij\rangle\,(t). \tag{26.26}$$

We have thus proved that the extra stresses are periodic with a period T.

In some instances, the absolute values of the oscillatory components of $v\,(y, t)$ at times t and $t + (T/2)$ are equal; or,

$$v\left(y, t + \frac{T}{2}\right) = -\,v\,(y, t) \tag{26.27}$$ [2]

is the condition being imposed on $v\,(y, t)$. Hence

$$\begin{aligned} \lambda\left(t + \frac{T}{2} - s\right) &= \frac{\partial}{\partial y}\int_0^s v\left(y, t + \frac{T}{2} - \sigma\right) d\sigma \\ &= -\frac{\partial}{\partial y}\int_0^s v\,(y, t-\sigma)\, d\sigma = -\,\lambda(t-s). \end{aligned} \tag{26.28}$$

Thus

$$\tilde{N}_i\left[\lambda\left(t + \frac{T}{2} - s\right)\right] = \tilde{N}_i\,[-\lambda^t(s)] = \tilde{N}_i\,[\lambda^t(s)], \quad i = 1, 2, \tag{26.29}$$

$$\tilde{\tau}\left[\lambda\left(t + \frac{T}{2} - s\right)\right] = -\,\tilde{\tau}\,[\lambda^t(s)], \tag{26.30}$$

2. For example, $v\,(y, t) = (ay\omega/h)\sin\omega t$ obeys (26.27).

or the shear stress functional has the property (26.30) if the velocity field does; the normal stress differences oscillate at twice the frequency if the velocity field obeys (26.27).

Note that, in general, one cannot say that if

$$v\,(y,\,t) = v_1\,(y) + v_2\,(y,\,t), \tag{26.31}$$

and $v_2\,(y,\,t)$ obeys (26.27), then (26.29) and (26.30) hold. In special fluids, like that obeying the finitely linear viscoelastic model (see Chap. 8), it is possible to make statements like (26.29) – (26.30).

As a second application, we turn to MWCSH. Suppose that 0 and t are two arbitrary positions on the time axis and let

$$\mathbf{C}_t(t-s) \equiv \mathbf{C}_0(0-s), \quad s \in [0,\,\infty). \tag{26.32}$$

Then, by the constitutive equation (25.4) of the incompressible simple fluid, the extra stresses at times 0 and t are identical:

$$\mathbf{T}_E\,(t) = \mathbf{T}_E\,(0). \tag{26.33}$$

Now, let the motion be in a MWCSH in the sense of Sec. 7, that is, let

$$\mathbf{C}_t(t-s) = e^{-s\mathbf{L}^T}\,e^{-s\mathbf{L}}, \tag{26.34}$$

where $\mathbf{L}$ is a constant tensor. In as much as $\mathbf{C}_t(t-s)$ is determined by $\mathbf{L}$ completely, a tensor-valued functional of $\mathbf{C}_t(t-s)$ must be dependent on $\mathbf{L}$. By virtue of (26.33), this functional of $\mathbf{C}_t(t-s)$ does not depend on t; however, it does depend on $\mathbf{L}$. So we express this relationship as

$$\mathbf{T}_E = \mathbf{f}(\mathbf{L}), \tag{26.35}$$

or that $\mathbf{T}_E$ is an ordinary function of $\mathbf{L}$. Moreover, since for any orthogonal tensor $\mathbf{Q}$,

$$\mathbf{Q}\,\mathbf{C}_t(t-s)\mathbf{Q}^T = e^{-s\mathbf{QL}^T\mathbf{Q}^T}\,e^{-s\mathbf{QLQ}^T}, \tag{26.36}$$

which may be verified by direct calculation, one has through (25.10) that

$$\mathbf{QT}_E\,\mathbf{Q}^T = \mathbf{f}\,(\mathbf{QLQ}^T), \tag{26.37}$$

or,

$$\mathbf{Qf}(\mathbf{L})\,\mathbf{Q}^T = \mathbf{f}\,(\mathbf{QLQ}^T). \tag{26.38}$$

Hence $\mathbf{f}(\mathbf{L})$ is an *isotropic function* of $\mathbf{L}$.

If we examine the following MWCSH, viz., simple shearing flow:

$$\dot{x} = \varkappa y, \qquad \dot{y} = \dot{z} = 0, \tag{26.39}$$

then we can conclude from the foregoing that [*cf.* (7.19)]:

$$\mathbf{T}_E = \mathbf{f}(\mathbf{L}), \tag{26.40}$$

where

$$[L\,\langle ij\rangle] = \begin{bmatrix} 0 & \varkappa & 0 \\ 0 & 0 & 0 \\ 0 & 0 & 0 \end{bmatrix}, \tag{26.41}$$

or that the components $\mathbf{T}_E$ depend on $\varkappa$ only. Turning to (26.6), (26.7) and (26.18) one has that the stresses corresponding to (26.39) are given by

$$T_E \langle 12 \rangle = \tau(\varkappa) = \mathop{\tilde{\tau}}_{s=0}^{\infty} (\varkappa s), \tag{26.42}$$

$$T_E \langle 11 \rangle - T_E \langle 22 \rangle = N_1(\varkappa) = \mathop{\tilde{N}_1}_{s=0}^{\infty} (\varkappa s), \tag{26.43}$$

$$T_E \langle 22 \rangle - T_E \langle 33 \rangle = N_2(\varkappa) = \mathop{\tilde{N}_2}_{s=0}^{\infty} (\varkappa s). \tag{26.44}$$

In (26.42) – (26.44), we have defined *shear stress* function $\tau(\varkappa)$ and *normal stress difference functions* $N_1(\varkappa)$ and $N_2(\varkappa)$ as arising from restricting the history $\lambda^t(s)$ to $\varkappa s$. From (26.22) – (26.23), we note that

$$N_i(\varkappa) = N_i(-\varkappa), \qquad i = 1, 2, \tag{26.45}$$

$$\tau(\varkappa) = -\tau(-\varkappa). \tag{26.46}$$

We call N_1, N_2 and τ the *viscometric material functions.*

What we have shown then is this: If, with respect to a local Cartesian coordinate system, the strain history has the form (26.3) with $\lambda^t(s) = -\varkappa s$, then with respect to that same basis, the stress components obey (26.5)–(26.10), (26.17) – (26.18) and (26.42) – (26.46). We shall write $N_i(\varkappa) = \overline{N}_i(\varkappa^2)$, $i = 1, 2$, when necessary.

Let us now consider the more difficult problem when the MWCSH obeys (*cf.* Sec. 8):

$$\mathbf{C}_t(t-s) = \mathbf{Q}(t)\,\mathbf{C}_0(0-s)\,\mathbf{Q}(t)^T. \tag{26.47}$$

Then, through (25.4) and (25.10), one has

$$\mathbf{T}_E(t) = \mathop{\boldsymbol{H}}_{s=0}^{\infty} [\mathbf{C}_t(t-s)] = \mathop{\boldsymbol{H}}_{s=0}^{\infty} (\mathbf{Q}(t)\,\mathbf{C}_0(0-s)\,\mathbf{Q}(t)^T)$$

$$= \mathbf{Q}(t) \mathop{\boldsymbol{H}}_{s=0}^{\infty} [\mathbf{C}_0(0-s)]\,\mathbf{Q}(t)^T = \mathbf{Q}(t)\,\mathbf{T}_E(0)\,\mathbf{Q}(t)^T. \tag{26.48}$$

If we now choose a rotating orthonormal basis $\mathbf{b}_i(t) = \mathbf{Q}(t)\,\mathbf{b}_i(0)$ as in (8.3), then with respect to this rotating basis

$$T_E \langle ij \rangle (t) = T_E \langle ij \rangle (0). \tag{26.49}$$

In other words, as long as one calculates the physical components of the strain history of a MWCSH with respect to the basis $\mathbf{b}_i(t)$, then the physical components of the extra stress tensor, when calculated with respect to this basis, are constant in time.

Next, consider a steady curvilineal flow (8.28), viz.,

$$v^1 = 0, \quad v^2 = v(x^1), \quad v^3 = w(x^1). \tag{26.50}$$

For this MWCSH, the strain history has the form

$$\mathbf{C}_t(t-s) = e^{-s\mathbf{L}_1(t)^T} e^{-s\mathbf{L}_1(t)}, \tag{26.51}$$

where $\mathbf{L}_1(t)$ is related to the velocity gradient $\mathbf{L}(t)$ through (8.19). Now, with respect to the basis (8.32), the components of $\mathbf{L}_1(t)$ are constant and are given by

$$[L_1\langle ij\rangle(t)] = \begin{bmatrix} 0 & 0 & 0 \\ \varkappa & 0 & 0 \\ 0 & 0 & 0 \end{bmatrix}, \tag{26.52}$$

that is, the flow has the local form of a simple shearing flow.

Define a new basis $\overline{\mathbf{b}}_i(t)$ such that [*cf.* (8.29)]:

$$\overline{\mathbf{b}}_1(t) = \mathbf{b}_2(t) = \alpha\mathbf{e}_2 + \beta\mathbf{e}_3,\ \overline{\mathbf{b}}_2(t) = \overline{\mathbf{b}}_1(t) = \mathbf{e}_1,\ \overline{\mathbf{b}}_3(t) = -\ \mathbf{b}_3(t) = \beta\mathbf{e}_2 - \alpha\mathbf{e}_3, \tag{26.53}$$

where α and β are given by $(8.31)_{2,3}$. With respect to this basis, $\mathbf{L}_1(t)$ has the form [*cf.* (26.41)] :

$$[L_1\langle ij\rangle(t)] = \begin{bmatrix} 0 & \varkappa & 0 \\ 0 & 0 & 0 \\ 0 & 0 & 0 \end{bmatrix}. \tag{26.54}$$

Thus, with respect to the basis $\overline{\mathbf{b}}_i(t)$, the extra stresses are given by [*cf.* (26.8) –(26.10)] :

$$\overline{T}_E\langle 11\rangle = \frac{1}{3}(2N_1 + N_2), \tag{26.55}$$

$$\overline{T}_E\langle 22\rangle = \frac{1}{3}(N_2 - N_1), \tag{26.56}$$

$$\overline{T}_E\langle 33\rangle = -\ \frac{1}{3}(N_1 + 2N_2), \tag{26.57}$$

$$\overline{T}_E\langle 12\rangle = \tau, \tag{26.58}$$

where N_1, N_2 and τ are functions of $\varkappa$, given by

$$\varkappa = \sqrt{(g_{22}v'^2 + g_{33}w'^2)/g_{11}}\,. \tag{8.31$_1$}$$

To express $\overline{\mathbf{b}}_i = \mathbf{Q}\,\tilde{\mathbf{e}}_i$, we note that $\mathbf{Q}$ must have the components

$$[Q\langle ij\rangle] = \begin{bmatrix} 0 & 1 & 0 \\ \alpha & 0 & \beta \\ \beta & 0 & -\alpha \end{bmatrix}. \tag{26.59}$$

Next, $T_E\langle ij\rangle$ must have the components

$$T_E\langle ij\rangle = Q\langle ik\rangle\,\overline{T}_E\langle kl\rangle\,Q\langle jl\rangle, \tag{26.60}$$

and thus[3]

$$T_E\langle 11\rangle = \overline{T}_E\langle 22\rangle = \frac{1}{3}(N_2 - N_1), \tag{26.61a}$$

$$T_E\langle 22\rangle = \alpha^2\,\overline{T}_E\langle 11\rangle + \beta^2\,\overline{T}_E\langle 33\rangle = \frac{1}{3}(2\alpha^2 - \beta^2)N_1 + \frac{1}{3}(\alpha^2 - 2\beta^2)N_2, \tag{26.61b}$$

$$T_E\langle 33\rangle = \beta^2\,\overline{T}_E\langle 11\rangle + \alpha^2\,\overline{T}_E\langle 33\rangle = \frac{1}{3}(2\beta^2 - \alpha^2)N_1 + \frac{1}{3}(\beta^2 - 2\alpha^2)N_2, \tag{26.61c}$$

$$T_E\langle 12\rangle = \alpha\,\overline{T}_E\langle 12\rangle = \alpha\tau, \tag{26.61d}$$

$$T_E\langle 13\rangle = \beta\,\overline{T}_E\langle 12\rangle = \beta\tau, \tag{26.61e}$$

$$T_E\langle 23\rangle = \alpha\beta\,(\overline{T}_E\langle 11\rangle - \overline{T}_E\langle 33\rangle) = \alpha\beta\,(N_1 + N_2). \tag{26.61f}$$

The nonlinearity of the material behavior is apparent here. For, while the velocity fields in (26.50) would individually yield no stress in the $\langle 23\rangle$ direction, their superposition leads to a stress in this direction.

Also, note that (26.61*a*) – (26.61*c*) imply that

$$T_E\langle 11\rangle - T_E\langle 23\rangle = \beta^2 N_2 - \alpha^2 N_1, \tag{26.62}$$

$$T_E\langle 22\rangle - T_E\langle 33\rangle = (\alpha^2 - \beta^2)(N_1 + N_2). \tag{26.63}$$[4]

Since in a simple shearing motion, the tensors $\mathbf{A}_1$ and $\mathbf{A}_2$ have the physical components given by

$$[A_1\langle ij\rangle] = \varkappa \begin{bmatrix} 0 & 1 & 0 \\ \cdot & 0 & 0 \\ \cdot & \cdot & 0 \end{bmatrix}, \quad [A_2\langle ij\rangle] = 2\varkappa^2 \begin{bmatrix} 0 & 0 & 0 \\ \cdot & 1 & 0 \\ \cdot & \cdot & 0 \end{bmatrix}, \tag{26.64}$$

the constitutive equation for $\mathbf{T}_E$ may be written as

$$\mathbf{T}_E = \frac{\tau(\varkappa)}{x}\mathbf{A}_1 + \frac{N_1(\varkappa) + N_2(\varkappa)}{\varkappa^2}\mathbf{A}_1^2 - \frac{N_1(\varkappa)}{2\varkappa^2}\mathbf{A}_2 - \frac{N_1(\varkappa) + 2N_2(\varkappa)}{3}\mathbf{1}$$
$$= \mathbf{T} + p\,\mathbf{1}, \tag{26.65}$$

and this satisfies the normalization condition $\operatorname{tr}\mathbf{T}_E = 0$.[5]

Now it may be felt that this normalization condition is unjustified. So, one can retrace all the earlier arguments and express T_E as

$$\mathbf{T}_E = \mathbf{T} + p\,\mathbf{1} = \frac{\tau(\varkappa)}{\varkappa}\mathbf{A}_1 + \frac{N_1(\varkappa) + N_2(\varkappa)}{\varkappa^2}\mathbf{A}_1^2 - \frac{N_1(\varkappa)}{2\varkappa^2}\mathbf{A}_2. \tag{26.66}$$

3. These results are equivalent to those in [1966 : 4].
4. In the works of COLEMAN and NOLL [1959 : 1; 2] σ_1 and σ_2 are used, where $\sigma_1 = N_2$, $\sigma_2 = N_1 + N_2$.
5. Such reductions were originally given by ERICKSEN [1958 : 2], who made use of RIVLIN'S earlier works [1956 : 4]. They also appear in [1958 : 1].

Of course the pressure field p associated with the form (26.65) is different from that associated with (26.66). But, as is obvious, the material functions $\tau(\varkappa)$, $N_1(\varkappa)$ and $N_2(\varkappa)$ are the same whether or not the normalization condition is used.

If, for example, one defined a new normal stress material function $N_3(\varkappa)$ through

$$N_3(\varkappa) = T_E\langle 11\rangle - T_E\langle 22\rangle, \tag{26.67}$$

we can rewrite the constitutive equation, without normalization, for $\mathbf{T}_E$ in matrix notation as

$$[T_E\langle ij\rangle] = \tau(\varkappa)\begin{bmatrix} 0 & 1 & 0 \\ . & 0 & 0 \\ . & . & 0 \end{bmatrix} - N_1(\varkappa)\begin{bmatrix} 0 & 0 & 0 \\ 0 & 1 & 0 \\ 0 & 0 & 0 \end{bmatrix} - N_3(\varkappa)\begin{bmatrix} 0 & 0 & 0 \\ . & 0 & 0 \\ . & . & 1 \end{bmatrix}. \tag{26.68}$$

For later use, we note that we may define three new functions $\overline{n}_1(\varkappa^2)$, $\overline{n}_2(\varkappa^2)$ and $\overline{n}_3(\varkappa^2)$ through

$$\overline{n}_i(\varkappa^2) = \overline{N}_i(\varkappa^2)/\varkappa^2, \quad i = 1, 2, 3. \tag{26.69}$$

Now $\overline{n}_i(\varkappa^2)$ are also even functions of $\varkappa$, as are the $\overline{N}_i(\varkappa^2)$. Hence, analyticity of N_i demands that

$$\left.\frac{d\overline{N}_i(\varkappa^2)}{d\varkappa}\right|_{\varkappa=0} = 0, \quad i = 1, 2, 3. \tag{26.70}$$

Moreover, non-Newtonian fluids exhibit the property that $N_i(0) = 0$ and thus near $\varkappa = 0$,

$$N_i(\varkappa^2) = \alpha_i\varkappa^2 + o(\varkappa^2), \quad i = 1, 2, 3, \tag{26.71}$$

and thus

$$\overline{n}_i(0) = \alpha_i, \quad \left.\frac{d\overline{n}_i(\varkappa^2)}{d\varkappa}\right|_{\varkappa=0} = 0. \tag{26.72}$$

Also, we define a *viscosity function* $\eta(\varkappa)$, which is even in $\varkappa$, through

$$\eta(\varkappa) = \tau(\varkappa)/\varkappa. \tag{26.73}$$

This has also the property:

$$\left.\frac{d\eta(\varkappa)}{d\varkappa}\right|_{\varkappa=0} = 0. \tag{26.74}$$

Also, non-Newtonian fluids exhibit a non-zero shear viscosity so that

$$\eta(0) = \eta_0 > 0. \tag{26.75}$$

Hence, near $\varkappa = 0$,

$$\eta(\varkappa) = \eta_0 + O(\varkappa^2). \tag{26.76}$$

Note that the inequality (25.11) leads to $\eta \geqq 0$ for all $\varkappa$, rather than (26.75).

Though some of the above comments need experimental confirmation (see Chap. 11), we have used known results to indicate here some properties of the viscometric material functions for use in Chap. 7.

Instead of (26.50), one can discuss[6] (non-steady) curvilineal flow, viz.,

$$v^1 = 0, \quad v^2 = v(x^1, t), \qquad v^3 = w(x^1, t), \tag{26.77}$$

with

$$\frac{\partial v}{\partial x^1} = f(x^1)\, q(x^1, t), \qquad \frac{\partial w}{\partial x^1} = h(x^1)\, q(x^1, t). \tag{26.78}$$

This will yield a stress system identical to (26.61*a*) – (26.61*f*), except that the N_1, N_2 and τ functions are replaced by the functionals $\tilde{N}_1$, $\tilde{N}_2$ and $\tilde{\tau}$ respectively. The argument of each functional is given by

$$\lambda^t(s) = k(t) - k(t-s), \qquad k(\tau) = \varkappa \int_0^\tau q(x^1, \sigma)\, d\sigma, \tag{26.79}$$

$$\varkappa = \sqrt{g_{11}^{-1}\,(g_{22}\, f^2 + g_{33}\, h^2)}, \quad \varkappa\alpha = f\sqrt{g_{22}\, g_{11}^{-1}}, \quad \varkappa\beta = h\sqrt{g_{33}\, g_{11}^{-1}}.$$

The crucial requirement in deducing this stress system is that α and β in (26.59) be functions of t at most, or better still, constants along the path of each particle. The assumed decomposition (26.78) assures us that the latter condition is met.

26 bis.

We now reexamine the state of stress near a surface of rest.[7] As shown by (3.31), the strain history has zero components in $\langle 23 \rangle$ direction and the $\langle 22 \rangle$ and $\langle 33 \rangle$ components are each identical to unity.

One can certainly find a basis to map $\mathbf{C}_t(t-s)$ to have the form

$$[C_t(t-s)\,\langle ij \rangle] = \begin{bmatrix} 1 & -\,\Omega^t(s) & 0 \\ \cdot & 1 + [\Omega^t(s)]^2 & 0 \\ \cdot & \cdot & 1 \end{bmatrix}. \tag{26.b1}$$

For this purpose, the form of $\mathbf{Q}$ given in (26.59) suffices, but α and β obey

$$\alpha^2 + \beta^2 = 1, \qquad \alpha U^t(s) - \beta V^t(s) = [\Omega^t(s)]^2, \tag{26.b2}$$

or α and β are each functions of $t - s$. So the *stress system* corresponding to (26.*b*1) *is not expressible* in terms of the material functionals $\tilde{N}_1$, $\tilde{N}_2$ and $\tilde{\tau}$. Of course, one may obtain such a representation if [*cf.* (26.78)]:

$$U^t(s) = f(\cdot)\, q^t(s), \qquad V^t(s) = h(\cdot)\, q^t(s), \tag{26.b3}$$

where $f(\cdot)$ and $h(\cdot)$ do not depend on $t - s$. Indeed, if the velocity field is given by (26.77) and (26.78), then in (26.*b*2) not only α and β become in-

6. NFTM, Sec. 107.
7. CASWELL [1967 : 6].

dependent of time, but also the *vortex lines are steady*[8]. Then $V^t(s) = 0$, $\alpha = 1$, $\beta = 0$, and the strain history is given by (3.36), or it is analogous to (26.3), and the stress system is expressible in terms of $\tilde{N}_1$, $\tilde{N}_2$ and $\tilde{\tau}$ as in (26.6), (26.7) and (26.18). Note that this modified stress system exists along the $\mathbf{i}_2$, $\mathbf{i}_1$ and $\mathbf{i}_3$ axes, which make (3.36) conform with (26.3).

If the *vorticity is itself steady*, then the stress system along the $\mathbf{i}_2$, $\mathbf{i}_1$, $\mathbf{i}_3$ axes is given by (26.42) – (26.44), with $\varkappa s$, replaced by ωs, $\omega = |\boldsymbol{\omega}|$. In other words, when the vorticity is steady, the stress system near a surface of rest is the viscometric stress system, regardless of the flow away from the surface of rest. For example, this is tantamount to saying that the steady flow in the boundary layer near a wall is viscometric.

For the more general situation represented by the strain history (3.31), it has been shown by CASWELL that the stress system may be written as

$$\mathbf{T} + p\,\mathbf{1} = \mathcal{S}_1(\Omega)\,\mathbf{n} \otimes \mathbf{n} + \mathcal{S}_2\,[\mathcal{N}_2(\Omega^t(s))\,(\boldsymbol{\Omega} \times \mathbf{n}) \otimes (\boldsymbol{\Omega} \times \mathbf{n})]$$
$$+ \mathcal{T}(\mathcal{M}(\Omega^t(s))\,[(\boldsymbol{\Omega} \times \mathbf{n}) \otimes \mathbf{n} + \mathbf{n} \otimes (\boldsymbol{\Omega} \times \mathbf{n})], \quad (26.b4)$$

where $\mathcal{N}_2[\Omega^t(s)]$ indicates the even, non-linear dependence of the functional $\mathcal{S}_2$ on $\Omega^t(s)$, and $\mathcal{S}_2$ is bilinear with respect to $(\boldsymbol{\Omega} \times \mathbf{n}) \otimes (\boldsymbol{\Omega} \times \mathbf{n})$; while $\mathcal{M}(\Omega^t(s))$ indicates the even, nonlinear dependence of the functional $\mathcal{T}$ on $\Omega^t(s)$, and this functional is linear in $(\boldsymbol{\Omega} \times \mathbf{n}) \otimes \mathbf{n}$.

For steady vortex lines, the functional $\mathcal{S}_2(\cdot)$ and $\mathcal{T}(\cdot)$ respectively become:

$$\mathcal{S}_2(\mathcal{N}_2(\Omega^t(s))\,\Omega^2(t))\,(\boldsymbol{\omega}^* \times \mathbf{n}) \otimes (\boldsymbol{\omega}^* \times \mathbf{n}), \quad (26.b5)$$
$$\mathcal{T}(\mathcal{M}(\Omega^t(s))\,\Omega(t))\,[(\boldsymbol{\omega}^* \times \mathbf{n}) \otimes \mathbf{n} + \mathbf{n} \otimes (\boldsymbol{\omega}^* \times \mathbf{n})].$$

The above formula is extremely elegant, and its use should be explored further.

27. Equations of Motion. General Solutions to a Few Problems

The Appendix to this chapter lists the equations of motion in Cartesian, cylindrical polar and spherical coordinates. In this section we shall list the solutions to these equations in a few interesting cases. Attention will be restricted to incompressible fluids only.

Generally speaking, we shall assume that body force density $\mathbf{b}$ is derivable from a potential χ :

$$\mathbf{b} = -\operatorname{grad} \chi. \quad (27.1)$$

Introducing a modified pressure ψ so that

$$\psi = p + \rho\chi, \quad (27.2)$$

the equations of motion become

$$-\nabla\psi + \operatorname{div} \mathbf{T}_E = \rho\mathbf{a}. \quad (27.3)$$

8. See CASWELL [1967 : 6] for a direct verification.

We now examine several cases, with and without body forces.

27.1. Homogeneous Time-Dependent Velocity Field. By a homogeneous velocity field, we mean the following one in a Cartesian coordinate system :

$$\dot{x}_i = L_{ij}(t)\, x_j + c_i(t), \quad L_{ii} = 0. \tag{27.4}$$

Here the velocity gradient is a function of time t only, and $c_i(t)$ is a time-dependent vector and thus (27.4) includes steady homogeneous velocity fields as a special case. The deformation gradient is given by

$$\mathbf{F}(t) = \exp\left(\int_0^t \mathbf{L}(\sigma)\, d\sigma\right) \mathbf{F}(0), \tag{27.5}$$

where 0 is some fixed time. Then

$$\mathbf{F}_t(t-s) = \exp\left(\int_0^{t-s} \mathbf{L}(\sigma)\, d\sigma\right) \exp\left(-\int_0^t \mathbf{L}(\sigma)\, d\sigma\right). \tag{27.6}$$

Hence the strain history $\mathbf{C}_t(t-s)$, $0 \leq s < \infty$, is a function of time only. The constitutive equation (25.4) then implies that the extra stresses are functions of time only. Thus the equations of motion reduce to

$$-\nabla \psi = \rho \mathbf{a}. \tag{27.7}$$

So the velocity field (27.4) is dynamically possible in an incompressible simple fluid if the acceleration field is irrotational[9], that is, if and only if $\mathbf{L}$ is symmetric.

As a special case, consider the unsteady relative extension problem of SLATTERY[10]. He has discussed the case when $L_{ij}(t)$ is diagonal and

$$\begin{aligned} L_{ii}(t) &= a_i f(t), \quad i = 1, 2, 3\,; \text{ no sum;} \\ L_{ij} &= 0, \quad i \neq j, \quad a_1 + a_2 + a_3 = 0. \end{aligned} \tag{27.8}$$

It is easy to see that this motion is dynamically possible, since $\mathbf{L}$ is symmetric and thus the acceleration field is irrotational.

27.2. Non-Steady Plane Flow. Let the velocity field be given by (26.1). From the calculations of Sec. 26, and the equations [(A6.1)—(A6.2)], the equations of motion reduce to

$$\begin{aligned} -\frac{\partial \psi}{\partial x} + \frac{\partial}{\partial y}\, T\langle xy\rangle &= \rho \frac{\partial v}{\partial t}, \\ -\frac{\partial \psi}{\partial y} + \frac{\partial}{\partial y}\, T_E\langle yy\rangle &= 0, \\ -\frac{\partial \psi}{\partial z} &= 0. \end{aligned} \tag{27.9}$$

---

9. This result is a special case of the result of COLEMAN and TRUESDELL [1965 : 6] for simple materials. However, our approach is slightly different.

10. [1964 : 22].

Now, $(27.9)_{2-3}$ yield that

$$\psi = T_E\langle yy\rangle + f(x, t), \tag{27.10}$$

where $f(x, t)$ is a function of x and t. But (27.10) is consistent with $(27.9)_1$ if and only if

$$f(x, t) = x\, g(t) + h(t). \tag{27.11}$$

Hence $(27.9)_1$ becomes

$$\frac{\partial}{\partial y}\, T\langle xy\rangle = \rho\, \frac{\partial v}{\partial t} + g(t), \tag{27.12}$$

so that the solution to the system (27.9) consists of assigning $g(t)$ and $h(t)$ in (27.11) and using $g(t)$ in (27.12) to find $v(y, t)$. This exact velocity distribution depends on the functional $T\langle xy\rangle = \tilde{\tau}[\lambda^t(s)]$, since its derivative appears in (27.12).

27.3. Non-Steady Helical Flow. In cylindrical coordinates, let the velocity field be given by the contravariant components

$$\dot{r} = 0, \qquad \dot{\theta} = \omega\,(r, t), \qquad \dot{z} = u\,(r, t). \tag{27.13}$$

The path lines will yield a relative deformation gradient which has mixed components depending on r and t. Moreover, the components of the metric tensor depend on r and this coordinate is constant along the path line of each particle. Thus the physical components of $\mathbf{C}_t(t - s)$ are functions of r and t only, with the result that all the extra stresses depend on r and t only.

From (A6.3)—(A6.4), the equations of motion are

$$\begin{gathered}
-\frac{\partial\psi}{\partial r} + \frac{\partial T_E\langle rr\rangle}{\partial r} + \frac{1}{r}\,(T_E\langle rr\rangle - T_E\langle\theta\theta\rangle) = -\,\rho r\omega^2,\\
-\frac{1}{r}\,\frac{\partial\psi}{\partial\theta} + \frac{\partial T\langle r\theta\rangle}{\partial r} + \frac{2}{r}\,T\langle r\theta\rangle = \rho r\,\frac{\partial\omega}{\partial t},\\
-\frac{\partial\psi}{\partial z} + \frac{\partial T\langle rz\rangle}{\partial r} + \frac{1}{r}\,T\langle rz\rangle = \rho\,\frac{\partial u}{\partial t}.
\end{gathered} \tag{27.14}$$

Equations $(27.14)_{2-3}$ imply that

$$\psi = -\,g(t)\,\theta - h(t)\,z + f(r, t). \tag{27.15}$$

Substituting this back into Eq. (27.14), we get

$$\left.\begin{aligned}
\frac{\partial}{\partial r}\,(r^2\, T\langle r\theta\rangle) &= \rho r^2\,\frac{\partial\omega}{\partial t} - r g(t),\\
\frac{\partial}{\partial r}\,(r\, T\langle rz\rangle) &= \rho r\,\frac{\partial u}{\partial t} - r h(t).
\end{aligned}\right\} \tag{27.16}$$

Because of (27.2) and (27.15), we must put

$$T\langle rr\rangle = -\,p + T_E\langle rr\rangle = g(t)\,\theta + h(t)\,z + k(r, t) + \rho\chi. \tag{27.17}$$

Hence,

$$\frac{\partial}{\partial r}\,T\langle rr\rangle = \frac{\partial k(r, t)}{\partial} + \rho\,\frac{\partial\chi}{\partial r}, \tag{27.18}$$

and thus $(27.14)_1$, which may be written as

$$-\rho\frac{\partial\psi}{\partial r}+\frac{\partial T_E\langle rr\rangle}{\partial r}+\frac{1}{r}\Big(T_E\langle rr\rangle-T_E\langle\theta\theta\rangle\Big)=-\rho r\omega^2, \qquad (27.19)$$

becomes

$$\frac{\partial k(r,t)}{\partial r}+\frac{1}{r}\Big(T_E\langle rr\rangle-T_E\langle\theta\theta\rangle\Big)=-\rho r\omega^2. \qquad (27.20)$$

Hence (27.16)-(27.17) and (27.20) are the solutions to the system (27.15)[11]. We now examine a special case of the above.

1°. Suppose now that ω and u are functions of r only. Then $(27.14)_{2,3}$ respectively become

$$-\frac{1}{r}\frac{\partial\psi}{\partial\theta}+\frac{\partial T\langle r\theta\rangle}{\partial r}+\frac{2}{r}T\langle r\theta\rangle=0, \qquad (27.b1)$$

$$-\frac{\partial\psi}{\partial z}+\frac{\partial T\langle rz\rangle}{\partial r}+\frac{1}{r}T\langle rz\rangle=0. \qquad (27.b2)$$

2°. If one also demands that the domain of the angle θ lies between $0\leqq\theta\leqq 2\pi$, then ψ must be independent of θ if it is to be single-valued; also ψ is linear in z. Hence the solutions to (27.*b*1) and (27.*b*2) are:

$$T\langle r\theta\rangle=M/2\pi r^2, \qquad M=\text{const.} \qquad (27.b3)$$

$$T\langle rz\rangle=-\frac{1}{2}cr+br^{-1}, \qquad (27.b4)$$

where b is a constant and

$$\frac{\partial\psi}{\partial z}=-c\ (\text{const.}), \qquad (27.b5)^{12}$$

3°. Now, from (27.15) it follows that

$$p=-cz+f(r)-\rho\chi. \qquad (27.b6)$$

Also, from (27.20), on noting that the dependence of $k(\cdot)$ is on r only we have

$$k(r)=\int\left\{\frac{1}{r}\Big(T\langle\theta\theta\rangle-T\langle rr\rangle\Big)-\rho r\omega^2\right\}dr+e, \qquad (27.b7)$$

where e is a constant. Hence from (27.17),

$$T\langle rr\rangle=\rho\chi+cz+e+\int\left\{\frac{1}{r}\Big(T_E\langle\theta\theta\rangle-T_E\langle rr\rangle\Big)-\rho r\omega^2\right\}dr. \qquad (27.b8)$$

27.4. Torsional Flow. In a cylindrical polar coordinate system, let the contravariant components of the velocity field be given by (see Fig. 27.1):

$$v^r=0, \qquad v^\theta=\omega(z), \qquad v^z=0. \qquad (27.21)$$

The extra stresses are obviously functions of r and z; they depend on r, because this coordinate enters into the physical components of the strain

11. NFTM, Sec. 112. Note that $k(r,t)=-f(r,t)+T_E\langle rr\rangle$.
12. In (27.17), we take $g(t)=0$, $h(t)=c$.

tensor through the metric tensor of the coordinate system.

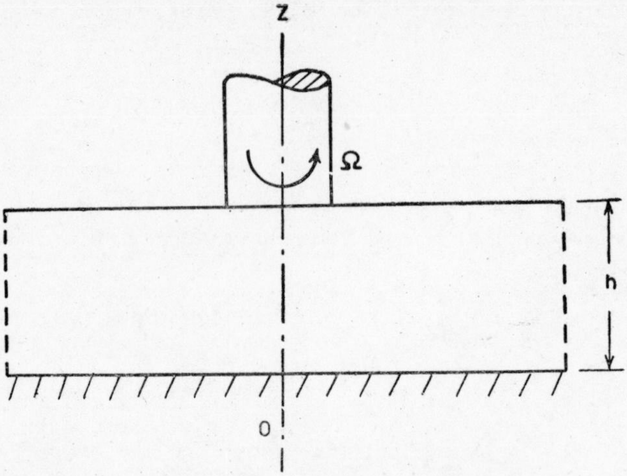

Fig. 27.1. Parallel plate system or torsional flow.

From the theory of Sec. 26, it follows that the stresses corresponding to (27.20) are such that $T_E\langle r\theta\rangle = T_E\langle rz\rangle = 0$; the other components of $\mathbf{T}_E$ depend on the shear rate $\varkappa$, given by

$$\varkappa = r\omega'(z). \qquad (27.22)^{13}$$

To verify these statements, take $z = x^1, \theta = x^2, r = x^3$ in (26.50). Then $\beta = 0$, (8.31) will yield (27.22), and (26.61a)-(26.61f) confirm the above claims.

The underlying equations of motion are obtained from (A6.3)-(A6.4) :

$$-\frac{\partial\psi}{\partial r} + \frac{\partial T_E\langle rr\rangle}{\partial r} + \frac{1}{r}\Big(T_E\langle rr\rangle - T_E\langle\theta\theta\rangle\Big) = -\rho r\omega^2,$$
$$-\frac{1}{r}\frac{\partial\psi}{\partial\theta} + \frac{\partial T_E\langle\theta z\rangle}{\partial z} = 0, \qquad (27.23)$$
$$-\frac{\partial\psi}{\partial z} + \frac{\partial T_E\langle zz\rangle}{\partial z} = 0.$$

Now in $(27.23)_2$, ψ has to be independent of θ if it is to be single-valued, since $0 \leq \theta \leq 2\pi$; and thus $T_E\langle\theta z\rangle$ is independent of z. This is possible if and only if $\varkappa$ in (27.21) is independent of z, since $T_E\langle\theta z\rangle$ is a function of $\varkappa$. Thus

$$\omega(z) = az + b, \qquad \varkappa = ar. \qquad (27.24)$$

Because of $(27.23)_3$, this condition implies that ψ is independent of z also, and hence, $(27.23)_1$ becomes

$$-\frac{d\psi}{dr} + \frac{\partial T_E\langle rr\rangle}{\partial r} = \frac{1}{r}\Big(T_E\langle\theta\theta\rangle - T_E\langle rr\rangle\Big) - \rho r\Big(az + b\Big)^2. \qquad (27.25)$$

But this equation demands that ψ depend on z also through $\int \rho r(az + b)^2\,dr$,

13. One can derive this result from (A1.44) also.

a contradiction. The only way to resolve this is to take $\rho = 0$. We cannot take $a = 0$, for otherwise we have a rigid motion.

Here, for the first time, one has an example of a flow which is possible if and only if the inertia terms are ignored, that is, if we put $\rho = 0$. Now, putting $\rho = 0$ is equivalent to assuming that $\psi = p$, so that the restrictions on ψ apply to p. But in any real fluid $\rho \neq 0$ and hence (27.24) is only approximately realizable in the laboratory.

27.5. Cone-and-Plate Flow. Let the contravariant components of a velocity field be given, in spherical coordinates by (see Fig. 27.2) :

$$\dot{r} = 0, \quad \dot{\theta} = 0, \quad \dot{\phi} = \omega(\theta). \tag{27.26}$$

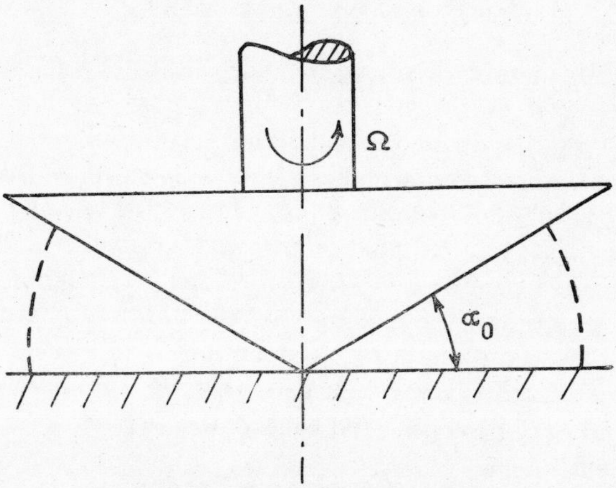

Fig. 27.2. Cone-and-plate system.

For this velocity field, the rate of shear $\varkappa$ is calculated as [see (A1.45)] :

$$\varkappa = \omega'(\theta) \sin\theta. \tag{27.27}$$

Put $\theta = x^1$, $\phi = x^2$ and $r = x^3$ in (26.50). Then $v(x^1) = \omega(\theta)$, $w(x^1) = 0$ and thus $\beta = 0$ in $(8.31)_3$. Since $g_{11} = r^2$, $g_{22} = r^2 \sin^2\theta$, $g_{33} = 1$, $(8.31)_1$ yields (27.27). Substitution of this information into (26.61*d*)-(21.61*f*) yields that $T_E\langle r\theta\rangle = T_E\langle r\phi\rangle = 0$, and $T_E\langle\theta\phi\rangle = \tau(\varkappa)$. Next, (26.62), shows us that $T_E\langle\theta\theta\rangle - T_E\langle\phi\phi\rangle = -N_1(\varkappa)$, while (26.63) implies $T_E\langle\phi\phi\rangle - T_E\langle rr\rangle = N_1(\varkappa) + N_2(\varkappa)$.

Assume that the body force acts in the direction $\theta = \pi$, with this axis pointing vertically downwards in the gravitational field of the earth. Then $b\langle\phi\rangle = 0$, but at each point (r, θ, ϕ) in the flow, $b\langle r\rangle$ and $b\langle\theta\rangle$ exist and depend on θ, with $b\langle r\rangle = -g\sin\theta$, $b\langle\theta\rangle = g\cos\theta$.

The equations of motion (A6.5), (A6.6) on recognizing that $\varkappa$ is a function of θ and hence that $\mathbf{T}_E$ is a function of θ become

$$-\frac{\partial p}{\partial r} - \frac{1}{r}[N_1(\varkappa) + 2N_2(\varkappa)] - \rho g \sin\theta = -\rho\omega^2 r \sin^2\theta,$$

$$-\frac{1}{r}\frac{\partial p}{\partial\theta} + \frac{1}{r}\frac{dT_E\langle\theta\theta\rangle}{d\theta} - \frac{1}{r}\cot\theta\, N_1(\varkappa) + \rho g\cos\theta = -\frac{\rho\omega^2 r \sin 2\theta}{2}, \tag{27.28}$$

$$-\frac{1}{r \sin \theta} \frac{\partial p}{\partial \phi} + \frac{1}{r} \frac{d\tau(\varkappa)}{d\theta} + \frac{2}{r} \cot \theta \, \tau(\varkappa) = 0.$$

From $(27.28)_1$ and $(27.28)_3$, one can derive that

$$p = h(\theta) + \phi \left(\frac{d\tau}{d\theta} + 2\,\tau \cot \theta \right) \sin \theta + \frac{\rho r}{2} \left(\omega^2 r \sin^2 \theta - 2\, g \sin \theta \right)$$
$$- (N_1 + 2\, N_2) \ln r$$
$$= h(\theta) + f(r, \theta, \phi) \text{ (say).} \tag{27.29}$$

Substitution of this into $(27.28)_2$ shows us that

$$\frac{dh}{d\theta} = \frac{dT_E\langle\theta\theta\rangle}{d\theta} - N_1 \cot \theta + \rho g r \cos\theta + \frac{\rho \omega^2 r^2 \sin 2\theta}{2} - \frac{\partial f}{\partial \theta}. \tag{27.30}$$

To satisfy (27.30), terms involving r and ϕ must drop out of the right side. The only way to do this is :

(i) take $\rho = 0$, that is, assume that the fluid has no inertia;

(ii) since dependence of p on ϕ makes it multi-valued in ϕ, take $\partial p/\partial \phi = 0$, so that $\partial f/\partial \theta$ will not contain ϕ, or reduce $f(r, \theta, \phi)$ to $-[N_1(\varkappa) + 2N_2(\varkappa)] \ln r$, that is, to a function of r and θ only :

(iii) to remove the dependence of $\partial f/\partial \theta$ on r, one may assume that $N_1(\varkappa) + 2N_2(\varkappa)$ is a constant in the domain of the flow and hence $\partial f/\partial \theta = 0$. We will now show that this assumption is not necessary, for we can make $\varkappa =$ const. across the gap, thereby automatically obtaining $N_1(\varkappa) + 2N_2(\varkappa) =$ const. To achieve this, let us solve the differential equation

$$\varkappa = \omega'(\theta) \sin\theta = \text{const.}, \quad \omega\left(\frac{\pi}{2}\right) = 0. \tag{27.31}$$

The boundary condition $\omega(\pi/2) = 0$ implies that the flat plate (at $\theta = \pi/2$) is at rest. The solution to (27.31) is[14]

$$\omega(\theta) = \varkappa \ln(\sin\theta) - \varkappa \ln(1 + \cos\theta). \tag{27.32}$$

If we assume that θ lies in the range $(\pi/2) - \alpha_0 \leqq \theta \leqq (\pi/2)$, where α_0 is very small ($< 4°$), then

$$\omega\left(\frac{\pi}{2} - \alpha_0\right) = \Omega = \varkappa \ln(\cos\alpha_0) - \varkappa \ln(1 + \sin\alpha_0) \doteq -\varkappa\,\alpha_0, \tag{27.33}$$

so that for a small cone angle[15], the rate of shear is $|\varkappa| = \Omega/\alpha_0$. Note that we get that $\varkappa$ is negative in (27.33) because $\omega(\theta)$ decreases from Ω at $\theta = \frac{\pi}{2} - \alpha_0$ to 0 at $\theta = \pi/2$.

The conclusion that $\varkappa =$ const. means that $(d/d\theta)\, \mathbf{T}_E = \mathbf{0}$. Hence $(27.28)_3$

14. YIN and PIPKIN [1970 : 31], Eq. (16.6).

15. If the cone angle is not small, (27.32) is incompatible with $(27.28)_3$ under the condition that $\varkappa =$ const. See Eq. (29.15) and the remarks therein.

reduces to the trivial equation $0 \approx 0$ because $\cot\theta \approx 0$, while $(27.28)_{1-2}$ yield

$$p = p_0 - [N_1(\varkappa) + 2N_2(\varkappa)] \ln r, \tag{27.32}$$

where p_0 is a constant. Also

$$T\langle rr\rangle = -p_R + \int_R^r \frac{1}{\xi}(N_1 + 2N_2)\, d\xi = -p_R + (N_1 + 2N_2)\ln(r/R),$$

$$\begin{aligned} T\langle\theta\theta\rangle &= (T\langle\theta\theta\rangle - T\langle\phi\phi\rangle + (T\langle\phi\phi\rangle - T\langle rr\rangle) + T\langle rr\rangle \\ &= T\langle rr\rangle + N_2(\varkappa), \end{aligned} \tag{27.33}$$

which will be useful in Chap. 11.

The above set of five examples indicates the general methodology of solving problems in nonlinear continuum mechanics. However, the method is fruitful only in semi-inverse problems and cannot be applied to boundary value problems of wider complexity.

Appendix: Equations of Motion in Curvilinear Coordinates, and Acceleration Fields

The objective here is to list the equations of motion in the Cartesian, cylindrical polar and spherical polar coordinate systems. The general form of the equations of motion is given by (17.21), and we have used the technique of App. A, Chap. 1.

Cartesian Coordinate System

$$\begin{aligned} \frac{\partial T\langle xx\rangle}{\partial x} + \frac{\partial T\langle xy\rangle}{\partial y} + \frac{\partial T\langle xz\rangle}{\partial z} + \rho\, b_x &= \rho\, a_x, \\ \frac{\partial T\langle xy\rangle}{\partial x} + \frac{\partial T\langle yy\rangle}{\partial y} + \frac{\partial T\langle yz\rangle}{\partial z} + \rho\, b_y &= \rho\, a_y, \\ \frac{\partial T\langle xz\rangle}{\partial x} + \frac{\partial T\langle yz\rangle}{\partial y} + \frac{\partial T\langle zz\rangle}{\partial z} + \rho\, b_z &= \rho\, a_z, \end{aligned} \tag{A6.1}$$

where $\mathbf{b} = (b_x, b_y, b_z)$ is the body force,

$$a_x = \frac{\partial v_x}{\partial t} + \frac{\partial v_x}{\partial x} v_x + \frac{\partial v_x}{\partial y} v_y + \frac{\partial v_x}{\partial z} v_z, \tag{A6.2}$$

with a_y and a_z obtainable by permutation.

Cylindrical Polar Coordinate System

$$\begin{aligned} \frac{\partial T\langle rr\rangle}{\partial r} + \frac{1}{r}\frac{\partial T\langle r\theta\rangle}{\partial\theta} + \frac{\partial T\langle rz\rangle}{\partial z} + \frac{1}{r}(T\langle rr\rangle - T\langle\theta\theta\rangle) + \rho b\langle r\rangle &= \rho a\langle r\rangle, \\ \frac{\partial T\langle r\theta\rangle}{\partial r} + \frac{1}{r}\frac{\partial T\langle\theta\theta\rangle}{\partial\theta} + \frac{\partial T\langle\theta z\rangle}{\partial z} + \frac{2}{r}T\langle r\theta\rangle + \rho b\langle\theta\rangle &= \rho a\langle\theta\rangle, \\ \frac{\partial T\langle rz\rangle}{\partial r} + \frac{1}{r}\frac{\partial T\langle\theta z\rangle}{\partial\theta} + \frac{\partial T\langle zz\rangle}{\partial z} + \frac{1}{r}T\langle rz\rangle + \rho\, b\langle z\rangle &= \rho a\langle z\rangle, \end{aligned} \tag{A6.3}$$

where

$$a\langle r\rangle = \frac{\partial v\langle r\rangle}{\partial t} + \frac{\partial v\langle r\rangle}{\partial r} v\langle r\rangle + \frac{v\langle\theta\rangle}{r}\frac{\partial v\langle r\rangle}{\partial\theta} + \frac{\partial v\langle r\rangle}{\partial z} v\langle z\rangle - \frac{v^2\langle\theta\rangle}{r},$$

$$a\langle\theta\rangle = \frac{\partial v\langle\theta\rangle}{\partial t} + \frac{\partial v\langle\theta\rangle}{\partial r} v\langle r\rangle + \frac{v\langle\theta\rangle}{r}\frac{\partial v\langle\theta\rangle}{\partial\theta} + \frac{\partial v\langle\theta\rangle}{\partial z} v\langle z\rangle + \frac{v\langle r\rangle v\langle\theta\rangle}{r}, \tag{A6.4}$$

$$a\langle z\rangle = \frac{\partial v\langle z\rangle}{\partial t} + \frac{\partial v\langle z\rangle}{\partial r} v\langle r\rangle + \frac{v\langle\theta\rangle}{r}\frac{\partial v\langle z\rangle}{\partial\theta} + \frac{\partial v\langle z\rangle}{\partial z} v\langle z\rangle .$$

Spherical Polar Coordinate System

$$\frac{\partial T\langle rr\rangle}{\partial r} + \frac{1}{r}\frac{\partial T\langle r\theta\rangle}{\partial\theta} + \frac{1}{r\sin\theta}\frac{\partial T\langle r\phi\rangle}{\partial\phi} + \frac{1}{r}[2T\langle rr\rangle - T\langle\theta\theta\rangle - T\langle\phi\phi\rangle + \cot\theta\, T\langle r\theta\rangle] + \rho b\langle r\rangle = \rho a\langle r\rangle,$$

$$\frac{\partial T\langle r\theta\rangle}{\partial r} + \frac{1}{r}\frac{\partial T\langle\theta\theta\rangle}{\partial\theta} + \frac{1}{r\sin\theta}\frac{\partial T\langle\theta\phi\rangle}{\partial\phi} + \frac{1}{r}[3T\langle r\theta\rangle + \cot\theta\,(T\langle\theta\theta\rangle - T\langle\phi\phi\rangle)] + \rho b\langle\theta\rangle = \rho a\langle\theta\rangle, \tag{A6.5}$$

$$\frac{\partial T\langle r\phi\rangle}{\partial r} + \frac{1}{r}\frac{\partial T\langle\theta\phi\rangle}{\partial\theta} + \frac{1}{r\sin\theta}\frac{\partial T\langle\phi\phi\rangle}{\partial\phi} + \frac{1}{r}[3T\langle r\phi\rangle + 2\cot\theta\, T\langle\theta\phi\rangle] + \rho b\langle\phi\rangle = \rho a\langle\phi\rangle,$$

where

$$a\langle r\rangle = \frac{\partial v\langle r\rangle}{\partial t} + \frac{\partial v\langle r\rangle}{\partial r} v\langle r\rangle + \frac{v\langle\theta\rangle}{r}\frac{\partial v\langle r\rangle}{\partial\theta} + \frac{v\langle\phi\rangle}{r\sin\theta}\frac{\partial v\langle\phi\rangle}{\partial\phi} - \frac{v^2\langle\theta\rangle + v^2\langle\phi\rangle}{r},$$

$$a\langle\theta\rangle = \frac{\partial v\langle\theta\rangle}{\partial t} + \frac{\partial v\langle\theta\rangle}{\partial r} v\langle r\rangle + \frac{v\langle\theta\rangle}{r}\frac{\partial v\langle\theta\rangle}{\partial\theta} + \frac{v\langle\phi\rangle}{r\sin\theta}\frac{\partial v\langle\theta\rangle}{\partial\phi} + \frac{v\langle r\rangle v\langle\theta\rangle}{r} - \frac{v^2\langle\phi\rangle\cot\theta}{r}, \tag{A6.6}$$

$$a\langle\phi\rangle = \frac{\partial v\langle\phi\rangle}{\partial t} + \frac{\partial v\langle\phi\rangle}{\partial r} v\langle r\rangle + \frac{v\langle\theta\rangle}{r}\frac{\partial v\langle\phi\rangle}{\partial\theta} + \frac{v\langle\phi\rangle}{r\sin\theta}\frac{\partial v\langle\phi\rangle}{\partial\phi} + \frac{v\langle r\rangle\, v\langle\phi\rangle}{r} + \frac{v\langle\theta\rangle v\langle\phi\rangle\cot\theta}{r}.$$

For incompressible materials, $\mathbf{T} = -p\,\mathbf{1} + \mathbf{T}_E$, so that the shear stress components of $\mathbf{T}$ and $\mathbf{T}_E$ are identical; but the normal stress components of $\mathbf{T}$ and $\mathbf{T}_E$ differ by the pressure p. So, whenever necessary, one may write

$$\frac{\partial T\langle ij\rangle}{\partial x^k} = -\frac{\partial p}{\partial x^k}\delta_{ij} + \frac{\partial T_E\langle ij\rangle}{\partial x^k}. \tag{A6.7}$$

CHAPTER 7

DYNAMICS OF MOTIONS WITH CONSTANT STRETCH HISTORY

At an earlier place (Chap. 2), we have explored the kinematics of MWCSH in detail and classified them into four categories. The motivation here is to undertake a study of their dynamics, that is, to try to discover those satisfying the equations of motion. This search is not systematic except in the case of viscometric flows. As far as these are concerned, it can be said that the present summary records almost all dynamically possible flows. At an earlier place (Chap. 2), we have explored the kinematics of MWCSH in detail and classified them into four categories. The motivation here is to undertake a study of their dynamics, that is, to try to discover those satisfying the equations of motion. This search is not systematic except in the case of viscometric flows. As far as these are concerned, it can be said that the present summary records almost all the dynamically possible flows, and since the shear stress (or equivalently the viscosity) function plays a crucial role here, we begin with a section on its importance, then move onto 'partially controllable' motions and finally discuss rectilinear flows. In the remaining sections, we choose an example of MWCSH-II, the orthogonal rheometer flow and the simple extensional flow and examine their dynamic feasibility.

28. Role of the Shear Stress Function in Viscometric Flows

In this section, we shall examine a few viscometric flows—the helical flow and its special cases (Couette and Poiseuille flow) and channel flow. It will be assumed that as in the torsional and cone-and-plate flow of Sec. 27, the contravariant components of the velocity fields are functions of one coordinate only. In those two cases, the compatibility of the equations of motion decided the forms of the velocity fields [see (27.24) and (27.32)]. Here we shall find that the equations of motion will decide the stresses, especially the shear stresses and that the shear stress function yields the velocity field.

We shall begin by using the cone-and-plate flow to determine $\tau(\varkappa)$ as a function of $\varkappa$ and assume that $d\tau(x)/d\varkappa \neq 0$. Then for a given $\tau(\varkappa)$, there is a unique value of $\varkappa$ and vice versa; or, mathematically speaking, $\varkappa = \tau^{-1}[\tau(\varkappa)]$ exists, and is unique. This will then be exploited in the sequel.

28.1. Cone-and-Plate Flow. In Sec. 27.5 of Chap. 6, it was shown that in the cone-and-plate flow, the dynamical equations are satisfied if inertia is neglected and if the angular gap is so small that the rate of shear $\varkappa$ is uniform across the gap. Let the torque needed to keep the bottom plate stationary be given by M in this apparatus. Then

$$M = 2\pi \int_0^R \tau(\varkappa)\, r^2 dr, \tag{28.1}$$

and since $\varkappa$ is a constant,

$$\tau(\varkappa) = 3M/2\pi R^3, \tag{28.2}$$

where R is the wetted radius of the bottom plate. If the cone rotates with an angular velocity Ω, and the angular gap is α ($< 4^\circ$), then $\varkappa = \Omega/\alpha$ in this apparatus. Knowing $\tau(\varkappa)$ and $\varkappa$, a graph may be constructed, yielding a relation between $\tau(\varkappa)$ and $\varkappa$, or $\eta(\varkappa)$ and $\varkappa$.

In practice, a finite number of observations of $\tau(\varkappa)$ and $\varkappa$ are obtained. The best curve fit between $\tau(\varkappa)$ and $\varkappa$ may then be obtained by numerical methods.

28.2. Helical Flow. Since the helical flow includes as its special cases the COUETTE flow, POISEUILLE flow and the axial flow between an annulus, we shall discuss the helical flow in detail. The first, fully analytic, successful attack on this problem was made by COLEMAN and NOLL[1], though RIVLIN[2] had attempted a solution earlier and obtained two coupled nonlinear differential equations to determine the angular velocity field $\omega(r)$ and the axial velocity field $u(r)$.

In steady helical flow, the coordinate system is cylindrical polar and

$$\dot{r} = 0, \qquad \dot{\theta} = \omega(r), \qquad \dot{z} = u(r) \tag{28.3}$$

are the contravariant components of the velocity field. Using (26.61d) and (26.61e), two of the non-zero shear stresses are :

$$T\langle r\theta\rangle = \alpha\tau(\varkappa), \qquad T\langle rz\rangle = \beta\tau(\varkappa), \tag{28.4}$$

$$\alpha = r\omega'(r)/\varkappa, \qquad \beta = u'(r)/\varkappa, \qquad \varkappa^2 = r^2\omega'^2 + u'^2. \tag{28.5}$$

Let the inner cylinder ($r = R_1$) rotate with a constant angular velocity Ω and let the outer circular cylindrical tube ($r = R_2$) be at rest. Then the boundary conditions on $\omega(r)$ due to the 'no slip' condition are :

$$\omega(R_1) = \Omega, \qquad \omega(R_2) = 0. \tag{28.6}$$

By the same assumption of 'no-slip' at the boundary, the function $u(r)$ is subject to the conditions :

$$u(R_1) = u(R_2) = 0. \tag{28.7}$$

From Eq. (27.b3) we know that

$$T\langle r\theta\rangle = M/2\pi r^2, \qquad M = \text{const.} \tag{28.8}$$

1. [1959 : 2].
2. [1956 : 4].

We can interpret M as the moment per unit height needed to keep the inner cylinder in motion. Also, (27.b4) tells us that

$$T\langle rz\rangle = -\frac{1}{2}cr + br^{-1}, \tag{28.9}$$

where b is a constant and $c = -\partial\psi/\partial z$. If the z-axis is vertically downwards, then the vertically downward force F, acting on the planes $z = z_I$ and $z = z_{II}$ is given by $(z_{II} > z_I)$:

$$F = \int_{R_1}^{R_2}\left\{T\langle zz\rangle\Big|_{z=z_{II}} - T\langle zz\rangle\Big|_{z=z_I}\right\}\times$$

$$\times\, 2\pi r\, dr + \int_{R_1}^{R_2}\int_{z_I}^{z_{II}} \rho \cdot 2\pi r\, dz\, dr, \tag{28.10}$$

where the latter integral is the weight of the mass of fluid in the annulus between the two planes. A straightforward calculation shows that

$$F = c\pi\left(R_2^2 - R_1^2\right)\left(z_{II} - z_I\right), \tag{28.11}$$

and yields a physical meaning to c as the vertically downward force per unit volume of the fluid in the domain of the flow.

We now indicate how c may be measured. Use (27.b8) for $T\langle rr\rangle$ to obtain

$$T\langle rr\rangle\Big|_{\substack{r=R_2\\ z=z_{II}}} - T\langle rr\rangle\Big|_{\substack{r=R_2\\ z=z_I}} = p\Big|_{\substack{r=R_2\\ z=z_I}} - p\Big|_{\substack{r=R_2\\ z=z_{II}}}$$

$$= c(z_{II}-z_I) + \rho\left(\chi\Big|_{z=z_{II}} - \chi\Big|_{z=z_I}\right). \tag{28.12}$$

Since $b\langle z\rangle = g = -\dfrac{d\chi}{dz}$, (28.12) reads

$$T\langle rr\rangle\Big|_{\substack{r=R_2\\ z=z_{II}}} - T\langle rr\rangle\Big|_{\substack{r=R_2\\ z=z_I}} = (c-\rho g)(z_{II}-z_I). \tag{28.13}$$

Thus, a measurement of the difference of two normal stresses on the outer cylinder at $z = z_{II}$ and $z = z_I$ yields the constant c readily.

To obtain b, theoretically, one proceeds to define

$$f^2(r) = \left(\frac{M}{2\pi r^2}\right)^2 + \left(\frac{b}{r} - \frac{cr}{2}\right)^2. \tag{28.14}$$

But

$$f^2(r) = \tau^2(\varkappa), \tag{28.15}$$

because of (28.4)—(28.5) and (28.8)—(28.9). Hence

$$\varkappa = \tau^{-1}[f(r)], \tag{28.16}$$

where τ^{-1} is the inverse of the function τ. Using (28.4), (28.5) and (28.9), one has :

$$u'(r) = \frac{\varkappa}{\tau(\varkappa)} T\langle rz\rangle = \frac{\tau^{-1}[f(r)]}{f(r)}\left(\frac{b}{r} - \frac{cr}{2}\right). \tag{28.17}$$

Since $u(R_1) = u(R_2) = 0$, (28.17) yields

$$0 = \int_{R_1}^{R_2} \frac{\tau^{-1}[f(r)]}{f(r)}\left(\frac{b}{r} - \frac{cr}{2}\right) dr, \tag{28.18}$$

which is to be solved to find b. This is not a very easy task, though (28.18) tells us that b can be expressed as a function of R_1, R_2, M and c. One could derive another formula for finding b from (28.14) as follows :

$$\left[b\,\frac{R_2+R_1}{R_2R_1} - \frac{c}{2}(R_2+R_1)\right]\left[b\,\frac{(R_2-R_1)}{R_2R_1} + \frac{c(R_2-R_1)}{2}\right]$$
$$= \tau^2(\varkappa)\Big|_{r=R_1} - \tau^2(\varkappa)\Big|_{r=R_2} - \frac{M^2(R_2^2-R_1^2)}{4\pi^2 R_2^2 R_1^2}. \tag{28.19}$$

In this formula, one needs the shear stress $\tau(\varkappa)$ on the walls $r = R_1$ and $r = R_2$ to determine b. It would be interesting to check whether (28.18) or (28.19) is easier to work with.

If one could suspend either the inner cylinder or the outer tube and note that the axial force on the cylinder or the tube per unit length is

$$T\langle rz\rangle\Big|_{r=R_1} \cdot 2\pi R_1 \qquad \text{or} \qquad T\langle rz\rangle\Big|_{r=R_2} \cdot 2\pi R_2,$$

then measurement of either value yields b immediately, since c is known by other measurements. This idea does not seem to be practical, however.

Using (28.4)—(28.5) and (28.8)—(28.9), one can record the following formulae :

$$u(r) = \int_{R_1}^{r} \frac{\tau^{-1}[f(\xi)]}{f(\xi)}\left(\frac{b}{\xi} - \frac{c\xi}{2}\right) d\xi, \tag{28.20}$$

$$\omega(r) = \frac{M}{2\pi}\int_{R_1}^{r} \frac{\tau^{-1}[f(\xi)]}{f(\xi)\xi^3}\, d\xi + \Omega_1, \tag{28.21}$$

with

$$0 = \Omega_1 + \frac{M}{2\pi}\int_{R_1}^{R_2} \frac{\tau^{-1}[f(\xi)]}{f(\xi)\xi^3}\, d\xi. \tag{28.22}$$

Next, the volume rate of flow Q is given by

$$Q = 2\pi\int_{R_1}^{R_2} u(r)\, r dr$$
$$= \pi\Big[r^2u(r)\Big]_{R_1}^{R_2} - \pi\int_{R_1}^{R_2} r^2u'(r)\, dr$$
$$= \pi\int_{R_1}^{R_2} \frac{\tau^{-1}[f(r)]}{f(r)}\left(\frac{cr^3}{2} - br\right) dr. \tag{28.23}$$

If the inner cylinder vanishes, that is, $R_1 \to 0$, then M and b both tend to zero. If $M = 0$, and $R_1 \neq 0$, then $b \neq 0$. We now examine two special cases of helical flow.

(a) Couette Flow

In this special case of helical flow, $u(r) \equiv 0$ and the inner cylinder rotates with a steady angular velocity Ω. Hence,

$$\varkappa = r\omega'(r), \qquad \tau(\varkappa) = T\langle r\theta\rangle = M/2\pi r^2,$$

and

$$\omega(r) = \Omega + \int_{R_1}^{r} \frac{\tau^{-1}(M/2\pi\xi^2)}{\xi}\, d\xi$$

$$= \Omega + \frac{1}{2}\int_{M/2\pi r^2}^{M/2\pi R_1^2} \frac{\tau^{-1}(\zeta)}{\zeta}\, d\zeta. \qquad (28.24)^3$$

Hence

$$\Omega = \frac{1}{2}\int_{M/2\pi R_1^2}^{M/2\pi R_2^2} \frac{1}{\zeta}\, \tau^{-1}(\zeta)\, d\zeta. \qquad (28.25)$$

We shall now indicate how this equation (28.24) can be solved graphically. If a curve τ vs $\varkappa$ is known (see Fig. 28.1), then in the range $M/2\pi R_2^2 \leqq \tau(r) \leqq M/2\pi R_1^2$ one can find $\varkappa(r)$ such that $\varkappa(R_2) \leqq \varkappa(r) \leqq \varkappa(R_1)$. Knowing $\varkappa(r)$

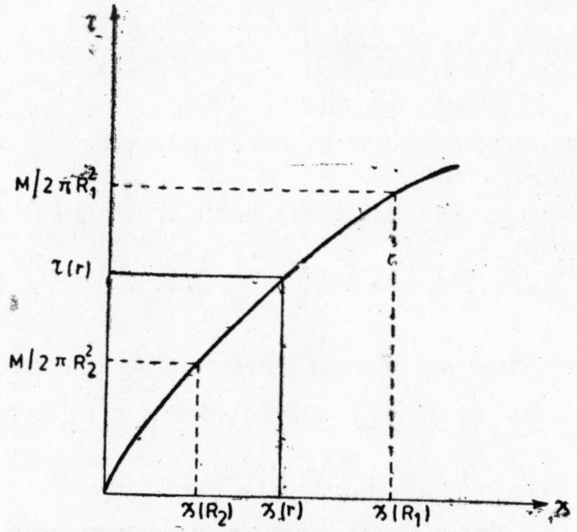

Fig. 28.1. Shear stress vs. shear rate.

$= r\, d\omega(r)/dr$, and using $\omega(R_1) = \Omega$ and $\omega(R_2) = 0$, one can perform numerical integration to find $\omega(r)$.

---

3. Coleman and Noll [1959 : 1]. An equivalent formula was derived earlier by Krieger and Elrod [1953 : 2]. The recent paper of Huang [1971 : 10] seems to be a re-derivation of these results. See Sec. 47 for the case of a rotating outer cylinder.

We now turn to the analytical approach of finding $\omega(r)$. Differentiation of (28.25) with respect to M gives

$$2M \frac{d\Omega(M)}{dM} = \tau^{-1}\left(\frac{M}{2\pi R_2^2}\right) - \tau^{-1}\left(\frac{M}{2\pi R_1^2}\right). \tag{28.26}$$

Eq. (28.26) is called the *Mooney equation.* Let

$$F(M) = 2M \frac{d\Omega(M)}{\alpha M}, \qquad \alpha = R_1^2 / R_2^2 . \tag{28.27}$$

Then, Eq. (28.26) becomes

$$F(M) = \tau^{-1}\left(\frac{\alpha M}{2\pi\, R_1^2}\right) - \tau^{-1}\left(\frac{M}{2\pi\, R_1^2}\right). \tag{28.28}$$

Replacing M by $\alpha^n M$ and summing from $n = 0$ to $n = N$, one has

$$\sum_{n=0}^{N} F(\alpha^n M) = \tau^{-1}\left(\frac{\alpha^{n+1}\, M}{2\pi\, R_1^2}\right) - \tau^{-1}\left(\frac{M}{2\pi\, R_1^2}\right). \tag{28.29}$$

Since $\alpha < 1$, $\lim_{n\to\infty} \alpha^{n+1} = 0$. Moreover τ^{-1} is continuous at $\varkappa = 0$ and $\tau^{-1}(0) = 0$, that is, the shear stress is zero if and only if the shear rate is zero. Hence

$$\tau^{-1}\left(\frac{M}{2\pi R_1^2}\right) = - \sum_{n=0}^{\infty} F(\alpha^n M). \tag{28.30}$$

In principle, we can obtain the rate of shear at the inner cylinder in the above manner, from which a curve between τ and $\varkappa$ may be obtained, and the latter used to find $\omega(r)$.

If $(R_2 - R_1)/R_1 \ll 1$, or the gap is very small, then (28.25) yields

$$\Omega = \frac{R_2 - R_1}{R_1} \cdot \tau^{-1}\left(\frac{M}{2\pi R_1^2}\right), \tag{28.31}$$

and this says that the rate of shear corresponding to a shear stress $M/2\pi R_1^2$ is $\Omega R_1/(R_2 - R_1)$. By varying Ω, one could obtain a curve between τ and $\varkappa$ quite readily.

(*b*) Poiseuille Flow

In this special case of helical flow, $\omega(r) \equiv 0$, and the fluid moves down a circular pipe ; or $R_1 = 0$, $R_2 = R$ and $u(R) = 0$ are the consequences of this flow description. Moreover, $b = 0$ in the formula (28.9) for $T\langle rz\rangle$, since the stress cannot be infinite at the center of the pipe.

Now, (28.9) tells us that

$$\varkappa = \frac{du}{dr} = \tau^{-1}\left(-\frac{1}{2}\, cr\right) = -\, \tau^{-1}\left(\frac{cr}{2}\right). \tag{28.32}$$

since $\tau(\varkappa)$ is an odd function of $\varkappa$. Thus

$$u(r) = \int_r^R \tau^{-1}\left(\frac{1}{2}c\zeta\right) d\zeta, \tag{28.33}$$[4]

and the volume flow rate Q [*cf.* (28.23)] becomes

$$Q = \pi \int_0^R \tau^{-1}\left(\frac{cr}{2}\right) r^2 dr = \frac{8\pi}{3}\int_0^{R/2} \zeta^2\tau^{-1}(\zeta)\, d\zeta. \tag{28.34}$$[4]

Differentiating (28.33) with respect to c leads to

$$\tau^{-1}\left(\frac{cR}{2}\right) = \frac{1}{\pi c^2 R^3}\frac{d}{dc}[c^3 Q(c)], \tag{28.35}$$

which yields $\varkappa$ as a function of c. Knowing this, a relation between τ and $\varkappa$ may be plotted, and used to find $u(r)$.

If $\tau(\varkappa) = \eta_0\varkappa$, where η_0 is a constant, then $\tau^{-1}(cr/2) = cr/2\eta_0$. Thus, for a fluid with constant viscosity

$$Q = \pi \int_0^R \frac{cr^3}{2\eta_0}\, dr = \frac{\pi c R^4}{8\eta_0}, \tag{28.36}$$

which is the famous HAGEN-POISEUILLE law of the incompressible Newtonian fluid theory. Indeed, we will now prove the converse that if $Q(c\,;R)$ is given by (28.36), then the viscosity must be a constant. Since (28.34) tells us that $Q = Q(c, R)$, let us rewrite it as

$$Q = Q(c, R) = \pi R^3 \int_0^1 \tau^{-1}\left(\frac{cR\xi}{2}\right)\xi^2 d\xi. \tag{28.37}$$

Consequently, if $Q \propto cR^4$, then $\tau^{-1}(cR\xi/2)$ must be linear in its argument, or,

$$\tau^{-1}\left(\frac{cR\xi}{2}\right) = \frac{1}{\eta_0}\frac{cR\xi}{2}, \qquad \eta_0 = \text{const.}, \tag{28.38}$$

and

$$\tau\left[\tau^{-1}\left(\frac{cR\xi}{2}\right)\right] = \frac{cR\xi}{2} = \tau\left(\frac{1}{\eta_0}\frac{cR\xi}{2}\right), \tag{28.39}$$

which implies that

$$\tau(\varkappa) = \eta_0\varkappa, \tag{28.40}$$

or the linearity of τ in $\varkappa$: or stated in words: *in the theory of simple fluids, the Hagen-Poiseuille formula* (28.36) *for efflux holds if and only if the viscosity function is constant.*[5]

28.3. Channel Flow[6]. In a Cartesian coordinate system, let the velocity

---

4. COLEMAN and NOLL [1959 : 1]. This formula is due to WEISSENBERG (*cf.* MARKOVITZ [1968 : 15]).

5. TRUESDELL [1962 : 11].

6. COLEMAN and NOLL [1959 : 1].

field be

$$\dot{x} = u(y), \qquad \dot{y} = \dot{z} = 0. \tag{28.41}$$

Let this rectilinear flow occur in a channel bounded by two pa allel, infinite plates (at rest) situated at $y = \pm h$.

It is trivial to verify that the modified pressures field $\psi = \psi(x, y)$ and [*cf.* (27.10)–(27.12)] :

$$\frac{\partial}{\partial y} T \langle xy \rangle - \frac{\partial \psi}{\partial x} = 0, \tag{28.42}$$

$$\frac{\partial}{\partial y} T_E \langle yy \rangle - \frac{\partial \psi}{\partial y} = 0.$$

Hence

$$\left.\begin{aligned} T\langle xy\rangle &= -cy + b, \\ \psi(x, y) &= -cx + T_E\langle yy\rangle + d, \end{aligned}\right\} \tag{28.43}$$

where b, c and d are constants. Now $\tau(\varkappa) = \tau(u')$ is an odd function of $\varkappa = u' = du/dy$ and thus $b = 0$ in (28.43). Hence

$$u'(y) = -\,\tau^{-1}(cy), \tag{28.44}$$

and using the adherence condition that $u(\pm h) = 0$, one obtains

$$u(y) = \int_h^y \tau^{-1}(c\xi)\, d\xi. \tag{28.45}$$

Here c is the modified pressure drop per unit length along the flow, and can be identified as the applied force per unit volume in the direction of the flow as well. The flow rate Q per unit width in the z-direction is given by :

$$\begin{aligned} Q = \int_{-h}^{h} u(y)\, dy &= -\int_{-h}^{h} yu'(y)\, dy = \int_{-h}^{h} y\,\tau^{-1}(cy)\, dy \\ &= \frac{2}{c^2}\int_0^{ch} \xi\,\tau^{-1}(\xi)\, d\xi. \end{aligned} \tag{28.46}$$

Since $Q = Q(c)$, we can differentiate $Q(c)$ with respect to c and obtain

$$\tau^{-1}(ch) = \frac{1}{2ch^2}\,\frac{d}{dc}\Big[\, c^2\, Q(c) \Big], \tag{28.47}$$

whence one can find the shear stress function if Q has been measured as a function of c.

REMARK. As we remarked at the beginning of this section, in Eqs. (28.20), (28.21), (28.24), (28.33) and (28.45), the shear stress function and its inverse played the crucial role in determining the velocity fields $\omega(r)$, $u(r)$ and $u(y)$. The normal stresses played no part at all. In each case, the normal stresses were such that in the equations of motion they were equivalent to the gradient of a scalar field. Such flows will be called *partially controllable*, following PIPKIN[7].

We shall make a detailed study of such velocity fields in the next section, where we shall admit motions with no inertia ($\rho = 0$) also into this class. In addition, we shall examine those flows (called *controllable*) in which the velocity fields are not determined by the shear stress function either, and which are fully compatible with the laws of dynamics in the presence of inertia.

29. Global Viscometric Flows

We have presented in Secs. 27 — 28 a few examples of dynamically possible viscometric flows. Some can be produced in the laboratory under the action of surface tractions in the presence of conservative body forces, while two can occur under zero body forces and zero inertia—in other words in a fluid with zero density. The reader will notice that no special forms of $N_1(\varkappa)$, $N_2(\varkappa)$ and $\tau(\varkappa)$ were used in deriving the dynamic feasibility of these flows, though $\tau(\varkappa)$ determined the velocity fields in some cases.

As PIPKIN[8] observed, the above flows are instructive because they reveal an interesting property : in each flow, the normal stresses are such that their divergence is the gradient of a scalar field P. To understand what is meant by this statement, let **a**, **b**, **c** represent the shear axes of a viscometric flow (Sec. 8), and let the constitutive equation be given by

$$\mathbf{T}_E = \bar{\eta}(\varkappa^2)\,\varkappa[\mathbf{a}\otimes\mathbf{b} + \mathbf{b}\otimes\mathbf{a}] - \bar{N}_1(\varkappa^2)\,\mathbf{b}\otimes\mathbf{b} - \\ - [\bar{N}_1(\varkappa^2) + \bar{N}_2(\varkappa^2)]\,\mathbf{c}\otimes\mathbf{c}. \tag{29.1}$$

By the existence of P, we mean that in a given flow for every choice of N_1 and N_2,

$$\operatorname{div}\{\bar{n}_1(\varkappa^2)\,\varkappa^2\mathbf{b}\otimes\mathbf{b} + [\bar{n}_1(\varkappa^2) + \bar{n}_2(\varkappa^2)]\,\varkappa^2\mathbf{c}\otimes\mathbf{c}\} = \nabla P. \tag{29.2}$$

We shall call such a flow *partially controllable.* The equation of motion now becomes

$$\rho\,\frac{d\mathbf{v}}{dt} = -\nabla(\psi + P) + \operatorname{div}\left\{\bar{\eta}(\varkappa^2)\,\varkappa[\mathbf{a}\otimes\mathbf{b} + \mathbf{b}\otimes\mathbf{a}]\right\}. \tag{29.3}$$

Then the *partially controllable* flow will be dynamically possible in a fluid with a density ρ and a viscosity function $\bar{\eta}(\varkappa^2)$ if (29.3) is satisfied. So follows the caution : *Partial controllability of a flow does not necessarily imply its dynamical feasibility.* Moreover, a dynamically feasible partially controllable flow may have a different velocity field from one fluid to another, depending on the viscosity function $\bar{\eta}(\varkappa^2)$, as the example of COUETTE flow suggests. Also, in the case of another partially controllable flow, viz., torsional flow, its dyna-

7. [1968 : 21].
8. [1968 : 21].

mic feasibility occurs under the condition $\rho = 0$. We have also got an example like simple shear (which is dynamically possible for $\rho \neq 0$) where the viscosity function does not determine the velocity field. So all these features of global viscometric flows have to be explored one by one. We shall summarize the deep results of PIPKIN[9] and YIN and PIPKIN[10] next, referring the reader to the original papers for full details.

29.1. Partial Controllability. In Sec. 8, we have summarized the conclusions of YIN and PIPKIN[11] and discussed some features of the **a**-, **b**-, **c**-lines. It is easy to see that demanding the left side of (29.2) to be irrotational for all n_1 and n_2 is to demand that the curl of the left side vanish. Hence one derives the following necessary and sufficient conditions[12, 13] on the shear rate $\varkappa$ and shear axes $\mathbf{e} = \mathbf{a}, \mathbf{b}$ and $\mathbf{c}$:

$$\begin{gathered}(\mathbf{e} \cdot \nabla\varkappa)\,(\mathbf{e}\times\nabla\varkappa) = \mathbf{0},\\ \nabla \times [\mathbf{e}\,(\mathbf{e} \cdot \nabla\varkappa)] + [\mathbf{e}\,(\nabla \cdot \mathbf{e}) + (\mathbf{e} \cdot \nabla)\mathbf{e}] \times \nabla\varkappa = \mathbf{0},\\ \nabla \times [\mathbf{e}\,(\nabla \cdot \mathbf{e}) + (\mathbf{e}\cdot\nabla)\mathbf{e}] = \mathbf{0}.\end{gathered} \tag{29.4}$$

Using these conditions and assuming $\nabla\varkappa \neq \mathbf{0}$, that is, a *non-uniform shear rate,* PIPKIN[14] showed that the $\varkappa =$ const. surfaces are either parallel planes or co-axial cylinders. Moreover, only the **b**-lines or **c**-lines are orthogonal to the $\varkappa =$ const. surfaces. So the $\varkappa =$ const. surfaces consist of either **a**- and **b**-lines or **a**- and **c**-lines. These lines form an orthogonal family of geodesics on $\varkappa =$ const. surface.

In addition, the slip surfaces which are formed by the **a**- and **c**-lines are *rigid* and *co-axial.* So there is a class of partially controllable viscometric flows with rigid, $\varkappa =$ const. surfaces, and there are other flows where $\varkappa =$ const. surfaces are not rigid. To illustrate these features, consider COUETTE flow. Here the **a**-line is azimuthal, the **b**-line radial and the **c**-line axial. The $\varkappa =$ const. surface $[\varkappa = r\, d\omega(r)/dr]$ is defined by a $r =$ const. (r = radial distance) surface. It coincides with the slip surface, and is hence rigid. In contrast, consider torsional flow. Here each $\varkappa =$ const. surface is a cylinder, but any vertical line on this surface will form a helix as the motion proceeds. So the $\varkappa =$ const. surface is not rigid and does not coincide with the slip surface which in this case is defined by $z =$ const.

Returning to the kinematical properties of partially controllable flows with non-uniform shear rate once again, we note that these flows are *steady with respect to some frame of reference.*[15] So PIPKIN's assumption[16] of this require-

9. [1968 : 21].
10. [1970 : 31].
11. [1970 : 31].
12. PIPKIN [1968 : 21], Eqs. (3.7)—(3.9).
13. YIN and PIPKIN [1970 : 31], Eqs. (10.4)—(10.6).
14. PIPKIN [1968 : 21].
15. YIN and PIPKIN [1970 : 31], Sec. 11.
16. [1968 : 21], pp. 87-88.

ment in his earlier analysis is redundant.

We now list *all partially controllable flows with a non-uniform shear rate*:

(i) Tangential sliding of parallel plane slip surfaces :

$$\mathbf{v} = u(y)\,\mathbf{i} + w(y)\,\mathbf{k}, \tag{29.5}$$

in a Cartesian coordinate system.

(ii) Axial translation, rotation and screw motions of co-axial circular cylindrical slip surfaces (e.g., helical flow). In cylindrical coordinates,

$$\mathbf{v} = r\omega(r)\,\mathbf{e}_\theta + u(r)\,\mathbf{e}_z. \tag{29.6}$$

(iii) Axial motion of fanned planes. In cylindrical coordinates,

$$\mathbf{v} = c\theta\,\mathbf{e}_z. \tag{29.7}$$

(iv) Screw motions of right helicoidal slip surfaces. In cylindrical coordinates,

$$\mathbf{v} = c(z - R\theta)\,(r\mathbf{e}_\theta + R\mathbf{e}_z). \tag{29.8}$$

The case $R = 0$ is the torsional flow.

If the *shear rate is uniform*, $\nabla\varkappa = \mathbf{0}$ everywhere and $(29.4)_{1-2}$ are satisfied trivially. Only $(29.4)_3$ remains to be exploited, along with the other kinematical properties of the **a**-, **b**- and **c**-lines. Using these, YIN and PIPKIN[17] showed that there are *only two partially controllable flows with uniform shear rate*. These are

(i) Rotation of conical slip surfaces about a common axis. In spherical coordinates :

$$\mathbf{v} = \varkappa r \sin\theta\,[\ln\sin\theta - \ln(1+\cos\theta)]\,\mathbf{e}_\phi\,, \tag{29.9}$$

encountered earlier in Sec. 27.

(ii) The motion with flexible slip surfaces defined by the motion (in Sec. 8):

$$\mathbf{x}(\mathbf{X}, t) = r_0(1 + \varkappa^2 t^2)^{-1}\,[\mathbf{a}(\alpha) - t\,\varkappa\,\mathbf{b}(\alpha)] + z_0\mathbf{k}. \tag{29.10}$$

As we have observed in Sec. 13, this motion is *intrinsically unsteady*.

29.2. Dynamic Feasibility. Whether the above six velocity fields are dynamically possible depends on : (i) whether the divergence of the shear stress in (29.3) is equivalent to the gradient of another scalar field for a given choice of $\bar{\eta}(\varkappa^2)$, and (ii) whether the acceleration field is irrotational or not.

Taking up (i) above, one can show that in the list (29.5) – (29.10), the flows in which the shear stress contributions are always 'irrotational' are given by:

(*a*) Steady simple shearing [*cf.* (29.5)]:

$$\mathbf{v} = \varkappa y\,\mathbf{i}. \tag{29.11}$$

(*b*) COUETTE flow with uniform shear rate [*cf.* (29.6)]:

$$\mathbf{v} = \varkappa r \ln(r/R)\,\mathbf{e}_\theta. \tag{29.12}$$

17. [1970 : 31].

(*c*) Axial motion of fanned planes (29.7).
(*d*) The helicoidal flow (29.8).
(*e*) The motion with flexible slip surfaces (29.10).

In the above list, flows (*a*) (*b*) and (*c*) have *irrotational acceleration fields.* So they are possible in every incompressible simple fluid, or are *completely controllable.* The flows (*d*) and (*e*) do not have such acceleration fields and hence can be maintained in a fluid with zero density, that is, $\rho = 0$.

What should be emphasised is that the velocity fields of the flows (*a*) to (*e*) above are *not* determined by the viscosity function $\bar{\eta}(\varkappa^2)$. Indeed, in the list (29.5) to (29.10), only (29.5) and (29.6) vary from one fluid to another. Taking up (29.5) first, it can be shown that

$$u(y) = a \int_{y_0}^{y} \frac{(y' - y_0)}{\tilde{\eta}(y')} dy' + u_0, \quad w(y) = b \int_{y_0}^{y} \frac{dy'}{\tilde{\eta}(y')} + w_0, \qquad (29.13)$$

where the equation

$$[\varkappa\bar{\eta}(\varkappa^2)]^2 = b^2 + a^2 (y - y_0^2), \qquad a, b = \text{const.}, \qquad (29.14)$$

determines $\eta = \tilde{\eta}(y)$, if $\bar{\eta}(\varkappa^2)$ is given. Concerning (29.6), it can be seen that if the domain is such that $0 \leqq \theta \leqq 2\pi$, then we have the helical flow of Sec. 28.2. If $0 \leqq \theta < 2\pi$, $u(r) = 0$, and $\omega(R_1) = \omega(R_2) = 0$, then one obtains the 'omega flow' of LOBO and OSMERS,[18] which has been used recently to measure N_1 at high rates of shear. Generally speaking, let $\partial\psi/\partial\theta = C$, and $\partial\psi/\partial z = D$. Then in (29.6),

$$\omega(r) = \int_R^r \varkappa(r') \frac{\sin\alpha\,(r')}{r'} dr' + \Omega_1, \quad u(r) = \int_R^r \varkappa(r') \cos\alpha\,(r')\, dr' + u_1, \qquad (29.15)$$

where

$$4[\varkappa\bar{\eta}(\varkappa^2)]^2 = (C + Er^{-2})^2 + (Dr + Fr^{-1})^2, \qquad (29.16)$$
$$2r^2 \varkappa\eta \sin\alpha = Cr^2 + E, \qquad 2r\varkappa\eta \cos\alpha = Dr^2 + F,$$

and C, D, E and F are constants. Here $(29.16)_1$ determines $\varkappa = \varkappa(r)$, while $(29.16)_{2,3}$ determine $\sin\alpha\,(r)$ and $\cos\alpha\,(r)$ respectively, and these can be used in (29.15) to find $\omega(r)$ and $u(r)$. We can identify E with M/π, D with $-c$ and F with $2b$ in (28.14), if (29.15) – (29.16) are to be compared with (28.20) – (28.22).

We shall refer the reader to YIN's thesis[19] to find the types of translations and rotations that can be added to the dynamically feasible viscometric flows (*a*) to (*e*) above to leave them dynamically feasible and viscometric. However, a partial list will be given here, from PIPKIN's work referred to above :

(i) Simple shearing flow (29.11) can be altered to

$$\mathbf{v} = u_0\mathbf{i} + v_0\mathbf{j} + w_0\mathbf{k} + \varkappa y\mathbf{i}. \qquad (29.17)$$

18. [1974 : *B*].
19. [1969 : 24].

(ii) Each one of the velocity fields (29.6) – (29.8) can be changed by an additional amount $\Omega\, r\, \mathbf{e}_\theta + U\mathbf{e}_z$.

(iii) Tangential sliding (29.5) can be altered to

$$\mathbf{v} = A\,\mathbf{i} + C\,\mathbf{k} + u(y)\,\mathbf{i} + w(y)\,\mathbf{k}. \tag{29.18}$$

The reader should consult Sec. 13 and convince himeself that the flows remain viscometric, after these additions. He should also check their dynamic feasibility.

One final remark: the flow (29.9) between conical surfaces is partially controllable, and has a uniform shear rate. Despite this condition, and despite the assumption that $\rho = 0$, *it is not dynamically feasible* unless one assumes that the cone angle $\alpha_0 = 0$. For, as we saw in Sec. 27.5, Eqs. $(27.28)_{1-2}$ are satisfied under the above conditions, but $(27.28)_3$ becomes $\partial p/\partial\varphi = 0$, and

$$\tau = A/\sin^2\theta, \tag{29.19}$$

where A is a constant, or τ is a function θ, and hence $\varkappa$ is, but $\varkappa$ has to be uniform, which is a contradiction. So θ = const., or the cone angle α_0 is zero. This explains the reason behind the additional assumption in Sec. 27.5 that α_0 is very small.

30. Uni-directional Rectilinear Motion of Simple Fluids

As we have just seen, the problem of rectilinear motion (motion with straight streamlines) of an incompressible simple fluid has been solved, subject to *partial controllability*. It was solved actually by PIPKIN[20], though YIN and PIPKIN's results[21] show that there are no rectilinear, partially controllable motions in addition to:

(i) fanned motion of axial planes, defined through (A, U = const.):

$$\mathbf{v} = (A\theta + U)\,\mathbf{e}_z, \quad 0 \leqq \theta < 2\pi, \quad r > 0, \tag{30.1}$$

in a cylindrical polar coordinate system;

(ii) rectilinear motion with concentric slip surfaces given by (U = const.):

$$\mathbf{v} = \left[U + \int_R^r \varkappa(\xi)\, d(\xi)\right]\mathbf{e}_z, \tag{30.2}$$

in a cylindrical polar coordinate system;

(iii) Tangential sliding of parallel plane slip surfaces, given in Cartesian coordinates by (A, C = const):

$$\mathbf{v} = u(y)\,\mathbf{i} + w(y)\,\mathbf{k} + A\mathbf{i} + C\mathbf{k}. \tag{30.3}$$

20. [1968 : 21].
21. [1970 : 31].

We can read off a number of instructive points from the above three flows:

(*a*) Rectilinear motion of an incompressible simple fluid occurs over a one-parameter family of slip surfaces. They may be parallel planes, concentric cylinders, or 'fanned planes'.

(*b*) On each slip surface, the speed is a constant[22].

(*c*) The flow (30.1) and the special case of (30.3) given by

$$\mathbf{v} = (A + \varkappa_0 y)\,\mathbf{i} + C\mathbf{k}, \qquad \varkappa_0 = \text{const.}, \tag{30.4}$$

occur under zero pressure gradient;

(*d*) However, the rate of shear is a non-uniform $\varkappa = -A/r$ in (30.1), while it is a constant $\varkappa_0$ in (30.4);

(*e*) If we examine the case of POISEUILLE flow, which is included in (30.2), then the rate of shear vanishes in the domain of the flow. The domain of the flow is bounded in the $z =$ const. plane in this case;

(*f*) The channel flow, contained in (30.3), also has a point at which the shear rate vanishes, but the cross-section normal to the flow is infinite;

(*g*) Finally, in Sec. 29 it was assumed that $\bar{\eta}(\varkappa^2)$ and $\bar{n}_2(\varkappa^2)$ were analytic functions of $\varkappa^2$ in the flow domain. Without this assumption, the conditions (29.4) cannot be derived, and so the whole analysis would break down at that point.

Having summarized the available information on rectilinear motions let us examine (ab initio) how to set up the problem of uni-directional rectilinear motion, and find the domain(s) in which it will occur. This question has recently been resolved by FOSDICK and SERRIN[23], who subjected the problem to quasi-linear elliptic theory and proved the uniqueness of PIPKIN's results relevant to this problem, along with providing counter examples for cases where one would wish to drop the requirements of analyticity of $\bar{n}_2(\varkappa^2)$, the vanishing of $\varkappa$ in the domain, etc. We shall return to this task shortly after setting up the relevant equations.

Consider a Cartesian coordinate system and a uni-directional velocity field given by

$$v_1 = v_2 = 0, \qquad v_3 = w(x, y), \tag{30.5}$$

in a bounded or unbounded domain Ω. It is trivial to verify that (30.5) is a viscometric flow and for an incompressible simple fluid, the extra stresses corresponding to (30.5) are given by [*cf.* (6.65)]:

$$T\langle xy\rangle = \bar{\eta}(\varkappa^2)\,\frac{\partial w}{\partial x}, \quad T\,\langle yz\rangle = \bar{\eta}(\varkappa^2)\frac{\partial w}{\partial y}, \tag{30.6}$$

22. The motion of fanned planes disproves the conjecture by TRUESDELL and NOLL, [in NFTM, ff. (117.22)], that lines of constant speeds are concentric circles or *parallel* straight lines. The word parallel is not in ERICKSEN's [1956 : 1] paper, which they quote.

23. [1973 : 1]. MAYNE [1968, 17] had attempted to solve this problem earlier but his analysis is incomplete. Neither of these two attempts makes the assumption of partial controllability, that is, the non-intervention of the normal stress in determining the velocity field.

$$T_E\langle xx\rangle = \bar{n}_2(\varkappa^2)\left(\frac{\partial w}{\partial x}\right)^2 - \frac{N_1 + 2N_2}{3},$$

$$T\langle xy\rangle = \bar{n}_2(\varkappa^2)\,\frac{\partial w}{\partial x}\,\frac{\partial w}{\partial y}, \tag{30.7}$$

$$T_E\langle yy\rangle = \bar{n}_2(\varkappa^2)\left(\frac{\partial w}{\partial y}\right)^2 - \frac{N_1 + 2N_2}{3}, \quad \varkappa^2 = \left(\frac{\partial w}{\partial x}\right)^2 + \left(\frac{\partial w}{\partial y}\right)^2. \tag{30.8}$$

Since the acceleration field is zero, the equations of motion yield the modified pressure ψ as:

$$\psi = 2cz + h(x, y). \tag{30.9}$$

The equation in the z-direction, viz.,

$$\frac{\partial\psi}{\partial z} = \frac{\partial\langle Txz\rangle}{\partial x} + \frac{\partial T\langle yz\rangle}{\partial y}, \tag{30.10}$$

leads to

$$\operatorname{div}\,(\bar{\eta}\,(\varkappa^2)\,Dw) = 2c, \quad Dw = \frac{\partial w}{\partial x}\,\mathbf{i} + \frac{\partial w}{\partial y}\,\mathbf{j}. \tag{30.11}$$ [24]

Assuming that $\chi = -gz$, the equation of motion in the x-direction is given by

$$\frac{\partial\psi}{\partial x} = \frac{\partial T_E\langle xx\rangle}{\partial x} + \frac{\partial T\langle xy\rangle}{\partial y}, \tag{30.12}$$

and this can be cast in the form

$$\frac{\partial w}{\partial x}\operatorname{div}\,(\bar{n}_2\,(\varkappa^2)\,Dw) = \frac{\partial p}{\partial x} - \bar{\eta}_2(\varkappa^2)\,\varkappa\,\frac{\partial\varkappa}{\partial x} + \\ + \frac{1}{3}\,\frac{\partial}{\partial x}[\bar{N}_1\,(\varkappa^2) + 2\bar{N}_2\,(\varkappa^2)]. \tag{30.13}$$

Similarly, the third equation leads to

$$\frac{\partial w}{\partial y}\operatorname{div}\,(\bar{n}_2\,(\varkappa^2)\,Dw) = \frac{\partial p}{\partial y} - \bar{n}_2\,(\varkappa^2)\,\varkappa\,\frac{\partial\varkappa}{\partial y} + \\ + \frac{1}{3}\,\frac{\partial}{\partial y}\,[\bar{N}_1\,(\varkappa^2) + 2\bar{N}_2\,(\varkappa^2)]. \tag{30.14}$$

Consequently, (30.13) and (30.14) are equivalent to

$$Dw\,\operatorname{div}\,(\bar{n}_2\,(\varkappa^2)\,Dw) = Df, \tag{30.15}$$

24. The equations (30.11) and (30.15) occur in CRIMINALE, *et al.* [1958 : 1], OLDROYD [1965 : 13], NFTM, (117.9)–(117.20), MAYNE [1968 : 17], PIPKIN [1969 : 18] and FOSDICK and SERRIN [1973 : 1]. We follow PIPKIN [1969 : 18]. In this section, D is the *two-dimensional operator, and* div *is also two-dimensional.*

whence

$$p(x, y, z) = (2c + \rho g)z + f(x, y) + \int_0^{\varkappa} n_2(\xi^2)\, \xi d\xi - \\ - \frac{1}{3} [\, \overline{N}_1(\varkappa^2) + 2\overline{N}_2(\varkappa^2)\,]. \qquad (30.16)$$

Thus we have to consider both (30.11) and (30.15) simultaneously, and they must be compatible if uni-directional rectilinear motion is to occur. In other words, let a fluid with prescribed properties $\bar{\eta}(\varkappa^2)$ and $\bar{n}_2(\varkappa^2)$ be given. Given c, the uniform pressure gradient, let us solve (30.11) and find $w(x, y)$. Then for this motion to be possible, we should be able to determine $f(x, y)$ from (30.15). On the other hand, given $f(x, y)$, one should be able to obtain $w(x, y)$ from (30.15) and satisfy (30.11). So we have two essentially independent equations to find $w(x, y)$. If we want the two equations to coalesce into one, we could demand *partial controllability*, which would turn the normal stress terms in the equations of motion into the gradient of a scalar P and yield us the types of streamlines that we are likely to encounter. Use of this in (30.11), gives us (30.1), (30.2), (30.3) with $A = u(y) = 0$, and (30.4).

Secondly, we could link $\bar{n}_2(\varkappa^2)$ to $\bar{\eta}(\varkappa^2)$ by demanding that $\bar{n}_2(\varkappa^2)$ be proportional to $\bar{\eta}(\varkappa^2)$. For example, let $\bar{\eta}(\varkappa^2) = \eta_0$, a constant and $\bar{n}_2(\varkappa^2) = \alpha$, another constant. Then $(30.11)_1$ reduces to $\eta_0 \Delta w = 2c$, $\Delta = (\partial^2/\partial x^2) + (\partial^2/\partial y^2)$, which is POISSON's equation in two dimensions and for which solutions exist in many bounded and unbounded domains, e.g. elliptical pipes. For such a fluid, (30.15) leads to

$$\alpha(Dw)\, \Delta w = Df = \alpha D(w\, \Delta w), \qquad (30.17)$$

implying that $f = \alpha\, w\Delta w$. So uni-directional rectilinear motion is possible in many domains for such a fluid[25]. Now, let us consider the more general case with $\bar{n}_2(\varkappa^2) = \alpha\bar{\eta}(\varkappa^2)$, where α is a constant. Then (30.15) becomes

$$Df = \alpha\, Dw \operatorname{div} (\bar{\eta}(\varkappa^2)\, Dw) = 2\alpha c\, Dw, \qquad (30.18)$$

or, to within a constant,

$$f(x, y) = 2\alpha c w\, (x, y) \qquad (30.19)$$

is the solution. Substitution of (30.19) into (30.16) leads to

$$p = (2c+\rho g)\, z + 2\alpha c w\, (x, y) + \alpha \int_0^{\varkappa} \bar{\eta}(\xi^2)\, \xi d\xi - \\ - \frac{1}{3}\Big[\, \overline{N}_1(\varkappa^2) + 2\alpha\bar{\eta}(\varkappa^2)\, \varkappa^2 \Big]. \qquad (30.20)$$

Now, had we used (26.66) instead of (26.65) in (30.5)—(30.8), we would have obtained

$$p = (2c+\rho g)\, z + 2\alpha c w(x, y) + \alpha \int_0^{\varkappa} \bar{\eta}(\xi^2)\, \xi d\xi. \qquad (30.21)^{26}$$

25. This idea is due to ERICKSEN [1956 : 1].

26. This equation, without the body force term, occurs in PIPKIN [1969 : 18]. See Eq. (3.6.9). In [1970 : 22], TANNER [see Eqs. (38)—(45)] has quoted this result of PIPKIN, when the pressure gradient $c = 0$ and inertia vanishes, that is, $\rho = 0$.

So, again, axially rectilinear motion is possible for such fluids in pipes of arbitrary cross-section. *In all cases where* $\bar{n}_2(\varkappa^2) = \alpha\bar{\eta}(\varkappa^2)$, $\eta(\cdot)$ *determines* $w(x, y)$. If $\bar{n}_2(\varkappa^2)$ is not proportional to $\bar{\eta}(\varkappa^2)$ then, (30.20) or (30.21) is not valid and the assumed motion may not be possible.

Indeed, ERICKSEN[27] showed from partial differential equation theory that there exist analytic functions $\bar{\eta}(\varkappa^2)$ and $\bar{n}_2(\varkappa^2)$ which make (30.11) and (30.15) independent of each other so that the solution $w(x, y)$ obtained from one need not satisfy the other. He conjectured that unless $\bar{n}_2(\varkappa^2)$ was proportional to $\bar{\eta}(\varkappa^2)$, only exceptional rectilinear motions occur—these being 'either rigid or such that the curves of constant speed are circles or straight lines'.

Immediately on the publication of this note, GREEN and RIVLIN[28] considered the problem of rectilinear flow through a pipe of elliptical cross-section for a fluid with $\bar{n}_1 = 0$ and $\bar{n}_2(\varkappa^2) = \alpha\bar{\eta}(\varkappa^2) + \delta\tilde{n}_2(\varkappa^2)$. Expanding in the perturbation parameter δ, these authors found that the flow is indeed non-rectilinear under a uniform pressure gradient, with a rectilinear flow pattern, superposed on four similar vortex-like flows in planes normal to the length of the tube (see Fig. 44.2). So one may thus assume that ERICKSEN's conjecture is true.

However, as FOSDICK and SERRIN have demonstrated, the situation is not so straightforward. For example, let

$$\bar{\eta} \equiv 1, \qquad \bar{n}_2(\varkappa^2) = K\varkappa^2 + L, \tag{30.22}$$

where K and L are constants. Then (30.11) and (30.15) admit the solution $c = 0$, and

$$w(x, y) = A(x^2 - y^2), \qquad f(x, y) = 8KA^2w^2(x, y), \tag{30.23}$$

where A is another constant. The flow domain Ω is unbounded, the curves of constant speeds are rectangular hyperbolae, and $Dw = 0$ at $(0, 0)$. So this disproves ERICKSEN's conjecture, or at least shows that the conjecture is not valid in unbounded domains.

As FOSDICK and SERRIN observed, in uni-directional rectilinear flow, one could have :

(i) $Dw = 0$ somewhere in Ω ;
(ii) the boundary $\partial\Omega$ may or may not be fixed ;[29]
(iii) Ω may be bounded or unbounded ;
(iv) the fluid may adhere to the boundary or not ;[29]
(v) the pressure gradient c in (30.11) may or may not vanish ;
(vi) for a constant viscosity fluid, the function $\bar{n}_2(\varkappa^2)$ may not be an analytic function of $\varkappa$; however, it may be differentiable in the flow domain considered.

---

27. [1956 : 1].
28. [1956 : 2]. They considered the problem for a REINER-RIVLIN fluid, for which $\mathbf{T}_E = \beta_1\mathbf{A}_1 + \beta_2\mathbf{A}_1^2$ where β_1 and β_2 are function of tr $\mathbf{A}_1^2$ and tr $\mathbf{A}_1^3$.
29. However, we shall assume that $\partial\Omega$ is fixed and that $w = 0$ on $\partial\Omega$.

We shall now summarize the findings of this work.

First of all, we shall show that (30.11) is a quasi-linear elliptic equation under some physically reasonable assumptions:

(i) the domain Ω is a connected open set; it may be bounded or unbounded, but it cannot be, for example, two disjointed pipes;

(ii) $\bar{\eta}$ and $\bar{n}_2$ are analytic functions of $\varkappa^2$ near $\varkappa = 0$;[30]

(iii) $\bar{\eta}(\varkappa^2)$ is of class C^2 in $\varkappa$ for all $\varkappa \geqq 0$;

(iv) $\tau(\varkappa) = \varkappa\bar{\eta}(\varkappa^2)$ is an increasing function of $\varkappa$, that is, $d\tau(\varkappa)/d\varkappa > 0$ for all $\varkappa \geqq 0$.

Now, carrying through the differentiation in (30.11) leads to

$$\eta \Delta w + 2\eta' w_\alpha\, w_\beta\, w_{\alpha\beta} = 2c, \qquad \alpha, \beta = 1, 2, \tag{30.24}$$

where $w_\alpha = \partial w/\partial x_\alpha$, $\eta' = d\bar{\eta}(\varkappa^2)/d\varkappa^2$. The characteristic quadratic form Q associated with (30.24) is

$$\begin{aligned} Q &= \eta\xi^2 + 2\eta' w_\alpha\, w_\beta\, \xi_\beta\, \xi_\beta \\ &= \eta\xi^2 + 2\eta'(w_1\xi_1 + w_2\xi_2)^2, \end{aligned} \tag{30.25}$$

where $\boldsymbol{\xi} = (\xi_1, \xi_2)$ is any non-vanishing vector, and $\xi = (\xi_1^2 + \xi_2^2)^{1/2}$. Since

$$(w_1\xi_1 + w_2\xi_2)^2 = (Dw \cdot \boldsymbol{\xi})^2 = \theta\varkappa^2\xi^2, \tag{30.26}$$

for some θ in $0 \leqq \theta \leqq 1$, (30.25) becomes

$$Q = \left[(1-\theta)\eta + \frac{d}{d\varkappa}(\eta\varkappa) \right] \xi^2. \tag{30.27}$$

By assumption (iv) above, $(d/d\varkappa)(\eta\varkappa) > 0$, which incidentally implies that $\eta > 0$. This result is stronger than the condition $\eta \geqq 0$ derived earlier by the non-negativity of stress power [see ff. (25.11)]. Thus $Q > 0$, which proves that the partial differential equation (30.24) is elliptic; it is quasi-linear because the coefficients of Δw and $w_{\alpha\beta}$ are functions of w_α. So any solution of (30.24) is *analytic* in Ω.

By the maximum principle[31] for such equations, if $w = 0$ on $\partial\Omega$ (that is, the fluid obeys the adherence condition and the boundary is fixed), then the solution is *unique*. Other consequences of this principle are: (i) if $c = 0$, then $w = 0$; (ii) if $c \neq 0$, $w = 0$ on $\partial\Omega$, then there must be some point in Ω at which $Dw = \mathbf{0}$; if $c < 0$, $w > 0$ in Ω and vice versa.

By exploiting the analyticity of the solution, and the equations (30.11) and (30.15) adoritly, FOSDICK and SERRIN proved

THEOREM 30.1. *If* $\bar{\eta}(\varkappa^2)$, Ω, *and* $n_2(\varkappa^2)$ *obey the conditions listed earlier,* $n_2(\varkappa^2) \neq$ const., $\bar{\eta}(\varkappa^2)$ *near* $\varkappa = 0$, *and if there is one point in* Ω *where* $Dw = 0$, *then* $w(x, y)$ *is radially symmetric, that is,* $w = w(\sqrt{x^2+y^2})$, *or plane symmetric, that is,* $w = w(x)$, *provided* $c \neq 0$.

---

30. This condition can be relaxed. See footnote on p. 317 of FOSDICK and SERRIN [1973 : 1].

31. COURANT and HILBERT [1962 : 5], pp. 322-324. For a delightful introduction to maximum principles, see PROTTER and WEINBERGER [1967 : 21].

We shall now examine the consequences of relaxing some of the conditions of the theorem. An example has already been listed in (30.22) which shows that the condition $c \neq 0$ is crucial. At least in that flow $Dw = \mathbf{0}$, at (0, 0). Suppose we now drop the condition $Dw = \mathbf{0}$ somewhere. Then we get simple shearing flow with $c = 0$ and $Dw \neq \mathbf{0}$ anywhere; this situation occurs in the axial motion of fanned planes also. Suppose we drop the requirement of analyticity of $n_2(\varkappa^2)$, but retain $c \neq 0$. Then, for the functions

$$\eta = 1, \qquad \bar{n}_2(\varkappa^2) = \frac{K}{\sqrt{\varkappa^2 - b}} + L, \tag{30.28}$$

one can find a rectilinear flow of the form

$$w(x, y) = Cx^2 + by, \qquad f = 2cLw. \tag{30.29}$$

The flow domain Ω is to be restricted to $x \neq 0$, with proper restrictions on x and b to ensure that $\bar{n}_2(\cdot)$ is differentiable in $|x| \geq \varepsilon$, $\varepsilon > 0$.

To sum up, all the conditions listed in the theorem are crucial. If anyone of them is relaxed, some kind of uni-directional rectilinear flow may occur for some fluid or the other, but this will not be a pipe flow in the usual sense of this term.

What the theorem says regarding pipe flow is this : if Ω is bounded and connected, $c \neq 0$, $w = 0$ on[32] $\partial\Omega$, and $n_2 \neq \alpha\eta$ near $\varkappa = 0$, then rectilinear axial flow will occur only in a circular pipe or in the annulus between two concentric cylinders. If Ω is unbounded, then channel flow will occur.[33]

In conclusion, it must be pointed out that the approaches of PIPKIN and that of FOSDICK and SERRIN are different. Had the latter authors assumed partial controllability, then conditions analogous to (29.4) would have arisen and the problem would have been solved along the lines of Sec. 29. However, these two authors assume the analyticity of $\bar{n}_2(\varkappa^2)$, rather than that this normal stress difference does not determine the velocity field. A consequence of this assumption is that their conclusion is valid only in the $\varkappa$-domain admitting the power series expansions of $\bar{\eta}(\varkappa^2)$ and $\bar{n}_2(\varkappa^2)$, and not outside. However, the procedure followed by PIPKIN does not suffer from this limitation.

31. Helical-Torsional Flow

In Sec. 8, a few of the kinematically possible MWCSH-II have been listed. The number is fairly small since the velocity field must depend on two spatial coordinates, be linear in, or be proportional to one of the coordinates, and the components of the metric tensor cannot change along the path

32. Note that the ellipticity of the equation (30.24) and $w = 0$ on $\partial\Omega$ imply that $Dw = \mathbf{0}$ somewhere in Ω.

33. MAYNE [1968 : 17] reached similar conclusions, but his analysis is not as thorough as that of FOSDICK and SERRIN [1973 : 1].

line of each particle. Under these restrictions, it seems that one is confined either to Cartesian coordinates where

$$\dot{x} = 0, \qquad \dot{y} = v(x), \qquad \dot{z} = w(y, x) \tag{31.1}$$

is the velocity field, with $w(y, x)$ linear in y, that is, $w(y, x) = \alpha y + h(x)$, or $w(y, x) = yh(x)$; or confined to cylindrical polar coordinates where the velocity field is

$$\dot{r} = 0, \qquad \dot{\theta} = \omega(r), \qquad \dot{z} = u(\theta, r), \tag{31.2}$$

with $u(\theta, r) = \alpha\theta + h(r)$ or $u(\theta, r) = \theta h(r)$. Of course, complementary to (31.2) is the following:

$$\dot{r} = 0, \qquad \dot{\theta} = \omega(r, z), \qquad \dot{z} = u(r), \tag{31.3}$$

$$\omega(r, z) = cz + h(r), \qquad \text{or} \qquad \omega(r, z) = zh(r).$$

It is the helical-torsional flow, viz.,

$$\dot{r} = 0, \qquad \dot{\theta} = \omega(r) + ez, \qquad \dot{z} = u(r), \quad e = \text{const.}, \tag{31.4}$$

which we shall consider here in detail.[34] For this velocity field one can obtain the physical components of the strain history through $\mathbf{C}_t(t-s) = e^{-s\mathbf{L}_1^T} e^{-s\mathbf{L}_1}$, where $\mathbf{L}_1$ has the matrix form

$$[L_1\langle ij\rangle] = \begin{bmatrix} 0 & 0 & 0 \\ r\omega' & 0 & er \\ u' & 0 & 0 \end{bmatrix}, \qquad \omega' = \frac{d\omega}{dr}, \qquad u' = \frac{du}{dr}. \tag{31.5}$$

The matrix form (31.5) is defined with respect to the natural basis of the cylindrical polar coordinate system. To be consistent with Eq. (7.23), $\mathbf{L}_1$ must be such that $L_1\langle 21\rangle$, $L_1\langle 31\rangle$ and $L_1\langle 32\rangle$ are non-zero, but the others are zero. The orthogonal matrix

$$[Q\langle ij\rangle] = \begin{bmatrix} 1 & 0 & 0 \\ 0 & 0 & 1 \\ 0 & 1 & 0 \end{bmatrix} \tag{31.6}$$

transforms the matrix form of $\mathbf{L}_1$ to read

$$[L_1^*\langle ij\rangle] = [(QL_1Q^T)\langle ij\rangle] = \varkappa \begin{bmatrix} 0 & 0 & 0 \\ l & 0 & 0 \\ m & n & 0 \end{bmatrix}, \tag{31.7}$$

with

$$\varkappa l = u', \qquad \varkappa m = r\omega', \qquad \varkappa n = er, \qquad l^2 + m^2 + n^2 = 1. \tag{31.8}$$

34. Huilgol [1971 : 12].

Corresponding to the form (31.7), let the material functions be defined by

$$\Sigma_1 = T^*\langle 22\rangle - T^*\langle 11\rangle, \qquad \Sigma_2 = T^*\langle 33\rangle - T^*\langle 11\rangle, \tag{31.9}$$

$$\tau_1 = T^*\langle 12\rangle,\ \tau_2 = T^*\langle 13\rangle, \qquad \tau_3 = T^*\langle 23\rangle, \tag{31.10}$$

where $\mathbf{T}^*$ is an isotropic function of $\mathbf{L}_1^*$, that is, for all constant orthogonal tensors Q,

$$\left.\begin{aligned} \mathbf{T}^* = \mathbf{QTQ}^T = \mathbf{f}(\mathbf{L}_1^*) = \mathbf{f}(\mathbf{QL}_1\mathbf{Q}^T), \\ \mathbf{Qf}(\mathbf{L}_1^*)\,\mathbf{Q}^T = \mathbf{f}(\mathbf{QL}_1^*\mathbf{Q}^T). \end{aligned}\right\} \tag{31.11}$$

It is impossible to reduce the number of the material functions in (31.9) *and* (31.10) *to four, or three.* On conversion to (r, θ, z) system, one obtains

$$\Sigma_1 = T\langle zz\rangle - T\langle rr\rangle, \qquad \Sigma_2 = T\langle \theta\theta\rangle - T\langle rr\rangle, \tag{31.12}$$

$$\tau_1 = T\langle rz\rangle,\ \tau_2 = T\langle r\theta\rangle, \qquad \tau_3 = T\langle \theta z\rangle. \tag{31.13}$$

All these material functions depend on $\varkappa, l, m, n$. We may note the following features of these functions obtained through the isotropy condition (31.11):

$$\begin{aligned} \Sigma_1(\varkappa, l, m, n) &= \Sigma_1(\varkappa, -l, -m, n) = \Sigma_1(\varkappa, -l, m, -n) \\ &= \Sigma_1(\varkappa, l, -m, -n); \end{aligned} \tag{31.14a}$$

$$\begin{aligned} \Sigma_2(\varkappa, l, m, n) &= \Sigma_2(\varkappa, -l, -m, n) = \Sigma_2(\varkappa, -l, m, -n) \\ &= \Sigma_2(\varkappa, l, -m, -n); \end{aligned} \tag{31.14b}$$

$$\begin{aligned} \tau_1(\varkappa, l, m, n) &= -\tau_1(\varkappa, -l, -m, n) = -\tau_1(\varkappa, -l, m, -n) \\ &= \tau_1(\varkappa, l, -m, -n); \end{aligned} \tag{31.14c}$$

$$\begin{aligned} \tau_2(\varkappa, l, m, n) &= -\tau_2(\varkappa, -l, -m, n) = \tau_2(\varkappa, -l, m, -n) \\ &= -\tau_2(\varkappa, l, -m, -n); \end{aligned} \tag{31.14d}$$

$$\begin{aligned} \tau_3(\varkappa, l, m, n) &= \tau_3(\varkappa, -l, -m, n) = -\tau_3(\varkappa, -l, m, -n) \\ &= -\tau_3(\varkappa, l, -m, -n). \end{aligned} \tag{31.14e}$$

Moreover, as $c \to 0$ or $n \to 0$, the flow (31.4) becomes the standard helical flow. Comparing with the set (26.61), we have

$$\Sigma_1(\varkappa, l, m, 0) = \alpha^2 N_1(\varkappa) - \beta^2 N_2(\varkappa), \tag{31.15a}$$

$$\Sigma_2(\varkappa, l, m, 0) = \alpha^2 N_2(\varkappa) - \beta^2 N_1(\varkappa), \tag{31.15b}$$

$$\tau_1(\varkappa, l, m, 0) = \beta\tau(\varkappa), \tag{31.15c}$$

$$\tau_2(\varkappa, l, m, 0) = \alpha\tau(\varkappa), \tag{31.15d}$$

$$\tau_3(\varkappa, l, m, 0) = \alpha\beta(N_1 + N_2), \tag{31.15e}$$

where N_1, N_2 and τ are the viscometric material functions. Note that $\alpha = r\omega'/\varkappa = m$ and $\beta = u'/\varkappa = l$ in the present notation.

For the helical-torsional flow, the extra stresses are all functions of r alone. Also, as in the torsional flow (*cf.* Sec. 26), the torsional term makes the inertia terms incompatible within the equations of motion. By assuming $\rho = 0$, omitting inertial and body force terms and using the modified pressure term $\psi = p$, we have the following equations of motion from (A6.3)–(A6.4):

$$-\frac{\partial\psi}{\partial r} + \frac{d}{dr} T^E \langle rr \rangle + \frac{1}{r}(T\langle rr \rangle - T\langle \theta\theta \rangle) = 0, \quad (31.16a)$$

$$\frac{d}{dr} T\langle r\theta \rangle + \frac{2}{r} T\langle r\theta \rangle = 0, \quad (31.16b)$$

$$-\frac{\partial\psi}{\partial z} + \frac{d}{dr} T\langle rz \rangle + \frac{1}{r} T\langle rz \rangle = 0. \quad (31.16c)$$

The solutions are given by:

$$\psi = -cz + h(r), \qquad \tau_2 = T\langle r\theta \rangle = M/2\pi r^2, \quad (31.17a)$$

$$\tau_1 = T\langle rz \rangle = -\frac{1}{2} cr + \frac{b}{r}, \qquad h'(r) = \frac{d}{dr} T_E \langle rr \rangle - \frac{\Sigma_2}{r}. \quad (31.17b)$$

We refer the reader to Sec. 28 for a discussion of the meaning of the constants c, b and M. The following apparatus suggested by OLDROYD[35], may be used to measure these three constants as well as the material functions τ_3, Σ_1 and Σ_2.

OLDROYD remarked that the POISEUILLE-torsional flow, that is, one with $\omega(r) = 0$ in (31.4), can be generated in a small region by rotating two porous disks, at different speeds, about a common axis placed along the axis of a circular pipe of approximately the same radius as the disks, so as to impose a torsional motion on the liquid flowing down the pipe. The helical-

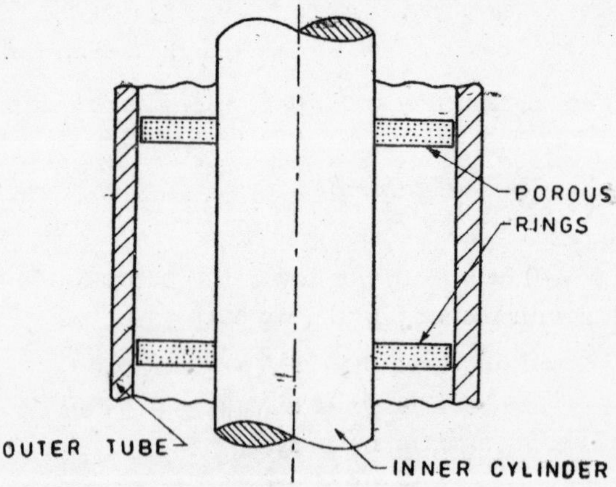

Fig. 31.1. Helical-torsional flow apparatus.

torsional flow can be generated in between two concentric cylinders by rotating two porous rings at different speeds in the annular space between the two cylinders, provided the width of each porous ring is almost equal to the annular space between the two cylinders (see Fig. 31.1).[36]

35. [1965 : 13].
36. HUILGOL [1971 : 12].

One can measure the constant c as in (28.11), and the constant b by suspending one of the two cylinders [(28.19) and ff.] and finally, M as in (28.8).

Next, the torque T on the porous ring is given by

$$T = 2\pi \int_{R_1}^{R_2} \tau_3 r^2 dr, \tag{31.18}$$

and, in principle, this yields τ_3.

Let the body force act in the z-direction only.[37] Then χ, the body force potential is given by $\chi = \chi(z)$. Thus $b_r = 0$. Then (31.16a) reads

$$\frac{\partial}{\partial r} T\langle rr\rangle = \frac{1}{r}\Big(T\langle\theta\theta\rangle - T\langle rr\rangle\Big) = \frac{1}{r}\Sigma_2, \tag{31.19}$$

from the definition in (31.12). Measuring the radial stress at the same height z, one derives

$$T\langle rr\rangle(R_2) - T\langle rr\rangle(R_1) = \int_{R_1}^{R_2} \frac{1}{r}\Sigma_2\, dr, \tag{31.20}$$

which determines Σ_2. Again, under this assumption of $\chi = \chi(z)$,

$$\frac{\partial}{\partial r} T\langle zz\rangle = \frac{\partial}{\partial r} T\langle rr\rangle + \frac{d}{dr}\Sigma_1, \tag{31.21}$$

from the definition of Σ_1. Since from $(31.17)_1$,

$$-p = \rho\chi - \psi = \rho\chi(z) + cz - h(r), \tag{31.22}$$

(31.19), (31.21) and (31.22) yield

$$T\langle zz\rangle = \rho\chi(z) + cz + \int_{\alpha}^{r}\left(\frac{1}{\xi}\Sigma_2 + \frac{d}{d\xi}\Sigma_1\right)d\xi, \qquad \alpha > 0. \tag{31.23}$$

Hence, at a fixed value of $z = z_0$,

$$T\langle zz\rangle(r) - T\langle zz\rangle(R_1) = \int_{R_1}^{r}\frac{1}{\xi}\Sigma_2 d\xi + \Sigma_1(r) - \Sigma_1(R_1), \tag{31.24}$$

where $\Sigma_1(r)$ means that the material function is evaluted at a distance r from the origin, using the values of $\varkappa$, l, m, n there. The thrust F on a porous ring is given by

$$F = 2\pi \int_{R_1}^{R_2} T\langle zz\rangle\, r dr, \tag{31.25}$$

and hence F leads to the determination of the function Σ_1.

There is at present no better method for the determination of these five functions. Indeed, the experiment listed here is not very useful either because the boundary conditions are not satisfied on the two cylindrical surfaces. Hence the determination of $\omega(r)$ and $u(r)$ is, physically speaking, meaningless. So, the analysis given here, while mathematically exact if the anticipated

---

37. By taking $\rho = 0$, this should be omitted. However its introduction here is for providing a hydrostatic correction.

flow can be produced in practice, is an approximation as far as reality is concerned. It seems that a realistic physical approximation would be to compute the stresses by assuming e to be small so that its powers can be neglected. One can then use these stresses in the equations of motion to solve for $\omega(r)$ and $u(r)$. Such an attempt is described elsewhere (Chap. 9), via the nearly viscometric theory.

For reference, the four RIVLIN-ERICKSEN tensors for the flow (31.4) will now be written in physical component form below :

$$[A_1\langle ij\rangle] = \begin{bmatrix} 0 & r\omega' & u' \\ . & 0 & er \\ . & . & 0 \end{bmatrix}, \quad [A_2\langle ij\rangle] = \begin{bmatrix} 2(r^2\omega'^2+u'^2) & eru' & 2er^2\omega' \\ , & 0 & 0 \\ . & . & 2e^2r^2 \end{bmatrix}, \tag{31.26}$$

$$[A_3\langle ij\rangle] = \begin{bmatrix} 6er^2\omega'u' & 0 & 3e^2r^2u' \\ . & 0 & 0 \\ . & . & 0 \end{bmatrix}, \quad [A_4\langle ij\rangle] = \begin{bmatrix} 6e^2r^2u'^2 & 0 & 0 \\ . & 0 & 0 \\ . & . & 0 \end{bmatrix}. \tag{31.27}$$

In the author's paper *loc. cit.*, the reader will find the solution to the flow (31.2), with $\dot{\theta} = \omega(r)$, and $\dot{z} = A\theta + u(r)$. It is shown there that such a flow is dynamically feasible, and that the intertial terms do not have to be neglected.

32. Maxwell Orthogonal Rheometer Flow[38]

As explained in Sec. 7, for the flow given by

$$\dot{x} = -\Omega y + \Omega\psi z, \quad \dot{y} = \Omega x, \quad \dot{z} = 0, \tag{32.1}$$

the strain history has the MWCSH form of a product of two exponential functions $\exp(-s\mathbf{L}^T)$ and $\exp(-s\mathbf{L})$, with the velocity gradient $\mathbf{L}$ given in matrix form by

$$[L\langle ij\rangle] = \begin{bmatrix} 0 & -\Omega & \Omega\psi \\ \Omega & 0 & 0 \\ 0 & 0 & 0 \end{bmatrix}. \tag{7.2}$$

Since the extra stress $\mathbf{T}_E$ is given by an isotropic function $\mathbf{f}(\cdot)$ of $\mathbf{L}$, we can apply the condition

$$\mathbf{Qf}(\mathbf{L})\,\mathbf{Q}^T = \mathbf{f}(\mathbf{QLQ}^T) \tag{32.2}$$

38. This section is based on HUILGOL [1969 : 8]. An examination of the flow in the rheometer by using a very special constitutive equation has been published by ABBOTT and WALTERS [1970 : 1]. Their results incorporate inertial effects, however. For a discussion, see Sec. 45.

for all constant orthogonal tensors **Q** and prove the following :

1°. The extra stress tensor can again be expressed in terms of five material functions. These cannot be reduced in number.

2°. By the theorem[39], proved in Sec. 9, the extra stress tensor is an isotropic function of the first two kinematical tensors $\mathbf{A}_1$ and $\mathbf{A}_2$. For, the first RIVLIN-ERICKSEN tensor $\mathbf{A}_1$ has the matrix form

$$[A_1\langle ij\rangle] = \begin{bmatrix} 0 & 0 & \Omega\psi \\ \cdot & 0 & 0 \\ \cdot & \cdot & 0 \end{bmatrix}, \tag{32.3}$$

and by the theorem[40] in Sec. 10, $\mathbf{A}_1$ has three distinct eigenvalues. Thus WANG's theorem assures us that

$$\mathbf{T}_E = \mathbf{f}\,(\mathbf{A}_1, \mathbf{A}_2). \tag{32.4}$$

Since the matrix of $\mathbf{A}_2$ is given by

$$[A_2\langle i\,j\rangle] = \begin{bmatrix} 0 & 0 & 0 \\ \cdot & 0 & -\Omega^2\psi \\ \cdot & \cdot & 2\Omega^2\psi^2 \end{bmatrix}, \tag{32.5}$$

each of the five material functions must be a function of $\Omega\psi$ and $\Omega^2\psi$.

3°. Defining the five material functions as $\mathcal{N}_1$, $\mathcal{N}_2$, $\mathcal{S}_1$, $\mathcal{S}_2$ and $\mathcal{S}_3$ as below, it follows from the objectivity condition applied to (32.4) that[41],

$$\begin{aligned} T\langle 11\rangle - T\langle 33\rangle &= \mathcal{N}_1(\Omega\psi, \Omega^2\psi) = \mathcal{N}_1(-\Omega\psi, -\Omega^2\psi) \\ &= \mathcal{N}_1(-\Omega\psi, \Omega^2\psi) = \mathcal{N}_1(\Omega\psi, -\Omega^2\psi)\,; \end{aligned} \tag{32.6a}$$

$$\begin{aligned} T\langle 22\rangle - T\langle 33\rangle &= \mathcal{N}_2(\Omega\psi, \Omega^2\psi) = \mathcal{N}_2(-\Omega\psi, -\Omega^2\psi) \\ &= \mathcal{N}_2(-\Omega\psi, \Omega^2\psi) = \mathcal{N}_2(\Omega\psi, -\Omega^2\psi)\,; \end{aligned} \tag{32.6b}$$

$$\begin{aligned} T\langle 12\rangle &= \mathcal{S}_1(\Omega\psi, \Omega^2\psi) = \mathcal{S}_1(-\Omega\psi, -\Omega^2\psi) \\ &= -\,\mathcal{S}_1(-\Omega\psi, \Omega^2\psi) = -\,\mathcal{S}_1(\Omega\psi, -\Omega^2\psi)\,; \end{aligned} \tag{32.6c}$$

$$\begin{aligned} T\langle 13\rangle &= \mathcal{S}_2(\Omega\psi, \Omega^2\psi) = -\,\mathcal{S}_2(-\Omega\psi, -\Omega^2\psi) \\ &= -\,\mathcal{S}_2(-\Omega\psi, \Omega^2\psi) = \mathcal{S}_2(\Omega\psi, -\Omega^2\psi)\,; \end{aligned} \tag{32.6d}$$

$$\begin{aligned} T\langle 23\rangle &= \mathcal{S}_3(\Omega\psi, \Omega^2\psi) = -\,\mathcal{S}_3(-\Omega\psi, -\Omega^2\psi) \\ &= \mathcal{S}_3(-\Omega\psi, \Omega^2\psi) = -\,\mathcal{S}_3(\Omega\psi, -\Omega^2\psi). \end{aligned} \tag{32.6e}$$

---

39. WANG [1965 : 20].
40. HUILGOL [1971 : 12].
41. Based on the idea that $\mathbf{A}_1$ and $\mathbf{A}_2$ in (32.3) and (32.5) are invariant under changing (i) ψ to $-\psi$ and the z axis to $-z$ axis ; (ii) Ω to $-\Omega$ and changing x to $-x$; (iii) ψ to $-\psi$, Ω to $-\Omega$ and y to $-y$.

We note from (32.3) and (32.5) that if we let $\Omega \to 0$ and $\psi \to \infty$ such that the product $\Omega\psi$ is infinite, then $\mathbf{A}_1$ and $\mathbf{A}_2$ would possess the forms as if the flow had been given by

$$\dot{x} = \Omega\psi z, \quad \dot{y} = \dot{z} = 0. \tag{32.7}$$

This is a viscometric flow. Hence

$$\mathcal{N}_1(\Omega\psi, 0) = N_1(\Omega\psi), \tag{32.8a}$$
$$\mathcal{N}_2(\Omega\psi, 0) = - N_2(\Omega\psi), \tag{32.8b}$$
$$\mathcal{S}_1(\Omega\psi, 0) = 0, \tag{32.8c}$$
$$\mathcal{S}_2(\Omega\psi, 0) = \tau(\Omega\psi), \tag{32.8d}$$
$$\mathcal{S}_3(\Omega\psi, 0) = 0. \tag{32.8e}$$

In (32.8), N_1 and N_2 and τ are the viscometric material functions.

Turning to the original motion (32.1), as far as the dynamical equations are concerned, the extra stresses are constants and their divergence is zero. But the motion (32.1) has an acceleration field given by

$$\mathbf{a} = - \Omega^2 x\, \mathbf{i} + \Omega^2(\psi z - y)\, \mathbf{j}. \tag{32.9}$$

Since curl $\mathbf{a} \neq \mathbf{0}$, the motion (32.1) is possible if and only if inertia is neglected. Neglecting inertia and body force terms[42], one is led to the solution that the pressure field is constant everywhere. So we may take $p = 0$, and thus $\mathbf{T} = \mathbf{T}_E$.

In the instrument described by MAXWELL and CHARTOFF[43] the forces in the x, y and z direction are measured on the top plate. These forces are :

$$F_x = \int T\langle xz\rangle\, da = \mathcal{S}_2\, A. \tag{32.10a}$$
$$F_y = \int T\langle yz\rangle\, da = \mathcal{S}_3\, A, \tag{32.10b}$$
$$F_z = \int T\langle zz\rangle\, da = \int T_E\langle zz\rangle\, da = T_E\langle zz\rangle\, A \tag{32.10c}$$
$$= -\frac{1}{3}(\mathcal{N}_1 + \mathcal{N}_2)\, A. \tag{32.10d}$$

The last result stems from (32.6a), (32.6b) and the normalization condition tr $\mathbf{T}_E = 0$. In (32.10), A is the area of contact of the specimen with the upper disk.

The above motion is the only known example of a homogeneous velocity field which may be approximately produced in the laboratory. However, the pleasure in discovering such a motion is short-lived, for the fact that the inertia terms have to be ignored means that Ω^2 has to be small, and so does $\Omega^2\psi$. In practice, this means a low anguiar velocity and a small eccentricity, which makes the flow physically uninteresting. On the other hand if one makes ψ small, but leaves Ω arbitrary, then as $\psi \to 0$ the flow becomes a rigid

42. Actually a constant body force in the z-direction can be absorbed into the pressure.
43. [1965 : 11].

motion, making the strain history close to the rest history. Such a case is dealt with by the linear theory and will be discussed in Sec. 34. Mathematically speaking, it is also possible to let $\Omega \to 0$, $\psi \to \infty$ such that $\Omega\psi$ is infinite. Then one can discuss the motion as a nearly viscometric flow (see Sec. 40), and compute the stresses.

Now, whatever approximations one may make to determine the stresses, the free (lateral) surface is open to the atmosphere and subject to a hydrostatic pressure externally, while the stress system in the fluid does not yield a normal stress vector. The secondary flows introduced by the inclusion of inertia and the non-conformity with the boundary condition have yet to be explored theoretically for the general fluid.[44]

For later use (in Sec. 34 ter), we note that the symmetries in (32.6) permit us to express the material functions as[45] :

$$\mathcal{N}_1(\Omega\psi, \Omega^2\psi^2) = \Omega^2\psi^2 f_{11}(\Omega^2, \Omega^2\psi^2), \tag{32.11a}$$

$$\mathcal{N}_2(\Omega\psi, \Omega^2\psi) = \Omega^2\psi^2 f_{22}(\Omega^2, \Omega^2\psi^2), \tag{32.11b}$$

$$\mathcal{S}_1(\Omega\psi, \Omega^2\psi) = \Omega^3\psi^2 f_{12}(\Omega^2, \Omega^2\psi^2), \tag{32.11c}$$

$$\mathcal{S}_2(\Omega\psi, \Omega^2\psi) = \Omega\psi f_{13}(\Omega^2, \Omega^2\psi^2), \tag{32.11d}$$

$$\mathcal{S}_3(\Omega\psi, \Omega^2\psi) = \Omega^2\psi f_{23}(\Omega^2, \Omega^2\psi^2). \tag{32.11e}$$

On taking the limit as $\psi \to \infty$ and $\Omega \to 0$, with $\Omega\psi$ finite, the formulae (32.11) become :

$$f_{11}(0, \Omega^2\psi^2) = \bar{n}_1(\Omega^2\psi^2), \quad f_{22}(0, \Omega^2\psi^2) = -\bar{n}_2(\Omega^2\psi^2), \tag{32.12a}$$

$$f_{12}(0, \Omega^2\psi^2) < \infty, \qquad f_{23}(0, \Omega^2\psi^2) < \infty, \tag{32.12b}$$

$$f_{13}(0, \Omega^2\psi^2) = \bar{\eta}\,(\Omega^2\psi^2). \tag{32.12c}$$

33. Simple Extensional Flow[46]

As analyzed in Secs. 7 — 8, this motion has the following form in Cartesian coordinates :

$$\left.\begin{aligned} \dot{x}_i &= a_i x_i \,; \qquad i = 1, 2, 3 \,;\ \text{no sum on } i \,; \\ a_1 &+ a_2 + a_3 = 0. \end{aligned}\right\}; \tag{33.1}$$

or, the following form

$$\dot{r} = -\frac{\alpha}{2}\,r, \quad \dot{\theta} = 0, \quad \dot{z} = \alpha z, \tag{33.2}$$

in cylindrical polar coordinates. The flow in (33.2) is also isochoric.

---

44. ABBOTT and WALTERS [1970 : 1] have carried out the analysis for a finitely linear theory, in the presence of inertia. See Sec. 45.

45. PIPKIN [1969 : 18], Eq. (5.3.4). Though he omits the use of condition (iii) in the footnote 41, these representations are correct. Note that $\Omega^3\psi^2$ is included in (32.11c) with the purpose of using $(32.6c)_{2-3}$ effectively and to demand that the $(32.12b)_1$ holds.

46. Based on COLEMAN and NOLL [1962 : 3].

For each of the above two flows, the strain history is determined completely by the respective $\mathbf{A}_1$ (*cf.* WANG's theorem[47] in Sec. 9). Hence

$$T_E = \mathbf{f}(\mathbf{A}_1), \tag{33.3}$$

where $\mathbf{f}(\cdot)$ is an isotropic function of $\mathbf{A}_1$. Thus the claim by COLEMAN and NOLL[48] that in simple extensional flow, the general simple fluid cannot be characterised by the RIVLIN-ERICKSEN fluid is incorrect[49].

For the velocity field (33.1), we have :

$$[A_1\langle ij\rangle] = \begin{bmatrix} a_1 & 0 & 0 \\ \cdot & a_2 & 0 \\ \cdot & \cdot & a_3 \end{bmatrix}, \quad \mathbf{a} = \frac{1}{2}\nabla[(a_1x_1)^2 + (a_2x_2)^2 + (a_3x_3)^2] \tag{33.4}$$
$$= -\frac{1}{2}\nabla\zeta,$$

where $\mathbf{a}$ is the acceleration field and ζ is its potential. Similarly, for the velocity field (33.2), we obtain

$$[A_1\langle ij\rangle] = \begin{bmatrix} -\alpha & 0 & 0 \\ \cdot & -\alpha & 0 \\ \cdot & \cdot & 2\alpha \end{bmatrix}, \quad \mathbf{a} = \frac{\alpha^2}{2}\nabla\left(\frac{r^2}{4} + z^2\right) = -\frac{\alpha^2}{2}\nabla\zeta. \tag{33.5}$$

So, in both cases the extra stresses are constants and each acceleration is derivable from a potential ζ. Hence the equations of motion are satisfied with the modified pressure ψ being given by

$$\psi = \rho\zeta + g(t), \tag{33.6}$$

where $g(t)$ is a function of time t alone. Defining the normal stresses $T\langle 11\rangle$, $T\langle 22\rangle$, $T\langle 33\rangle$ to be T_1, T_2, T_3 respectively and assuming that the isotropic pressure term is absorbed into $g(t)$, one has, with the summation convention :

$$T_i = \frac{1}{2}\rho(a_jx_j)^2 + \rho\chi + a_ih + a_i^2\, l + g(t), \quad i = 1, 2, \tag{33.7}$$

for the flow (33.1). In (33.7), h and l are functions arising from the expansion of the relation (33.3). From the proof given, for example, by SERRIN[50], it follows that the function in (33.3) may be expanded as

$$\mathbf{T}_E = \alpha_0\mathbf{1} + h\,\mathbf{A}_1 + l\,\mathbf{A}_1^2\,, \tag{33.8}$$

where α_0, h and l are analytic functions of the three invariants of $\mathbf{A}_1$. Ab-

47. [1965 : 20].
48. [1962 : 3].
49. For a full treatment of such fluids, see Chap. 8.For the present, by RIVLIN-ERICKSEN fluids, we mean those obeying an isotropic constitutive equation of the form $\mathbf{T}_E = \mathbf{f}(\mathbf{A}_1, \mathbf{A}_2, \ldots, \mathbf{A}_N)$, $N = 1, 2, 3, \ldots M$, and M is finite.
50. [1959 : 6].

sorbing α_0 into $g(t)$, one derives (33.7) from (33.8). Since tr $\mathbf{A}_1 = 0$, h and l are thus functions of the other two invariants of $\mathbf{A}_1$, or functions of $\sum_{i=1}^{3} a_i^2$ and $\sum_{i=1}^{3} a_i^3$.

For this case of the flow (33.2), one has that T_r, T_θ and T_z are the three normal stresses, and

$$T_r = T_\theta = \frac{\rho\alpha^2}{2}\left(\frac{r^2}{4} + z^2\right) + \rho\chi - \alpha h + \alpha^2 l + g(t), \tag{33.9}$$

$$T_z = \frac{\rho\alpha^2}{2}\left(\frac{r^2}{4} + z^2\right) + \rho\chi + 2\alpha h + 4\alpha^2 l + g(t), \tag{33.10}$$

where h and l are functions of α^2 and α^3.

Suppose we consider a simple fluid with the shape of a circular cylinder of radius $R(t)$ and length $L(t)$, and suppose that this fluid body has been subject to the extensional flow (33.2). It is obvious from the velocity field (33.2) that the shape of the fluid body will always be a circular cylinder. Now

$$\alpha z = \dot{z} = \frac{d}{dt} L(t), \tag{33.11}$$

and thus

$$\alpha = \frac{d}{dt} \ln L = -2\frac{d}{dt} \ln R. \tag{33.12}$$

At time t, let the cylinder lie between the planes

$$z = 0, \quad z = L = L(t). \tag{33.13}$$

By (33.10), the normal stresses in the z-direction at these cross-sections are given by

$$\left.\begin{aligned} T_z(0) &= \frac{\rho\alpha^2 r^2}{8} + \rho\chi + 2\alpha h + 4\alpha^2 l + g(t), \\ T_z(L) &= \frac{\rho\alpha^2}{2}\left(\frac{r^2}{4} + L^2\right) + \rho\chi + 2\alpha h + 4\alpha^2 l + g(t). \end{aligned}\right\} \tag{33.14}$$

On the lateral surface of the cylinder, the normal stress is

$$T_r(R) = \frac{\rho\alpha^2}{2}\left(\frac{R^2}{4} + z^2\right) + \rho\chi - \alpha h + \alpha^2 l + g(t). \tag{33.15}$$

If one assumes that $\chi = 0$ (body force to be zero), and adjusts $g(t)$ so that

$$T_r(R) = 0 \quad \text{at} \quad z = L/2, \tag{33.16}$$

then one finds,

$$T_r(R) = \frac{\rho\alpha^2}{2}\left(z^2 - \frac{L^2}{4}\right), \tag{33.17}$$

$$T_z(0) = \frac{\rho\alpha^2}{8}(r^2 - R^2 - L^2) + 3\alpha h + 3\alpha^2 l, \tag{33.18}$$[51]

$$T_z(L) = \frac{\rho\alpha^2}{8}(r^2 - R^2 + 3L^2) + 3\alpha h + 3\alpha^2 l. \tag{33.19}$$[51]

If inertia is neglected, we obtain

$$T_r(R) = 0, \quad T_z(0) = T_z(L) = 3\alpha h + 3\alpha^2 l. \tag{33.20}$$

We note that the stress power inequality (25.11) reads

$$\alpha(T_z - T_r) \geqq 0, \tag{33.21}$$

or, from (33.10), that

$$\alpha(3\alpha h + 3\alpha^2 l) \geqq 0. \tag{33.22}$$

Thus, when inertia is neglected, $T_z(0) = T_z(L)$ have the same sign as the logarithmic extension rate α. Hence, a steady elongation requires 'pulling rather than pushing', and this is a consequence of the dissipation inequality.[52]

If inertia is not neglected, and normal tractions as specified by (33.17) not supplied, then for $0 \leqq z < L/2$, the cylinder will tend to bulge, but for $L/2 < z \leqq L$, it will tend to contract.

COLEMAN and NOLL[53] have also analyzed the case of the extension of a rectangular parallelopiped by using the velocity field (33.1).

REMARKS. 1. The reader can refer to Sec. 27.1 and derive the stresses in the non-steady extensional flow of SLATTERY[54] to be functionals of $\sum_{i=1}^{3} a_i^2 f^2(t)$, and $\sum_{i=1}^{3} a_i^3 f^3(t)$, that is, dependent on time only, and easily reverify the dynamic feasibility of the motion (27.8).

2. ZAHORSKI[55] has conjectured that at low rates of shear q, the viscometric functions $\eta(\cdot)$, $N_1(\cdot)$ and $N_2(\cdot)$ could approximate the extensional viscosity η_e, defined by $\eta_e = T\langle zz\rangle/q$, when q is the logarithmic extension rate, through

$$\eta_e(q) = 3\eta(q) + \frac{\sqrt{3}}{2}\,\frac{N_1(q) + 2N_2(q)}{q}. \tag{33.23}$$

This relation should be subjected to experimental verification.

51. There is a misprint in Eqs. $(4.9)_{2-3}$ of COLEMAN and NOLL [1962 : 3], for the terms involving R^2 are missing.
52. COLEMAN and NOLL [1962 : 3].
53. [1962 : 3].
54. [1964 : 22].
55. [1971 : 31].

APPENDIX : THE REINER-RIVLIN PARADOX[1]

In MWCSH, if the velocity gradient **L** has a zero material derivative, we have that

$$\mathbf{C}_t(t-s) = e^{-s\mathbf{L}^T}\, e^{-s\mathbf{L}}. \tag{A7.1}$$

The extra stresses are then an isotropic function of **L**, that is,

$$\mathbf{T}_E = f(\mathbf{L}), \qquad \mathbf{Q}\mathbf{f}(\mathbf{L})\mathbf{Q}^T = \mathbf{f}(\mathbf{Q}\mathbf{L}\,\mathbf{Q}^T). \tag{A7.2}$$

We have used this in Chaps. 6 and 7. Now suppose we regard $\mathbf{T}_E = \mathbf{f}(\mathbf{L})$, that is, that the stress is an isotropic function of the velocity gradient only. Then by the objectivity arguments of Chap. 5 (see p. 176 below),

$$\mathbf{T}_E = \mathbf{f}(\mathbf{D}), \quad 2\mathbf{D} = \mathbf{L} + \mathbf{L}^T. \tag{A7.3}$$

Hence by the well-known representation theorem[2],

$$\mathbf{T}_E = \alpha_0\mathbf{1} + \alpha_1\mathbf{D} + \alpha_2\mathbf{D}^2. \tag{A7.4}$$

This is the so-called REINER-RIVLIN constitutive equation.

Consider the two MWCSH :

Simple shear : $\dot{x} = \Omega\psi z, \qquad \dot{y} = \dot{z} = 0.$ (A7.5)

MAXWELL rheometer flow : $\dot{x} = -\,\Omega y + \Omega\psi z,\ \dot{y} = \Omega x,\ \dot{z} = 0.$ (A7.6)

In both motions, if Ω and ψ are constants, the respective velocity gradients have zero material derivatives. But *in both simple shear and the Maxwell rheometer flow,*

$$[D\langle ij\rangle] = \Omega\psi \begin{bmatrix} 0 & 0 & 1 \\ . & 0 & 0 \\ . & . & 0 \end{bmatrix}. \tag{A7.7}$$

So, Eq. (A7.4) tells us that the stresses in these two MWCSH must be the same. But, as we have seen already, the extra stresses are not. This is the 'paradox'.

We can resolve it in the following way[3]. In the constitutive relation

$$\mathbf{T}_E = \mathop{\mathcal{H}}_{s=0}^{\infty} [\mathbf{C}_t\,(t-s)], \tag{A7.8}$$

no additional reductions due to objectivity are possible, except for the following one :

$$\mathbf{Q} \mathop{\mathcal{H}}_{s=0}^{\infty} [\mathbf{C}_t(t-s)]\, \mathbf{Q}^T = \mathop{\mathcal{H}}_{s=0}^{\infty} \mathbf{Q}\,[\mathbf{C}_t(t-s)\,\mathbf{Q}^T], \tag{A7.9}$$

1. So termed by PIPKIN [1969 : 18]. In unravelling the 'paradox', it is sufficient to consider the cases when $\dot{\mathbf{L}} = \mathbf{0}$.
2. SERRIN [1959 : 6], Sec. 59.
3. PIPKIN [1969 : 19] does it by saying that the histories corresponding to (A7.5) and (A7.6) are different and that the assumption (A7.2) is not equivalent to (A7.3).

valid for all contant orthogonal tensors **Q**. So (*A*7.8) follows from (*A*7.9) because in MWCSH

$$\mathbf{Q}\mathbf{C}_t(t-s)\,\mathbf{Q}^T = e^{-s(\mathbf{QLQ}^T)^T}\,e^{-s(\mathbf{QLQ}^T)}, \qquad (A7.10)$$

or **L** determines $\mathbf{C}_t(t-s)$, and $\mathbf{QLQ}^T$ determines $\mathbf{QC}_t(t-s)\mathbf{Q}^T$. However, (*A*7.3) does not follow from (*A*7.9). Hence the paradox is resolved.

As we have just seen the paradox, if it can be so called, arises if one misunderstands the real meaning of the objectivity restriction on a constitutive equation. There are at least two procedures for avoiding such an error. One is to write the constitutive equation in a MWCSH in terms of the tensors $\mathbf{A}_1$, $\mathbf{A}_2$, and $\mathbf{A}_3$. Since these three tensors are not identical for the flows (*A*7.5) and (*A*7.6), the stresses cannot be the same. A second method, which works when $\dot{\mathbf{L}} = \mathbf{O}$, is to note that any function of $\mathbf{A}_1$ and $\mathbf{A}_2$ is a function of $\mathbf{A}_1$ and **W**, the spin tensor.[4] Since the stresses in the motions (*A*7.5) and (*A*7.6) are functions of the respective $\mathbf{A}_1$ and $\mathbf{A}_2$ only and since the **W** tensors for the two flows are not identical, the stresses must be different.

4. Giesekus [1962 : 6], Eq. (44). This paper considers constitutive equations depending on $\mathbf{A}_1, \ldots, \mathbf{A}_N$ when $\dot{\mathbf{L}} = \mathbf{O}$, which implies that

$$\dot{\mathbf{A}}_n = \mathbf{O}, \quad \mathbf{A}_{n+1} = \mathbf{A}_n\mathbf{A}_1 + \mathbf{A}_1\mathbf{A}_n + \mathbf{A}_n\mathbf{W} - \mathbf{W}\mathbf{A}_n, \quad n = 1, 2, \ldots .$$

CHAPTER 8

Other Constitutive Equations and Responses in Some Flows

So far, we have developed the constitutive equation for a simple fluid and made use of the incompressible simple fluid to solve specific problems belonging to MWCSH. While these form a fascinating class of flows, clearly they do not exhaust the totality of strain histories. Even in MWCSH it becomes obvious that the boundary conditions and/or the inertia terms cannot be incorporated accurately into the solutions of the equations of motion[1]. To deal with a wider class of strain histories, there is only one realistic option open to us. This is the prescription of a more specific constitutive equation, for the simple fluid rather than that

$$\mathbf{T} = -p(\rho)\,\mathbf{1} + \mathop{\tilde{\tilde{H}}}_{s=0}^{\infty} (\mathbf{G}(s);\,\rho).$$

Now these approximations have to be meaningful and cannot be arbitrarily chosen. They have to be derived from the original constitutive equations and the processes of approximation involved must be clearly stated so that one knows precisely what assumptions one is making in deriving the end result.

To obtain a class of approximations, COLEMAN and NOLL[2] introduced the concept of fading memory and showed how linear viscoelasticity, finite linear viscoelasticity[3] and 'second-order viscoelasticity' fit into the general theory of simple materials and, in particular, simple fluids. These theories (integral models) are based on the assumption that the 'norm' of $\mathbf{C}_t(t-s)$ is very close to the norm of the rest history, viz., $\mathbf{C}_t(t-s) = \mathbf{1}$, $s \in [0, \infty)$.

These two authors also developed[3] an approximation scheme for obtaining the stress in a simple material when the given strain history is replaced by the same history, executed at a slower rate. The 'magnitude of slowness' dictates the types of constitutive equations which arise. For simple fluids, such equations are known as first-order (Newtonian), second-order, third-order fluids, etc. In fact, these approximations had been introduced earlier by GREEN and

---

1. However such deviations are dealt with under perturbations. See Chaps. 9 and 10 for the techniques of dealing with them.

2. [1960 : 1], [1961 : 1], [1961 : 2], [1961 : 4]. Such approximations had been constructed earlier by GREEN and RIVLIN [1957 : 1], but this work does not list the sufficient conditions under which expansions are possible.

3. LODGE [1956 : 3] has constructed such a theory earlier by using a molecular model. His book [1964 : 4] is a detailed treatment of such a constitutive relation.

RIVLIN[4] and applied by LANGLOIS and RIVLIN[5] in 1959, and COLEMAN and NOLL's treatment[6] provides an asymptotic status for these fluids in the hierarchy of simple fluids.

In view of the work of WINEMAN and PIPKIN[7], who proved rigorously that the constitutive functional of any incompressible simple fluid may be expressed as a sum of five linear functionals, the derivation of the results of COLEMAN and NOLL[8] on first- and second-order viscoelasticity is trivial. However, we shall present the derivation in the original spirit here, for it is instructive. We shall discuss WANG's theories[9] briefly as well, since these seem to offer a more realistic manner of looking at approximate constitutive relations.

Practically speaking, these approximate models provide no meaningful prescriptions for the stresses in strain histories differing from the rest history. Indeed, their responses in viscometric flows are poor. However, the finitely linear theory is very useful in providing inertia corrections for new rheometers[10] which have been developed to measure the complex viscosity $\eta^*(\omega)$. The second-order viscoelasticity theory has been used to predict the sign of the first normal stress difference in viscometric flows and serves as an example of simple fluid for which certain results valid in the additive functional theory do not hold[11].

The so-called order fluids have been useful in steady flows which are close to the Newtonian velocity fields. In such situations, these models indicate the effects of normal stresses on the streamlines and the extra forces needed to maintain a desired motion. But the second-order fluid, the simplest of such models, has severe drawbacks in non-steady flow problems and this has caused its use to be abandoned in initial value problems, for example. The Appendix to this chapter examines this question.

The best approximation to emerge so far is the BKZ fluid[12], which has been placed on a different theoretical setting by RIVLIN and SAWYERS[13]. We shall study the latter model in detail here.

Finally, in this chapter we discuss two constitutive relations of OLDROYD[14] and examine how they may be approximated by the order fluids or integral models.

---

4. [1957 : 1].
5. [1959 : 4].
6. [1960 : 1].
7. [1964 : 27].
8. [1961 : 1].
9. [1965 : 17], [1965 : 18].
10. E.g. MAXWELL rheometer. See ABBOTT and WALTERS [1970 : 1]. For other such rheometers, see [1971 : 1 ; 6], as well as Chaps. 10 and 11.
11. See Sec. 39.
12. [1963 : 1].
13. [1971 : 21].
14. [1958 : 4].

34. Histories of Small Norm. Finite Linear and Second-Order Viscoelasticity.[15] Static Continuation and Stress Relaxation[16]

The constitutive equation of a compressible simple fluid is given by

$$\mathbf{T} = -p(\rho)\,\mathbf{1} + \overset{\infty}{\underset{s=0}{\tilde{H}}}\,(\mathbf{G}(s);\,\rho), \tag{34.1}$$

where $\mathbf{G}(s) = \mathbf{C}_t(t-s) - \mathbf{1}$, and the density $\rho = \rho(t)$ acts as a parameter. We shall postulate that

$$\overset{\infty}{\underset{s=0}{\tilde{H}}}\,(\mathbf{0}\,;\,\rho) = \mathbf{0}, \tag{34.2}$$

or that in a state of rest, the stress is hydrostatic. Let us define the norm of $\mathbf{G}(s)$ through

$$\|\,\mathbf{G}(s)\,\| = \int_0^\infty |\,\mathbf{G}(s)\,|^2\, h^2(s)\, ds, \tag{34.3}$$

where $h^2(s) = [h(s)]^2$ and

(i) $|\,\mathbf{G}(s)^2\,| = \mathrm{tr}\{[\mathbf{G}(s)^2\}$;

(ii) $h(s)$, called an *influence function,* is a positive monotonically decreasing function of s, $h(0) = 1$, and which goes to zero fast enough as $s \to \infty$ so that the norm in (34.3) is finite. For example, *if* $h(s) > 0$ *for* $s \in [0, \infty)$ and

$$\lim_{s\to\infty} s^r h(s) = 0 \tag{34.4}$$

monotonically in s, *we say that* $h(s)$ *is an influence function of order* r. Note that $h(s) = (s+1)^{-p}$, $p > r$, satisfies (34.4), so does $h(s) = e^{-\beta s}$, $\beta > 0$; $e^{-\beta s}$ is an influence function of infinite order.

Now, the collection of all $\mathbf{G}(s)$, with a finite norm is not a vector space[17], because $\mathbf{1} + \mathbf{G}(s)$ must always be positive-definite. So we imbed the set $\mathcal{S}_1 = \{\mathbf{G}(s) : \|\,\mathbf{G}(s)\,\| < \infty\}$ in the set $\mathcal{S}$ consisting of all symmetric tensor-valued functions $\mathbf{J}(s)$ of s over $[0, \infty)$ with a finite norm. This set is a vector space, and forms a complete normed space.[18].

In this space any two histories $\mathbf{G}_1(s)$ and $\mathbf{G}_2(s)$ which differ on a set of measure zero, have the same norm $\|\cdot\|$. Hence (34.3) defines an equivalence class of histories and therefore to speak of $\mathbf{G}(s)$ as a point in the normed space is not quite correct. However, we shall continue to do so, with the understanding that we are discussing equivalence classes. In addition, we

15. Based on COLEMAN and NOLL [1961 : 1].
16. COLEMAN and NOLL [1962 : 4].
17. For a definition, see e.g. RUDIN [1966 : 17], p. 33.
18. RUDIN [1966 : 17], p. 95.

shall turn this space $\mathcal{S}$ of histories into a HILBERT space by defining an inner product through

$$[\mathbf{J}_1(s), \mathbf{J}_2(s)] = \int_0^\infty | \mathbf{J}_1(s)\ \mathbf{J}_2(s) |\ h^2(s)\ ds, \tag{34.5}$$

where

$$| \mathbf{J}_1(s)\ \mathbf{J}_2(s) | = \mathrm{tr}\{\mathbf{J}_1(s)\ \mathbf{J}_2(s)\}. \tag{34.6}$$

One could obtain a metric for this space through

$$d[\mathbf{J}_1(s), \mathbf{J}_2(s)] = \| \mathbf{J}_1(s) - \mathbf{J}_2(s) \|. \tag{34.7}$$

Note that this metric ensures that the history space is HAUSDORFF, that is, if $\mathbf{J}_1(s) \neq \mathbf{J}_2(s)$, in other words, if their norms are different, then we can choose two open sets $\mathcal{U}$ and $\mathcal{V}$ belonging to the space $\mathcal{S}$ such that $\mathbf{J}_1(s) \in \mathcal{U}$, $\mathbf{J}_2(s) \in \mathcal{V}$ but $\mathcal{U} \cap \mathcal{V} = \emptyset$. We shall postulate that :

The constitutive functional yields bounded stresses for all histories in $\mathcal{S}$.

Now consider any difference history $\mathbf{G}(s)$ quite close to $\mathbf{0}$, and let us approximate the response of the constitutive equation to this history as :

$$\underset{s=0}{\overset{\infty}{\tilde{\boldsymbol{H}}}} (\mathbf{G}(s)\ ;\ \rho) = \underset{s=0}{\overset{\infty}{\tilde{\boldsymbol{H}}}} (\mathbf{0}\ ;\ \rho) + \delta \underset{s=0}{\overset{\infty}{\tilde{\boldsymbol{H}}}} (\mathbf{0}\ ;\ \rho \mid \mathbf{G}(s))$$
$$+ \underset{s=0}{\overset{\infty}{\boldsymbol{R}}} (\mathbf{G}(s)\ ;\ \rho), \tag{34.8}$$

where the first FRECHET derivative of $\underset{s=0}{\overset{\infty}{\tilde{\boldsymbol{H}}}} (\mathbf{G}(s)\ ;\ \rho)$ about the zero history $\mathbf{0}$ is the continuous functional $\delta \underset{s=0}{\overset{\infty}{\tilde{\boldsymbol{H}}}} (\mathbf{0}\ ;\ \rho \mid \mathbf{G}(s))$, linear in $\mathbf{G}(s)$, and the error term $\underset{s=0}{\overset{\infty}{\boldsymbol{H}}} (\cdot\ ;\ \cdot)$ goes to zero faster than $\| \mathbf{G}(s) \|$ as $\| \mathbf{G}(s) \| \to 0$. Generally speaking, the first FRECHET derivative is defined, at arbitrary $\mathbf{G}(s)$, through

$$\delta \underset{s=0}{\overset{\infty}{\tilde{\boldsymbol{H}}}} (\mathbf{G}(s)\ ;\ \rho \mid \mathbf{g}(s)) = \frac{\partial}{\partial \alpha} \underset{s=0}{\overset{}{\tilde{\boldsymbol{H}}}} (\mathbf{G}(s) + \alpha \mathbf{g}(s)\ ;\ \rho) \Big|_{\alpha=0}. \tag{34.9}$$

To find a theory of finite linear viscoelasticity, by which we mean a linear theory of stress response about a state of rest, let us examine the linear functional in (34.8) closely. By virtue of the theorem in HILBERT space theory[19], which states that every continuous linear functional in a HILBERT space can be written as an inner product, we have, for example, that the component $T_E\langle 11 \rangle$ of the stress is given in Cartesian coordinates by

$$\delta \underset{s=0}{\overset{\infty}{\tilde{\boldsymbol{H}}}} (\mathbf{0}\ ;\ \rho \mid \mathbf{G}(s)) = \int_0^\infty \mathrm{tr}\{\boldsymbol{\Upsilon}_{11}(s\ ;\ \rho)\ \mathbf{G}(s)\}\ h^2(s)\ ds. \tag{34.10}$$

19. RUDIN [1966 : 17], p. 80.

Note that $\delta \overset{\infty}{\underset{s=0}{\tilde{\boldsymbol{H}}_{11}}}(\cdot\,;\,\cdot\,|\,\cdot)$ is a linear functional of $\mathbf{G}(s)$ and $\boldsymbol{\Upsilon}_{11}(s\,;\,\rho)$ is an element of $\mathcal{S}$ which is uniquely determined[20]. This means that once we prescribe $h(s)$ and the linear functional, the latter is given by (34.10) with $\boldsymbol{\Upsilon}_{11}(s\,;\,\rho)$ *always the same*, whatever $\mathbf{G}(s)$ is. Collecting all the six components of $\delta\,\tilde{\boldsymbol{H}}$ together, we have

$$\delta \overset{\infty}{\underset{s=0}{\tilde{\boldsymbol{H}}_{ij}}}(\mathbf{0}\,;\,\rho\,|\,\mathbf{G}(s)) = \int_0^\infty \Upsilon_{ijkl}(s\,;\,\rho)\,\mathbf{G}_{kl}(s)\,h^2(s)\,ds. \tag{34.11}$$

Note that each $\boldsymbol{\Upsilon}_{ij}$ is a symmetric tensor with components Υ_{ijkl} and the trace operation is incorporated into the summation convention. Hence we may write (34.10) as

$$\delta \overset{\infty}{\underset{s=0}{\tilde{\boldsymbol{H}}}}(\mathbf{0}\,;\,\rho\,|\,\mathbf{G}(s)) = \int_0^\infty \underline{\underline{\Gamma}}(s\,;\,\rho)\,[\mathbf{G}(s)]\,ds, \tag{34.12}$$

where $\underline{\underline{\Gamma}}(s\,;\,\rho)$ is a fourth-order tensor with components $\Upsilon_{ijkl}(s\,;\,\rho)$, and it operates on the symmetric second-order tensor $\mathbf{G}(s)$, yielding a symmetric second-order tensor.

Since each $\boldsymbol{\Upsilon}_{ij}(s\,;\,\rho)$ appearing in (34.10) has a finite norm, we have, for example,

$$\|\,\boldsymbol{\Upsilon}_{11}(s)\,\| = \int_0^\infty \{\Upsilon_{11kl}(s\,;\,\rho)\,\Upsilon_{11kl}(s\,;\,\rho)\}\,h^2(s)\,ds < \infty. \tag{34.13}$$

If we write the stress tensor T_{ij} as $T_1, T_2, T_3, T_4, T_5, T_6$ for $T\langle 11\rangle$, $T\langle 22\rangle$, $T\langle 33\rangle$, $T\langle 12\rangle$; $T\langle 23\rangle$, $T\langle 31\rangle$ respectively, with similar notation for $\mathbf{G}(s)$, one can write (34.12) as

$$\begin{bmatrix} T_1 \\ T_2 \\ T_3 \\ T_4 \\ T_5 \\ T_6 \end{bmatrix} = \begin{bmatrix} \Gamma_{11} & \Gamma_{12} & \Gamma_{13} & \Gamma_{14} & \Gamma_{15} & \Gamma_{16} \\ \Gamma_{21} & \Gamma_{22} & \cdot & \cdot & \cdot & \Gamma_{26} \\ \Gamma_{31} & & & \cdot & & \cdot \\ \cdot & & & \cdot & & \cdot \\ \cdot & & & & \cdot & \cdot \\ \Gamma_{61} & \Gamma_{62} & \cdot & \cdot & \cdot & \Gamma_{66} \end{bmatrix} \begin{bmatrix} G_1 \\ G_2 \\ G_3 \\ G_4 \\ G_5 \\ G_6 \end{bmatrix}, \tag{34.14}$$

where $\Gamma_{11} = \Upsilon_{1111}$, etc. Then (34.13) is equivalent to

$$\int_0^\infty |\,\boldsymbol{\Gamma}(s\,;\,\rho)\,|^2\,h^2(s)\,ds < \infty, \tag{34.15}$$

where

$$\boldsymbol{\Gamma}(s)\,;\,\rho\,|^2 = \sum_{i,j} \Gamma_{ij}^2. \tag{34.16}$$

20. RUDIN [1966 : 17], p. 80.

Next, a simple fluid is always isotropic. Thus

$$\mathbf{Q}\,\delta \overset{\infty}{\underset{s=0}{\tilde{H}}}(\mathbf{0}\,;\rho \mid \mathbf{G}(s))\,\mathbf{Q}^T = \delta \overset{\infty}{\underset{s=0}{\tilde{H}}}(\mathbf{0}\,;\rho \mid \mathbf{Q}\,\mathbf{G}(s)\,\mathbf{Q}^T) \tag{34.17}$$

for all constant orthogonal tensors **Q**. Using this in (34.11) and using the linearity in $\mathbf{G}(s)$ of the linear functional leads us to the following :

$$\delta \overset{\infty}{\underset{s=0}{\tilde{H}}}(\mathbf{0}\,;\rho \mid \mathbf{G}(s)) = \int_0^\infty [\lambda(s\,;\rho)\,\{\mathrm{tr}\,\mathbf{G}(s)\}\,\mathbf{1} + \\ + \mu(s\,;\rho)\,\mathbf{G}(s)]\,ds. \tag{34.18}$$

In incompressible materials, **T** is determined upto a pressure p only ; thus absorbing the term involving **1** into $p\mathbf{1}$, we have the constitutive equation of an incompressible fluid, obeying the theory of finite linear viscoelasticity :

$$\mathbf{T}_E = \mathbf{T} + p\mathbf{1} = \int_0^\infty \mu(s)\,\mathbf{G}(s)\,ds. \tag{34.19}$$

In defining (34.19), we have dropped the remainder term in (34.8) which goes to zero faster than $\|\mathbf{G}(s)\|$ as $\mathbf{G}(s) \to \mathbf{0}$, and of course omitted the dependence on ρ. We call (34.19) a theory of finite linear viscoelasticity, because the *finite strain measure* is used and the equation is *linear* in the strain history.

To obtain a second-order theory of viscoelasticity, the functional $\overset{\infty}{\underset{s=0}{\tilde{H}}}(\mathbf{G}(s)\,;\rho)$ is assumed to be twice differentiable about the zero history **0**, so that

$$\overset{\infty}{\underset{s=0}{\tilde{H}}}(\mathbf{G}(s)\,;\rho) = \delta \overset{\infty}{\underset{s_1,\,s_2=0}{\tilde{H}}}((\mathbf{0}\,;\rho) \mid \mathbf{G}(s)) + \delta^2 \overset{\infty}{\underset{s_1,\,s_2=0}{\tilde{H}}}(\mathbf{0}\,;\rho \mid \mathbf{G}(s_1), \mathbf{G}(s_2)) + \\ + \overset{\infty}{\underset{s=0}{R'}}(\mathbf{G}(s)\,;\rho), \tag{34.20}$$

where the remainder term goes to zero faster than $\|\mathbf{G}(s)\|^2$ as $\mathbf{G}(s) \to \mathbf{0}$.

The second FRÉCHET derivative $\delta^2 \overset{\infty}{\underset{s=0}{\tilde{H}}}(\mathbf{G}(s)\,;\rho \mid \mathbf{g}(s), \mathbf{h}(s))$ is defined by

$$\delta \overset{\infty}{\underset{s=0}{\tilde{H}}}(\mathbf{G}(s)\,;\rho \mid \mathbf{g}(s), \mathbf{h}(s)) = \frac{\partial^2}{\partial\alpha\,\partial\beta} \overset{\infty}{\underset{s=0}{\tilde{H}}}(\mathbf{G}(s) + \alpha\mathbf{g}(s) + \beta\mathbf{h}(s)\,;\rho)\Big|_{\substack{\alpha=0\\ \beta=0}} \\ = \delta^2 \overset{\infty}{\underset{s=0}{\tilde{H}}}(\mathbf{G}(s)\,;\rho \mid \mathbf{h}(s), \mathbf{g}(s)). \tag{34.21}$$

While the linearity and continuity of the functional $\delta \overset{\infty}{\underset{s=0}{\tilde{H}}}(\cdot\,;\cdot \mid \cdot)$, assures its boundedness[21], such a statement is not true in general for the second deri-

21. This is a general result in HILBERT space theory. See e.g. AKHIEZER and GLAZMAN [1966 : 1], p. 39.

vative $\delta^2 \overset{\infty}{\underset{s=0}{\tilde{H}}}(\cdot\,;\cdot|\cdot,\cdot)$. However, we shall assume its boundedness and state its properties :

(i) $\delta^2 \overset{\infty}{\underset{s=0}{\tilde{H}}}(\mathbf{G}(s)\,;\,\rho\mid \mathbf{g}(s), \mathbf{h}(s))$ is symmetric in $\mathbf{g}(s)$ and $\mathbf{h}(s)$, which is implied in (34.21) ;

(ii) this derivative is bilinear ; in other words, if we keep $\mathbf{g}(s)$ fixed, then

$$\delta^2 \overset{\infty}{\underset{s=0}{\tilde{H}}}(\mathbf{G}(s)\,;\,\rho\mid \mathbf{g}(s), \quad \alpha\mathbf{h}_1(s) + \beta\mathbf{h}_2(s))$$

$$= \alpha\,\delta^2 \overset{\infty}{\underset{s=0}{\tilde{H}}}(\mathbf{G}(s)\,;\,\rho\mid \mathbf{g}(s), \quad \mathbf{h}(s)) + \beta\,\delta^2 \overset{\infty}{\underset{s=0}{\tilde{H}}}(\mathbf{G}(s)\,;\,\rho\mid \mathbf{g}(s), \mathbf{h}_2(s)).$$

Similarly, the derivative is linear in $\mathbf{g}(s)$ for a fixed $\mathbf{h}(s)$.

Property (ii) and the boundedness imply that the second derivative is continuous in $\mathbf{g}(s)$ and $\mathbf{h}(s)$[22]. Recalling (34.20), since $\delta^2 \overset{\infty}{\underset{s_1,\,s_2=0}{\tilde{H}}}(\mathbf{0}\,;\,\rho\mid \mathbf{G}(s_1), \mathbf{G}(s_2))$ is a bilinear functional, by the representation theorem for such functionals[23],

$$\delta^2 \overset{\infty}{\underset{s_1,\,s_2=0}{\tilde{H}}}(\mathbf{0}\,;\,\rho\mid \mathbf{G}(s_1), \mathbf{G}(s_2)) = (\mathcal{L}\mathbf{G}(s_1), \mathbf{G}(s_2)), \tag{34.22}$$

where $\mathcal{L}[\,\cdot\,]$ is a linear operator, and the inner product in (34.22) is in the sense of (34.5) — (34.6).

An application of the previously used representation theorem for the linear functionals in (34.22), the isotropy of the material and the fact that the second FRECHET derivative is symmetric in $\mathbf{G}(s_1)$ and $\mathbf{G}(s_2)$ lead to the following expansion :[24]

$$\begin{aligned}\delta^2 \overset{\infty}{\underset{s_1,\,s_2=0}{H}}(\mathbf{0}\,;\,\rho\mid \mathbf{G}(s_1), \mathbf{G}(s_2)) = & \int_0^\infty\!\!\int_0^\infty \{\gamma_1(s_1, s_2\,;\,\rho)\,[\operatorname{tr}\mathbf{G}(s_1)]\,[\operatorname{tr}\mathbf{G}(s_2)] + \\ & + \gamma_2(s_1, s_2\,;\,\rho)\operatorname{tr}[\mathbf{G}(s_1)\,\mathbf{G}(s_2)]\}\,ds_1\,d_2\,\mathbf{1} + \\ & + \int_0^\infty\!\!\int_0^\infty \{\beta(s_1, s_2\,;\,\rho)\,[\operatorname{tr}\mathbf{G}(s_1)]\,\mathbf{G}(s_2) + \\ & + \alpha(s_1, s_2\,;\,\rho)\,\mathbf{G}(s_1)\,\mathbf{G}(s_2)\}\,ds_1\,ds_2.\end{aligned} \tag{34.23}$$

In (34.23), to preserve the symmetry in $\mathbf{G}(s_1)$ and $\mathbf{G}(s_2)$, we demand that $\alpha(s_1, s_2 : \rho) = \alpha(s_2, s_1\,;\,\rho)$. The function $\beta(s_1, s_2\,;\,\rho)$ is also symmetric, since its skew symmetric part can be shown to be zero[25] as follows. Calculate $\mathbf{G}(s)$

22. AKHIEZER and GLAZMAN [1966 ; 1], p. 40.
23. *Ibid.*, p. 42.
24. NFTM, Eq. (37.9) ; COLEMAN and NOLL [1962 : 4], Eq. (6.2*a*).
25. Restrictions on $\alpha(s_1, s_2\,;\,\rho)$ and $\beta(s_1, s_2\,;\,\rho)$ were first reported by COLEMAN and NOLL [1961 : 1] and [1962 : 4].

for a simple shear flow $\dot{x} = \varkappa y$, $\dot{y} = \dot{z} = 0$. Then the component $\mathbf{G}_\alpha(s)\langle yy\rangle = s_\alpha^2\,\varkappa = \text{tr}\,\mathbf{G}(s_\alpha)$, $\alpha = 1, 2$. Hence

$$\int_0^\infty \int_0^\infty (\beta(s_1, s_2\,;\,\rho) - \beta(s_2, s_1\,;\rho))\,\varkappa^4\, s_1^2\, s_2^2\, ds_1 ds_2 = 0,$$

which implies

$$\beta(s_1, s_2\,;\,\rho) = \beta(s_2, s_1\,;\,\rho).$$

For incompressible fluids, absorbing the terms involving **1** into $-p\mathbf{1}$, omitting dependence on ρ, and neglecting the error term in (34.21), we obtain the *constitutive equation for second-order viscoelasticity* as:

$$\mathbf{T}_E = \mathbf{T} + p\mathbf{1} = \int_0^\infty \mu(s)\,\mathbf{G}(s)\,ds + \int_0^\infty \int_0^\infty \{\beta(s_1, s_2)\,[\text{tr}\,\mathbf{G}(s_1)]\,\mathbf{G}(s_2) + \alpha(s_1, s_2)\,\mathbf{G}(s_1)\,\mathbf{G}(s_2)\}\,ds_1 ds_2. \quad (34.24)$$

We now record a refinement due to to PIPKIN[26], which is applicable to incompressible fluids. Since, in such materials, $\det \mathbf{C}_t(t - s) = \det[\mathbf{1} + \mathbf{G}(s)] = 1$, by the expansion of this determinant[27] one has

$$\text{tr}\,\mathbf{G}(s) + [\text{tr}\,\mathbf{G}(s)]^2 - \text{tr}\,\mathbf{G}^2(s) + \tfrac{2}{3}\{2\,\text{tr}\,\mathbf{G}^3(s) - 3[\text{tr}\,\mathbf{G}(s)]\,[\text{tr}\,\mathbf{G}^2(s)] + (\text{tr}\,\mathbf{G}(s))^3\} = 0. \quad (34.25)$$

Hence $\text{tr}\,\mathbf{G}(s)$ is always expressible in terms of second and higher-order invariants. Thus (34.24) may be shortened to[26]

$$\mathbf{T}_E = \int_0^\infty \mu(s)\,\mathbf{G}(s)\,ds + \int_0^\infty \int_0^\infty \alpha(s_1, s_1)\,\mathbf{G}(s_1)\mathbf{G}(s_2)\,ds_1 ds_2. \quad (34.26)$$

PIPKIN[26] has also shown how to obtain a third-order expansion. This is obtained by adding

$$\int_0^\infty \int_0^\infty \int_0^\infty \{\beta\{s_1, s_2, s_3)\,\mathbf{G}(s_1)\,\text{tr}\,[\mathbf{G}(s_2)\,\mathbf{G}(s_3)] + \gamma(s_1, s_2, s_3)\,\mathbf{G}(s_1)\,\mathbf{G}(s_2)\,\mathbf{G}(s_3)\}\,ds_1\,ds_2\,ds_3 \quad (34.27)$$

to (34.26). Since (34.27) is essentially the third FRÉCHET derivative

$$\delta^3 \underset{s_1,\, s_2\, s_3 = 0}{\overset{\infty}{\tilde{H}}} (0\,;\,\rho \mid \mathbf{G}(s_1), \mathbf{G}(s_2), \mathbf{G}(s_3)),$$

assumed to be bounded and multilinear, that is, linear in each $\mathbf{G}(s_i)$, $i = 1, 2, 3$ whenever the other two $\mathbf{G}(s_i)$ are fixed, and symmetric in the $\mathbf{G}(s_i)$, restrictions on the kernels $\beta(s_1, s_2, s_3)$ and $\gamma(s_1, s_2, s_3)$ arise. These are, however, not recorded here.

26. [1964 : 18].
27. This is derived by writting the CAYLEY-HAMILTON expansion for $\mathbf{1} + \mathbf{G}(s)$ and taking the trace of this expansion, which gives the determinant.

We now offer some comments on the principle of fading memory :

(i) Demanding that the order r of the influence function obey $r > \frac{1}{2}$ yields simply the continuity of the constitutive functional in $\mathbf{G}(s)$, the continuity being with respect to the norm (34.3). This restriction on $h(s)$ yields the *weak principle of fading memory*[28], and the stress relaxation property ;

(ii) Demanding that $r > n + \frac{1}{2}$, where n is the number of times one wishes the constitutive functional to be FRÉCHET differentiable, yields *the strong principle of fading memory*[29], and the integral approximations (see also Sec. 35) ;

(iii) It has already been remarked that the set of histories $\mathbf{G}(s)$ does not form a vector space. We had to extend the domain of the constitutive functional to $\mathbf{J}(s)$ to obtain the expansions (34.18) and (34.23). For incompressible fluids, $\mathbf{G}(s)$ obeys the additional restriction det $[\mathbf{G}(s)+\mathbf{1}] = 1$. Since the domain of stress functional of incompressible fluids is restricted to the class of $\mathbf{G}(s)$ such that (*a*) $\| \mathbf{G}(s) \|$ is finite, (*b*) $\mathbf{G}(s)+\mathbf{1}$ is positive-definite and symmetric, and (*c*) det $[\mathbf{G}(s)+\mathbf{1}] = 1$, we have to extend this domain to apply to all $\mathbf{J}(s)$. One way of doing this is to demand that[30]

$$\overset{\infty}{\underset{s=0}{\tilde{H}}} (\mathbf{J}(s)) \equiv \overset{\infty}{\underset{s=0}{\tilde{H}}} ([\det (\mathbf{J}(s) + \mathbf{1})]^{-1} [\mathbf{J}(s) + \mathbf{1}] - \mathbf{1}). \qquad (34.28)$$

This extension of the functional to all of $\mathcal{S}$ is necessary to use the representation theorem for linear functionals, for example. If we assume that such an extension (34.28) obeys the strong principle of fading memory, all the previous results are valid.

At this juncture, it is worthwhile to point out that the above formulation of strong fading memory is motivated by the desire to obtain finite linear viscoelasticity as the first approximation. As a consequence, this type of fading memory leads to the following unphysical characteristic of the simple fluid : if the fluid particle has been at rest for an infinitely long time, and suddenly one imposes a jump in $\mathbf{A}_1$ and no other kinematical variable, the stress in the fluid would not change[31]. Had one defined the simple fluid constitutive equation[32] to be a functional of $\mathbf{G}(s)$ *and* $\mathbf{A}_1, \ldots, \mathbf{A}_N$, that is,

$$\mathbf{T}_E = \overset{\infty}{\underset{s=0}{\bar{\bar{H}}}} (\mathbf{G}(s) ; \mathbf{A}_1(t), \ldots, \mathbf{A}_N(t)), \qquad (34.29)$$

the above objection cannot be made. We shall revert to this point in Sec. 35.

28. The terminology is due to WANG [1965 : 17].
29. This terminology is due to WANG [1965 : 18].
30. COLEMAN and NOLL [1960 : 1].
31. OLDROYD [1965 : 13] seems to have been the first to state this explicitly.
32. GREEN and RIVLIN [1960 : 2], Eq. (6.13).

A second objection to the above principle of fading memory was raised by WANG[33], who pointed out that under this formulation, a Newtonian fluid does not have fading memory. Physically speaking, it does; its memory is 'catastrophic'. Because a Newtonian fluid is a special case of the so-called RIVLIN-ERICKSEN fluid, we shall defer the discussion of this point also to Sec. 35, showing there that (34.29) is equivalent to WANG's theory.

We shall now return to the COLEMAN and NOLL's principle of fading memory and establish some consequences of it.

34.1. Static Continuation and Stress Relaxation. In Sec. 6, we discussed the form taken by the history $\mathbf{C}_t(t-s)$ when the material is held in its current configuration for a time $\delta = t - t_0 > 0$ from an instant t_0. It was shown there that

$$\mathbf{C}_t(t-s) = \begin{cases} \mathbf{1}, & 0 \leqq s \leqq \delta, \\ \mathbf{C}_{t_0}(t_0 - s + \delta), & \delta < s < \infty. \end{cases} \tag{6.4}$$

Consequently, in *static continuation*,

$$\mathbf{G}(s) = \begin{cases} \mathbf{0}, & 0 \leqq s \leqq \delta, \qquad (34.30) \\ \mathbf{C}_{t_0}(t_0 - s + \delta) - \mathbf{1}, & \delta < s < \infty. \qquad (34.31) \end{cases}$$

Clearly, as $\delta \to \infty$, the material would have been held in its current state for an infinitely long time and it would have been subjected to a zero difference history, except perhaps at an infinitely long time in the past. Intuitively, we expect the stresses to be zero, on account of relaxation. Here we see, mathematically, that in static continuation

$$\lim_{\delta \to \infty} \| \mathbf{G}(s) \| = 0. \tag{34.32}$$

Or, as $\delta \to \infty$, $\| \mathbf{G}(s) \| = 0$ and by the continuity of the stress functional (that is, the weak principle of fading memory), it follows that in *static continuation*:

$$\lim_{\delta \to \infty} \ \underset{s=0}{\overset{\infty}{\tilde{H}}} (\mathbf{G}(s); \rho) = \mathbf{0}. \tag{34.33}$$

We have shown therefore that the stresses relax in static continuation[34].

34.2. A Bounded Jump in Strain implies a Bounded Stress Jump. Suppose, as in (6.10)—(6.12), there is a sudden jump in strain history at time t_0. Then for $t > t_0$,

$$\begin{aligned} | \mathbf{C}_t(t-s) |^2 &= \operatorname{tr} \{\mathbf{F}_t(t_0)^T \mathbf{C}_{t_0}(t_0-s) \mathbf{F}_t(t_0) \mathbf{F}_t(t_0)^T \mathbf{C}_{t_0}(t_0-s) \mathbf{F}_t(t_0)\} \\ &= | \mathbf{C}_{t_0}(t_0-s) \mathbf{B}_t(t_0) |^2, \ \mathbf{B}_t(t_0) = \mathbf{F}_t(t_0) \mathbf{F}_t(t_0)^T \qquad (34.34) \\ &\leqq | \mathbf{C}_{t_0}(t_0-s) |^2 \, | \mathbf{B}_t(t_0) |^2. \qquad (34.35) \end{aligned}$$

33. [1965 : 18].
34. COLEMAN and NOLL [1962 : 4].

Thus $|\mathbf{G}(s)|^2 \leqq |\mathbf{C}_{t_0}(t_0-s)|^2 \, |\mathbf{B}_t(t_0)|^2$ and provided the difference history $\mathbf{G}_0(s)$, corresponding to $\mathbf{C}_{t_0}(t_0-s)$, belonged to $\mathcal{S}$ and $\mathbf{B}_t(t_0)$ is bounded, the norm of $\mathbf{G}(s)$, defined through (34.3), will then be bounded. Hence the stress jump which occurs due to the strain jump must also be bounded, because of our postulate on p. 162. Physically speaking, this is translated as saying that the simple fluid behaves like an elastic solid when subjected to a sudden jump in strain history†. This concept becomes transparent if one were to look at the constitutive equation of linearly viscoelastic fluid (34.*b*4), for example.

We note in passing that all of the expansions can also be derived from the work of WINEMAN and PIPKIN[35]. Their representation theorem for the constitutive equation of an incompressible simple fluid states that

$$\mathbf{T}_E = \sum_{\beta=1}^{5} \underline{\underline{\mathcal{L}}}^{\beta}\{\mathbf{f}^{(\beta)}\,;\, I_1, \ldots, I_6\}, \tag{34.36}$$

where

$$I_1 = \operatorname{tr} \mathbf{G}(s)\,; \qquad I_2 = \operatorname{tr}\{\mathbf{G}(s_1)\,\mathbf{G}(s_2)\}, \ldots$$
$$I_6 = \operatorname{tr}\{\mathbf{G}(s_1)\,\mathbf{G}(s_2)\ldots\mathbf{G}(s_6)\}, \tag{34.37}$$
$$2\mathbf{f}^{(\beta)} = [\mathbf{G}(s_1)\ldots\mathbf{G}(s_\beta) + \mathbf{G}(s_\beta)\ldots\mathbf{G}(s_1)], \tag{34.38}$$

and $\underline{\underline{\mathcal{L}}}^{(\beta)}$, for each β, is a linear functional of $\mathbf{f}^{(\beta)}$, with $I_1, \ldots, I_6$ acting as parameters.

Basically what (34.36) says is that in the incompressible material, there are at most five independent material functionals—the two normal stress differences and the three shear stresses. For reasons of economy of space and because (34.36) is useful conceptually rather than for practical problem solving, we shall not develop (34.36) any further here. The only use so far of the expansion like (34.36) is due to CARROLL[36], who showed that certain deformations possible in isotropic finitely elastic materials were also tractable in isotropic simple materials.

We shall examine the relation between the linearly viscoelastic fluid and the simple fluid theory next.

34bis. Relation to Linear Viscoelasticity

We first recapitulate briefly the material from Sec. 5.

Let $\mathbf{F}(t)$ be the deformation gradient at a particle at time t, computed with respect to a fixed reference configuration. Let us assume that the displacement gradient $\mathbf{H}(t) = \mathbf{F}(t) - \mathbf{1}$ is very small in magnitude, that is, let $\varepsilon = \sup_{\tau} |\mathbf{H}(\tau)| \ll 1$. Then

$$\mathbf{F}_t(\tau) = \mathbf{1} + \mathbf{H}(\tau) - \mathbf{H}(t) + 0(\varepsilon^2), \qquad -\infty < \tau < \infty. \tag{34.b1}$$

†This result was derived in a different way by LEIGH [1968 : *A*]. See also [1966 : 13] and [1966 : 15].

35. [1964 : 27], Eq. (3.12), p. 205. By the expansions, we mean not only those due to COLEMAN and NOLL, but also due to GREEN and RIVLIN [1957 : 1].

36. [1967 : 5]. See also FOSDICK [1968 : 8].

Hence, with $2\mathbf{E}(\tau) = \mathbf{H}(\tau) + \mathbf{H}(\tau)^T$, one has

$$\mathbf{C}_t(\tau) = \mathbf{1} + 2[\mathbf{E}(\tau) - \mathbf{E}(t)] + \mathbf{0}(\varepsilon^2). \tag{34.b2}$$

Thus, in the linear theory of viscoelasticity, where it is assumed that $|\mathbf{H}(t)|$ is very small, one has that

$$\mathbf{G}(s) = 2[\mathbf{E}(t-s) - \mathbf{E}(t)] + \mathbf{0}(\varepsilon^2). \tag{34.b3}$$

The classical theory of incompressible linearly viscoelastic fluids is based on

$$\mathbf{T}_E = 2G(0)\,\mathbf{E}(t) + \int_0^\infty 2\dot{G}(s)\,\mathbf{E}(t-s)\,ds, \tag{34.b4}$$

where $\dot{G}(s) = \dfrac{dG(s)}{ds}$ and $G(s)$ is called the *relaxation modulus.* Noting that $\lim_{s\to\infty} G(s) = 0$, (34.$b$4) is equivalent to

$$\mathbf{T}_E = \int_0^\infty 2\dot{G}(s)\,[\mathbf{E}(t-s) - \mathbf{E}(t)]\,ds = -\int_0^\infty G(s)\,\frac{d\mathbf{G}(s)}{ds}\,ds. \tag{34.b5}$$

Comparing (34.b5) with (34.19), we learn that they are the same, except that $(34.b5)_1$ is obtained by replacing $\mathbf{G}(s)$ in (34.19) with $\mathbf{G}(s)$ in (34.b3) without the error term $O(\varepsilon^2)$ and putting

$$\dot{G}(s) = \mu(s), \qquad G(s) = -\int_s^\infty \mu(\sigma)\,d\sigma. \tag{34.b6}$$

Thus, COLEMAN and NOLL's results[37] accord a proper place for the linear theory in the sequence of approximations to the general theory of simple fluids.

For use below, we note that the *shear storage modulus* $G'(\omega)$ and the *shear loss modulus* $G''(\omega)$ of the linear theory are given by

$$G'(\omega) = \omega \int_0^\infty G(s)\sin\omega s\,ds = \omega\eta''(\omega), \tag{34.b7}$$

$$G''(\omega) = \omega \int_0^\infty G(s)\cos\omega s\,ds = \omega\eta'(\omega), \tag{34.b8}$$

where $\eta'(\omega)$ is called the *dynamic shear viscosity.* Moreover, we anticipate that $G(s) > 0$, and that $\dot{G}(s) < 0$, as well as that

$$\lim_{s\to\infty} s^2 G(s) = 0, \tag{34.b9}$$

or $G(s)$ goes to zero rapidly as s becomes very large. In the next section we shall discuss some consequences of the above assumptions and relate the behaviour of a viscoelastic fluid, obeying the theory of finite linear viscoelasticity, in a MAXWELL orthogonal rheometer with (34.b7)—(34.b8). We also connect (34.b7)—(34.b8) with two other rheometrical systems studied by

37. [1961 : 1].

JONES and WALTERS[38] and ABBOTT and WALTERS[39].

For future use, we define the *complex viscosity* $\eta^*(\omega)$ as

$$\eta^*(\omega) = \eta'(\omega) - i\eta''(\omega) = \int_0^\infty G(s)\, e^{-i\omega s} ds \tag{34.b10}$$

$$= \frac{i}{\omega} \int_0^\infty \mu(s)\, [1 - e^{-i\omega s}]\, ds.$$

34 ter. Response in MWCSH and other Special Flows

For a fluid obeying the finite linear theory (34.19),

$$\mathbf{T}_E = \int_0^\infty \mu(s)\, \mathbf{G}(s)\, ds.$$

This constitutive relation will now be used to record its response in some flows.

(i) *Simple shear with* $\dot{x} = \varkappa y$, $\dot{y} = \dot{z} = 0$. The viscometric material functions are

$$\bar{N}_1(\varkappa^2) = T\langle xx\rangle - T\langle yy\rangle = -\varkappa^2 \int_0^\infty \mu(s)\, s^2 ds\,; \tag{34.C1}$$

$$\bar{N}_2(\varkappa^2) = T\langle yy\rangle - T\langle zz\rangle = \varkappa^2 \int_0^\infty \mu(s)\, s^2 ds\,; \tag{34.C2}$$

$$\tau(\varkappa) = T\langle xy\rangle = -\varkappa \int_0^\infty \mu(s)\, s ds = \varkappa\eta_0. \tag{34.C3}$$

Thus, for this fluid,

$$\overline{N_1}(\varkappa^2) = -\overline{N_2}(\varkappa^2) \tag{34.C4}$$

for all values of $\varkappa$. Also the viscosity η_0 is a constant and

$$\eta_0 = -\int_0^\infty s\mu(s)\, ds = \int_0^\infty G(s)\, ds > 0, \tag{34.C5}$$

from (34.*b*6) and (34.*b*9).

(ii) *Maxwell orthogonal rheometer flow, with* $\dot{x} = -\Omega y + \Omega\psi z$, $\dot{y} = \Omega x$, $\dot{z} = 0$. The strain history is given by (7.45) and to ensure the norm of $\|\mathbf{G}(s)\|$ to be small enough so that the theory of finite linear viscoelasticity applies, we take ψ to be small and neglect terms of order ψ^2. Then

$$T_E\langle xx\rangle = 0, \quad T_E\langle yy\rangle = 0, \quad T\langle xy\rangle = 0\,; \tag{34.C6}$$

$$T\langle xz\rangle = -\psi \int_0^\infty \mu(s) \sin \Omega s ds\,; \tag{34.C7}$$

38. [1969 : 10]. Instrument called Kepes balance rheometer. See also WALTER's [1970 : 2].
39. [1970 : 2]. Flow between eccentric rotating cylinders.

$$T\langle yz\rangle = -\psi \int_0^\infty \mu(s)\,(1 - \cos\Omega s)\,ds\,; \tag{34.C8}$$

$$T_E\langle zz\rangle = 2\psi^2 \int_0^\infty \mu(s)\,(1 - \cos\Omega s)\,ds \approx 0. \tag{34.C9}$$

Using $\mu(s) = dG(s)/ds$ in (34.*C*7), applying the RIEMANN-LEBESGUE lemma[40], one obtains through integration by parts, the following result :

$$T\langle xz\rangle = \psi\Omega \int_0^\infty G(s)\cos\Omega s\,ds = \psi G''(\Omega). \tag{34.C10}$$

Hence a measurement of $T\langle xz\rangle$, at very small ψ and arbitrary Ω in this rheometer, yields the dynamic viscosity $\eta'(\Omega)$ or the modulus $G''(\Omega)$[41]. Similarly,

$$T\langle yz\rangle = \psi G'(\Omega). \tag{34.C11}$$

We now relate the formulae (34.*C*10) and (34.*C*11) to (32.11*d*) and (32.11*e*) respectively. Since ψ is very small and terms of order ψ^2 are neglected,

$$\Omega\psi f_{13}(\Omega^2, 0) = \psi G''(\Omega) = \Omega\psi\,\eta'(\Omega), \tag{34.C12}$$

$$\Omega^2\psi f_{23}(\Omega^2, 0) = \psi G'(\Omega) = \Omega\psi\,\eta''(\Omega). \tag{34.C13}$$

By definition $\eta'(\Omega)$ and $\eta''(\Omega)/\Omega$ are both even functions of Ω; this is clearly reflected in the above equations[42].

There are two other situations in which the shear storage and loss modulii can be measured by performing experiments in which a parameter is kept very small. In (2.22) – (2.30) and (3.7) – (3.8), we have already discussed the flow in the balance rheometer, also called the 'Kepes apparatus.' Using the strain history (3.8) in the constitutive relation (34.19), the shear stresses $T\langle r\theta\rangle$ and $T\langle r\phi\rangle$ in this rheometer are given by[43]

$$T\langle r\theta\rangle = \frac{3\alpha\lambda r_1^3}{r^3}\int_0^\infty \frac{dG(s)}{ds}\,[\sin\phi\sin\Omega s + {} + \cos\phi\,(\cos\Omega s - 1)]\,ds, \tag{34.C14}$$

$$T\langle r\phi\rangle = \cos\theta\,\frac{3\alpha\lambda r_1^3}{r^3}\int_0^\infty \frac{dG(s)}{ds}\,[\cos\phi\sin\Omega s + {} + \sin\phi\,(1 - \cos\Omega s)]\,ds. \tag{34.C15}$$

In view of the relations (34.*b*7) and (34.*b*8), (34.*C*14) and (34.*C*15) can be

40. E.g. CHURCHILL [1963 : 2].
41. HUILGOL [1969 : 8]. The relations (34.*C*10) and (34.*C*11) also appear in ABBOTT and WALTERS [1970 : 1].
42. PIPKIN [1969 : 18], Eq. (5.3.10).
43. JONES and WALTERS [1969 : 10].

written as

$$\left.\begin{aligned} T\langle r\theta\rangle &= -\frac{3\alpha\lambda\Omega r_1^3}{r^3}[\eta'(\Omega)\sin\phi - \eta''(\Omega)\cos\phi], \\ T\langle r\phi\rangle &= -\frac{3\alpha\lambda\Omega r_1^3}{r^3}\cos\theta[\eta'(\Omega)\cos\phi + \eta''(\Omega)\sin\phi] \end{aligned}\right\}. \tag{34.C16}$$

If the couples about x, y, z axes, fixed in the sphere of radius r_1, are respectively denoted by M_x, M_y, M_z, we have:

$$M_x = -r_1^3\int_0^{2\pi}\int_0^{\pi}(T\langle r\theta\rangle\sin\phi + T\langle r\phi\rangle\cos\phi\cos\theta)\sin\theta\, d\theta\, d\phi,$$

$$M_y = r_1^3\int_0^{2\pi}\int_0^{\pi}(T\langle r\theta\rangle\cos\phi - T\langle r\phi\rangle\sin\phi\cos\theta)\sin\theta\, d\theta d\phi,$$

$$M_z = r_1^3\int_0^{2\pi}\int_0^{\pi} T\langle r\phi\rangle\sin^2\theta\, d\theta\, d\phi. \tag{34.C17}$$ [44]

In (34.C17), the stresses are evaluated at $r = r_i$. Substitution of (34.C16) into (34.C17) leads to

$$M_x = 8\pi\lambda\alpha r_1^3\,\Omega\eta'(\Omega), \qquad M_y = 8\pi\lambda\alpha r_1^3\,\Omega\eta''(\Omega), \qquad M_z = 0. \tag{34.C18}$$

Thus these couples yield us $\eta'(\Omega)$ and $\eta''(\Omega)$ directly.

For the isochoric flow between two eccentric cylinders in rotation[45] (see Secs. 2 and 3), we use the strain history given by (3.10) in (34.19) to obtain the extra stresses as:

$$\begin{aligned} T_E\langle rr\rangle &= 2\alpha\Omega\eta^*(\Omega)\,F'e^{i\theta}, \quad \Omega\eta^*(\Omega) = G''(\Omega) - iG'(\Omega), \\ T_E\langle\theta\theta\rangle &= -2\alpha\Omega\eta^*(\Omega)\,F'e^{i\theta}, \\ T\langle r\theta\rangle &= i\alpha\Omega\eta^*(\Omega)\,(rF''+F')\,e^{i\theta} \\ T\langle rz\rangle &= T\langle\theta z\rangle = T_E\langle zz\rangle = 0. \end{aligned} \tag{34.C19}$$

Neglecting inertia and body forces, the above stresses can be used in the equations of motion (A6.3)—(A6.4) with an assumed pressure function

$$p = p_0 + p_1(r)\,\alpha\Omega\eta^*(\Omega)\,e^{i\theta}, \tag{34.C20}$$

to obtain the two equations:

r-direction:
$$F'' + \frac{3}{r}F' = dp_1/dr, \tag{34.C21$_1$}$$

---

44. To verify (34.C17), take $\mathbf{n} = \mathbf{e}_r$ in the spherical coordinate system. Compute $\mathbf{t} = \mathbf{Tn} = T\langle rr\rangle\,\mathbf{e}_r + T\langle r\theta\rangle\,\mathbf{e}_\theta + T\langle r\phi\rangle\,\mathbf{e}_\phi$; express $\mathbf{e}_r$, $\mathbf{e}_\theta$, $\mathbf{e}_\phi$ in terms of **i, j, k.** Then

$$M_x = \int\int (yt_z - zt_y)\,dA,$$

etc. Also, note that $T\langle rr\rangle$ contributes zero moment because it is radial.

45. ABBOTT and WALTERS [1970 : 2].

θ-direction : $$r^2F''' + 4rF'' = p_1. \qquad (34.C21)_2$$

Eliminating p_1 in (34.C21), we obtain that $F(r)$ obeys a fourth-order equation:

$$r^4F^{(iv)} + 6r^3F''' + 3r^2F'' - 3rF' = 0. \qquad (34.C22)$$

Noting the conditions on $F(r)$ from kinematical considerations [see (2.36)], viz.

$$F(r_1) = F'(r_1) = F'(r_2) = 0, \qquad F(r_2) = 1, \qquad (34.C23)$$

we have

$$F(r) = Ar^2 + B \ln r + \frac{C}{r^2} + D, \qquad p_1(r) = 8Ar - \frac{2B}{r}, \qquad (34C24)$$

$$A = \frac{2(r_1^2 - r_2^2)}{r_1^3 r_2^3 \Delta}, \qquad B = \frac{4(r_2^4 - r_1^4)}{r_1^3 r_2^3 \Delta}, \qquad C = \frac{2(r_2^2 - r_1^2)}{r_1 r_2 \Delta}, \qquad (34.C25)$$

$$D = \frac{4(r_1^4 - r_2^4) \ln r_1 - 2(r_2^2 - r_1^2)}{r_1^3 r_2^3 \Delta}, \qquad (34.C26)$$

$$r_1^3 r_2^3 \Delta = 4\left[(r_2^4 - r_1^4) \ln \frac{r_2}{r_1} - (r_2^2 - r_1^2)^2 \right].$$

Let **n** be the unit normal from the inner cylinder into the fluid. In (r, θ, z) coordinates, $\mathbf{n} = \{1, 0, 0\}$. Thus, on the cylinder at $r = r_1$, the traction vector **t** is

$$\mathbf{t} = T\langle rr \rangle \, \mathbf{e}_r + T\langle r\theta \rangle \, \mathbf{e}_\theta \qquad (34.C27)$$

$$= (T\langle rr \rangle \cos\theta - T\langle r\theta \rangle \sin\theta)\, \mathbf{i} + (T\langle rr \rangle \sin\theta + T\langle r\theta \rangle \cos\theta)\, \mathbf{j}. \qquad (34.C28)$$

Moreover, these stresses are evaluated at $r = r_1$ and since $F'(r_1) = 0$ from (34.C23), $T\langle rr \rangle(r_1) = -p_1(\Upsilon_1) = -p_1(r_1)\, \alpha\Omega\eta^*(\Omega)\, e^{i\theta} - p_0$. Therefore, the *forces F_x and F_y in the x and y directions on the cylinder* are :

$$(F_x - iF_y) = Lr_1 \int_0^{2\pi} [-p(r_1) - i\, T\langle r\theta \rangle]\, e^{-i\theta}\, d\theta$$

$$= -\pi\, a\Omega\eta^*(\Omega)\, L[\Upsilon_1^3\, F'''(r_1) + 3\Upsilon_1^2\, F''(r_1)], \qquad (34.C29)$$

where L is the height of the liquid column.

Thus the MAXWELL orthogonal rheometer, the KEPES apparatus and the eccentric cylindrical rotational viscometer permit one to measure the complex dynamic viscosity $\eta^*(\Omega)$ at low values of Ω without subjecting the material to sinusoidal oscillations.[46]

46. Additional examples and a theorem due to ABBOTT, BOWEN and WALTERS [1971 : 1] on this matter are given in Chap. 10.

(iii) *Simple extensional flow with* $\dot{x} = a_1 x$, $\dot{y} = a_2 y$, $\dot{z} = a_3 z$, *and* $a_1 + a_2 + a_3 = 0$. The extra stresses are :

$$T_E \langle ii \rangle = \int_0^\infty \mu(s)\,(e^{-2a_i s} - 1)\,ds, \qquad i = 1, 2, 3 \text{ ; no sum.} \tag{34.C30}$$

We shall discuss this result in connection with elongational viscosity in Sec. 49.

For the fluid obeying the theory of second-order viscoelasticity (34.24), the only result of interest is the following. In *simple shear*, this fluid is described by the material functions

$$N_1(\varkappa) = -\varkappa^2 \int_0^\infty \mu(s)\,s^2 ds - \varkappa^4 \int_0^\infty \int_0^\infty [\alpha(s_1, s_2) + \beta(s_1, s_2)]\,s_1^2\,s_2^2\,ds_1\,ds_2 \text{ ;} \tag{34.C31}$$

$$N_2(\varkappa) = \varkappa^2 \int_0^\infty \mu(s)\,s^2 ds + \varkappa^2 \int_0^\infty \int_0^\infty \{\alpha(s_1, s_2)\,[1 + \varkappa^2\,s_1 s_2] + + \varkappa^2 \beta(s_1, s_2)\}\,ds_1 ds_2 \text{ ;} \tag{34.C32}$$

$$T(\varkappa) = -\varkappa \int_0^\infty \mu(s)\,s ds - \varkappa^3 \int_0^\infty \int_0^\infty [\alpha(s_1, s_2) + \beta(s_1, s_2)]\,s_1^2\,s_2\,ds_1 ds_2. \tag{34.C33}$$

Thus, $N_1(\varkappa) \neq -N_2(\varkappa)$ for this fluid.

If $\varkappa$ is so small that terms involving $\varkappa^4$ are neglible compared to $\varkappa^2$, then

$$N_1(\varkappa) = -\varkappa^2 \int_0^\infty \mu(s)\,s^2\,ds \tag{34.C34}$$

$$= 2\varkappa^2 \int_0^\infty s\,G(s)\,ds, \tag{34.C35}$$

from which we learn that $N_1(\varkappa) > 0$ at low values of $\varkappa$[47]. Not only this, one notes from (34.*b*8) that

$$\lim_{\omega \to 0} \eta'(\omega) = \lim_{\omega \to 0} \int_0^\infty G(s) \cos \omega s\,ds = \eta_0, \tag{34.C36}$$

and from (34.*b*7) that

$$\lim_{\omega \to 0} G'(\omega) = 0, \qquad \lim_{\omega \to 0} \frac{d\,G'(\omega)}{d\omega} = 0, \tag{34.C37}$$

$$\lim_{\omega \to 0} \frac{d^2\,G'(\omega)}{d\omega^2} = 2 \int_0^\infty s\,G(s)\,ds. \tag{34.C38}$$

Hence,

$$\lim_{\varkappa \to 0} \frac{N_1(\varkappa)}{2\varkappa^2} = \lim_{\omega \to 0} \frac{1}{2} \frac{d^2\,G'(\omega)}{d\omega^2} = \lim_{\omega \to 0} \frac{G'(\omega)}{\omega^2}. \tag{34.C39}$$ [47]

47. An equivalent statement was published by COLEMAN and MARKOVITZ [1964 : 6], Eq. (2.24).

As COLEMAN and MARKOVITZ[48] have remarked, (34.C39) is true only in the theory of simple fluids. Thus, a test of this relation can be made if independent measursments of $N_1(\varkappa)$ and $G'(\omega)$ are available. These measurements of $N_1(\varkappa)$ will be dealt with in detail later in Sec. 48. In this section, we have listed three ways of measuring $G'(\omega)$ by methods other than subjecting a material to sinusoidal oscillations. Despite this, it is very doubtful whether (34.C39) can ever be checked, because experimental errors seem to dominate the measurements of $G'(\omega)$ at low values of ω.

35. Rivlin-Ericksen Fluids[49] and the Retardation Theorem[50]

In Sec. 12, it was shown that the velocity gradient $\mathbf{L}$ is not objective. In an objective motion [*cf.* (12.5)], the velocity gradient $\mathbf{L}$ appears as

$$\mathbf{L}^* = \mathbf{QLQ}^T + \dot{\mathbf{Q}}\mathbf{Q}^T = \mathbf{QDQ}^T + \mathbf{QWQ}^T + \dot{\mathbf{Q}}\mathbf{Q}^T . \tag{35.1}$$

Since $\dot{\mathbf{Q}}\mathbf{Q}^T$ is skew (8.12), and $\mathbf{QWQ}^T$ is skew, it follows that the observer $\mathcal{O}^*$ can rotate himself or choose $\dot{\mathbf{Q}} = -\mathbf{QW}$ so that $\mathbf{L}^* = \mathbf{QDQ}^T$; $\mathbf{L}^* - \mathbf{L}^{*T} = \mathbf{0}$ in his frame of reference. So a constitutive equation of the form

$$\mathbf{T} = \mathbf{g}(\mathbf{L}), \tag{35.2}$$

where $\mathbf{g}(\cdot)$ is a tensor-valued function of $\mathbf{L}$, becomes [*cf.* (22.8)]

$$\mathbf{T}^* = \mathbf{g}(\mathbf{L}^*) = \mathbf{g}(\mathbf{QDQ}^T) ; \tag{35.3}$$

or,

$$\mathbf{T}^* = \mathbf{QTQ}^T = \mathbf{g}(\mathbf{QDQ}^T), \qquad \forall\ \mathbf{Q} \in \mathcal{O}, \tag{35.4}$$

which implies that $\mathbf{T} = \mathbf{g}(\mathbf{D}) = \tilde{\mathbf{g}}(\mathbf{A}_1)$, where $\mathbf{A}_1$ is the first RIVLIN-ERICKSEN tensor.

In Sec. 12, we showed further that an observer can choose $\dot{\mathbf{Q}}$ and $\ddot{\mathbf{Q}}$ such that in his frame of reference $\mathbf{L}_2^*$, the gradient of the acceleration field, is symmetric, that is, $\mathbf{L}_2^* = \mathbf{L}_2^{*T}$. Since $\mathbf{L}_2^*$ is a linear combination of $\mathbf{A}_1^*$ and $\mathbf{A}_2^*$, these results imply that if we begin with a constitutive equation depending on $\mathbf{L}$ and $\mathbf{L}_2$, it must reduce to an isotropic function of $\mathbf{A}_1$ and $\mathbf{A}_2$. Proceeding analogously, it is possible to find higher derivatives of $\mathbf{Q}(t)$ to ensure that a constitutive equation of the form

$$\mathbf{T} = \mathbf{g}(\mathbf{L}_1, \mathbf{L}_2, \ldots, \mathbf{L}_n) \tag{35.5}$$

must reduce to an isotropic function of the type

$$\mathbf{T} = \mathbf{f}(\mathbf{A}_1, \mathbf{A}_2, \ldots, \mathbf{A}_n). \tag{35.6}$$

48. COLEMAN and MARKOVITZ [1964 : 6], Eq. (2.34).
49. RIVLIN and ERICKSEN [1955 : 4].
50. COLEMAN and NOLL [1960 : 1] : COLEMAN [1971 : 8].

The equation (35.6) is that of a fluid and not a solid because $\mathbf{T} = \mathbf{0}$ in a state of rest. We say that *n-th order incompressible Rivlin-Ericksen fluid* is described by

$$\mathbf{T}_E = \mathbf{f}(\mathbf{A}_1, \mathbf{A}_2, \ldots, \mathbf{A}_n), \qquad n < \infty. \tag{35.7}$$ [51]

If $n = 1$, the first-order RIVLIN-ERICKSEN fluid is identical to the Stokesian fluid.

Since each $\mathbf{A}_r$ is of dimension t^{-r}, where t is the unit of time, one may expand[51] (35.7) in full as follows :

$$\mathbf{T}_E = \mathbf{T}_E^{(1)} + \mathbf{T}_E^{(2)} + \mathbf{T}_E^{(3)} + \ldots, \tag{35.8}$$

in which $\mathbf{T}_E^{(r)}$ is of dimension r and each term in $\mathbf{T}_E^{(r)}$ is composed of the products of the traces of $\mathbf{A}_i$ and the $\mathbf{A}_i = 1, 2, \ldots, r$, with each product of dimension r. Thus (on absorbing the term involving $\mathbf{1}$ into $-p\mathbf{1}$) :

$$\begin{aligned}
\mathbf{T}_E^{(1)} &= \eta_0 \mathbf{A}_1,\\
\mathbf{T}_E^{(2)} &= \beta \mathbf{A}_1^2 + \gamma \mathbf{A}_2,\\
\mathbf{T}_E^{(3)} &= \alpha_0(\operatorname{tr} \mathbf{A}_1^2)\mathbf{A}_1 + \alpha_1(\mathbf{A}_1\mathbf{A}_2 + \mathbf{A}_2\mathbf{A}_1) + \alpha_2\mathbf{A}_3, \qquad (35.9)^{52}\\
\mathbf{T}_E^{(4)} &= [\,\alpha_3 \operatorname{tr}(\mathbf{A}_1\mathbf{A}_2) + \alpha_4 \operatorname{tr} \mathbf{A}_1^3\,]\,\mathbf{A}_1 + \alpha_5(\operatorname{tr} \mathbf{A}_1^2)\,\mathbf{A}_1^2 +\\
&\quad + \alpha_6(\operatorname{tr} \mathbf{A}_1^2)\,\mathbf{A}_2 + \alpha_7(\mathbf{A}_1^2\mathbf{A}_2 + \mathbf{A}_2\mathbf{A}_1^2) + \alpha_8\mathbf{A}_2 +\\
&\quad + \alpha_9(\mathbf{A}_1\mathbf{A}_3 + \mathbf{A}_3\mathbf{A}_1) + \alpha_{10}\mathbf{A}_4, \text{ etc.}
\end{aligned}$$

In (35.9), the coefficients $\eta_0, \ldots, \alpha_{10}$ are all constants and use has been made of the fact that in incompressible materials all motions obey $\operatorname{tr} \mathbf{A}_1 = 0$, $\operatorname{tr} \mathbf{A}_1^2 = \operatorname{tr} \mathbf{A}_2$. However, if one wants to obtain a complete expansion of the constitutive relation (35.6), it is possible to do so in the following way[53] :

(i) Let φ_{ij} be a symmetric tensor, which is objective. Multipy (35.6) by φ_{ij} and form the scalar product $F = \varphi_{ik} T_{kj}$. Now F is a scalar, invariant under the full orthogonal group.

(ii) By HILBERT's theorem[54], if F is a polynominal, then there is a finite

---

51. Representation theorems or functions of the type (35.7) were derived by RIVLIN [1955 : 3], SPENCER and RIVLIN [1959 : 8], [1959 : 9], [1960 : 4] and SMITH [1960 : 3]. Review articles have been published by RIVLIN and SMITH [1969 : 20], and SPENCER [1971 : 24].

52. In compresible materials $\operatorname{tr} \mathbf{A}_2 = \operatorname{tr} \mathbf{A}_1^2$, and $\operatorname{tr} \mathbf{A}_n$ $(n = 2, 3, \ldots)$ can be expressed as a polynomial in traces of products of lower order $\mathbf{A}_n$ (RIVLIN [1962 : 8]). These two facts have been used in (35.9).

53. PIPKIN and RIVLIN [1959 : 5].

54. See WEYL [1946 : 1].

integrity basis for it in terms of the various invariants of φ_{ij}, $A_{ij}^{(1)}$, $A_{ij}^{(2)}, \ldots, A_{ij}^{(n)}$ and their products.

(iii) Out of this integrity basis, consider the terms linear in φ_{ij}, i.e., of the form $\varphi_{ik} f_{kj}$, where f_{kj} is a polynomial in δ_{lm}, $A_{lm}^{(1)}, \ldots, A_{lm}^{(n)}$ and their products, with scalar-valued coefficients being polynomials in the invariants of $\mathbf{A}_1, \mathbf{A}_2, \ldots, \mathbf{A}_n$ and their products.

(iv) Then, since φ_{ij} is arbitrary, $T_{ij} = f_{ij}$.

(v) Since the integrity basis is also a function basis[55], the f_{ij} so obtained is a single-valued expansion provided the coefficients are single-valued in their variables. If the expansion is to be differentiable, it does not seem possible to drop the assumption of polynomiality.

For example, a full expansion of an isotropic function $\mathbf{h}(\mathbf{A}_1, \mathbf{A}_2)$ reads :

$$\begin{aligned}\mathbf{h}(\mathbf{A}_1, \mathbf{A}_2) = {} & \beta_0 \mathbf{1} + \beta_1 \mathbf{A}_1 + \beta_2 \mathbf{A}_1^2 + \beta_3 \mathbf{A}_2 + \beta_4 \mathbf{A}_2^2 + \beta_5(\mathbf{A}_1\mathbf{A}_2 + \mathbf{A}_2\mathbf{A}_1) + \\ & + \beta_6(\mathbf{A}_1^2 \mathbf{A}_2 + \mathbf{A}_2 \mathbf{A}_1^2) + \beta_7(\mathbf{A}_1 \mathbf{A}_2^2 + \mathbf{A}_2^2 \mathbf{A}_1) + \\ & + \beta_8 (\mathbf{A}_1^2 \mathbf{A}_2^2 + \mathbf{A}_2^2 \mathbf{A}_1^2),\end{aligned} \qquad (35.10)^{56}$$

where the β_j, $j = 0, 1, \ldots, 8$, are analytic functions of the invariants :

$$\begin{aligned}&\operatorname{tr} \mathbf{A}_1, \operatorname{tr} \mathbf{A}_1^2, \operatorname{tr} \mathbf{A}_1^3, \operatorname{tr} \mathbf{A}_2, \operatorname{tr} \mathbf{A}_2^2, \operatorname{tr} \mathbf{A}_2^3, \\ &\operatorname{tr} \mathbf{A}_1 \mathbf{A}_2, \operatorname{tr} \mathbf{A}_1^2 \mathbf{A}_2, \operatorname{tr} \mathbf{A}_1 \mathbf{A}_2^2, \operatorname{tr} \mathbf{A}_1^2 \mathbf{A}_2^2 .\end{aligned} \qquad (35.11)^{56}$$

One can also consider other types of expansions of the constitutive relation (35.7). Let us suppose that the $\mathbf{A}_n$ have been determined for a *steady velocity field* $\mathbf{v}$ so that $\dot{\mathbf{A}}_n = 0$. If $\mathbf{v}$ is replaced by $\varepsilon\mathbf{v}$, where ε is a constant, the computation of the new set of RIVLIN-ERICKSEN $\tilde{\mathbf{A}}_n$ tensors shows that

$$\tilde{\mathbf{A}}_n = \varepsilon^n \mathbf{A}_n, \qquad n = 1, 2, \ldots \qquad (35.12)$$

Thus if (35.7) is expanded to an order ε, ε^2, ε^3 and ε^4 respectively, the tensors $\mathbf{T}_E^{(1)}, \mathbf{T}_E^{(2)}, \mathbf{T}_E^{(3)}$ and $\mathbf{T}_E^{(4)}$ in (35.9) are respectively of the order ε, ε^2, ε^3 and ε^4.

Next, let us change the length scale of a given motion so that the new length $\bar{l}$ is related to the old length l by $\varepsilon\bar{l} = l$, where ε is a small positive fixed number. Change the time scale t also so that $\varepsilon\bar{t} = t$. Then the velocities are equal, that is, $\bar{\mathbf{v}} = \mathbf{v}$, but $\bar{\mathbf{L}} = \varepsilon\mathbf{L}$, $\bar{\mathbf{A}}_1 = \varepsilon\mathbf{A}_1$, or $\bar{\mathbf{A}}_r = \varepsilon^r \mathbf{A}_r$, $r = 1, 2, \ldots, n$. Therefore these changes in length and time scales again yield expansions in terms of $\mathbf{T}_E^{(1)}, \mathbf{T}_E^{(2)}$, etc.

55. WINEMAN and PIPKIN [1964 : 27].
56. RIVLIN [1955 : 3].

Perturbations about a ground state can be examined on replacing $\mathbf{v}$ by $\mathbf{v} + \varepsilon\mathbf{u}$, where ε is small. Then the new set of RIVLIN-ERICKSEN tensor $\hat{\mathbf{A}}_n$ is related to the old set $\mathbf{A}_n$ by

$$\hat{\mathbf{A}}_n = \mathbf{A}_n + \mathbf{0}(\varepsilon), \tag{35.13}$$

where $\mathbf{0}(\varepsilon)$ means terms of order ε are retained and those of higher order are ignored. The expansion of (35.7) for the $\tilde{\mathbf{A}}_n$ in (35.13) can be carried through to an order ε. Since this approach is meaningful in the case of stability problems, which are not discussed in this book the consideration of such an approach shall not be attempted.

We shall now concentrate on the position of the RIVLIN-ERICKSEN fluid in the theories of simple fluids with fading memory. First we examine their position with respect to the COLEMAN-NOLL (hereafter abridged to C-N) theory (Sec. 34).

1°. Since the RIVLIN-ERICKSEN fluids do not possess the gradual stress relaxation property and do not obey the C-N principle of fading memory, *they are not special cases of the simple fluid* within that theory of fading memory, To put it another way, the DIRAC delta function is not an influence function of the type mentioned in Sec. 34. Thus, the first-order RIVLIN-ERICKSEN fluid is not a simple fluid with C-N fading memory, and so nor are the others.

2°. However, in many flows, e.g. MWCSH and where $\mathbf{C}_t(t-s)$ has a finite expansion in terms of the $\mathbf{A}_n$, the simple fluid is indistinguishable from the RIVLIN-ERICKSEN fluid of order n.

3°. The above situation induces one to ask : suppose that the strain history is approximated by its TAYLOR series with the remainder term. Does there exist a RIVLIN-ERICKSEN constitutive equation which approximates the stress in the simple fluid?

To answer this, we now examine the retardation theorem of COLEMAN and NOLL[57] and provide a more direct, but heuristic, proof of it.

Given a history $\mathbf{C}_t(t-s)$. Let $0 < \alpha \leqq 1$ be a real number. We say that $\mathbf{C}_t(t-\alpha s)$ is a *retarded history*, because replacing s by αs is equivalent to 'performing the original history at a slower rate'. Let $\mathbf{C}_t(t-s)$ possess a TAYLOR series of the form

$$\mathbf{C}_t(t-s) = \mathbf{1} - s\mathbf{A}_1 + \frac{1}{2!}\, s^2\mathbf{A}_2 - \ldots + (-1)^n \frac{s^n\mathbf{A}_n}{n} + \boldsymbol{R}(s^n), \tag{35.14}$$

where $\boldsymbol{R}(s^n)$ is the remainder term which goes to zero faster than s^n as $s \to 0$, or

$$\lim_{s\to 0} \frac{\| \boldsymbol{R}(s^n) \|}{s^n} = 0.$$

57. [1960 : 1]. This theorem makes precise the point 3° above, by a change in the *time scale only.* As we have just shown. a simultaneous change in length and time scale also yields an expansion in terms of RIVLIN-ERICKSEN tensors since the TAYLOR series of the altered motion has the form (35.17).

It is trivial to verify that

$$\mathbf{C}_t(t - s) = \mathbf{1} - \alpha s \mathbf{A}_1 + \frac{1}{2!}\, \alpha^2 s^2 \mathbf{A}_2 - \ldots + (-1)^n \frac{s^n \mathbf{A}_n}{n!} + R(\alpha^n s^n), \quad (35.15$$

where $\| R(\alpha^n s^n) \|$ is $\mathbf{o}(\alpha^n s^n)$. We introduce the notation

$$\overset{\alpha}{\mathbf{A}}_n = \alpha^n \mathbf{A}_n, \qquad n = 1, 2, \ldots \quad (35.16)$$

and write

$$\mathbf{C}_t(t - \alpha s) = \mathbf{1} - s\overset{\alpha}{\mathbf{A}}_1 + \frac{1}{2!}\, s^2 \overset{\alpha}{\mathbf{A}}_2 - \ldots + \frac{(-1)^n s^n \overset{\alpha}{\mathbf{A}}_n}{n!} + R(\alpha^n s^n). \quad (35.17)$$

Let us now treat (37.17) as the strain history of a simple fluid particle, *for a fixed n and α.* Then, by using the principle of strong fading memory

$$\mathbf{T}_E = \overset{\infty}{\underset{s=0}{H}} (\mathbf{C}_t(t - \alpha s)) = \overset{\infty}{\underset{s=0}{H}} \left(\mathbf{1} - s\, \overset{\alpha}{\mathbf{A}}_1 + \ldots + \frac{(-1)^n s^n}{n\,!}\, \overset{\alpha}{\mathbf{A}}_n \right) +$$
$$+ \delta \overset{\infty}{\underset{s=0}{H}} \left(\mathbf{1} - s\overset{\alpha}{\mathbf{A}}_1 + \ldots + (-1)^n \frac{s^n}{n!}\, \overset{\alpha}{\mathbf{A}}_n \,|\, R(\alpha^n s^n) \right) +$$
$$+ \mathbf{o}(\| R(\alpha^n s^n) \|), \quad (35.18)$$

where $\delta H(\cdot \mid \cdot)$ is the FRÉCHET derivative of $H(\,\cdot\,)$ about the history $1 - s\overset{\alpha}{\mathbf{A}}_1 + \ldots + (-1)^n \frac{s^n}{n!}\, \overset{\alpha}{\mathbf{A}}_n$. The first term in (35.18) is a functional of this strain history and this strain history is completely determined by the $\overset{\alpha}{\mathbf{A}}_1$, $r = 1, 2, \ldots, n$. Hence (35.18) reduces to

$$\mathbf{T}_E = \mathbf{f}(\overset{\alpha}{\mathbf{A}}_1, \ldots, \overset{\alpha}{\mathbf{A}}_n) + \delta \overset{\infty}{\underset{s=0}{H}} (\cdot \mid R(\alpha^n s^n)) + \mathbf{o}(\| R(\alpha^n s^n) \|), \quad (35.19)$$

where $\mathbf{f}(\,\cdot\,)$ is isotropic function of its arguments. By the representation theorem for such functions, we have

$$\mathbf{f}(\overset{\alpha}{\mathbf{A}}_1, \ldots, \overset{\alpha}{\mathbf{A}}_n) = \mathbf{f}^{(n)}(\overset{\alpha}{\mathbf{A}}_1, \ldots, \overset{\alpha}{\mathbf{A}}_n) + R^1(\alpha^n), \quad (35.20)$$

where $\mathbf{f}^{(n)}(\,\cdot\,)$ is utmost of the order α^n in its expansion and $R^1(\alpha^n)$ is such that $\lim\limits_{\alpha \to 0} \frac{\| R^1(\alpha^n) \|}{\alpha^n} = 0$. By the linearity of the FRÉCHET derivative and the fact that the remainder term $\| R(\alpha^n s^n) \| = o(\alpha^n)$, we conclude that

$$\mathbf{T}_E = \mathbf{f}^{(n)} (\overset{\alpha}{\mathbf{A}}_1, \ldots, \overset{\alpha}{\mathbf{A}}_n) + \mathbf{o}(\alpha^n), \quad (35.21)$$

where $\| \mathbf{o}(\alpha^n) \|$ gose to zero faster than α^n as $\alpha \to 0$. If we take $n = 1$, we have

$$\mathbf{T}_E = \mathbf{f}^{(1)} (\overset{\alpha}{\mathbf{A}}_1) + \mathbf{o}(\alpha) = \tilde{\eta}_0\, \overset{\alpha}{\mathbf{A}}_1 + \mathbf{o}(\alpha)\,; \qquad \tilde{\eta}_0 = \text{const.} \quad (35.22)$$

Hence the Newtonian fluid and the simple fluid have the same stress in a retarded motion to within an error term which goes to zero faster than α as $\alpha \to 0$. Let us put $n = 2$ in (35.21). We get

$$\mathbf{T}_E = \tilde{\eta}_0 \mathbf{A}_1^{\alpha} + \tilde{\beta}(\mathbf{A}_1^{\alpha})^2 + \tilde{\gamma} \mathbf{A}_2^{\alpha} + o(\alpha^2) ; \qquad \tilde{\eta}_6, \tilde{\beta}, \tilde{\gamma}, \text{ constants.} \tag{35.23}$$

Thus the simple fluid is described by (35.23) in retarded flows to within an error term which goes to zero faster than α^2 as $\alpha \to 0$. We call the fluid described by the constitutive relation

$$\mathbf{T}_E = \eta_0 \mathbf{A}_1 + \beta \mathbf{A}_1^2 + \gamma \mathbf{A}_2, \tag{35.24}$$

the *second-order fluid.* Similarly, we call the constitutive relation

$$\mathbf{T}_E = \eta_0 \mathbf{A}_1 + \beta \mathbf{A}_1^2 + \gamma \mathbf{A}_2 + \alpha_0(\operatorname{tr} \mathbf{A}_1^2)\mathbf{A}_1 + \\ + \alpha_1(\mathbf{A}_1\mathbf{A}_2 + \mathbf{A}_2\mathbf{A}_1) + \alpha_2\mathbf{A}_3 \tag{35.25}$$

the *third-order fluid*, and the fluid described by [*cf.* (35.9)]

$$\mathbf{T}_E = \mathbf{T}_E^{(1)} + \mathbf{T}_E^{(2)} + \mathbf{T}_E^{(3)} + \mathbf{T}_E^{(4)} \tag{35.26}$$

the *fourth-order fluid.*

We wish to emphasize the difference between an n-th-order fluid and an n-th-order RIVLIN-ERICKSEN fluid at this point. For example, *a second-order fluid* is given by (34.24), while a *second-order Rivlin-Ericksen fluid* obeys (35.10)—(35.11). Of course, in retarded flows, (35.24) approximates (35.10)—(35.11) to an error $o(\alpha^2)$.

Having considered the position of RIVLIN-ERICKSEN fluids in the context of C-N theory of fading memory, we shall discuss WANG's theories[58] in a heuristic manner. According to WANG, the constitutive functional $\underset{s=0}{\overset{\infty}{\boldsymbol{H}}} (\mathbf{C}_t(t-s))$ is defined over the set of positive-definite strain histories, $\mathbf{C}_t(t-s)$, $s \in [0, \infty)$. This set is not a vector space, and thus the topology used is not that of (34.3). It is what is known as a topology of compact convergence, in which convergence is defined by uniform convergence of the strain history and its first p derivatives, the number p depending on the order of the material, which we define next :

> *A simple fluid is said to be of order p ($p \geqq 0$) if p is the largest integer for which the constitutive functional $\underset{s=0}{\overset{\infty}{\boldsymbol{H}}} (\mathbf{C}_t(t-s))$ depends explicitly on $(d^p/ds^p)\, \mathbf{C}_t(t-s)$ for some $s \in [0, \infty)$.*

Under this definition, the simple fluid theory of order 0 leads to finite linear and second-order viscoelasticity theories, provided one assumes differentiability. RIVLIN-ERICKSEN fluids are included, for we can define them

58. [1965 : 18].

through $\underset{s=0}{\overset{\infty}{\boldsymbol{H}}}(1; \mathbf{A}_1, \ldots, \mathbf{A}_n)$, and thus (34.29) is also included.

Let $\mathbf{D}_t(\tau)$ be a strain history, and let $\mathbf{D}_t^{(n)}(\tau) = \frac{d^n}{d\tau^n}\mathbf{D}_t(\tau)$, $n = 0, 1, \ldots, p$ be its first $(p+1)$ derivatives. In this notation, the rest history has $(p+1)$ derivatives given by $\mathbf{1}^{(0)} \equiv \mathbf{1}$, $\mathbf{1}^{(n)} \equiv \mathbf{0}$, $n = 1, 2, \ldots, p$. Consider the class of all those $\mathbf{D}_t(\tau)$, which satisfy $\mathbf{D}_t^{(n)}(\tau) - \mathbf{1}^{(n)}| < \delta$ for $0 \leqq n \leqq p$, where δ is some positive number. Compute $\underset{-\infty}{\overset{t}{\boldsymbol{H}}}(\mathbf{D}_t(\tau))$ and confine attention only to those $\mathbf{D}_t(\tau)$ for which $\left|\underset{-\infty}{\overset{t}{\boldsymbol{H}}}(\mathbf{D}_t(\tau))\right|$ is bounded. Then consider the set, defining λ_0:

$$\lambda_0 = \sup\left\{\lambda : \lim_{\delta\to 0+} \sup\left\{\left|\underset{-\infty}{\overset{t}{\boldsymbol{H}}}(\mathbf{D}_t(\tau))\right| : \left|\mathbf{D}_t^{(n)}(\tau) - \mathbf{1}^{(n)}\right| < \delta \right.\right.$$
$$\left.\left. \text{for all } 0 \leqq n \leqq p, \text{ when } \tau \in [t-\lambda, t]\right\} < \infty \right\}. \quad (35.27)$$

As $\delta \to 0$, let $\left|\underset{-\infty}{\overset{t}{\boldsymbol{H}}}[\mathbf{D}_t(\tau)]\right|$ attain a bounded supremum somewhere on $\tau \in [t-\lambda, t]$. We call the upper bound λ_0 of such an λ, the *time of sentinence.* By the very definition of the RIVLIN-ERICKSEN fluid, its time of sentinence is 0, provided its constitutive equation be continuous. For general simple fluids, the time of sentinence λ_0 is a measure of the '*major*' *memory* of the fluid. By this we mean that if $\mathbf{D}_t^{(n)}(\tau)$ is close to $\mathbf{1}^{(n)}$, $0 \leqq n \leqq p$, in $\tau \in [t - \lambda_0, t]$, the stress will be very close to $\mathbf{0}$ (that at rest history), whatever the history $\mathbf{D}_t(\tau)$ may have been in $\tau \in (-\infty, t - \lambda_0]$. This latter period is the '*minor*' *memory* of the fluid.

Now consider the stress $\underset{-\infty}{\overset{t}{\boldsymbol{H}}}[\mathbf{D}_t(\tau)]$ again such that $|\mathbf{D}_t^{(n)}(\tau) - \mathbf{1}^{(n)}| < \delta$, $0 \leqq n \leqq p$, and $\tau \in [t - \lambda, t]$. Let $\lambda \to \infty$. Then what is the largest value δ_0 of that δ for which $\left|\underset{-\infty}{\overset{t}{\boldsymbol{H}}}(\mathbf{D}_t(\tau))\right|$ is bounded over $\tau \in (-\infty, t]$? We shall call δ_0, defined by

$$\delta_0 = \sup\left\{\delta : \lim_{\lambda\to\infty} \left\{\left|\underset{-\infty}{\overset{t}{\boldsymbol{H}}}(\mathbf{D}_t(\tau))\right| : \left|\mathbf{D}_t^{(n)}(\tau) - \mathbf{1}^{(n)}\right| < \delta \right.\right.$$
$$\left.\left. \text{for all } 0 \leqq n \leqq p \text{ when } \tau \in [t - \lambda, t]\right\} < \infty \right\}, \quad (35.28)$$

the *grade of sentinence.* What this says is that if we choose $\delta < \delta_0$, then the stresses will be close to $\mathbf{0}$ (that at rest history), and otherwise if $\delta > \delta_0$. So, any $\delta < \delta_0$ yields a *minor deviation* and $\delta > \delta_0$ yields a *major deviation.* For a RIVLIN-ERICKSEN fluid, whenever $\mathbf{A}_1, \ldots, \mathbf{A}_n$ are bounded, the stresses are bounded. So δ is infinite and thus $\delta_0 = \infty$ for this fluid. In general, every fluid has thus a pair (λ_0, δ_0) defining its properties. In fluids with memory, λ_0 is usually finite while δ_0 is infinite. WANG has defined and interlaced the above ideas with weak and strong principles of fading memory in his paper to which the reader is referred.

From the point of view of applications, WANG's theory for simple fluids of order 0 yields constitutive equations similar to those we have discussed here in Secs. 34 and 35, but for materials of order $p \geqq 1$, it removes the objections raised by OLDROYD, listed in Sec. 34. The more recent papers by COLEMAN and MIZEL[59] have investigated theories of fading memory in L_p spaces and for finding stress responses in retarded flows[60] with discontinuities in velocity gradients, etc., and are thus extensions of the C-N theory of Sec. 34. However, we shall not discuss these theories here, for they do not yield any new results for our applications, these being more useful in the context of shock waves, etc.

We now recall the approximate constitutive equations and note their properties in viscometric flows.

	Second Order	*Third Order*	*Fourth Order*
$\tau(\varkappa)$	$\eta_0\varkappa$	$\eta_0\varkappa + 2(\alpha_0 + \alpha_1)\varkappa^3$	$\eta_0\varkappa + 2(\alpha_0 + \alpha_1)\varkappa^3$
$N_1(\varkappa)$	$-2\gamma\varkappa^2$	$-2\gamma\varkappa^2$	$-2\gamma\varkappa^2 - 4\varkappa^4(\alpha_6 + \alpha_7 + \alpha_8)$
$N_2(\varkappa)$	$(\beta + 2\gamma)\varkappa^2$	$(\beta + 2\gamma)\varkappa^2$	$(\beta + 2\gamma)\varkappa^2 + 2\varkappa^4(\alpha_5 + 2\alpha_6 + 2\alpha_7 + 2\alpha_8)$.

The reader should compare these properties with those of the OLDROYD model (Sec. 37). One sees that the 'order fluids' have very poor predictions. However, their utility lies in the tractability of a problem developed by using the order fluid constitutive equations.

Next, note that in viscometric flows, if the rate of shear $\varkappa$ is small (that is, if $\varkappa$ is replaced by $\alpha\varkappa$), the *second-order* fluid (35.24) approximates the stress in a simple fluid to an error $\mathbf{o}(\alpha^2)$. If one computes the stresses to $\mathbf{o}(\alpha^2)$ by using the *second-order viscoelasticity* theory (34.24), one obtains that [*cf.* (34.*C*35)]

$$N_1(\alpha\varkappa) = 2\alpha^2\varkappa^2 \int_0^\infty sG(s)\,ds + o(\alpha^2). \tag{35.29}$$

Comparing $N_1(\alpha\varkappa) = -2\gamma\alpha^2\varkappa^2$, for the second-order fluid, with (35.28), one concludes that

$$\gamma > 0, \qquad (35.30)[61]$$

implying that $N_1 > 0$ at low rates of shear.

59. [1968 : 6], [1968 : 7].
60. COLEMAN [1971 : 8].
61. COLEMAN and MARKOVITZ [1964 : 6].

Before we conclude this section, we offer some comments on the meaning behind the approximate constitutive relations and what one may expect from them. First of all, the approximations are in the nature of 'kinematics'. By this we mean that given a kinematically feasible motion and a simple fluid, we can find an explicit constitutive equation by a change in time scale (and/or length scale). *A priori* there is no guarantee that the original motion is dynamically possible in the general simple fluid. Similarly, there is no guarantee that the altered motion will be possible in the n-th order fluid, even if the n-th order fluid approximates the stress to an error term of order $\mathbf{o}(\alpha^n)$. Interestingly enough, even if the original motion is dynamically possible in the general simple fluid, *it may not be possible in* the r-th *order fluid.* While this may sound odd, the n-th order fluid is not a special case of the simple fluid with C-N fading memory.

While examples of such flows are hard to give, it is possible to construct velocity fields which are plausible in the second-order fluid (for a given set of boundary conditions), but these are impossible in the Newtonian fluid.[62] Borrowing from finite elasticity we note that version of a sphere is possible in finite incompressible isotropic elasticity theory, but not in the linear elasticity.

What we say is this : one need not be surprised if it turns out that these order models are inadequate models. For example, the second-order fluid implies that in the initial value problem of cessation of steady simple shearing the velocity field will become unbounded instead of decaying to zero (see Appendix). One way out of such dilemmas is to keep a close watch on the error terms and use a higher-order model or integral models when the error terms exceed the assumed bounds. But the difficulty encountered here is really analytical. Numerical solutions will probably pick up absurdities straightaway, but analytical methods are not simple enough to tackle the nonlinear partial differential equations which arise when these differential models are used. For example, in plane creeping flows of second-order fluids, the stream function ψ obeys[63] :

$$\Delta^2\psi + \frac{\gamma}{\eta_0}\,\mathbf{v}\cdot\nabla(\Delta^2\psi) = 0. \tag{35.31}$$

The nonlinear terms are of higher order (in derivatives) than the linear (elliptic) term, and *a priori* bounds on solutions of (35.31) are not available. So, we cannot look at a given problem, form the differential equation for an order model and decide whether or not to use this model. Though it may be felt that in steady flows which are close to the Newtonian solution, order fluids may be used, it is worth noting that non-uniqueness of the solution to

62. See Sec. 41, Chap. 10 for details.
63. TANNER [1966 : 19]. See also Sec. 41, Chap. 10.

the problem may arise there[64], and the fluid model may be experimentally unacceptable (see Appendix).

Actually this discussion leads one to consider the classification of flow fields into categories and to delineate the conditions under which one should choose a particular constitutive equation. This question is taken up in the Introduction to Chap. 10 in somewhat more detail.

36. Additive Functionals: Rivlin and Sawyer's Generalization[65] of the Theory of BKZ Fluid

In 1963, BERNSTEIN, KEARSLEY and ZAPAS[66] suggested that the constitutive equation for viscoelastic fluid may be derived by the following argument.

The stress at time t is influenced by the strain history upto time t; the effect of any configuration at time $\tau < t$ on the stress at time t is equivalent to treating the configuration at time τ as the preferred or unstressed configuration of an elastic solid, and computing the stress at time t by the rules of isotropic finite elasticity, with a stored energy function. In finite elasticity the strain measure is computed with respect to the preferred configuration and denoted by

$$\mathbf{B}(t, \tau) = \mathbf{B}_\tau(t) = \mathbf{F}_\tau(t)\, \mathbf{F}_\tau(t)^T. \tag{36.1}$$

We have, intentionally, made explicit the dependence on τ. This strain measure is called the *left Cauchy-Green strain tensor*[67]. In isotropic, finite hyper-elasticity one assumes the existence of a stored energy function $U = U[\mathbf{B}_\tau(t)]$, which on account of the fact that the invariants I, II and III of $\mathbf{B}_\tau(t)$ determine the eigenvalues of $\mathbf{B}_\tau(t)$ uniquely, becomes

$$U = U(I, II, III). \tag{36.2}$$

The invariants are:

$$\begin{aligned} I &= \operatorname{tr} \mathbf{B}_\tau(t), \\ II &= \frac{1}{2}\left\{\operatorname{tr} \mathbf{B}^2_\tau(t) - I^2\right\}, \\ III &= \det \mathbf{B}_\tau(t). \end{aligned} \tag{36.3}$$

In incompressible materials, $\det \mathbf{B}_\tau(t) = 1$, and hence U becomes a function of I and II, or equivalently,

$$U = U(I_1, I_2), \tag{36.4}$$

$$\left.\begin{aligned} I_1 &= \operatorname{tr} \mathbf{B}_\tau(t), \\ I_2 &= \operatorname{tr} \mathbf{B}^{-1}_\tau(t) \end{aligned}\right\}. \tag{36.5}$$

64. See Sec. 41, Chap. 10, and Appendix to this chapter.
65. [1971 : 21].
66. [1963 : 1].
67. See, e.g. NFTM, Sec. 23.

In isotropic finite elasticity, the stress at time t is given by

$$\mathbf{T}_E(t) = 2\frac{\partial U}{\partial I_1}\mathbf{B}_\tau(t) - 2\frac{\partial U}{\partial I_2}\mathbf{B}_\tau(t)^{-1}. \tag{36.6}$$

For the elastic fluid, we must make certain changes listed below:

(i) $U = U(I_1, I_2, t - \tau)$, because viscoelastic fluids exhibit fading memory and hence the contribution to the stress at time t, from the configuration at time τ, must decrease as $t - \tau$ gets larger;

(ii) while we assume that a formula of the type (36.6) holds, we postulate that the stress at time t in an incompressible fluid is given by

$$\mathbf{T}_E(t) = 2\int_0^\infty \left\{ \frac{\partial U(I_1, I_2, s)}{\partial I_1}\mathbf{B}_{t-s}(t) - \right.$$
$$\left. - \frac{\partial U(I_1, I_2, s)}{\partial I_2}\mathbf{B}_{t-s}^{-1}(t) \right\} ds. \tag{36.7}$$

Since $\mathbf{B}_\tau(t) = \mathbf{C}_t^{-1}(\tau)$, one may write (36.5) and (36.7) in a form more useful in fluid flow:

$$\mathbf{T}_E(t) = 2\int_0^\infty \left\{ \frac{\partial U(I_1, I_2, s)}{\partial I_1}\mathbf{C}_t^{-1}(t-s) - \right.$$
$$\left. - \frac{\partial U(I_1, I_2, s)}{\partial I_2}\mathbf{C}_t(t-s) \right\} ds, \tag{36.8}$$

$$I_1 = \operatorname{tr}\mathbf{C}_t^{-1}(t-s), \qquad I_2 = \operatorname{tr}\mathbf{C}_t(t-s), \qquad \det\mathbf{C}_t(t-s) = 1. \tag{36.9}$$

We call (36.8) *the constitutive equation of the incompressible BKZ Fluid.* Wherever we talk of the BKZ fluid in this book, we mean (36.8) – (36.9).

We shall now show how a BKZ type constitutive equation can be derived from a potential functional[68]. Let $W(t)$ be a scalar-valued functional of $\mathbf{F}(t-s)^{-1}$, $s \in [0, \infty)$, that is,

$$W(t) = \underset{s=0}{\overset{\infty}{\mathcal{W}}}(\mathbf{F}(t-s)^{-1}). \tag{36.10}$$

Objectivity demands that for all orthogonal functions $\mathbf{Q}(\cdot)$,

$$\underset{s=0}{\overset{\infty}{\mathcal{W}}}(\mathbf{F}(t-s)^{-1}) = \underset{s=0}{\overset{\infty}{\mathcal{W}}}(\mathbf{F}(t-s)^{-1}\mathbf{Q}(t-s)^T), \tag{36.11}$$

while symmetry demands that

$$\underset{s=0}{\overset{\infty}{\mathcal{W}}}(\mathbf{F}(t-s)^{-1}) = \underset{s=0}{\overset{\infty}{\mathcal{W}}}(\mathbf{H}^{-1}\mathbf{F}(t-s)^{-1}), \tag{36.12}$$

where $\mathbf{H}$ is a constant tensor belonging to the symmetry group for a fluid

---

68. BERNSTEIN, *et al.* [1964 : 3], [1965 : 2].

$\mathbf{H} \in \mathcal{U}$, the unimodular group. Choose $\mathbf{H} = |\det \mathbf{F}(t)| \, \mathbf{F}(t)^{-1}$. Then (36.12) becomes

$$W(t) = \mathcal{W}\left((v(t), \int_0^\infty \tilde{U}(\mathbf{F}_{t-s}(t), s)\, ds \right) \quad \text{(say)}. \tag{36.13}$$

Here $v = 1/\rho$ in the specific volume and beeause of (16.7), $\rho_R v = |\det \mathbf{F}(t)|$, and $\tilde{U}$ is a scalar-valued function of its arguments. We denote the integral in the parenthesis in (36.13) by $S(t)$, and define

$$-p = \frac{\partial \mathcal{W}}{\partial v}, \quad T = \frac{\partial \mathcal{W}}{\partial S}, \quad (F_{t-s}(t))_{im} = \frac{\partial x_i(t)}{\partial \xi_m(t-s)},$$
$$F_{i\alpha}(t) = \frac{\partial x_i(t)}{\partial X_\alpha}. \tag{36.14}$$

Then

$$\frac{\partial W(t)}{\partial F_{i\alpha}(t)} = -pv \frac{\partial X_\alpha}{\partial x_i(t)} + T \int_0^\infty \frac{\partial \tilde{U}}{\partial F_{im}} \frac{\partial X_\alpha}{\partial \xi_m(t-s)}\, ds$$
$$= v \frac{\partial X_\alpha}{\partial x_j(t)} \left[-p\delta_{ij} + \rho T \int_0^\infty \frac{\partial \tilde{U}}{\partial F_{im}} F_{im}\, ds \right]. \tag{36.15}$$

Multiplying through by $\rho F_{j\alpha}(t)$, we can define the terms in the brackets on the right-hand side as the stress tensor and then use symmetry and objectivity principles to recast it in the form (36.7).

One may question the derivation of the stress tensor via the 'potential' $W(t)$ for a viscoelastic substance. Though such results are common in theories of thermodynamics[69], it seems desirable to obtain a BKZ type fluid theory without any energetic arguments. We shall do so in what follows[70].

If we reexamine (36.8) – (36.9), and concentrate on the case when

$$\mathbf{C}_t(t-s) = \mathbf{C}_1(s) + \mathbf{C}_2(s), \tag{36.16}$$

where $\mathbf{C}_1(s)$ and $\mathbf{C}_2(s)$ are defined on two intervals of s in $0 \leqq s < \infty$ such that these time intervals are disjoint, then we notice that (36.8) implies that

$$\mathbf{T}_E(\mathbf{C}_t(t-s)) = \mathbf{T}_E(\mathbf{C}_1(s)) + \mathbf{T}_E(\mathbf{C}_2(s)). \tag{36.17}$$

In other words, we are looking for a 'superposition' type of representation for the constitutive functional of the incompressible simple fluid (25.4) :

$$\mathop{\mathbf{\mathit{H}}}_{s=0}^{\infty}(\mathbf{C}_t(t-s)) = \mathop{\mathbf{\mathit{H}}}_{s=0}^{\infty}(\mathbf{C}_1(s)) + \mathop{\mathbf{\mathit{H}}}_{s=0}^{\infty}(\mathbf{C}_2(s)), \tag{36.18}$$

when $\mathbf{C}_1(\cdot)$ and $\mathbf{C}_2(\cdot)$ *are defined on disjoint time intervals on the s-axis.*

Functionals such as those in (36.18) are called *additive*. Representation theorems, ensuring single integral forms such as (36.8), exist for additive

69. COLEMAN [1964 : 4]; BERNSTEIN, KEARSLEY and ZAPAS [1964 : 3].
70. RIVLIN and SAWYERS [1971 : 21].

functionals[71] of the type (36.18). According to this theorem, Eq. (36.18) is equivalent to

$$\mathbf{T}_E(t) = \int_0^\infty \mathbf{f}(\mathbf{C}_t(t-s), s)\, ds, \tag{36.19}$$

where $\mathbf{f}(\cdot\,,\cdot)$ is a *function* of its arguments. By the application of the isotropy condition to (36.19) and incorporating the **1** term into $-p\mathbf{1}$, we have

$$\mathbf{T} + p\mathbf{1} = \int_0^\infty \Big\{\alpha_{-1}(I_1, I_2, s)\,\mathbf{C}_t^{-1}(t-s) - \alpha_1(I_1, I_2, s)\,\mathbf{C}_t(t-s)\Big\} ds, \tag{36.20}$$

where α_j, $j = \pm 1$ are functions of (I_1, I_2, s), with I_1 and I_2 defined by (36.9). We call the constitutive equation (36.20), with (36.9), the *additive functional theory*. Perhaps it should be called the RIVLIN-SAWYERS fluid.

In the sequel, we shall use the addltive functional theory instead of the BKZ fluid, whenever necessary. A reversion to the BKZ model can always be made by putting[72]

$$\alpha_{-1} = 2\frac{\partial U}{\partial I_1}, \qquad \alpha_1 = 2\frac{\partial U}{\partial I_2}.$$

We now record the response of the additive functional theory (36.20) [and (36.9)] in some flows.

(i) *Simple shear,* with $\dot{x} = \varkappa y$, $\dot{y} = \dot{z} = 0$. For this flow, one obtains:

$$\begin{aligned} \tau(\varkappa) &= \int_0^\infty (\alpha_{-1} + \alpha_1)\,\varkappa s\, ds, \\ \bar{N}_1(\varkappa^2) &= \int_0^\infty (\alpha_{-1} + \alpha_1)\,\varkappa^2 s^2 ds, \\ \bar{N}_2(\varkappa^2) &= -\int_0^\infty \alpha_1\,\varkappa^2 s^2 ds, \end{aligned} \tag{36.21}$$

with α_{-1} and α_1 being functions of $I_1 = I_2 = 3 + \varkappa^2 s^2$ and s.

(ii) In *simple extension,* with $\dot{x}_i = a_i x_i$ ($i = 1, 2, 3$; no sum), and $a_1 + a_2 + a_3 = 0$, the extra stresses are :

$$T_E\langle x_i x_i\rangle = \int_0^\infty \Big[\alpha_{-1}\exp(2a_i s) - \alpha_1 \exp(-2a_i s)\Big] ds, \quad i = 1, 2, 3;\ \text{no sum}; \tag{36.22}$$

and the α_j, $j = \pm 1$, are functions of $I_1 = e^{2a_1 s} + e^{2a_2 s} + e^{2a_3 s}$, $I_2 = e^{-2a_1 s} + e^{-2a_2 s} + e^{-2a_3 s}$ and s.

We now relate the additive functional theory to linear viscoelasticity. If

71. See FRIEDMAN and KATZ [1966 : 7] for other references and a nice discussion. Additive functionals were discovered independently by MARTIN and MIZEL [1964 : 16] and CHACON and FRIEDMAN [1965 : 3]. In these papers, it is demanded that $\mathbf{C}_1(s)\,\mathbf{C}_2(s) = \mathbf{0}$, requiring $\mathbf{C}_1(\cdot)$ and $\mathbf{C}_2(\cdot)$ to be positive definite, which is equivalent to $\mathbf{C}_1(\cdot)$ and $\mathbf{C}_2(\cdot)$ being defined on disjoint sets.

72. Many of the results obtained by BERNSTEIN [1966 : 2] for the BKZ fluid can be adapted for the additive functional theory. However, we omit the details.

we make the usual assumptions about the displacements (*cf.* Sec. 5), then

$$\mathbf{C}_t(t-s) = \mathbf{1} + 2[\mathbf{E}(t-s) - \mathbf{E}(t)] + \mathbf{0}(\varepsilon^2)\,; \tag{34.b2}$$

whence

$$\mathbf{C}_t^{-1}(t-s) = \mathbf{1} - 2[\mathbf{E}(t-s) - \mathbf{E}(t)] + \mathbf{0}(\varepsilon^2). \tag{36.23}$$

Since incompressibility demands that det $\mathbf{C}_t(t-s) = 1$, we have $I_1 = I_2 = 3$. Under these restrictions, the additive functional theory becomes

$$\mathbf{T}_E = \int_0^\infty 2(\alpha_{-1} + \alpha_1)[\mathbf{E}(t-s) - \mathbf{E}(t)]\, ds, \tag{36.24}$$

where we have neglected the error terms $\mathbf{0}(\varepsilon^2)$. Note that in (36.24), α_j, $j = \pm 1$, are functions of s only. Comparison with (34.*b*5) shows that

$$\alpha_{-1}(3, 3, s) + \alpha_1(3, 3, s) = -\frac{dG(s)}{ds}. \tag{36.25}$$

For the *lineal flow*, $x = v(y, t)$, $\dot{y} = \dot{z} = 0$, the strain history $\mathbf{C}_t(t-s)$ is given by (26.3) and hence

$$I_1 = I_2 = 3 + [\lambda(t-s)]^2. \tag{36.26}$$

If $\lambda(t-s) = \varkappa s + \varepsilon f(\omega, t, s)$, where ε is very small so that terms of order ε^2 are neglected, then

$$\alpha_j(I_1, I_2, s) = \alpha_j(3 + \varkappa^2 s^2, 3 + \varkappa^2 s^2, s) + 2\varepsilon\left(\frac{\partial \alpha_j}{\partial I_1} + \frac{\partial \alpha_j}{\partial I_2}\right)\varkappa s\, f(\omega, t, s) + O(\varepsilon^2),$$
$$j = \pm 1, \tag{36.27}$$

where $\partial\alpha_j/\partial I_1$, $\partial\alpha_j/\partial I_2$ are evaluated at $I_1 = I_2 = 3 + \varkappa^3 s^2$. We shall need the formulae (36.27) in Sec. 39.

A deliberate attempt has been made here to emphasize the additive functional theory, for it is a mechanical theory, and does not involve any thermodynamic reasonings, since thermodynamics is beyond the scope of this text.

REMARK. In the literature, there are two empirical single integral models, which look like the additive functional theory, except for the important difference that the coefficients of $\mathbf{C}_t(t-s)^{-1}$ and $\mathbf{C}_t(t-s)$ depend on the history of the second invariant of $\mathbf{A}_1$, that is, on tr $\mathbf{A}_1^2\,(t-s)$, $s \in [0, \infty)$. Such models have been studied extensively by BIRD and his coworkers[73], and by BOGUE and WHITE[74].

Let $\mathfrak{x}$ be the fixed position of a particle in a convected coordinate system (*cf.* App. A to Chap. 5). BIRD and CARREAU postulate that the contravariant convected components $\hat{T}_E^{ij}$ of the stress tensor $\mathbf{T}_E$ are given by

$$\hat{T}_E^{ij} = -\int_{-\infty}^{t} m[t-t', II(t')]\Big[\Big(1 + \frac{E}{2}\Big)[\hat{g}^{ij}(t') - \hat{g}^{ij}(t)] + \\ + \frac{E}{2}\,\hat{g}^{ir}(t)\,\hat{g}^{js}(t)\,[\hat{g}_{rs}(t') - \hat{g}_{rs}(t)]\Big]\, dt'. \tag{36.28}$$

73. For the latest version, see BIRD and CARREAU [1968 : 1].
74. BOGUE and WHITE [1970 : 4].

Here $E = E[II(t')]$, but we shall assume it to be a constant ε, following BIRD and CARREAU; $II\,(t') = \text{tr}\,\mathbf{A}_1^2\,(t')$ at the particle. $m(\cdot\,,\cdot)$ is the memory function defined by

$$m[t-t', II(t')] = \sum_{p=1}^{\infty} \frac{\eta_p}{\lambda_{2p}^2} \frac{e^{-(t-t')/\lambda_{2p}}}{1 + \frac{1}{2}\lambda_{1p}^2\, II\,(t')}, \tag{36.29}$$

where η_p, λ_{1p} and λ_{2p} are constants. Finally, $\hat{g}^{ij}\,(\tau)$ and $\hat{g}_{ij}\,(\tau)$ are respectively the contravariant and covariant convected components of the metric tensor at ξ at time τ. Using the configuration at time as the reference configuration, and as well the transformation rules of OLDROYD in Eqs. (A5.3) — (A5.4), one can write the constitutive equation (36.28) in *fixed coordinates* as:

$$T_E^{ij} = -\int_0^{\infty} m(s, II(t-s))\Big[\Big(1+\frac{\varepsilon}{2}\Big)\Big(g^{mn}(t-s)\frac{\partial x^i}{\partial x'^m}\frac{\partial x^j}{\partial x'^n} - g^{ij}\,(t) +$$
$$+ \frac{\varepsilon}{2} g^{ir}\,(t)\, g^{js}\,(t)\,\Big(g_{mn}\,(t-s)\frac{\partial x'^m}{\partial x^r}\frac{\partial x'^n}{\partial x^s} - g_{rs}\,(t)\Big)\Big]\,ds. \tag{36.30}$$

Here $\mathbf{x}' = \mathbf{x}'(\mathbf{x}, t, t-s)$ is the position, in space at time $t-s$, occupied by the particle at $\mathbf{x}$ at time t. Now, in direct notation,

$$g_{mn}\,(t-s)\frac{\partial x'^m}{\partial x^r}\frac{\partial x'^n}{\partial x^s} \text{ are components of } \mathbf{C}_t(t-s),$$

$$g^{mn}\,(t-s)\frac{\partial x^i}{\partial x'^m}\frac{\partial x^j}{\partial x'^n} \text{ are components of } \mathbf{C}_t(t-s)^{-1}.$$

Choosing a Cartesian coordinate system in space, one rewrites (36.30) as:

$$\mathbf{T}_E = -\int_0^{\infty} m[s, II(t-s)]\Big[\Big(1+\frac{\varepsilon}{2}\Big)\Big(\mathbf{C}_t(t-s)^{-1} - \mathbf{1}\Big)+$$
$$+\frac{\varepsilon}{2}\Big(\mathbf{C}_t(t-s) - \mathbf{1}\Big)\Big]\,ds. \tag{36.31}$$

One can absorb the terms involving $\mathbf{1}$ into the pressure term and thus compare (36.31) with the additive functional theory. BIRD and CARREAU have imposed additional restrictions on η_p, λ_{1p} and λ_{2p} through:

$$\eta_p = \eta_0 \frac{\lambda_{1p}}{\sum_{p=1}^{\infty}\lambda_{1p}}, \quad \lambda_{1p} = \lambda_1\left(\frac{1+n_1}{p+n_1}\right)^{\alpha_1},$$

$$\lambda_{2p} = \lambda_2\left(\frac{1+n_2}{p+n_1}\right)^{\alpha_2}. \tag{36.32}$$

Hence, the unknown constants of the theory are: η_0, λ_1, λ_2, n_1, n_2, α_1 and α_2.

The model proposed by BOGUE and WHITE can be obtained from (36.31) by replacing the infinite series in the memory function (36.29) $m[t - t', II(t')]$ by

$$\sum_{i=1}^{\infty} \frac{G_i}{\lambda_{(\text{eff})_i}} e^{-(t-t')/\lambda_{(\text{eff})_i}}. \tag{36.33}$$

Here G_i are some constants and the 'relaxation times' $\lambda_{(\text{eff})_i}$ are defined through

$$\lambda^{-1}_{(\text{eff})_i} = \lambda_i^{-1} + \frac{a \int_0^{t-t'} | II(\xi) |^{\frac{1}{2}} d\xi}{t - t'}, \tag{36.34}$$

where λ_i and a are some other constants.

37. A Study of the Oldroyd 6 and 8 Constant Models

We now begin a study of a theoretical model of OLDROYD, by exaamining the constitutive equation (A5.22), viz,

$$\mathbf{T}_E + \frac{\mathscr{D}\,\mathbf{T}_E}{\mathscr{D}t} + \frac{\mu_0}{2}(\text{tr }\mathbf{T}_E)\,\mathbf{A}_1 + \frac{\mu_1}{2}(\mathbf{T}_E\,\mathbf{A}_1 + \mathbf{A}_1\,\mathbf{T}_E) + \frac{\nu_1}{2}[\text{tr }(\mathbf{T}_B\,\mathbf{A}_1)]\mathbf{1}$$
$$= \eta_0\left[\mathbf{A}_1 + \lambda_2\frac{\mathscr{D}\,\mathbf{A}_1}{\mathscr{D}t} - \mu_2\,\mathbf{A}_1^2 + \frac{\nu_2}{2}(\text{tr }\mathbf{A}_1^2)\,\mathbf{1}\right] \tag{37.1}$$

in steady simple shear $\dot{x} = \varkappa y$, $\dot{y} = \dot{z} = 0$. The viscometric functionals are[75]:

$$\tau(\varkappa) = \varkappa\,\eta_0\,\frac{1 + \sigma_2\,\varkappa^2}{1 + \sigma_1\,\varkappa^2} \equiv \varkappa\,\eta(\varkappa^2), \tag{37.2}$$

$$\overline{N}_1(\varkappa^2) = 2(\lambda_1\,\eta(\varkappa^2) - \lambda_2\,\eta_0)\,\varkappa^2, \tag{37.3}$$

$$\overline{N}_2(\varkappa^2) = [(\mu_1 - \lambda_1)\,\eta(\varkappa^2) + (\lambda_2 - \mu_2)\,\eta_0]\,\varkappa^2, \tag{37.4}$$

where

$$\sigma_1 = \lambda_1^2 + \mu_0\left(\mu_1 - \frac{3}{2}\nu_1\right) - \mu_1(\mu_1 - \nu_1), \tag{37.5}$$

$$\sigma_2 = \lambda_1\,\lambda_2 + \mu_0\left(\mu_2 - \frac{3}{2}\nu_2\right) - \mu_1(\mu_2 - \nu_2). \tag{37.6}$$

Let us demand that $\overline{\eta}\,(\varkappa^2)$ be a positive-valued, decreasing function of $\varkappa$. To achieve this, the sufficient conditions on η_0, σ_1 and σ_2 are:

$$\eta_0 > 0, \qquad \sigma_1 > \sigma_2 > 0. \tag{37.7}$$

Moreover, let $\tau(\varkappa)$ be an increasing function of $\varkappa$, that is, let $(d/d\varkappa)\,\tau(\varkappa) \geqq 0$. Then we have

$$\frac{1 + \sigma_2\varkappa^2}{1 + \sigma_1\varkappa^2} + \frac{(1 + \sigma_1\varkappa^2)\,\sigma_2 - (1 + \sigma_2\varkappa^2)\,\sigma_1}{(1 + \sigma_1\varkappa^2)^2}\,2\varkappa^2 \geqq 0; \tag{37.8}$$

75. OLDROYD [1958 : 4].

that is, the desired inequality is

$$3\sigma_2 \geqq \sigma_1 > 0. \tag{37.9}$$

Combining (37.7) and (37.9) we obtain[76]

$$3\sigma_2 \geqq \sigma_1 > \sigma_2 \geqq \frac{1}{3}\sigma_1 > 0. \tag{37.10}$$

Demanding that $\overline{N}_1(\varkappa^2) > 0$ means that one has the inequality $\lambda_1 (1+\sigma_2\varkappa^2)/(1+\sigma_1\varkappa^2) - \lambda_2 > 0$, or

$$\lambda_1 > \lambda_2, \qquad \lambda_1\sigma_2 \geqq \lambda_2\sigma_1, \tag{37.11}$$

and as well that $\overline{N}_2(\varkappa^2) < 0$, leads us to

$$(\mu_1 - \lambda_1)(1 + \sigma_2\varkappa^2) + (\lambda_2 - \mu_2)(1 + \sigma_1\varkappa^2) < 0; \tag{37.12}$$

or, let us take

$$0 < \mu_1 - \lambda_1 < \lambda_2 - \mu_2, \tag{37.13}$$

$$(\mu_1 - \lambda_1)\sigma_2 < (\lambda_2 - \mu_2)\sigma_1. \tag{37.14}$$

If, as OLDROYD[77] has suggested, we take $\mu_1 = \lambda_1$, $\mu_2 = \lambda_2$, then the second normal stress difference will be zero, and hence we no not adopt these restrictions. On noting that (37.14) is a consequence of (37.7) and (37.13), we can say that the inequalities we need are (37.10), (37.11) and (37.13). We obtain one more restriction next. In COUETTE flow (Sec. 28), OLDROYD[78] has shown that the stress $T\langle zz\rangle$ has the leading term :

$$2\eta_0\,\sigma_1\sigma_2^{-2}\,(2\mu_1\sigma_2 - \lambda_1\sigma_2 - 2\mu_2\sigma_1 + \lambda_2\sigma_1)\,b^2r^{-4}, \tag{37.15}$$

and this determines whether the liquid climbs up the inner cylinder or not[79]. Using (37.14) we see that the term in (37.15) will be positive, thereby leading to the rod climbing effect, if

$$\mu_1\sigma_2 \geqq \mu_2\sigma_1 > 0. \tag{37.16}$$

Interestingly, no direct restrictions on μ_0, ν_1 and ν_2 arise. Having derived these inequalities, we note some of the nice rheological properties inherent in the model (37.1) :

1°. The viscosity is a decreasing function of $\varkappa$, going from η_0 (at $\varkappa = 0$) to $\eta_0\sigma_2/\sigma_1$ as $\varkappa \to \infty$.

2°. Unequal normal stresses are exhibited in viscometric flows.

3°. If at time $t = 0$, the motion is stopped so that $\mathbf{v}$ and $\mathbf{A}_1$ are zero, then (37.1) reduces to

$$\mathbf{T}_E + \lambda_1 \frac{\partial \mathbf{T}_E}{\partial t} = \mathbf{0}, \qquad t > 0. \tag{37.17}$$

Since $\lambda_1 > 0$, stress relaxation occurs and $\mathbf{T}_E(t) \to \mathbf{0}$ as $t \to \infty$. So the fluid model exhibits this important property.

76. OLDROYD [1958 : 4].
77. [1961 : 4], Eq. (15).
78. [1958 : 4], p. 284.
79. It should be noted that such a remark is purely qualitative. See Sec. 44 for an exact investigation of the rod climbing effect.

4°. The claim by TANNER and SIMMONS[80] that this model with $\mu_0 = 0$ may exhibit instabilities similar to those predicted by second-order fluids is not true if one accepts that $\overline{N}_2(\varkappa^2) < 0$, which seems to be the case experimentally. Their analysis is based on $\overline{N}_2(\varkappa^2) > 0$.

We now suggest a method for reducing the constitutive equation (37.1) to an algebraic equation in MWCSH. In these flows, both $\mathcal{D}\mathbf{T}_E/\mathcal{D}t$ and $\mathcal{D}\mathbf{A}_1/\mathcal{D}t$ can be written respectively as :

$$\frac{\mathcal{D}\mathbf{T}_E}{\mathcal{D}t} = \mathbf{T}_E(\mathbf{W} - \mathbf{Z}) - (\mathbf{W} - \mathbf{Z})\,\mathbf{T}_E\,, \qquad (37.18)^{81}$$

$$\frac{\mathcal{D}\mathbf{A}_1}{\mathcal{D}t} = \mathbf{A}_1(\mathbf{W} - \mathbf{Z}) - (\mathbf{W} - \mathbf{Z})\,\mathbf{A}_1, \qquad (37.19)^{81}$$

where $\mathbf{Z}$ is defined by (37.1). Hence $(8.11)_2$ can be solved quite readily for $\mathbf{T}_E$ which can be used in the equations of motion.

For more general circumstances, GODDARD and MILLER[82] have shown that one can obtain an integral for the co-rotational derivative $\mathcal{D}\mathbf{T}_E\,/\,\mathcal{D}t$ in terms of matrizants. However, this exprssion does not seem to facilitate the task of simplifying (37.1) to any appreciable extent.

Hence, in more complex steady flow problems, or where the flow is not steady, the constitutive equations form a set of 6 simultaneous linear partial differential equations to be solved for thc stresses, which are coupled with the three equations of motion. So we have nine coupled equations from which the three velocity components are to be derived, the consistency of which has not been explored. It seems therefore that the versatility[83] of the model (37.1) is lost to a large extent by the algebraic complexity thus introduced. For example, even GRIFFITHS, JONES and WALTERS[84], have employed the third-order equation (35.25) in discussing a complicated flow arising due to edge effects, despite the fact that this was a steady flow.

Attempts have thus been made by some authors to express the OLDROYD equation in terms of the n-th order fluids, $n = 1, 2, 3, \ldots,$ to within an error term. We turn to a study of these next. In 1964, NIILER and PIPKIN[85] studied the equation

$$\mathbf{T}_E + \lambda_1 \frac{\delta\mathbf{T}_E}{\delta t} - 2\varkappa_1(\mathbf{T}_E\mathbf{A}_1 + \mathbf{A}_1\mathbf{T}_E) + \mu_0\mathbf{A}_1(\operatorname{tr}\mathbf{T}_E) + \nu_1\mathbf{1}\operatorname{tr}(\mathbf{T}_E\mathbf{A}_1)$$

$$= 2\mu[\mathbf{A}_1 + \lambda_2\mathbf{A}_2 - 4\varkappa_2\mathbf{A}_1^2 + \nu_2\mathbf{1}(\operatorname{tr}\mathbf{A}_1^2)] \quad (37.20)$$

---

80. [1967 : 24].

81. To be compared with OLDROYD [1958 : 4], p. 283 and NFTM, Eq. (119.19).

82. [1966 : 8], Eq. (18) or (19), These authors call the irrotational derivative the JAUMANN derivative.

83. WALTERS [1970 : 29], p. 599, calls attention to this versatility.

84. [1969 : 6].

85. [1964 : 17]. Our notation is different from their notation.

in the lineal flow problem

$$u_1 = u_1(x_2, t), \qquad u_2 = u_3 = 0, \tag{37.21}$$

and showed that in non-dimensional form, the stress $\mathbf{T}'_E$ in (37.20) is given by

$$\begin{aligned}\mathbf{T}'_E &= \mathbf{B}_1 - \varepsilon^2(\lambda'_1 - \lambda'_2)\,\mathbf{B}_2 + 4\varepsilon^3 Q(\varkappa'_1 - \varkappa'_2)\,\mathbf{B}_1^2 + \varepsilon^4 \lambda'_1(\lambda'_1 - \lambda'_2)\,\mathbf{B}_2 - \\ &\quad - 2\varepsilon^5 Q[2\lambda'_1(\varkappa'_1 - \varkappa'_2) + \varkappa'_1(\lambda'_1 - \lambda'_2]\,(\mathbf{B}_1\mathbf{B}_2 + \mathbf{B}_2\mathbf{B}_1) - \\ &\quad - \varepsilon^6 \lambda'^2_1(\lambda'_1 - \lambda'_2)\,\mathbf{B}_4 + \varepsilon^6 Q^2[4(\lambda'_1 + 2\varkappa'_1 - \mu'_0)(\varkappa'_1 - \varkappa'_2) + \\ &\quad + (2\lambda'_1 + 3\mu'_0 - 4\varkappa'_1)(\nu'_1 - \nu'_2) + 2\mu'_0(\lambda'_1 - \lambda'_2)]\,\mathbf{B}_1 \operatorname{tr} \mathbf{B}_1^2 + \\ &\quad + \mathbf{1}(\ldots) + O(\varepsilon^7).\end{aligned} \tag{37.22}$$

In (37.20), the operator $\delta/\delta t$ is defined as in (A5.11) and with t representing a characteristic time, one has

$$\lambda_\alpha = \lambda'_\alpha t, \quad \varkappa_\alpha = \varkappa'_\alpha t, \quad \nu_\alpha = \nu'_\alpha t, \quad \mu_0 = \mu'_0 t, \quad \alpha = 1, 2, \tag{37.23}$$

with A representing an amplitude, ω a frequency, ρ the density, and

$$\mathbf{T}'_E = \frac{1}{2A\mu\omega k}\mathbf{T}_E; \qquad k^2 = \rho\omega/2\mu. \tag{37.24}$$

Also, $\varepsilon^2 = \omega t$, and $Q^2 = \rho A^2/2\mu t$. Moreover,

$$\mathbf{A}_n = Ak\,\omega^n\,\mathbf{B}_n, \qquad n = 1, 2, \ldots . \tag{37.25}$$

For this problem of lineal flow, by using ths RIVLIN-ERICKSEN constitutive relation (35.7), it can be shown that[86]

$$\begin{aligned}\mathbf{T}'_E &= \mathbf{B}_1 + \varepsilon^2\beta_2\mathbf{B}_2 + 2Q\varepsilon^3\beta_{11}\mathbf{B}_1^2 + \varepsilon^4\beta_3\mathbf{B}_3 + 2\varepsilon^5 Q\beta_{12}(\mathbf{B}_1\mathbf{B}_2 + \mathbf{B}_2\mathbf{B}_1) + \\ &\quad + \varepsilon^6(\beta_4\mathbf{B}_4 + Q^2\mathbf{B}_{1,11}\,\mathbf{B}_1 \operatorname{tr}\mathbf{B}_1^2) + O(\varepsilon^7),\end{aligned} \tag{37.26}$$

where we have derived (37.26) from

$$\begin{aligned}\frac{\mathbf{T}_E}{2\mu} &= \Sigma\,\alpha_n\mathbf{A}_n + \Sigma\,\Sigma\,\alpha_{mn}(\mathbf{A}_m\mathbf{A}_n + \mathbf{A}_n\mathbf{A}_m) + \\ &\quad + \Sigma\,\Sigma\,\Sigma[\alpha_{p,\,mn}\,\mathbf{A}_p \operatorname{tr}(\mathbf{A}_m\,\mathbf{A}_n) + \alpha_{mnp}(\mathbf{A}_m\,\mathbf{A}_n\,\mathbf{A}_p + \\ &\quad + \mathbf{A}_p\,\mathbf{A}_n\,\mathbf{A}_m)] + \ldots .\end{aligned} \tag{37.27}$$

Thus, α_n, α_{mn}, $\alpha_{1,11}$ are related to β_n, β_{mn}, $\beta_{1,11}$ through

$$\alpha_n = \beta_n\,t^{n-1}, \qquad \alpha_{mn} = \beta_{mn}\,t^{m+n-1}, \qquad \alpha_{1,11} = \beta_{1,11}\,t^2. \tag{37.28}$$

86. NIILER and PIPKIN [1964 : 17], Eq. (4.5).

Comparing (37.22) and (37.27), and neglecting the term involving **1** in (37.22) because it can be absorbed into the pressure term, we obtain

$$\alpha_2 = -(\lambda_1 - \lambda_2), \quad \alpha_3 = \lambda_1(\lambda_1 - \lambda_2), \quad \alpha_4 = -\lambda_1^2(\lambda_1 - \lambda_2)$$

$$\alpha_{11} = 2(\varkappa_1 - \varkappa_2), \quad \alpha_{12} = -2\lambda_1(\varkappa_1 - \varkappa_2) - \varkappa_1(\lambda_1 - \lambda_2), \tag{37.29}$$

$$\alpha_{1,11} = 4(\lambda_1 + 2\varkappa_1 - \varkappa_0)(\varkappa_1 - \varkappa_2) + (2\lambda_1 + 3\mu_0 - 4\varkappa_1)(\nu_1 - \nu_2) + 2\mu_0(\lambda_1 - \lambda_2).$$

So the OLDROYD coefficients in Equation (37.20) are related to the RIVLIN-ERICKSEN coefficients in (37.27) by (37.29) as well as the fact that they obey $\alpha_4 = \alpha_3^2 / \alpha_2$ when $\lambda_1 \neq \lambda_2$.

This comparitive study of NIILER and PIPKIN is not specialized in that it is not limited to a single problem, though it may appear that way. However, the comparison is not valid for all flows. So we shall study a more useful comparison by following WALTERS[37] and setting the two constants $\nu_1 = \nu_2 = 0$ in (37.1), and examine the simple constitutive relation (the 6 constant model):

$$\mathbf{T}_E + \lambda_1 \frac{\delta}{\delta t} \mathbf{T}_E + \frac{1}{2}\mu_0 \mathbf{A}_1(\operatorname{tr} \mathbf{T}_E) - \frac{\mu_1}{2}(\mathbf{T}_E \mathbf{A}_1 + \mathbf{A}_1 \mathbf{T}_E) = \eta_0[\mathbf{A}_1 + \lambda_2 \mathbf{A}_2 - \mu_2 \mathbf{A}_1^2], \tag{37.30}$$

with a view to obtaining a relation between the coefficients of this fluid and those of the first, second, and third-order fluids of Sec. 35.

Steady Flow. In this case, $(\partial \mathbf{v}/\partial t) = 0$ and we may assume that $\partial \mathbf{T}_E / \partial t = \mathbf{0}$. Then

$$\frac{\delta}{\delta t} \mathbf{T}_E = (\nabla \mathbf{T}_E)\mathbf{v} + \mathbf{T}_E \mathbf{L} + \mathbf{L}^T \mathbf{T}_E, \tag{37.31}$$

and is thus of higher order in the velocity $\mathbf{v}$ than $\mathbf{L}$.

Also, if $\mathbf{v} = O(\varepsilon)$, then $\mathbf{A}_n = O(\varepsilon^n)$, $n = 1, 2, \ldots$. Thus to the first order in ε, (37.30) is approximated by

$$\mathbf{T}_E^{(1)} = \eta_0 \mathbf{A}_1. \tag{37.32}$$

Substitution of $\mathbf{T}_E^{(1)}$ into (37.30) leads to

$$(\mathbf{T}_E^{(1)} + \mathbf{T}_E^{(2)}) + \lambda_1 \frac{\delta}{\delta t} \mathbf{T}_E^{(1)} - \frac{\mu_1}{2}(\mathbf{T}_E^{(1)} \mathbf{A}_1 + \mathbf{A}_1 \mathbf{T}_E^{(1)}) = \eta_0(\mathbf{A}_1 + \lambda_2 \mathbf{A}_2 - \mu_2 \mathbf{A}_1^2). \tag{37.33}$$

On noting that $\delta \mathbf{A}_1/\delta t = \mathbf{A}_2$, a rearrangment of (37.33), leads to

$$\mathbf{T}_E^{(2)} = \eta_0[(\lambda_2 - \lambda_1)\mathbf{A}_2 + (\mu_1 - \mu_2)\mathbf{A}_1^2]. \tag{37.34}$$

---

37. [1970 : 29].

Similar manipulations lead us to

$$\mathbf{T}_E^{(3)} = \eta_0\lambda(\lambda_1 - \lambda_2)\,\mathbf{A}_3 + \frac{\eta_0}{2}[\mu_0\{\lambda_1 - \lambda_2) - (\mu_1 - \mu_2\} + $$
$$+ \mu_1(\mu_1 - \mu_2) + \lambda_1(\mu_1 - \mu_2)]\,(\operatorname{tr}\mathbf{A}_1^2)\,\mathbf{A}_1 +$$
$$+ \frac{\eta_0}{2}[\mu_1\lambda_2 - 3\mu_1\lambda_2 + 2\lambda_1\mu_1](\mathbf{A}_1\mathbf{A}_2 + \mathbf{A}_2\mathbf{A}_1). \quad (37.35)$$

Comparison with (35.25) shows that the constants of the third-order fluid and the OLDROYD model (37.30) are related through :

$$\eta_0 = \eta_0,$$
$$\beta = \eta_0(\mu_1 - \mu_2),$$
$$\gamma = \eta_0(\lambda_2 - \lambda_1),$$
$$\alpha_0 = \frac{\eta_0}{2}[\mu_0\{(\lambda_1 - \lambda_2) - (\mu_1 - \mu_2)\} +$$
$$+ \mu_1(\mu_1 - \mu_2) + \lambda_1(\mu_1 - \mu_2)], \quad (37.36)$$
$$\alpha_1 = \frac{\eta_0}{2}(\mu_1\lambda_2 - 3\mu_1\lambda_1 + 2\lambda_1\mu_2),$$
$$\alpha_2 = \eta_0\lambda_1(\lambda_1 - \lambda_2).$$

Unsteady Flow. In these problems, since the n-th-order fluid does not have the relaxation property, it seems reasonable to compare the OLDROYD model (37.30) with theories of finite linear viscoelasticity and second-order viscoelasticity. For this purpose, let (37.30) be written in convected coordinates as

$$\pi'_{ik} + \lambda_1 \frac{D}{Dt}\pi'_{ik} + \mu_0\pi'^{j}_{j}\gamma^{(1)}_{ik} - \mu_1(\pi'^{j}_{i}\gamma^{(1)}_{jk} + \pi'^{j}_{k}\gamma^{(1)}_{ji})$$
$$= 2\eta_0[\gamma^{(1)}_{ik} + \lambda_2\frac{D}{Dt}\gamma^{(1)}_{ik} - 2\mu_2\gamma^{(1)j}_{i}\gamma^{(1)}_{jk}]. \quad (37.37)$$

Since the flow is unsteady, the convected derivative $D\Upsilon/Dt$ is of the same order as Υ. Hence, to the first order in ε, if $\mathbf{v} = O(\varepsilon)$, (37.37) is :

$$\pi'_{ik} + \lambda_1\frac{D}{Dt}\pi'_{ik} = 2\eta_0\left[\gamma^{(1)}_{ik} + \lambda_2\frac{D\gamma^{(1)}_{ik}}{Dt}\right]. \quad (37.38)$$

Equivalently,

$$\pi'_{ik} = 2\eta_0\frac{\lambda_2}{\lambda_1}\gamma^{(1)}_{ik} + 2\eta_0\frac{\lambda_1 - \lambda_2}{\lambda_1^2}\int_0^\infty e^{-s/\lambda_1}\gamma^{(1)}_{ik}(t-s)\,ds. \quad (37.39)$$

In fixed coordinates, we have

$$\mathbf{T}_E = \eta_0\frac{\lambda_2}{\lambda_1}\mathbf{A}_1 - 2\eta_0\frac{\lambda_1 - \lambda_2}{\lambda_1^2}\int_0^\infty e^{-s/\lambda_1}\frac{d}{ds}\mathbf{G}(s)\,ds. \quad (37.40)$$

where we have used

$$\frac{d}{d\tau}\,\mathbf{C}_t(\tau) = \frac{d}{d\tau}\,(\mathbf{C}_t(\tau)-\mathbf{1}) = -\frac{d}{ds}\,(\mathbf{C}_t(t-s)-\mathbf{1}) = -\frac{d}{ds}\,\mathbf{G}(s). \tag{37.41}$$

[88]

We can convert (37.40) to have the form of finite linear viscoelasticity if we put

$$\mathbf{T}_E = -\int_0^\infty k_1(s)\,\frac{d\,\mathbf{G}(s)}{ds}\,ds,$$

$$k_1(s) = \eta_0\,\frac{\lambda_2}{\lambda_1}\,\delta(0) + 2\eta_0\,\frac{\lambda_1-\lambda_2}{\lambda_1^2}\,e^{-s/\lambda_1} \tag{37.42}$$

where $\delta(\,\cdot\,)$ is the DIRAC delta function.

WALTERS[89] has also shown that working up to second-order in ε leads to

$$\mathbf{T}_E = -\int_0^\infty k_1(s)\,\frac{d}{ds}\,\mathbf{G}(s)\,ds + \int_0^\infty\int_0^\infty k_2(s_2, s_3)\Big[\frac{d}{ds_2}\,\mathbf{G}(s_2)\,\frac{d}{ds_3}\,\mathbf{G}(s_3) + \frac{d}{ds_3}\,\mathbf{G}(s_3)\,\frac{d}{ds_2}\,\mathbf{G}(s_2)\Big]\,ds_2\,ds_3, \tag{37.43}$$

$$k_2(s_2, s_3) = \begin{cases} 0, & 0 \leqq s_3 < s_2, \\ \dfrac{2\eta_0(\mu_1\lambda_2 - \mu_2\lambda_1)}{\lambda_1^2}\,e^{-s_2/\lambda_1}\,\delta(s_3 - s_2) + \dfrac{2\eta_0\mu_1(\lambda_1-\lambda_2)\,e^{-s_3/\lambda_1}}{\lambda_1^2}, & s_3 \geqq s_2. \end{cases} \tag{37.44}$$

We have recorded in this section a systematic procedure by which the OLDROYD constitutive equations may be expressed as closely as we desire by an n-th-order fluid in steady flow, or expressed as special type of integral models in non-steady flows.

These approximation schemes are motivated by a desire[90] to replace the OLDROYD model by explicitly constitutive equations, for the OLDROYD models are arbitrarily chosen and not derived as a sequence of equations with increasing complexity. However, the approximation procedure is complicated because none of the order models nor the theories of finitely linear and second-order viscoelasticity have material functions comparable to those of the OLDROYD fluid in viscometric flows. But at deviations large enough from the state of rest in non-steady flows, the OLROYD model with a single relaxation time cannot be a good descriptor of viscoelastic liquids. So, if one accepts the reasoning that for 'large motions' one cannot use the OLDROYD model, and believes that it is useful at small deformations, one may represent it in terms

88. Since $\mathbf{C}_t(\tau) = \mathbf{F}_t(\tau)^T\,\mathbf{F}_t(\tau)$, $(d/d\tau)\,\mathbf{C}_t(\tau) = \mathbf{F}_t(\tau)^T\,\mathbf{A}_1(\tau)\,\mathbf{F}_t(\tau)$. This is used by WALTERS [1970 : 29], in his Eq. (14).

89. [1970 : 29], Eqs. (36)—(37).

90. This point of view is not shared by WALTERS [1970 : 29], for example.

of 'order' fluids and theories of finitely linear and second-order viscoelasticity. Thus, the OLDROYD fluids occupy a place in between the RIVLIN-ERICKSEN fluids and the additive functional theory.

Appendix : Instability, Uniqueness and Nonexistence of Solutions to Flows of Second-order Fluids

Consider the constitutive equation of the second-order fluid (35.24), subject to the restrictions

$$\eta_0 > 0, \qquad \gamma < 0. \tag{A8.1}$$

Suppose that this fluid undergoes a steady shearing motion

$$\dot{x} = \varkappa y, \qquad \dot{y} = \dot{z} = 0, \tag{A8.2}$$

between two infinite parallel plates at $y = 0$ and $y = h$. The plate at $y = 0$ is at rest always ; and let the plate at $y = h$ be brought to rest also at time $t = 0$. Then ($A8.2$) is replaced by the velocity field :

$$\dot{x} = v(y, t), \qquad \dot{y} = \dot{z} = 0\,; \qquad v(0, t) = v(h, t) = 0, \qquad t > 0. \tag{A8.3}$$

From Eq. (27.12), we know that the differential equation governing the motion is

$$\frac{\partial}{\partial y} T\langle xy\rangle = \rho \frac{\partial v}{\partial t} + g(t). \tag{A8.4}$$

The constitutive equation (35.24) implies that

$$T\langle xy\rangle = \eta_0 \frac{\partial v}{\partial y} + \gamma \frac{\partial^2 v}{\partial t\, \partial y}, \tag{A8.5}$$

and by assuming that there exists no pressure gradient in the direction of motion, that is, $g(t) \equiv 0$, ($A8.4$) becomes

$$\eta_0 \frac{\partial^2 v}{\partial y^2} + \gamma \frac{\partial^3 v}{\partial y\, \partial t\, \partial y} = \rho \frac{\partial v}{\partial t}. \tag{A8.6}$$

This is a linear partial differential equation of third order and by separation of variables, the unique solution, satisfying the boundary conditions ($A8.3$), is

$$v(y, t) = \sum_{n=1}^{\infty} A_n e^{\mu_n t} \sin \lambda_n y, \tag{A8.7}$$

where

$$\lambda_n = \frac{n\pi}{h}, \qquad \mu_n = -\frac{\eta_0 \lambda_n^2}{\rho + \gamma\lambda_n^2}, \qquad n = 1, 2, \ldots \tag{A8.8}$$

Since $\gamma < 0$, as n gets very large, $\mu_n > 0$. In fact if $\lambda_n^2 > \rho / |\gamma|$, $\mu_n > 0$. Thus the velocity field in ($A8.7$) will grow without bounds and will not decay to zero. This is the major conclusion of COLEMAN, DUFFIN and MIZEL[91].

---

91. [1965 : 5].

For the Nowtonian fluid $\gamma = 0$, $\mu_n < 0$ and hence the velocity decays to zero as $t \to \infty$. However, for the second-order fluid, whatever may be the magnitude of γ, so long as it is negative, the velocity field grows without bounds. This is a very unsatisfactory feature of the second-order fluid theory.

If the boundary conditions (*A*8.3) are replaced by

$$v(0, t) = 0, \qquad \left[\eta_0 \frac{\partial v(y, t)}{\partial y} + \gamma \frac{\partial^2 v(y, t)}{\partial t\, \partial y}\right]_{y=h} = 0, \tag{A8.9}$$

that is, there are no shear stresses on the top plate from $t \geqq 0$, then the solution is again given by (*A*8.7), but instead of (*A*8.8), we have

$$\lambda_n = (2n - 1)\,\pi/2h,$$

$$\mu_n^2 = -\eta_0(2n - 1)^2\pi^4/[4\rho h^2 + \gamma\pi^2(2n - 1)^2]. \tag{A8.10}$$

If we take $\gamma < 0$ as usual, then whenever

$$(2n - 1)^2 > \frac{4\rho h^2}{\pi^2 \mid \gamma \mid}, \tag{A8.11}$$

the solution is again unbounded, but unique.

COLEMAN, DUFFIN and MIZEL proved further that if the boundary conditions (*A*8.3) are obeyed and if for some integer n,

$$h = n\pi\sqrt{\mid \gamma \mid/\rho}, \tag{A8.12}$$

and if

$$\int_0^h [\eta_0\, \rho v(y, 0) - \gamma g_0] \sin (y\sqrt{\rho/\mid \gamma \mid})\, dy \neq 0, \tag{A8.13}$$

with $g_0 = g(0)$ in Eq. (*A*8.4), *then there is no solutiou* $v(y, t)$ of

$$\eta_0 \frac{\partial^2 v}{\partial y^2} + \gamma \frac{\partial^3 v}{\partial y\, \partial t\, \partial y} = \rho \frac{\partial v}{\partial t} + g(t). \tag{A8.14}$$

Later on, COLEMAN and MIZEL[92] proved that if at time $t = 0$, the velocity field $\mathbf{v(x)} = u(\mathbf{x}, 0)\, \mathbf{i} + v(\mathbf{x}, 0)\, \mathbf{j} + w(\mathbf{x}, 0)\, \mathbf{k}$ is of the lineal type with $u(0, 0) = u(h, 0)$ and $u(y, 0)$ obeys (*A*8.13), then for any solution of the differential equation

$$\rho\dot{\mathbf{v}} = \operatorname{div} [\eta_0\, \mathbf{A}_1 + \beta\, \mathbf{A}_1^2 + \gamma\, \mathbf{A}_2] - g(t)\, \mathbf{i}, \tag{A8.15}$$

one of the following equations must fail to hold at some point in the fluid :

$$\frac{\partial^2}{\partial t \partial z} u \bigg|_{t=0} = 0, \qquad \frac{\partial^2 v}{\partial t}\bigg|_{t=0} = 0, \qquad \frac{\partial w}{\partial t}\bigg|_{t=0} = 0. \tag{A8.16}$$

In other words, the flow cannot remain a lineal flow for a critical value of h unless the initial velocity profile minus $\gamma g_0/\eta_0\rho$ is orthogonal to $\sin (y\sqrt{\rho / \mid \gamma \mid})$. Moreover, this result asserts that if the solution of the form $\mathbf{v} = u(y, t)\, \mathbf{i}$

92. [1966 : 5]. Here $g_0 = \partial_x\psi$, evaluated at $t = 0$, $x = y = z = 0$, and ψ is the modified pressure.

exists at all, there are various factors at work to destroy this flow.

This appendix illustrates why the second-order fluid is a poor model to use in initial value problems. It may even be questioned whether use of this model in steady flows is justifiable. Though the model will be used in this book to understand the qualitative behaviour of viscoelastic liquids, the answer to this question has to be in the negative as we shall show next. In MWCSH, for this second-order fluid, the stress power tr $(\mathbf{T}_E \mathbf{A}_1)$ can be shown to be given by

$$\mathrm{tr}\,(\mathbf{T}_E \mathbf{A}_1) = \eta_0 \,\mathrm{tr}\, \mathbf{A}_1^2 + (\beta + \gamma)\,\mathrm{tr}\, \mathbf{A}_1^3 . \qquad (A8.17)$$

Now tr $(\mathbf{T}_E \mathbf{A}_1) \geqq 0$ if and only if

$$\eta_0 \geqq 0, \qquad \beta + \gamma = 0. \qquad (A8.18)$$ [93]

In other words, the second-order fluid satisfies the requirement of non-negative stress power in MWCSH if and only if ($A8.18$) is satisfied. In terms of normal stress differences, ($A8.18$) demands that $N_2 = -\frac{1}{2} N_1$. It is doubtful whether such a fluid exists in the physical world, for in polymeric liquids tested so far ($A8.18$) is not obeyed. Thus we conclude that the second-order fluid is a bad model.

Interestingly, the conditions ($A8.18$) have been shown by Hills[94] to lead to a unique solution for various boundary data provided the first and second gradients of the velocity field remain bounded.

93. This condition was derived, in a different way, by Green and Laws [1967 : 11] to be sufficient for this fluid to sustain irreversible thermodynamic processes. It seems that Noll was also aware of it in 1965. See footnote on p. 115 of [1965 : 5].

94. [1967 : 13].

CHAPTER 9

PERTURBATION ABOUT AN ARBITRARY GROUND STATE. NEARLY VISCO-METRIC FLOWS

We now revert to the theory of the general simple fluid and will study the material response of such a fluid to arbitrarily small disturbances (to be made precise below) superposed on a given motion. For the sake of simplicity, we suppose that the given motion is the viscometric flow. One could, with relevant modifications, apply the arguments here to perturbations of MWCSH-II, or simple extension, etc. We shall be content with flows close to viscometricity because these are fairly simple to analyze and adequate experimental data are available to compare theoretical predictions with measurements. Also, this class of nearly viscometric flows permits us to discover which integral model is a good approximation to the simple fluid and, in fact, indicates that the additive functional theory (36.20) is superior to other integral models in representing the simple fluid.

Finally, the nearly viscometric flow theory has been worked out in elegant detail by PIPKIN and OWEN[1] and applied by PIPKIN[2] and others[3] to small oscillations superposed on steady shearing, and by the author[4] to MWCSH, treating two MWCSH as nearly viscometric flows. So we have a wealth of material to draw upon to illustrate the principles and practice of such a theory.

Logically, the theory of nearly viscometric flows in this chapter is independent of the approximations in Chap. 8. However, we have presented the various explicit constitutive models in Chap. 8, enabling us to make a comparison between some of them in Sec. 39.3.

38. Nearly Viscometric Flows: Theory[5]

To recapitulate, let the viscometric flow be $\dot{x} = \varkappa y$, $\dot{y} = \dot{z} = 0$, and let the strain history be given by

$$°\mathbf{C} = \mathbf{1} - s\mathbf{A}_1 + \frac{1}{2} s^2\mathbf{A}_2. \tag{38.1}$$

1. [1967 : 20].
2. [1968 : 20].
3. [1971 : 4], [1972 : 2].
4. [1970 : 14].
5. This is based completely on PIPKIN and OWEN [1967 : 20].

We introduce the vectors $\mathbf{a}^1$, $\mathbf{a}^2$ and $\mathbf{a}^3$ which are respectively the unit vector along the flow, in the direction of the gradient, and normal to both these directions so that $\mathbf{a}^1$, $\mathbf{a}^2$ and $\mathbf{a}^3$ constitute a right-handed orthonormal triad. It is also known that

$$2\varkappa^2 = \operatorname{tr} \mathbf{A}_1^2 = \operatorname{tr} \mathbf{A}_2. \tag{38.2}$$

Moreover, in tensor product notation

$$\mathbf{A}_1 = \varkappa(\mathbf{a}^1 \otimes \mathbf{a}^2 + \mathbf{a}^2 \otimes \mathbf{a}^1), \qquad \mathbf{A}_2 = 2\varkappa^2 \mathbf{a}_2 \otimes \mathbf{a}_2. \tag{38.3}$$

Now, let $\mathbf{C}$ be another relative strain history close to $°\mathbf{C}$, so that the difference

$$\mathbf{E} = \mathbf{E}(t - s, t) = \mathbf{E}_t(t - s) = \mathbf{C} - °\mathbf{C} \tag{38.4}$$

has a very small norm in the HILBERT space of histories (see Chap. 8). Then the constitutive equation (25.4) and its FRÉCHET differentiability yield

$$\underset{s=0}{\overset{\infty}{\mathbf{\mathit{H}}}} (\mathbf{C}) - \underset{s=0}{\overset{\infty}{\mathbf{\mathit{H}}}} (°\mathbf{C}) = \delta \underset{s=0}{\overset{\infty}{\mathbf{\mathit{H}}}} (°\mathbf{C} \mid \mathbf{E}) + \mathbf{o}(\| \mathbf{E} \|), \tag{38.5}$$

or, that the difference in the stresses due to the perturbation (38.4) is given by the linear functional $\delta \underset{s=0}{\overset{\infty}{\mathbf{\mathit{H}}}} (°\mathbf{C} \mid \mathbf{E})$. So in all, there are 81 linear functionals, for one can write this functional as

$$\delta \mathbf{T}^E = \delta \underset{s=0}{\overset{\infty}{\mathbf{\mathit{H}}}} (°\mathbf{C} \mid \mathbf{E}) = \delta \underset{s=0}{\overset{\infty}{\mathbf{\mathit{H}}}} (\varkappa, \mathbf{a}^2(t) \mid \mathbf{E})$$

$$= \mathbf{a}^i(t) \otimes \mathbf{a}^j(t)\, \delta S_{ijkl}\, (\varkappa \mid E_{kl}). \qquad (38.6)^6$$

The justification for (38.6) stems from the fact that $°\mathbf{C}$ is completely determined by $\varkappa$ and the $\mathbf{a}^i(t)$, $i = 1, 2, 3$; moreover $\delta \underset{s=0}{\overset{\infty}{\mathbf{\mathit{H}}}} (°\mathbf{C} \mid \mathbf{E})$ is a linear functional (see Chap. 8) of $\mathbf{E}$ and is isotropic in the following manner:

$$\mathbf{Q}\, \delta \underset{s=0}{\overset{\infty}{\mathbf{\mathit{H}}}} (\mathbf{C}° \mid \mathbf{E})\, \mathbf{Q}^T = \delta \underset{s=0}{\overset{\infty}{\mathbf{\mathit{H}}}} (\mathbf{Q}\, °\mathbf{C}\mathbf{Q}^T \mid \mathbf{Q}\, \mathbf{E}\mathbf{Q}^T), \quad \forall\, \mathbf{Q} \in \mathcal{O}. \tag{38.7}$$

Now, $\delta S_{ijkl} = \delta S_{jikl}$ by the symmetry of the stress tensor, and since $E_{kl} = E_{lk}$, we can assume $\delta S_{ijkl} = \delta S_{ijlk}$ without loss of generality. These two steps reduce the number of independent functionals to 36. The assumption (25.7) that for all strain histories $\mathbf{C}$, $\operatorname{tr} \underset{s=0}{\overset{\infty}{\mathbf{\mathit{H}}}} (\mathbf{C}) = 0$ implies that

$$\delta S_{iijk} = 0. \tag{38.8}$$

We shall revert to this later.

6. We shall use δS_{ijkl} as the notation for the linear functionals arising from perturbations about viscometricity.

Consider the components $\delta T^E_{\alpha\beta} = \delta S_{\alpha\beta kl}(\varkappa \mid E_{kl})$, where $\alpha, \beta = 1, 2$. Under the change of frame such that $\mathbf{a}^3(t) \to -\mathbf{a}^3(t)$, $\delta T^E_{\alpha\beta}$ are not altered: but $E_{13} = E_{31}$ and $E_{23} = E_{32}$ become $-E_{13}$ and $-E_{23}$ respectively. Also, $°\mathbf{C}$ is not altered by this change of $\mathbf{a}^3$ to $-\mathbf{a}^3$. Hence, with $\alpha, \beta, \gamma, \delta = 1, 2$, we have

$$\begin{aligned}\delta T^E_{\alpha\beta} &= \delta S_{\alpha\beta\gamma\delta}(\varkappa \mid E_{\gamma\delta}) + \delta S_{\alpha\beta3\gamma}(\varkappa \mid E_{3\gamma}) + \\ &\quad + \delta S_{\alpha\beta\gamma3}(\varkappa \mid E_{\gamma3}) + \delta S_{\alpha\beta33}(\varkappa \mid E_{33}) \\ &= \delta S_{\alpha\beta\gamma\delta}(\varkappa \mid E_{\gamma\delta}) + \delta S_{\alpha\beta3\gamma}(\varkappa \mid -E_{3\gamma}) + \\ &\quad + \delta S_{\alpha\beta\gamma3}(\varkappa \mid -E_{\gamma3}) + \delta S_{\alpha\beta33}(\varkappa \mid E_{33}). \qquad (38.9)\end{aligned}$$

Since $E_{3\gamma}$ are arbitrary, we must have

$$\delta S_{\alpha\beta3\gamma} = \delta S_{\alpha\beta\gamma3} = 0. \qquad (38.10)$$

Similar arguments, where we note that replacing $\varkappa$ by $-\varkappa$ and $\mathbf{a}^1$ by $-\mathbf{a}^1$ does not alter $°\mathbf{C}$, and neither do the replacements of $\varkappa$ by $-\varkappa$ and $\mathbf{a}^2$ by $-\mathbf{a}^2$ change $°\mathbf{C}$, lead us to the following:

$$\delta S_{3\alpha\beta\gamma} = \delta S_{\alpha3\beta\gamma} = \delta S_{\alpha333} = \delta S_{3\alpha33} = \delta S_{33\alpha3} = \delta S_{333\alpha} = 0. \qquad (38.11)$$

Now (38.10) is equivalent to 6 independent restraints on δS_{ijkl}, while in (38.11) one has 10 more, so, in all the number of independent functionals has now been reduced to 20. The interesting part of (38.10) – (38.11) is that the linear functionals with an odd number of subscripts equal to 3 are zero.

Let us recall (38.8) and observe that this now reads:

$$\delta S_{11\alpha\beta} + \delta S_{22\alpha\beta} + \delta S_{33\alpha\beta} = 0, \qquad \alpha, \beta = 1, 2, \qquad (38.12)$$

and

$$\delta S_{1133} + \delta S_{2233} + \delta S_{3333} = 0. \qquad (38.13)$$

Thus (38.12) – (38.13) are equivalent to four restrictions on the linear functionals. So, ultimately there are 16 independent linear functionals.

We have not yet explored the limitation on the δS_{ijkl} due to the incompressibility of the fluid. Since this demands that $\det \mathbf{C} = 1$, one can write

$$1 = \det(°\mathbf{C} + \mathbf{E}) = \det °\mathbf{C} \det(\mathbf{1} + °\mathbf{C}^{-1}\mathbf{E}). \qquad (38.14)$$

Since $\det °\mathbf{C} = 1$, (38.14) is equivalent to

$$1 = \det(\mathbf{1} + °\mathbf{C}^{-1}\mathbf{E}). \qquad (38.15)$$

But the expansion of the right-hand side leads to

$$\det(\mathbf{1} + °\mathbf{C}^{-1}\mathbf{E}) = 1 + \operatorname{tr} °\mathbf{C}^{-1}\mathbf{E} + o(\| \mathbf{E} \|), \qquad (38.16)$$

where $(o(\| \mathbf{E} \|)/\| \mathbf{E} \|) \to 0$ as $\| \mathbf{E} \| \to 0$. Comparing (38.15) and (38.16), one notices that

$$\operatorname{tr}(°\mathbf{C}^{-1}\mathbf{E}) = o(\| \mathbf{E} \|). \qquad (38.17)$$

In other words, to the order of approximation being considered, that is, to a linearity in **E**, tr $({}^{\circ}\mathbf{C}^{-1}\,\mathbf{E}) = 0$. Thus, *the perturbation* **E** *is isochoric if and only if*

$$\operatorname{tr}\,({}^{\circ}\mathbf{C}^{-1}\,\mathbf{E}) = 0. \tag{38.18}$$

Now, (38.18) demands that **E** be orthogonal to the space spanned by ${}^{\circ}\mathbf{C}^{-1}$, or in incompressible materials, the perturbation **E** cannot be in the subspace spanned by ${}^{\circ}\mathbf{C}^{-1}$. So, consistency of the constitutive equation demands that the stress perturbation generated by such strain histories must be zero—to be exact, of order $o\,(\|\,\mathbf{E}\,\|)$. However, we shall put such stresses to be zero.

So, if $\mathbf{E} = f(s)\,{}^{\circ}\mathbf{C}^{-1}$, where $f(s)$ is an arbitrary function, then

$$\delta S_{ijkl}\,(\varkappa \mid f(s)\,{}^{\circ}C^{-1}_{kl}) = 0. \tag{38.19}$$

But

$${}^{\circ}C^{-1}_{13} = {}^{\circ}C^{-1}_{31} = {}^{\circ}C^{-1}_{23} = {}^{\circ}C^{-1}_{32} = 0,$$

while

$${}^{\circ}C^{-1}_{11} = 1 + \varkappa^2 s^2, \quad {}^{\circ}C^{-1}_{12} = {}^{\circ}C^{-1}_{21} = \varkappa s, \quad {}^{\circ}C^{-1}_{22} = {}^{\circ}C^{-1}_{33} = 1.$$

Thus (38.19) reads

$$\delta S_{ijkk}\,(\varkappa \mid f(s)) + 2\varkappa\,\delta S_{ij12}\,(\varkappa \mid sf(s)) + \varkappa^2\delta S_{ij11}\,(\varkappa \mid s^2 f(s)) = 0. \tag{38.20}$$

We can rewrite (38.20) as

$$\delta S_{\alpha\beta 33}\,(\varkappa \mid f(s)) = -\,\delta S_{\alpha\beta\gamma\gamma}\,(\varkappa \mid f(s)) - \\ -\,2\varkappa\,\delta S_{\alpha\beta 12}\,(\varkappa \mid sf(s)) - \varkappa^2\delta S_{\alpha\beta 11}\,(\varkappa \mid s^2 f(s)). \tag{38.21}$$

Also, from (38.12), (38.13) and (38.20) one obtains

$$\delta S_{3333}\,(\varkappa \mid f(s)) = -\,\delta S_{\alpha\alpha\beta\beta}\,(\varkappa \mid f(s)) - 2\varkappa\,\delta S_{\alpha\alpha 12}\,(\varkappa \mid s f(s)) - \\ -\,\varkappa^2\delta S_{\alpha\alpha 11}\,(\varkappa \mid s^2 f(s)). \tag{38.22}$$

Now (38.21) and (38.22) are equivalent to 3 independent restrictions on δS_{ijkl}. Thus there are only 13 linearly independent linear functionals: $\delta S_{\alpha\beta\gamma\delta}$, of which there are 9, and $\delta S_{3\alpha 3\beta}$, of which there are 4.

It is intuitively obvious that the 13 linear functionals listed above must be related to the basic viscometric functions $\tau(\varkappa) = \varkappa\eta(\varkappa)$, $N_1(\varkappa)$ and $N_2(\varkappa)$. Following CRIMINALE, *et al.*[7], the constitutive equation in viscometric flows

7. [1958 : 1].

may be written as [*cf.* (26.65)]:

$$\mathbf{T}_E = \eta(\varkappa)\,\mathbf{A}_1 + n_2(\varkappa)\,\mathbf{A}_1^2 + \frac{n_1(\varkappa)}{2}(2\mathbf{A}_1^2 - \mathbf{A}_2) - \frac{1}{6}[n_1(\varkappa) + 2n_2(\varkappa)]\,\mathbf{1}\,\mathrm{tr}\,\mathbf{A}_1^2$$
$$= \eta(\varkappa)\,\varkappa(\mathbf{a}^1 \otimes \mathbf{a}^2 + \mathbf{a}^2 \otimes \mathbf{a}^1) + [n_1(\varkappa) + n_2(\varkappa)]\varkappa^2\,\mathbf{a}^1 \otimes \mathbf{a}^1 + n_2(\varkappa)\,\varkappa^2\mathbf{a}^2 \otimes \mathbf{a}^2$$
$$- \frac{\varkappa^2}{3}[n_1(\varkappa) + 2n_2(\varkappa)]\mathbf{1}, \qquad (38.23)^8$$

with

$$N_\alpha(\varkappa) = \varkappa^2 n_\alpha(\varkappa), \qquad \alpha = 1, 2. \qquad (38.24)$$

Now, let $\mathbf{C} = {}^\circ\mathbf{C} + \mathbf{E}$ be also *viscometric*, with a shear rate very close to $\varkappa$ and with its own orthonormal triad of shear axes very close to $\mathbf{a}^i(t)$. Then the strain history $\mathbf{E}$ will include a perturbation due to the shear rate as well as due to the rotation of the axes—note that the perturbed triad can be obtained from the old one by rotation. To facilitate the calculation, let us introduce a parameter ξ so that

$$\varkappa = \varkappa(\xi), \qquad \mathbf{a}^i(t) = \mathbf{a}^i(\xi), \qquad (38.25)$$

and let ${}^\circ\mathbf{C} = {}^\circ\mathbf{C}(\xi_0)$, and $\mathbf{C}(\xi) = {}^\circ\mathbf{C}(\xi_0) + {}^\circ\mathbf{C}(\xi_0)'d\xi$, where the prime denotes differentiation with respect to ξ. Then by the orthonormality of $\{\mathbf{a}^i(\xi)\}$, we have

$$\frac{d}{d\xi}\begin{bmatrix}\mathbf{a}^1\\ \mathbf{a}^2\\ \mathbf{a}^3\end{bmatrix} = \begin{bmatrix}0 & \omega_3 & -\omega_2\\ -\omega_3 & 0 & \omega_1\\ \omega_2 & -\omega_1 & 0\end{bmatrix}\begin{bmatrix}\mathbf{a}^1\\ \mathbf{a}^2\\ \mathbf{a}^3\end{bmatrix}. \qquad (38.26)$$

Thus, in symbolic notation,

$${}^\circ\mathbf{C}(\xi') = \left(\frac{\partial}{\partial\varkappa}{}^\circ\mathbf{C}\right)\varkappa' - \varkappa s\frac{d}{d\xi}(\mathbf{a}^1 \otimes \mathbf{a}^2 + \mathbf{a}^2 \otimes \mathbf{a}^1) + s^2\varkappa^2\frac{d}{d\xi}(\mathbf{a}^2 \otimes \mathbf{a}^2). \qquad (38.27)$$

Now

$$\left(\frac{\partial}{\partial\varkappa}{}^\circ\mathbf{C}\right)\varkappa' = \left[-s(\mathbf{a}^1 \otimes \mathbf{a}^2 + \mathbf{a}^2 \otimes \mathbf{a}^1) + 2s^2\varkappa\,\mathbf{a}^2 \otimes \mathbf{a}^2\right]\varkappa'. \qquad (38.28)$$

For example, using (38.26), one obtains

$$\frac{d}{d\xi}(\mathbf{a}^1 \otimes \mathbf{a}^2) = \mathbf{a}^{1\prime} \otimes \mathbf{a}^2 + \mathbf{a}^1 \otimes \mathbf{a}^{2\prime} = \omega_3\mathbf{a}^2 \otimes \mathbf{a}^2 - \omega_2\mathbf{a}^3 \otimes \mathbf{a}^2$$
$$+ \omega_1\mathbf{a}^1 \otimes \mathbf{a}^3 - \omega_3\mathbf{a}^1 \otimes \mathbf{a}^1. \qquad (38.29)$$

Hence

$${}^\circ\mathbf{C}(\xi)' = [-s(\mathbf{a}^1 \otimes \mathbf{a}^2 + \mathbf{a}^2 \otimes \mathbf{a}^1) + 2s^2\varkappa\,\mathbf{a}^2 \otimes \mathbf{a}^2]\,\varkappa' +$$
$$+ \omega_1[-s\varkappa(\mathbf{a}^1 \otimes \mathbf{a}^3 + \mathbf{a}^3 \otimes \mathbf{a}^1) + s^2\varkappa^2(\mathbf{a}^2 \otimes \mathbf{a}^3 + \mathbf{a}^3 \otimes \mathbf{a}^2)]$$
$$+ \omega_2[s\varkappa(\mathbf{a}^2 \otimes \mathbf{a}^3 + \mathbf{a}^3 \otimes \mathbf{a}^2)] + \qquad (38.30)$$
$$+ \omega_3[2s\varkappa(\mathbf{a}^1 \otimes \mathbf{a}^1 - \mathbf{a}^2 \otimes \mathbf{a}^2) - s^2\varkappa^2(\mathbf{a}^1 \otimes \mathbf{a}^2 + \mathbf{a}^2 \otimes \mathbf{a}^1)].$$

8. In the notation of PIPKIN and OWEN, $N_1 = \varkappa^2\nu$, $N_2 = \varkappa^2(\varphi - \nu)$.

Now, if $°\mathbf{C}_0 = °\mathbf{C}(\xi_0)$ and $\mathbf{C}(\xi) = °\mathbf{C} + °\mathbf{C}'\,d\xi$, then

$$\mathbf{T}_E\ [\mathbf{C}(\xi)] \doteq \mathbf{T}_E\ [°\mathbf{C}(\xi_0)] + \delta \overset{\infty}{\underset{s=0}{H}} (°\mathbf{C}(\xi_0) \mid °\mathbf{C}(\xi_0)'\,d\xi). \tag{38.31}$$

Hence

$$\left.\frac{d\mathbf{T}_E}{d\xi}\right|_{\xi=\xi_0} = \delta \overset{\infty}{\underset{s=0}{H}} (°\mathbf{C}(\xi_0) \mid °\mathbf{C}(\xi_0)'). \tag{38.32}$$

By using $(38.23)_2$ for $\mathbf{T}_E$ one has that

$$\begin{aligned}\frac{d\mathbf{T}_E}{d\xi} = \varkappa'\Big[&(\mathbf{a}^1 \otimes \mathbf{a}^2 + \mathbf{a}^2 \otimes \mathbf{a}^1)\frac{d[\varkappa\,\eta(\varkappa)]}{d\varkappa} + \mathbf{a}^1 \otimes \mathbf{a}^1 \frac{d}{d\varkappa}[N_1(\varkappa) + N_2(\varkappa)]\\ &+ \mathbf{a}^2 \otimes \mathbf{a}^2 \frac{dN_2(\varkappa)}{d\varkappa} - \frac{1}{3}\mathbf{1}\frac{d}{d\varkappa}[N_1(\varkappa) + 2\,N_2(\varkappa)]\Big]\\ &+ \omega_1\,[\eta(\varkappa)\,\varkappa\,(\mathbf{a}^1 \otimes \mathbf{a}^3 + \mathbf{a}^3 \otimes \mathbf{a}^1) + N_2(\varkappa)\,(\mathbf{a}^2 \otimes \mathbf{a}^3 + \mathbf{a}^3 \otimes \mathbf{a}^2)]\\ &- \omega_2\,[\eta(\varkappa)\,\varkappa\,(\mathbf{a}^2 \otimes \mathbf{a}^3 + \mathbf{a}^3 \otimes \mathbf{a}^2)\\ &+ (N_1(\varkappa) + N_2(\varkappa))\,(\mathbf{a}^1 \otimes \mathbf{a}^3 + \mathbf{a}^3 \otimes \mathbf{a}^1)]\\ &+ \omega_3\,[2\eta(\varkappa)\,\varkappa\,(\mathbf{a}^2 \otimes \mathbf{a}^2 - \mathbf{a}^1 \otimes \mathbf{a}^1) + N_1(\varkappa)\,(\mathbf{a}^1 \otimes \mathbf{a}^2 + \mathbf{a}^2 \otimes \mathbf{a}^1)].\end{aligned} \tag{38.33}$$

Using (38.31), and the previous results (38.10) and (38.11) on the linear functionals δS_{ijkl} we obtain

$$\begin{aligned}\delta T^E_{\alpha\beta} &= 2\varkappa'\{-\,\delta S_{\alpha\beta 12}\,(\varkappa \mid s) + \varkappa\,\delta S_{\alpha\beta 22}\,(\varkappa \mid s^2)\}\\ &\quad + 2\,\varkappa\omega_3\{\delta S_{\alpha\beta 11}\,(\varkappa \mid s) - \delta S_{\alpha\beta 22}\,(\varkappa \mid s) - \varkappa\,\delta S_{\alpha\beta 12}\,(\varkappa \mid s^2)\};\end{aligned} \tag{38.34}$$

$$\begin{aligned}\delta T^E_{3\alpha} &= 2\,\varkappa\omega_1\{-\,\delta S_{3\alpha 13}\,(\varkappa \mid s)\\ &\quad + \varkappa\,\delta S_{3\alpha 23}\,(\varkappa \mid s^2)\} + 2\,\varkappa\omega_2\,\delta S_{3\alpha 23}(\varkappa \mid s);\end{aligned} \tag{38.35}$$

$$\begin{aligned}\delta T^E_{33} &= 2\,\varkappa'\,\{-\,\delta S_{3312}\,(\varkappa \mid s) + \varkappa\,\delta S_{3322}\,(\varkappa \mid s^2)\}\\ &\quad + 2\,\varkappa\omega_3\{\delta S_{3311}\,(\varkappa \mid s) - \delta S_{3322}\,(\varkappa \mid s) - \varkappa\,\delta S_{3312}\,(\varkappa \mid s^2)\}.\end{aligned} \tag{38.36}$$

Comparing (38.33) with (38.34) — (38.36), we read off the *consistency relations*:

$$\varkappa'\ \&\ \delta T^E_{12}\ :\ \frac{d[\varkappa\,\eta(\varkappa)]}{d\varkappa} = -\,2\,\delta S_{1212}(\varkappa \mid s) + \varkappa\,\delta S_{1222}(\varkappa \mid s^2); \tag{38.37}$$

$$\varkappa'\ \&\ \delta T^E_{11}\ :\ \frac{d[2N_1(\varkappa) + N_2(\varkappa)]}{d\varkappa} = -\,6\,\delta S_{1112}(\varkappa \mid s) + 6\,\varkappa\,\delta S_{1122}(\varkappa \mid s^2); \tag{38.38}$$

$$\varkappa'\ \&\ \delta T^E_{22}\ :\ \frac{d[N_2(\varkappa) - N_1(\varkappa)]}{d\varkappa} = -\,6\,\delta S_{2212}(\varkappa \mid s) + 6\,\varkappa\,\delta S_{2222}(\varkappa \mid s^2); \tag{38.39}$$

$$\omega_1\ \&\ \delta T^E_{31}\ :\ \eta(\varkappa) = -\,2\,\delta S_{3113}(\varkappa \mid s) + 2\,\varkappa\,\delta S_{3123}(\varkappa \mid s^2); \tag{38.40}$$

$$\omega_1\ \&\ \delta T^E_{32}\ :\ N_2(\varkappa) = -\,2\,\varkappa\,\delta S_{3213}(\varkappa \mid s) + 2\,\varkappa^2\,\delta S_{3223}(\varkappa \mid s^2); \tag{38.41}$$

$$\omega_2 \ \& \ \delta T^E_{31} \ : \ N_1(\varkappa) + N_2(\varkappa) = -\, 2\, \varkappa\, \delta S_{3123}(\varkappa \mid s); \tag{38.42}$$

$$\omega_2 \ \& \ \delta T^E_{32} \ : \ \eta(\varkappa) = -\, 2\, \delta S_{3223}(\varkappa \mid s); \tag{38.43}$$

$$\omega_3 \ \& \ \delta T^E_{11} \ : \ -\, \eta(\varkappa) = \delta S_{1111}(\varkappa \mid s) - \delta S_{1122}(\varkappa \mid s) - \varkappa\, \delta S_{1112}(\varkappa \mid s^2); \tag{38.44}$$

$$\omega_3 \ \& \ \delta T^E_{22} \ : \ \eta(\varkappa) = \delta S_{2211}(\varkappa \mid s) - \delta S_{2222}(\varkappa \mid s) - \varkappa\, \delta S_{2212}(\varkappa \mid s^2); \tag{38.45}$$

$$\omega_3 \ \& \ \delta T^E_{12} \ : \ N_1(\varkappa) = 2\, \varkappa\, \delta S_{1211}(\varkappa \mid s) - 2\, \varkappa\, \delta S_{1222}(\varkappa \mid s) - 2\, \varkappa^2\, \delta S_{1212}(\varkappa \mid s^2). \tag{38.46}$$

Using (38.12) and adding (38.44) and (38.45) one derives that

$$0 = \delta S_{3311}(\varkappa \mid s) - \delta S_{3322}(\varkappa \mid s) - \varkappa\, \delta S_{3312}(\varkappa \mid s^2). \tag{38.47}$$

Again using (31.12) and adding (38.38) and (38.39), we obtain

$$-\, \frac{d[N_1(\varkappa) + 2\, N_2(\varkappa)]}{d\varkappa} = -\, 6\, \delta S_{3312}(\varkappa \mid s) + 6\varkappa\, \delta S_{3322}(\varkappa \mid s). \tag{38.48}$$

Thus we have 10 consistency relations among the 13 independent linear functionals, and two redundant ones in (38.47) — (38.48).

Finally, in closing this section, we note that we may write the linear functionals as integrals:

$$\delta S_{ijkl}(\varkappa \mid E_{kl}(t - s, t)) = \int_0^\infty S_{ijkl}(\varkappa, s)\, E_{kl}(t - s, t)\, ds, \tag{38.49}$$

where the viscometric flow kernels $S_{ijkl}(\varkappa, s)$ are expected to possess properties similar to the linear stress relaxation moduli. These properties will be assumed in the next section.

39. Small Oscillations Superposed on Large

39.1. Asymptotic Relations. In a fixed Cartesian coordinate system, let the velocity field be given by[9]

$$\dot{x} = \varkappa y + (Ay\omega/h) \cos \omega t, \qquad \dot{y} = \dot{z} = 0, \tag{39.1}$$

or, by (2.11)

$$\dot{x} = \varkappa y, \quad \dot{y} = 0, \quad \dot{z} = (Ay\omega/h) \cos \omega t. \tag{39.2}$$

The flow in (39.1) is said to involve in-line oscillations, while that in (39.2) to concern transverse oscillations. In both flows, $\varkappa$ is the rate of steady shear; A is the amplitude and ω is the frequency of the superposed oscillation, while h is the gap width across which this oscillation acts.

Assuming that A is so small that linearizations in A are permissible, the

9. The perturbations (39.1) and (39.2) are equivalent to an oscillatory displacement field $(Ay/h) \sin \omega t$.

path lines corresponding to (39.1) and (39.2) may be used to determine the two strain histories:

in-line: $$\mathbf{C}_{\parallel} = {}^{\circ}\mathbf{C} + \mathbf{E}_{\parallel}, \tag{39.3}$$

transverse: $$\mathbf{C}_{\perp} = {}^{\circ}\mathbf{C} + \mathbf{E}_{\perp}, \tag{39.4}$$

where the components of $^{\circ}\mathbf{C}$ are given by (5.15), while those of $\mathbf{E}_{\parallel}$ and $\mathbf{E}_{\perp}$ have the form:

$$[E_{\parallel}(t-s,t)\langle ij\rangle] = \begin{bmatrix} 0 & \frac{A}{h} f(\omega,t,s) & 0 \\ \cdot & -\frac{2A\,s\varkappa}{h} f(\omega,t,s) & 0 \\ \cdot & \cdot & 0 \end{bmatrix}, \tag{39.5}$$

$$[E_{\perp}(t-s,t)\langle ij\rangle] = \begin{bmatrix} 0 & 0 & 0 \\ \cdot & 0 & \frac{A}{h} f(\omega,t,s) \\ \cdot & \cdot & \cdot \end{bmatrix} \tag{39.6}$$

where

$$f(\omega,t,s) = \sin\omega t\,(\cos\omega s - 1) - \cos\omega t \sin\omega s. \tag{39.7}$$

It is trivial to verify that both (39.5) and (39.6) satisfy the incompressibility condition for nearly viscometric flows, viz., $\mathrm{tr}\,{}^{\circ}\mathbf{C}^{-1}\mathbf{E}_{\parallel} = \mathrm{tr}\,{}^{\circ}\mathbf{C}^{-1}\mathbf{E}_{\perp} = 0$. If we let A be very small, then $\|\mathbf{E}\|$ is very small, and thus (39.1)—(39.2) can be considered as nearly viscometric flows in the sense of Sec. 38.

From Eqs. (38.10)—(38.11) we note that for the in-line case, δT^E_{11}, δT^E_{22}, δT^E_{33} and δT^E_{12} are not zero, and for the transverse case δT^E_{13} and δT^E_{23} are not zero. Hence to the first order in A, the in-line case superposes perturbations on the normal stress differences and the shear stress of the original flow, while in the transverse case shear stress perturbations in directions normal to the main stream occur.

Guided by this, we suppose that the perturbations given by

$$\tau_{\parallel} = \frac{A\sin\omega t}{h} G'_{\parallel}(\varkappa,\omega) + \frac{A\cos\omega t}{h} G''_{\parallel}(\varkappa,\omega), \tag{39.8}$$

$$N_{1\parallel} = \frac{A\sin\omega t}{h} N'_{1\parallel}(\varkappa,\omega) + \frac{A\cos\omega t}{h} N''_{1\parallel}(\varkappa,\omega), \tag{39.9}$$ [10]

$$N_{2\parallel} = \frac{A\sin\omega t}{h} N'_{2\parallel}(\varkappa,\omega) + \frac{A\cos\omega t}{h} N''_{2\parallel}(\varkappa,\omega), \tag{39.10}$$ [10]

10. From (26.29), it is true that $N_{1\parallel}$ and $N_{2\parallel}$ have components with frequencies proportional to 2ω. However, these components are proportional to A^2 and thus are not included here.

where τ is the in-line oscillatory shear stress in the xy direction and normal stress differences are associated with $T\langle xx\rangle - T\langle yy\rangle$ and $T\langle yy\rangle - T\langle zz\rangle$ respectively. Similarly, we suppose that

$$\tau_{\perp} = \frac{A \sin \omega t}{h} G'_{\perp}(\varkappa, \omega) + \frac{A \cos \omega t}{h} G''_{\perp}(\varkappa, \omega) \tag{39.11}$$

is the transverse oscillatory shear stress δT^{E}_{32}, since this component is easy to measure. In conformity with linear viscoelasticity, we call $G'_{\|}(\varkappa, \omega)$, etc., the *dynamic moduli* and those moduli which are multiplicands of sin ωt (that is, out of phase with velocity) as the *elastic* or *storage moduli*, and the others the *loss moduli*. We also define two viscosities:

$$\eta'_{\|}(\varkappa, \omega) = G''_{\|}(\varkappa, \omega)/\omega, \tag{39.12}$$

$$\eta'_{\perp}(\varkappa, \omega) = G''_{\perp}(\varkappa, \omega)/\omega. \tag{39.13}$$

A straightforward comparison between the formulae in Sec. 38 and (39.8)—(39.11) shows that

$$G'_{\|}(\varkappa, \omega) + i\, G''_{\|}(\varkappa, \omega)$$
$$= 2 \int_0^\infty \{S_{1222}(\varkappa, s)\, \varkappa s - S_{1212}(\varkappa, s)\}\,(1 - e^{-i\omega s})\, ds, \tag{39.14}$$

where

$$e^{-i\omega s} = \cos \omega s - i \sin \omega s. \tag{39.15}$$

Next,

$$N'_{1\|}(\varkappa, \omega) + i\, N''_{1\|}(\varkappa, \omega) = 2 \int_0^\infty \{S_{2212}(\varkappa, s) - S_{1112}(\varkappa, s) + \varkappa s\,(S_{1122}(\varkappa, s)$$
$$- S_{2222}(\varkappa, s))\}\,(1 - e^{-i\omega s})\, ds, \tag{39.16}$$

$$N'_{2\|}(\varkappa, \omega) + i\, N''_{2\|}(\varkappa, \omega) = 2 \int_0^\infty \{S_{3312}(\varkappa, s) - S_{2212}(\varkappa, s) + \varkappa s\,(S_{2222}(\varkappa, s)$$
$$- S_{3322}(\varkappa, s))\}\,(1 - e^{-i\omega s})\, ds, \tag{39.17}$$

$$G'_{\perp}(\varkappa, \omega) + i\, G''_{\perp}(\varkappa, \omega) = -2 \int_0^\infty S_{2323}(\varkappa, s)\,(1 - e^{-i\omega s})\, ds. \tag{39.18}$$

If we compare (38.43) and (39.18), we see that

$$\lim_{\omega \to 0} \frac{G''_{\perp}(\varkappa, \omega)}{\omega} = \lim_{\omega \to 0} \eta'_{\perp}(\varkappa, \omega) = \eta(\varkappa). \tag{39.19}$$ [11]

11. PIPKIN [1968 : 20], Eq. (9.8); MARKOVITZ [1969 : 14], Eq. (47).

Moreover, a comparison of (38.37) and (39.14) shows that

$$\lim_{\omega\to 0} \frac{G''_{\|}(\varkappa, \omega)}{\omega} = \lim_{\omega\to 0} \eta'_{\|}(\varkappa, \omega) = \frac{d\,\tau(\varkappa)}{d\varkappa}. \tag{39.20}$$ [12]

Intuitively, Eqs. (39.19) and (39.20) are to be expected: in one case, the viscosity is that existing at the base shear rate $\varkappa$, while in the other case the base shear rate is itself being altered and hence the viscosity is the derivative of $\tau(\varkappa)$.

From (38.38), (38.39) and (39.16) we obtain

$$\lim_{\omega\to 0} \frac{N''_{1\|}(\varkappa, \omega)}{\omega} = \frac{d\,N_1(\varkappa)}{d\varkappa}, \tag{39.21}$$ [13]

and from (38.39), (38.48) and (39.17) we read off

$$\lim_{\omega\to 0} \frac{N''_{2\|}(\varkappa, \omega)}{\omega} = \frac{d\,N_2(\varkappa)}{d\varkappa}. \tag{39.22}$$ [14]

Incidentally, for the flow field (39.2), the shear stress perturbation δT^E_{31} is given by

$$\delta T^E_{31} = -\frac{2A}{h}\int_0^\infty S_{3123}(\varkappa, s)\,\{(1-\cos\omega s)\sin\omega t + \cos\omega t \sin\omega s\}\,ds. \tag{39.23}$$

Hence, from (38.42) we have that

$$\lim_{\omega\to 0} \frac{\delta T^E_{31}}{\omega\cos\omega t} = [N_1(\varkappa) + N_2(\varkappa)]/\varkappa, \tag{39.24}$$ [15]

but δT^E_{31} is not easy to measure and hence we do not dwell at length on (39.23)—(39.24).

We now turn our attention to ultrasonic limits, that is, find the behaviour of the moduli as $\omega \to \infty$.

In the theory of linear viscoelasticity, for the deformation field

$$x = X + \frac{Ay}{h}\sin\omega t, \qquad y = Y, \qquad z = Z, \tag{39.25}$$

the shear stress is given by [*cf.* (34.*b*5)]

$$\tau = -\int_0^\infty \dot{G}(s)\,\{(1-\cos\omega s)\sin\omega t + \sin\omega s\cos\omega t\}\,ds. \tag{39.26}$$

12. PIPKIN [1968 : 20], Eq. (9.7); MARKOVITZ [1969 : 14], Eq. (36).
13. Derived for BKZ fluids by BERNSTEIN [1969 : 1].
14. Derived for BKZ fluids by BERNSTEIN, HUILGOL and TANNER [1972 : 2].
15. PIPKIN [1968 : 20], Eq. (9.9). In deriving (39.24) from (39.23), an application of L'HOSPITAL's rule is essential.

Hence, consistency demands that as $\varkappa \to 0$,

$$G''(\omega) = \bar{G}''_{\parallel}(0, \omega) = -\int_0^\infty \dot{G}(s) \sin \omega s \, ds$$

$$= -2 \int_0^\infty S_{1212}(0, s) \sin \omega s \, ds = G''_{\perp}(0, \omega). \tag{39.27}$$

Now, suppose that $S_{1212}(\varkappa, s)$ is such that

(i) it is integrable over $[0, \infty)$;

(ii) $(1/\omega)\, S_{1212}(\varkappa, s)\, e^{i\omega s} \to 0$ as $\omega \to \infty$;

(iii) $\int_0^\infty \left[\frac{\partial}{\partial s} S_{1212}(\varkappa, s)\right] e^{i\omega s}\, ds$ is finite;

then by the RIEMANN-LEBESGUE lemma[16],

$$\lim_{\omega \to \infty} \int_0^\infty S_{1212}(\varkappa, s) \sin \omega s \, ds = 0, \tag{39.28}$$

and

$$\lim_{\omega \to \infty} \int_0^\infty \left[\frac{\partial}{\partial s} S_{1212}(\varkappa, s)\right] \sin \omega s \, ds = 0. \tag{39.29}$$

Then, integration by parts yields the following:

$$\int_0^\infty S_{1212}(\varkappa, s) \sin \omega s \, ds = \frac{1}{\omega} S_{1212}(\varkappa, 0)$$

$$+ \frac{1}{\omega} \int_0^\infty \left[\frac{\partial}{\partial s} S_{1212}(\varkappa, s)\right] \cos \omega s \, ds. \tag{39.30}$$

Next,

$$\int_0^\infty S_{1212}(0, s) \sin \omega s \, ds = \frac{1}{\omega} S_{1212}(0, 0)$$

$$+ \frac{1}{\omega} \int_0^\infty \left[\frac{\partial}{\partial s} S_{1212}(0, s)\right] \cos \omega s \, ds. \tag{39.31}$$

Thus, applying the assumption (iii) to (39.30) and (39.31), one notes that

$$\lim_{\omega \to \infty} \frac{\int_0^\infty S_{1212}(\varkappa, s) \sin \omega s \, ds}{\int_0^\infty S_{1212}(0, s) \sin \omega s \, ds} \neq 1, \tag{39.32}$$

unless

$$S_{1212}(\varkappa, 0) = S_{1212}(0, 0); \tag{39.33}$$

if (39.33) is true, then the limit in (39.32) is equal to 1.

For the additive functional constitutive equation (36.20), a straightforward

16. E.g., see CHURCHILL [1963 : 2], p. 87. Wherever such integrals occur in the sequel, assumptions equivalent to (i) - (iii) above have been made to ensure that limits of the type (39.28) obtain.

calculation shows that $S_{1222}(\varkappa, s) = 0$ and

$$2\, S_{1212}(\varkappa, s) = (\alpha_{-1} + \alpha_1) + 2\, \varkappa^2 s^2 \left(\frac{\partial \alpha_{-1}}{\partial I_1} + \frac{\partial \alpha_{-1}}{\partial I_2} + \frac{\partial \alpha_1}{\partial I_1} + \frac{\partial \alpha_1}{\partial I_2} \right), \quad (39.34)$$

where (36.27) has been used. Hence, for this fluid

$$S_{1212}(\varkappa, 0) = S_{1212}(0, 0), \quad (39.35)$$

since by putting $s = 0$ in (39.34), we obtain that

$$S_{1212}(\varkappa, 0) = \frac{1}{2} (\alpha_{-1} + \alpha_1)_{I_1 = I_2 = 3}\,, \quad (39.36)$$

for all values of $\varkappa$. Thus retracing the steps, we note that for the additive functional theory,

$$\lim_{\omega \to \infty} \frac{\eta'_{\|}(\varkappa, \omega)}{\eta'(\omega)} = \lim_{\omega \to \infty} \frac{G''_{\|}(\varkappa, \omega)}{G''(\omega)} = 1, \quad (39.37)^{17}$$

because of the equality expressed by (39.35).

Similar calculations show that for the additive functional fluid the following is also true :

$$\lim_{\omega \to \infty} \frac{\eta'_{\perp}(\varkappa, \omega)}{\eta'(\omega)} = \lim_{\omega \to \infty} \frac{G''_{\perp}(\varkappa, \omega)}{G'(\omega)} = 1. \quad (39.38)^{18,19}$$

Moreover, for this fluid, we have just shown that :

$$\lim_{\omega \to \infty} \omega\, G''_{\|}(\varkappa, \omega) = \lim_{\omega \to \infty} \omega\, G''(\omega) = (\alpha_{-1} + \alpha_1)_{I_1 = I_2 = 3}\,. \quad (39.39)$$

But from Eq. (39.27), we may write (39.39) as [*cf.* (36.25)] :

$$\lim_{\omega \to \infty} \omega^2\, \eta'_{\|}(\varkappa, \omega) = \lim_{\omega \to \infty} \omega^2\, \eta'(\omega) = -\, \dot{G}(0), \quad (39.40)^{18}$$

where $\dot{G}(\tau)$ is the derivative of the relaxation modulus of the linear theory at time τ. The limit in (39.40) holds for $\omega^2 \eta'_{\perp}(\varkappa, \omega)$ also, since $2S_{2323}(\varkappa, s) = (\alpha_{-1} + \alpha_1)$, $I_1 = I_2 = 3 + \varkappa^2 s^2$, for this fluid.

So, if on a log-log plot one plots $\eta'_{\|}(\varkappa, \omega)$ or $\eta'_{\perp}(\varkappa, \omega)$ against ω, with one curve for each fixed $\varkappa$, these curves must merge with that of $\eta'(\omega)$ as ω becomes large ; moreover, the following equation

$$\lim_{\omega \to \infty} [2 \ln \omega + \ln \eta'_{\perp}(\varkappa, \omega)] = \ln[-\, \dot{G}(0)], \quad (39.41)$$

implies that for large values of ω, the slope of the graph must become -2. Figure 39.1 below shows a reproduction of SIMMONS' results[20] along with a

17. BERNSTEIN and HUILGOL [1971 : 4] obtained this relation for the BKZ fluid (36.8). See also [1974 : 1].
18. BERNSTEIN and HUILGOL [1971 : 4] for the BKZ fluid (36.8). See also [1974 : 1].
19. TANNER and WILLIAMS [1971 : 28] for the BKZ fluid (36.8).
20. SIMMONS [1968 : 22].

line of slope -2. The agreement between the prediction of the additive

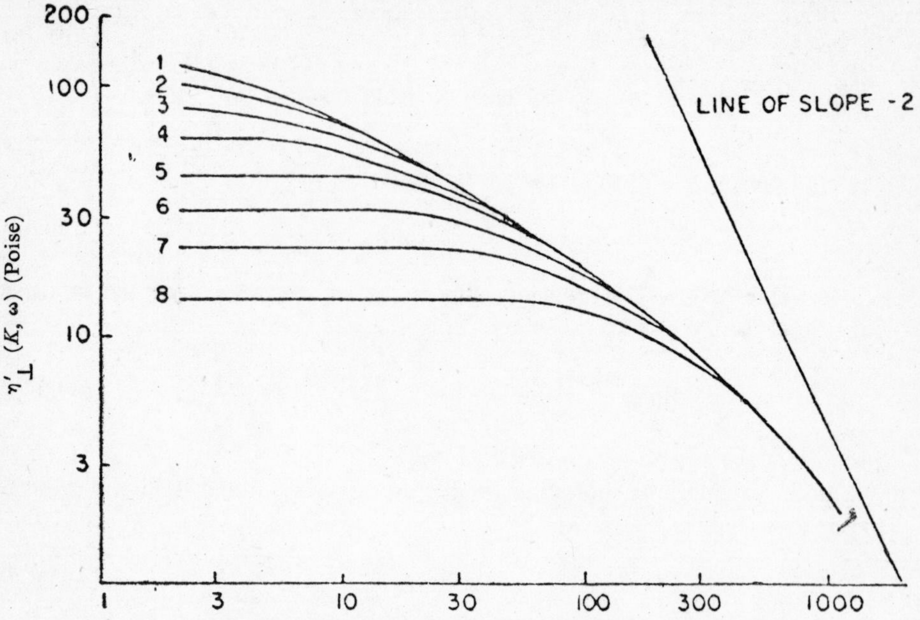

Fig. 39.1. Effect of steady shearing on dynamic viscosity of 8.54% p.i.b. in cetane. From SIMMONS. 1-$\varkappa$=0.2 sec^{-1}, 2-$\varkappa$=6,36, 3-$\varkappa$=12.7, 4-$\varkappa$=25.4, 5-$\varkappa$=50.9, 6-$\varkappa$=102, 7-$\varkappa$=204, 8-$\varkappa$=407.

functional theory and experimental evidence is excellent. We shall examine the foregoing from a more general point of view next.

It is possible to treat the stress perturbations in this section as represented by an integral of the type[21]

$$\int_0^\infty h(\varkappa s, s)\, e^{i\omega s}\, ds \tag{39.42}$$

for Eq. (36.20). If we put $\varphi(s) = h(\varkappa s, s)$ then under suitable conditions, the following asymptotic expansion may be obtained :

$$\int_0^\infty \varphi(s)\, e^{i\omega s}\, ds \sim \frac{i}{\omega}\varphi(0) + \left(\frac{i}{\omega}\right)^2 \varphi'(0) + \ldots + \left(\frac{i}{\omega}\right)^n \varphi^{(n-1)}(0), \tag{39.43}$$

where $\varphi'(0) = \left.\frac{d\varphi}{ds}\right|_{s=0}, \quad \varphi^{(n-1)}(0) = \left.\frac{d^{n-1}\varphi}{ds^{n-1}}\right|_{s=0}, \quad n = 1, 2, \ldots$

So, if in the asymptotic expansion of

$$f(\varkappa, \omega) = \int_0^\infty h(\varkappa s, s)\, e^{i\omega s}\, ds, \tag{39.44}$$

$h(0, 0) \neq 0$, then we recover (39.37) and (39.38) for

$$\lim_{\omega \to \infty} \frac{f(\varkappa, \omega)}{f(0, \omega)} = 1. \tag{39.45}$$

21. BERNSTEIN, HUILGOL and TANNER [1972 : 2] considered the BKZ fluid (36.8).

Using the above asymptotic expansion as a guide, it can be shown that for the additive functional theory[22]

$$\lim_{\omega\to\infty}\left[\omega^3 \frac{\partial G'_{\parallel}(\varkappa, \omega)}{\partial\omega}\right] = \lim_{\omega\to\infty}\left[\omega^3 \frac{\partial G'_{\perp}(\varkappa, \omega)}{\partial\omega}\right]$$

$$= \lim_{\omega\to\infty}\left[\omega^3 \frac{\partial G'(\omega)}{\partial\omega}\right] = 2\ddot{G}(0); \quad (39.46)$$

$$\lim_{\omega\to\infty} \omega^3 N''_{1_{\parallel}}(\varkappa, \omega) = 4\varkappa\, \ddot{G}(0)\,; \quad (39.47)$$

$$\lim_{\omega\to\infty} \omega^3 \frac{\partial N'_{1_{\parallel}}(\varkappa, \omega)}{\partial\omega} = 4\varkappa\, \ddot{G}(0)\,; \quad (39.48)$$

$$\lim_{\omega\to\infty} \omega^3 \frac{\partial N'_{2_{\parallel}}(\varkappa, \omega)}{\partial\omega} = 4\varkappa\, \alpha_1\Big|_{I_1=I_2=3}. \quad (39.49)$$

$$\lim_{\omega\to\infty} \omega^3 N''_{2_{\parallel}}(\varkappa, \omega) = 4\varkappa \frac{d\alpha_1}{ds}\Big|_{I_1=I_2=3}. \quad (39.50)$$

It can also be shown that for the additive functional theory[23]

$$G'_{\parallel}(\varkappa, \omega) + i\,G''_{\parallel}(\varkappa, \omega) = G'_{\perp}(\varkappa, \omega) + i\,G''_{\perp}(\varkappa, \omega)$$

$$+ \varkappa' \frac{\partial}{\partial\varkappa}\left[G'_{\perp}(\varkappa, \omega) + i\,G''_{\perp}(\varkappa, \omega)\right] \quad (39.51)$$

for all values of ω This claim can be experimentally confirmed or denied now by subjecting the same fluid to both in-line and transverse oscillations and measuring the two oscillatory shear stresses δT^E_{12} and δT^E_{23}. At the time of writing no results are available.

39.2. Methods for Measuring Oscillatory-Shear Stresses. We shall now suggest the conceptual framework for measuring δT^E_{12} and δT^E_{23}. Suppose that the liquid is contained between two concentric coaxial cylinders and that the outer one rotates at a fixed angular velocity. For various values of this angular velocity, we obtain different values of $\varkappa$, the base rate of shear. Now superpose on this motion an oscillating flow due to the oscillation of the inner cylinder. If the inner cylinder performs azimuthal oscillation then the dynamic shear stress on it is $\delta T_E\langle 12\rangle$; if it executes longitudinal oscillations, it is $\delta T_E\langle 23\rangle$.

The WEISSENBERG rheogoniometer, for example, may be used to measure

22. The calculations in [1972: 2] are for the BKZ fluid. We record the above by suitable modifications.

23. In [1972: 1], BERNSTEIN derived (39.51) for the BKZ fluid.

the stress $\delta T_E \langle 12 \rangle$. If the rheogoniometer is set up in a cone-and-plate configuration and while the cone is retained by a torsional wire, the bottom plate is subjected to oscillations superposed on a steady velocity, one can thereby measure $\delta T_E \langle 12 \rangle$.

BOOIJ[24], KATAOKA and UEDA[25], KUROIWA and NAKAMURA[26], OSAKI, *et al.*[27], WALTERS and JONES[28] measured the in-line shear stress dynamic moduli. But their observations are neither at very low nor at very high frequencies so that a comparison between theoretical predictions and experimental data is not possible. For transverse oscillations, commercial instruments are not available and SIMMONS built his own apparatus to measure the dynamic moduli. TANNER and WILLIAMS[29] repeated SIMMONS' experiments and confirmed his observations (see Fig. 39.2). But no one, according to author's knowledge

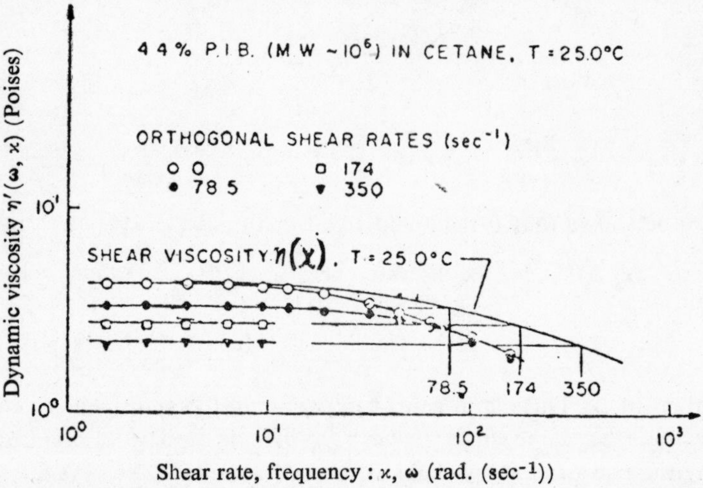

Fig 39.2. Effect of steady orthogonal shear, $\varkappa$ on the real Part, η'. of the complex dynamic viscosity for a 4.4% polyisobutylene sample.

has used the same liquid and conducted both types of experiments. This is a severe handicap to the theoretician.

Finally, in some of the above papers, the graphical representations show that the curves of $G'_{\|}(\varkappa, \omega)$ for various values of $\varkappa$ coalesce with the curve of $G'(\omega)$ as ω becomes very large, while in others, the curves of $G'_{\|}(\varkappa, \omega)$ are parallel to $G'(\omega)$ for different values of ω (see Figs. 39.3 – 39.4). The theory

24. [1966 : 3], [1970 : 5].
25. [1969 : 11]
26. [1967 : 16].
27. [1965 : 14].
28 [1970 : 30]. See also JONES and WALTERS [1970 : 14].
29. [1971 : 28]. These authors have removed the discrepancy between theory and experiment noted by SIMMONS. He had found that $\lim_{\omega \to 0} \eta'(\varkappa, \omega) > \eta(\varkappa)$.

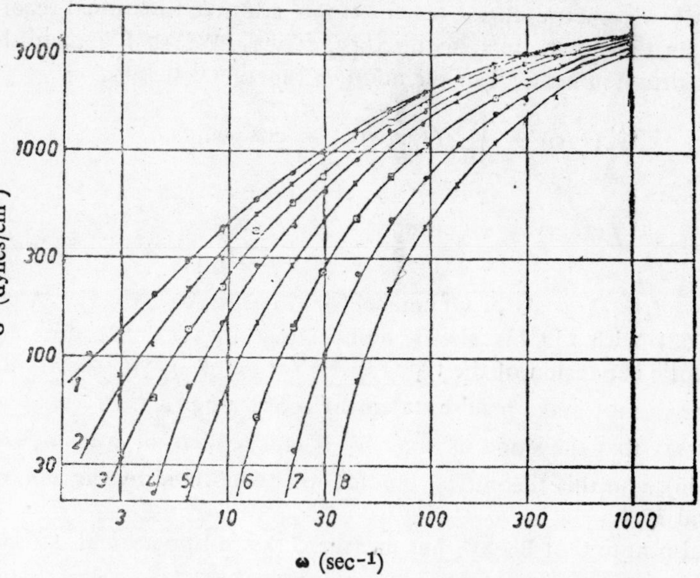

Fig. 39.3. Effect of steady orthogonal shearing of rate $\varkappa$ on G' for 8.54% polyisobutylene in cetane at 25°C and .08 oscillatory shear amplitude. $\varkappa=0$, 2-$\varkappa=6.36$ sec^{-1}, 3-$\varkappa=12.7$, 4-$\varkappa=25.4$, 5-$\varkappa=50.9$, 6-$\varkappa=102$, 7-$\varkappa=407$. From SIMMONS [1968 : 22].

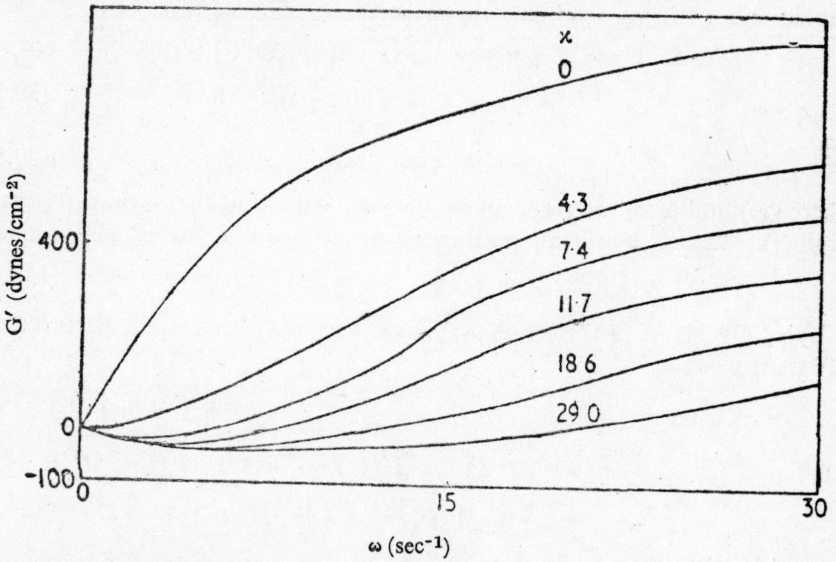

Fig. 39.4. Experimental (G', ω) curves for a 4% solution of polyacrylamide. From JONES and WALTERS [1971 : 14].

of nearly vicometric flows, even if the additive functional theory is used, cannot shed any definite light on this matter. We shall explain the reasons for this situation next. For the additive functional theory,

$$G'_{\parallel}(\varkappa, \omega) = \int_0^{\infty} f(\varkappa s, s)\,(1 - \cos \omega s)\, ds, \tag{39.52}$$

where

$$f(\varkappa s, s) = \alpha_{-1} + \alpha_1 + 2\varkappa^2 s^2 \sum_{i=-1,1} \left(\frac{\partial \alpha_i}{\partial I_1} + \frac{\partial \alpha_i}{\partial I_2} \right), \tag{39.53}$$

and $I_1 = I_2 = 3 + \varkappa^2 s^2$. Of course, for $G'(\omega)$, take $f(0, s)$. A comparison of (39.52) with (39.43) shows immediately that (39.52) does not have an asymptotic expansion of the type (39.43) for large ω. It is for this reason that one cannot make precise statement concerning $G'_{\parallel}(\varkappa, \omega)$ as $\omega \to \infty$, but we can say that the slope of $G''_{\parallel}(x, \omega)$ is independent of ω as $\omega \to \infty$ because of (39.46), and this theoretical prediction is confirmed by the curves in Figs. 39.3 and 39.4

We also note that BOOIJ[30] has measured the in-line normal stress dynamic moduli $N'_{1_{\parallel}}(\varkappa, \omega)$ and $N''_{1_{\parallel}}(\varkappa, \omega)$. However, before one can make realistic comparison, data should be collected at very low and very high frequencies, and it is for this reason that we have not cited the results of BOOIJ.

39.3. Small Displacement on Large. We will now set down equations analogous to (38.34) – (38.36) by using the strain history arising from small displacements superposed on a viscometric flow [*cf.* (5.23)] :

$$\begin{aligned} \mathbf{C} = {}^{\circ}\mathbf{C} - {}^{\circ}\mathbf{C}\, \nabla\mathbf{u}(t) - \nabla\mathbf{u}(t)^T\, {}^{\circ}\mathbf{C} - \{\nabla\, {}^{\circ}\mathbf{C}\}\, \mathbf{u}(t) \\ + {}^{\circ}\mathbf{F}^T\, \{\nabla\mathbf{u}(t-s) + \nabla\mathbf{u}(t-s)^T\}\, {}^{\circ}\mathbf{F}. \end{aligned} \tag{39.54}$$

Let

$$\nabla\mathbf{u}(\tau) = \boldsymbol{\varepsilon}(\tau) + \boldsymbol{\omega}(\tau), \tag{39.55}$$

where $\boldsymbol{\varepsilon}(\tau)$ and $\boldsymbol{\omega}(\tau)$ are respectively the symmetric and skew-symmetric parts of $\nabla\mathbf{u}(\tau)$. Let e_{ijk} be the alternating tensor in three dimensions, and define

$$\Omega_{ij}(\tau) = \varepsilon_{23}(\tau)\, e_{1ij} + \varepsilon_{13}(\tau)\, e_{2ij} - \varepsilon_{12}(\tau)\, e_{3ij}. \tag{39.56}$$

Using (39.54) – (39.56), PIPKIN[31] has shown that the stress perturbation $\delta\mathbf{T}^E(t)$ is given by

$$\begin{aligned} \delta\mathbf{T}^E(t) = \varkappa[\varepsilon_{11}(t) - \varepsilon_{22}(t)]\, \frac{d\, {}^{\circ}\mathbf{T}^E(\varkappa)}{d\varkappa} - \{\nabla\, {}^{\circ}\mathbf{T}^E(\varkappa)\}\, \mathbf{u}(t) + \\ + [\boldsymbol{\omega}(t) + \boldsymbol{\Omega}(t)]\, {}^{\circ}\mathbf{T}^E(\varkappa) - {}^{\circ}\mathbf{T}^E(\varkappa)\, [\boldsymbol{\omega}(t) + \boldsymbol{\Omega}(t)] + \\ + \int_0^{\infty} \mathbf{T}(\varkappa, s)\, \{\boldsymbol{\varepsilon}(t) - \boldsymbol{\varepsilon}(t-s)\}\, ds, \end{aligned} \tag{39.57}$$

30. [1968 : 3], [1970 : 5].

31. [1968 : 20]. Reader's attention is drawn to Sec. 5 where $\Delta\mathbf{u}(t)$ and $\Delta\mathbf{u}(t-s)$, etc., are defined.

where $°\mathbf{T}^E$ is the viscometric stress corresponding to °C. In indicial notation,

$$\delta T^E_{ij}(t) = \varkappa[\varepsilon_{11}(t) - \varepsilon_{22}(t)] \frac{d°T^E_{ij}(\varkappa)}{d\varkappa} - °T^E_{ij,k}(\varkappa)\, u_k(t) +$$

$$+ [\omega_{ik}(t) + {}_i\Omega_k(t)]\, °T^E_{kj}(\varkappa) - °T^E_{ij}(\varkappa)\, [\omega_{kj}(t) + \Omega_{kj}(t)]$$

$$+ \int_0^\infty T_{ijkl}(\varkappa, s)\, \{\mathcal{E}_{kl}(t) - \mathcal{E}_{kl}(t - s)\}\, ds. \qquad (39.58)$$

The kernel $T_{ijkl}(\varkappa, s)$ is related to the viscometric kernel $S_{ijkl}(\varkappa, s)$ by

$$T_{ijkl}(\varkappa, s) = -2S_{ijpq}(\varkappa, s)\, °F_{kp} °F_{lq}. \qquad (39.59)$$

We believe in the traditional spirit of fluid mechanics, that is, that velocity fields are to be used to solve the problems. This explains why we have placed more emphasis on computing $\mathbf{C}_t(t - s)$ through the velocity field rather than the displacement field, and have thus used the former approach in computing the dynamic viscosities.

39.4. Comparison of the Additive Functional Theory with the Empirical Theories Depending on the Second Invariant of $\mathbf{A}_1$. As remarked in the introduction to this chapter, the experimental results concerning dynamic viscosities tend to support the additive functional theory. The reader must note that the additive functional theory not only predicts that, for example, both $\eta'_{\|}(\varkappa, \omega)$ and $\eta'(\omega) \to 0$ as $\omega \to \infty$, but also that $\omega^2 \eta'_{11}$ and $\omega^2 \eta'$ tend to the same limit as $\omega \to \infty$, independent of $\varkappa$. This theoretical prediction finds support in Fig. 39.1, as remarked before.

Consider the following special case of the BIRD-CARREAU model (36.31), with a memory function[32]

$$m[t - t, \mathrm{II}(t')] = \frac{\eta_0}{\sum_{n=1}^{\infty} \lambda_n} \sum_{n=1}^{\infty} \frac{e^{-(t-t')/\lambda_n}}{\lambda_n \left[1 + \frac{1}{2} c^2 \lambda_n^2\, \mathrm{II}(t')\right]}. \qquad (39.60)$$

For such a model, the dynamic viscosity $\eta'_{\|}(\varkappa, \omega)$ is given by

$$\eta'_{\|}(\varkappa, \omega) = \frac{\eta_0}{\Sigma \lambda_n} \sum_{n=1}^{\infty} \frac{\lambda_n[1 + \lambda_n^2\omega^2 + c^2\lambda_n^2 \varkappa^2 (3\lambda_n^2 \omega^2 - 1)]}{(1 + \lambda_n^2 \omega^2)^2 (1 + c^2\lambda_n^2 \varkappa^2)^2}, \qquad (39.61)$$

while $\eta'(\omega)$ is obtained by putting $\varkappa = 0$ in (39.61). Now, as $\omega \to \infty$, both $\eta'_{\|}(\varkappa' \omega)$ and $\eta'(\omega) \to 0$. However,

$$\lim_{\omega \to \infty} \omega^2 \eta'_{\|}(\varkappa, \omega) = \frac{\eta_0}{\Sigma \lambda_n} \sum_{n=1}^{\infty} \frac{1 + 3c^2\lambda_n^2 \varkappa^2}{\lambda_n(1 + c^2\lambda_n^2 \varkappa^2)^2}, \qquad (39.62)$$

32. MACDONALD and BIRD [1966 : 11]. See Eqs. (2) and (6).

or $\omega^2\eta'_{\|}$ depends on $\varkappa$ as $\omega \to \infty$. In other words, a graph of $\eta'_{\|}$ vs ω for large ω would not be asymptotic to a line of slope -2, as in the additive functional theory. It seems, therefore, that experimental results call for a revision of theories with memory functions relying on the history of the second invariant[33] of $\mathbf{A}_1$.

40. Motions with Constant Stretch History as Nearly Viscometric Flows

40.1. Poiseuille-torsional Flow. Consider the following velocity field in the contravariant component form:

$$v^r = 0, \quad v^\theta = ez, \quad v^z = u(r). \tag{40.1}$$

This is a MWCH-II and was discussed in Sec. 8 from a kinematical viewpoint and in Sec. 31 from dynamic considerations. In the latter section it was noted that the motion (40.1), called the *Poiseuille-torsional flow*, is partially controllable, for the equations of motion are satisfied if and only if the inertia terms are ignored. Moreover, the boundary conditions are not satisfied, on the cylindrical surfaces. So we shall treat (40.1) as a nearly viscometric flow[34] by letting e be small so that higher powers of e can be ignored and compute the nearly viscometric stresses. With respect to the natural basis of (r, θ, z) coordinate system, the strain history corresponding to (40.1) is given by

$$\mathbf{C} = {}^\circ\mathbf{C} + \mathbf{E}, \tag{40.2}$$

where

$$[{}^\circ C\langle ij\rangle] = \begin{bmatrix} 1+\varkappa^2 s^2 & 0 & -\varkappa s \\ . & 1 & 0 \\ . & . & 1 \end{bmatrix}, \qquad \varkappa = \frac{du}{dr}, \tag{40.3}$$

$$[E\langle ij\rangle] = e\begin{bmatrix} 0 & s^2 r\varkappa/2 & 0 \\ . & 0 & -sr \\ . & . & 0 \end{bmatrix}. \tag{40.4}$$

Since tr ${}^\circ\mathbf{C}^{-1}\mathbf{E} = 0$, the flow is nearly viscometric if e is quite small, which we assume. If we make a local change of axes such that z-axis becomes $\bar{x}$, r-axis becomes $\bar{y}$ and θ becomes $\bar{z}$, then we have a local Cartesian coordinate system, with $\mathbf{C}_0$ having the familiar form ${}^\circ C\langle\bar{x}\bar{x}\rangle = {}^\circ C\langle\bar{z}\bar{z}\rangle$

33. See also the footnote in Sec. 48.7, and [1974 : 1], which re-examines this question in light of MACDONALD's findings [1974 : 3] and shows that the $B-C$ theory is still not a good approximation to the simple fluid.

34. HUILGOL [1970 : 14] has discussed the more difficult case of helical-torsional flow in this manner. The present discussion is based on this paper.

$= 1$, $°C\langle \bar{y}\,\bar{y}\rangle = 1 + s^2\varkappa^2$, $°C\langle \bar{y}\,\bar{x}\rangle = °C\langle \bar{x}\,\bar{y}\rangle = -\varkappa s$, the others being zero. Next $E\langle \bar{y}\,\bar{z}\rangle = s^2\, er\,\varkappa/2$, $E\langle \bar{x}\,\bar{z}\rangle = -ser$. Thus the perturbations in stresses which are non-zero are $\delta\, T_E\langle \bar{x}\,\bar{z}\rangle = \delta\, T_E\langle \theta\, z\rangle$ and $\delta\, T_E\langle \bar{y}\,\bar{z}\rangle = \delta\, T_E\langle r\,\theta\rangle$. These are given by

$$\delta T_E\langle \bar{x}\,\bar{z}\rangle = 2\,\delta S_{1313}\,[\varkappa \mid -ser] + 2\,\delta S_{1323}\,[\varkappa \mid s^2\, er\varkappa/2]\,, \tag{40.5}$$

$$\delta T_E\langle \bar{y}\,\bar{z}\rangle = 2\,\delta S_{2313}\,[\varkappa \mid -ser] + 2\,\delta S_{2323}\,[\varkappa \mid s^2\, er\varkappa/2]. \tag{40.6}$$

Compare (40.5) with (38.40), getting

$$\delta T_E\langle \theta z\rangle = er\,\{\eta(\varkappa) - \varkappa\,\delta S_{1323}\,[\varkappa \mid s^2]\}, \tag{40.7}$$

while (40.6) and (38.41) yield the following:

$$\delta T_E\langle r\theta\rangle = er\,\left\{\frac{N_2(\varkappa)}{\varkappa} - \varkappa\,\delta S_{2323}\,[\varkappa \mid s^2]\right\}. \tag{40.8}$$

Now the other stresses $T_E\langle rr\rangle$, $T_E\langle \theta\theta\rangle$, $T_E\langle zz\rangle$ and $T\langle rz\rangle$ have the same values as in (31.9) — (31.14), obtained by putting $\varkappa = u'$, $m = 0$, $l = 1$. Therefore

$$\tau_1 = T\langle rz\rangle = -\frac{1}{2}\, cr, \tag{40.9}$$

from (31.17*b*), with $b = 0$, since the origin of the coordinate system is in the flow domain now. But $\tau_1 = \tau(\varkappa)$, where $\tau(\cdot)$ is the viscometric shear stress function. Thus [see (28.33)]

$$u(r) = \int_r^R \tau^{-1}\left(\frac{1}{2}\, c\xi\right) d\xi, \tag{40.10}$$

or the velocity field in the POISEUILLE-torsional flow is completely known.

It has been shown by the author that for the helical-torsional flow, the application of the nearly viscometric flow theory leads to two coupled nonlinear differential equations for $u(r)$ and $\omega(r)$:

$$\eta u' + \frac{er^2\,\omega'}{\varkappa^2}\,\bar{\eta}_2(\varkappa^2)\,(r^2\,\omega'^2 - u'^2) = \frac{b}{r} - \frac{cr}{2}, \tag{40.11}$$

$$\eta r\omega' + \frac{eru'}{\varkappa^2}\,\bar{n}_2(\varkappa^2)\,(u'^2 - r^2\,\omega'^2) + \frac{eru'\, n_1(\varkappa^2)}{2} = \frac{M}{2\pi r^2}\,. \tag{40.12}$$

Here $\varkappa^2 = r^2\,\omega'^2 + u'^2$, $\eta = \eta(\varkappa)$ the viscometric viscosity function, and $\bar{n}_1(\cdot)$ and $\bar{n}_2(\cdot)$ are defined in (26.69). The solutions to (40.11) — (40.12) are unknown at present.

The question of the magnitudes of the linear functionals in (40.7) and (40.8) is open, though it was argued[35] on the basis of certain equivalences between the simple fluid and the BKZ fluid that the following relationship is

35. [1970 : 14].

valid for all simple fluids:

$$\delta S_{2323}\,[\varkappa \mid s^2] = -\frac{1}{2\varkappa^2}\,N_1(\varkappa) = -\frac{1}{2}\,n_1(\varkappa). \tag{40.13}$$

It was conjectured also that for all simple fluids

$$\delta S_{1323}\,[\varkappa \mid s^2] = -\varkappa \int_0^\infty \frac{\partial U}{\partial I_1}\, s^3\, ds, \qquad (40.14)^{36}$$

where $U(I_1, I_2, s)$ is the BKZ fluid potential and $\partial U/\partial I_1$ is evaluated at $I_1 = I_2 = 3 + \varkappa^2 s^2$. So far, no evidence is available to confirm (40.13) and (40.14). It is interesting that (40.13) and (40.14) cannot be derived by the relations in Sec. 38.

40.2. Maxwell Orthogonal Rheometer Flow. The strain history for the MAXWELL orthogonal rheometer flow, viz.

$$\dot{x} = -\,\Omega y + \Omega \psi z, \quad \dot{y} = \Omega x, \quad \dot{z} = 0 \tag{40.15}$$

can be written as a nearly viscometric flow by the following artifact: let ε be very small and $\Omega = O(\varepsilon)$ and $1/\psi = O(\varepsilon)$ such that $\Omega\,\psi = O\,(1)$, $\Omega^2\psi^2 = O(1)$, $\Omega^2\psi = O(\varepsilon)$, $\Omega^3\psi = o(\varepsilon)$.

The total strain history corresponding to (40.15) has the form

$$[C\langle ij\rangle] = [C_t(t-s)\,\langle ij\rangle] = \begin{bmatrix} 1 & 0 & -\psi \sin \Omega s \\ \cdot & 1 & -\psi(1-\cos\Omega\, s) \\ \cdot & \cdot & 1 + 2\psi^2(1-\cos\Omega s) \end{bmatrix}. \tag{40.16}$$

Expanding $\psi \sin \Omega s$ and $\psi(1 - \cos \Omega s)$ as a power series in Ω to the order $O(\varepsilon)$, we have that

$$\mathbf{C} = {}^{\circ}\mathbf{C} + \mathbf{E}, \tag{40.17}$$

where

$$[{}^{\circ}C\langle ij\rangle] = \begin{bmatrix} 1 & \cdot & -s\varkappa \\ \cdot & 1 & 0 \\ \cdot & \cdot & 1 + \varkappa^2 s^2 \end{bmatrix}, \quad \varkappa = \Omega\psi, \tag{40.18}$$

$$[E\langle ij\rangle] = \begin{bmatrix} 0 & 0 & 0 \\ \cdot & \cdot & -\,\varkappa\Omega s^2/2 \\ \cdot & \cdot & 0 \end{bmatrix}. \qquad (40.19)^{37}$$

36. There is a misprint in Eq. (58) [1970: 14].

37. In [1970 : 14] HUILGOL retained terms of order $\varkappa\Omega^2$ and suggested that these be ignored in computing the stresses. The calculations reported here are in line with assigning a parameter a very small magnitude and maintaining linearity in it (*cf.* Secs. 39 and 40.1) and hence correct the analysis there.

Now, $\| \mathbf{E} \|$ is small if Ω is small, and tr $°\mathbf{C}^{-1}\,\mathbf{E} = 0$ ensures the satisfaction of isochoricity. The stress perturbations corresponding to (40.19) are given by

$$\delta T_E \langle xy \rangle = 2\,\delta S_{1323}\,[\varkappa \mid -\varkappa\,\Omega s^2/2] = -\,\Omega\varkappa \int_0^\infty S_{1323}\,(\varkappa, s)\, s^2\, ds, \quad (40.20)$$

$$\delta T_E \langle yz \rangle = 2\,\delta S_{2323}\,[\varkappa \mid -\varkappa\,\Omega s^2/2] = -\,\Omega\varkappa \int_0^\infty S_{2323}\,(\varkappa, s)\, s^2\, ds. \quad (40.21)$$

The above two integrals should be compared with (40.13) and (40.14) to obtain the estimates for their magnitudes.

No experimental results are available to compare with the predictions contained in (40.7), (40.8), (40.20) and (40.21) at the time of writing.

CHAPTER 10

PERTURBATIONS FOR SIMPLE CONSTITUTIVE EQUATIONS

In the last chapter, the perturbations about a fairly simple base flow, e. g., simple shear, were examined, but the constitutive equation used was one of utmost generality. Here we shall concentrate on flows which are complex, but the constitutive relations are of a lower order of generality.

There are a number of perturbation schemes which will be employed in this chapter. The first concerns the so-called slow viscoelastic flow analysis. Here it is assumed that the velocity field is given by $\mathbf{v} = \varepsilon\, \mathbf{v}_1 + \ldots + \varepsilon^n\, \mathbf{v}_n$, where ε is a small positive parameter. This velocity field is substituted into the RIVLIN-ERICKSEN fluid constitutive equation (35.9) and expansions to order ε, ε^2, ε^3, ε^4, etc., are obtained. Such equations will be called *slow flow constitutive equations* (abridged to SFCE). Since, for example, the second-order SFCE based on $\varepsilon\, \mathbf{v}_1 + \varepsilon^2\, \mathbf{v}_2$ is of order ε^2 only, it is not identical to that of a second-order fluid (35.24), which would be of order ε^4, when $\varepsilon\, \mathbf{v}_1 + \varepsilon^2\, \mathbf{v}_2$ is used there. So one has to distinguish carefully between SFCE's and order fluids, and a solution to an order ε^2 is not necessarily the solution to the same problem using a second-order fluid. However, there are cases, such as those in Secs. 41 and 42, where the solutions to the order ε^2 are identical to those for a second-order fluid. In Sec. 43, this is not the case. So we have examples illustrating this vital difference. Moreover in Sec. 42, one sees that the respective solutions to order ε^3 and ε^4 are not those for the third- and fourth-order fluids either.

The perturbation scheme of these three sections is an extension of the work of RIVLIN[1] for second-order elasticity. This extension to viscoelastic liquids was done by LANGLOIS and RIVLIN[2]. In a series of papers, these authors applied the perturbation scheme to flows in non-circular tubes and other geometries[3], and we shall discuss the problem of pipe flow in Sec. 42.

In Sec. 41, we concentrate on the case when the first-order solution is identical to the second-order solution. This occurs whenever the GIESEKUS-PIPKIN[4] theorem holds, and use of the identical natures of the two flows has been made in the so-called 'pressure hole error' analysis by TANNER and

1. [1953 : 3].
2. [1959 : 4].
3. [1962 : 9], [1963 : 4-6], [1964 : 12].
4. [1963 : 3], [1969 : 18], [1972 : 13].

PIPKIN[5] as well as KEARSLEY[6]. We pay special attention to this matter by discussing some questions of uniqueness.

Also, TANNER's work on the free surface flow[7] has revived interest in slow viscoelastic flows down tilted troughs. In fact, the experimental results in this paper establish beyond doubt that the second normal stress difference $N_2(\varkappa)$ is negative, at low shear rates at least. This analysis is paralleled along the earlier work of WINEMAN and PIPKIN[8] and we summarize this work in Sec. 43.

In Sec. 44, the perturbation scheme employed is the domain perturbation method of JOSEPH[9]. This has been recently employed by JOSEPH, FOSDICK and BEAVERS[10] to solve the problem of finding the free surface in COUETTE flow. These authors have incorporated the influence of surface tension as well and because of the usefulness of this method, we have summarized it in Sec. 44.

Earlier in the text (Sec. 34 ter), the measurement of the complex viscosity by the MAXWELL orthogonal rheometer and by the eccentric cylinder rotational viscometer has been discussed. The analysis was based on the constitutive relation of finite linear viscoelasticity, and neglected inertial effects and edge conditions (that is, finiteness of the dimensions). In Sec. 45, we shall record, by using the calculations of WALTERS and his colleagues[11], the perturbation techniques by which corrections can be adduced to incorporate inertia (in MAXWELL rheometer) and end conditions (rotational viscometer).

In the final section the initial value problem of the initiation of rectilinear flow in a tube is set up[12] for a fluid obeying the constitutive equation of finite linear viscoelasticity. It is solved for the case of a circular pipe under a special form for the relaxation function[13]. Certain aspects of the flow under a periodic pressure gradient are discussed.

At this juncture a few comments are in order. The first concerns the constitutive equation which must be employed in the solution of a given problem. One can summarize the conclusions as below[14] :

(i) if the motion is steady and slow, e. g. if the driving force is small, then the RIVLIN-ERICKSEN fluids suffice ; for example, one could solve the pipe flow problems by using such models. It is the second normal stress difference which causes departure from rectilinear flow and induces transverse flow and such effects are easily explained by these fluids ;

---

5. [1969 : 23].
6. [1970 : 15].
7. [1970 : 22].
8. [1966 : 20].
9. [1967 : 15].
10. [1973 : 7], [1973 : 8].
11. [1970 : 1], [1971 : 6].
12. NORMAN and PIPKIN [1967 : 19].
13. WATERS and KING [1971 : 29].
14. See METZNER, *et al.* [1966 : 13], PIPKIN [1966 : 15], HUILGOL [1973 : 5].

(ii) again if the motion is slow and steady, is spatially periodic as well and is being used to measure the complex viscosity, then obviously the finite linear viscoelasticy model is to be used ;

(iii) if an initial value problem is to be solved then integral models should be used. Such models are to be employed when the fluid particle undergoes short but fast deformation, or when it experiences a spatially steady but pathwise dependent sudden acceleration.

However, in MWCSH, the RIVLIN-ERICKSEN fluids suffice, and here there is no limit on the driving force or the quesion of spatial periodicity. So the problem of which constitutive equation to use must be resolved on a more realistic measure of 'forces', 'strain history', etc. One such approach was outlined by REINER[15], who tried to mingle theology with rheology and invented the DEBORAH number, based on the 'observation' of the Prophetess DEBORAH that 'the mountains flowed before the Lord'[16]. REINER claimed that in the eyes of the Lord, whose observation time span is infinite, the process of mountain forming (flow) is finite. Thus, if the ratio (time of relaxation/time of observation) is very small, the material behaves like a fluid, while the converse would imply a solid-like behavior.

The above idea was taken up by METZNER[17], *et al.*, who designated the ratio of a characteristic relaxation time θ to a characteristic residence time T, as the DEBORAH number. If θ/T is small, one has a fluid-like behavior and if it is large, solid-like features occur. They claimed, in essence, that if θ/T is small, one should use the order fluids, and in the opposite case, one should use integral models.

It seems that a more precise approach[18] would be based on the idea that the rate at which the strain history is experienced by the particle is the parameter defining the type of equation to use. Let us call

$$\mathcal{T}(\tau) = \int_0^\infty \{\operatorname{tr} \mathbf{C}_\tau(\tau - s)\}\, g(s)\, ds \qquad (*)$$

the *duration of execution of the strain history*, where $g(s)$ is a suitable non-dimensional influence function, analogous to $h(s)$ in Sec. 34, so that $\mathcal{T}(\tau)$ exists for all meaningful strain histories[19]. To prescribe the strain history upto time t, one could define $\mathbf{C}_\tau(\tau - s)$ as a function $\mathbf{C}[\mathcal{T}(\tau)]$ and the length $0 \leqq \mathcal{T} \leqq \tilde{\mathcal{T}}$, where $\tilde{\mathcal{T}} = \mathcal{T}(t)$. If one were to postulate that the constitutive functional depends on $\mathbf{C}[\mathcal{T}(\tau)]$ and $\tau \in (-\infty, t]$, one would obtain a

15. [1964 : 21].

16. The King James Version reads : 'The mountains melted from before the Lord', while the New English Bible has 'Mountains shook in fear before the Lord', Judges, 5 : 5.

17. [1966 : 13].

18. HUILGOL [1973 : 5]. A revised version will appear in *Trans. Soc. Rheol.* **19** (1975).

19. A similar measure was introduced by PIPKIN and RIVLIN [1965 : 15] to define the arc length of the strain path.

rate-dependent fluid theory[20]. Dropping such a dependenc rate-independent theory, considered by PIPKIN and RIVLI context of solids. We shall not pursue this matter any further

Since in a MWCSH, $\mathcal{T}(t)$ is independent of t, it is perhaps use the departure of $d\mathcal{T}(t)/dt$ from zero as equivalent to the DEBO N_{De}. If $\dot{\mathcal{T}}(t) = 0$, then $N_{De} = 0$, and in this case RIVLIN-ERICK us suffice. It $\dot{\mathcal{T}}(t) \neq 0$, then integral models ought to be employed. M tivated by this and in order to avoid negative values for N_{De}, we shall define the DEBORAH number through

$$N_{De} = \sup_{-\infty < \tau \leqslant t} \left| \frac{d}{d\tau} \mathcal{T}(\tau) \right|. \qquad (**)$$

Some comments on this definition follow next :

(i) N_{De} is an objective quantity, since $\mathcal{T}(t)$ is one.

(ii) In small amplitude oscillatory motions, it is known that integral models are to be employed, and not RIVLIN-ERICKSEN fluids. If one were to justify this choice on the basis of the norm of the history (close to 1) and smoothness (as smooth as one pleases), it would be difficult to do so. However, by taking $g(s) = e^{-\beta s}$, $\beta > 0$, one can show that $N_{De} \sim A\ \omega/\beta$, where A is the 'amplitude factor'. Thus an explicit numerical value is obtained and it does not require that the residence time be determined, which must be done if the approach of METZNER, *et al.* is followed. Moreover, in the latter case, it is difficult to justify why the residence time, in an oscillatory motion, is finite rather than infinite.

(iii) ASTARITA[21] defined the DEBORAH number as

$$N_{De} = \theta \left(\left| \frac{d}{dt} \left(\tfrac{1}{2} \operatorname{tr} \mathbf{A}_1^2 \right)^{1/2} \right| \right)^{1/2},$$

where θ is the relaxation time. In oscillatory motions, this becomes zero, a difficulty avoided in our definition.

(iv) It is also clear that the definition given here yields results which are better than those obtained by the amplitude A vs. $\omega\theta$ diagram of PIPKIN[22]. This diagram suggests that if $\omega\theta$ is small, i.e., if ω/β is small, one should use order fluids and if $\omega\theta$ is large, one should use integral models. Not only does the present definition agree with this, it suggests that in die swell integral models should be used. This latter conclusion is not obvious from PIPKIN's theory.

(v) In passing, it may be remarked that the concept of natural time, intro-

---

20. Such a theory is in much better agreement with fading memory than the rate-independent theory. See NFTM, p. 402. OWEN and WILLIAMS [1968 : 18] show that the idea of rate-independence in NFTM and that of PIPKIN and RIVLIN are equivalent.

21. [1967 : 2], Eq. (72).

22. [1972 : 13], p. 133.

duced by TRUESDELL[23], is not sufficient to delineate the situation when order fluids or integral models must be used. This coneept is equivalent to the retardation of a given flow and is thus not fully adequate for the present purpose.

To conclude, one can suggest that if $N_{De} \simeq 0$, order fluids suffice and, if N_{De} is large, integral models must be used in solving flow problems. Though this procedure is not directly applied in what follows, the reader may convince himself that the examples chosen in this chapter obey this criterion.

41. Creeping Flow : Theorems of Giesekus, Pipkin and Tanner and Questions of Uniqueness

Let us now assume that the velocity field is given by

$$\mathbf{v} = \varepsilon\,\mathbf{v}_1 + \varepsilon^2\,\mathbf{v}_2 + \ldots + \varepsilon^n\,\mathbf{v}_n, \tag{41.1}$$

where ε is a very small positive number, and that the flow is *steady*. We shall call such motions *slow viscoelastic* or *creeping flows*. Using (41.1) in the constitutive equation (35.9), we obtain that to an error $o(\varepsilon^r)$, $r = 1, 2, 3, 4$, the extra stresses are[24] :

$$\mathbf{T}_E = \varepsilon\,\overset{(1)}{\mathbf{S}}_E + \varepsilon^2\,\overset{(2)}{\mathbf{S}}_E + \varepsilon^3\,\overset{(3)}{\mathbf{S}}_E + \varepsilon^4\,\overset{(4)}{\mathbf{S}}_E + o(\varepsilon^4), \tag{41.2a}$$

$$\overset{(1)}{\mathbf{S}}_E = \eta_0\,\mathbf{A}, \tag{41.2b}$$

$$\overset{(2)}{\mathbf{S}}_E = \eta_0\,\mathbf{D} + \beta\,\mathbf{A}^2 + \gamma\,\mathbf{B}, \tag{41.2c}$$

$$\overset{(3)}{\mathbf{S}}_E = \eta_0\,\mathbf{F} + \beta(\mathbf{AD} + \mathbf{DA}) + \gamma\,\mathbf{E} + \alpha_0(\mathrm{tr}\,\mathbf{B})\,\mathbf{A} + + \alpha_1(\mathbf{AB} + \mathbf{BA}) + \alpha_2\mathbf{C}, \tag{41.3}$$

$$\begin{aligned}\overset{(4)}{\mathbf{S}}_E = {} & \{\alpha_3\,\mathrm{tr}\,(\mathbf{AB} + \mathbf{BA}) + \alpha_4\,\mathrm{tr}\,\mathbf{A}^3\}\,\mathbf{A} + \\ & + \alpha_5(\mathrm{tr}\,\mathbf{A}^2)\mathbf{A}^2 + \alpha_6(\mathrm{tr}\,\mathbf{A}^2)\,\mathbf{B} + \alpha_7(\mathbf{A}^2\mathbf{B} + \mathbf{BA}^2) + \\ & + \alpha_8\mathbf{B}^2 + \alpha_9(\mathbf{AC} + \mathbf{CA}) + \alpha_{10}\mathbf{W}.\end{aligned} \tag{41.4}$$

In (41.2) – (41.4), we have put

$$\begin{aligned}\mathbf{A} &= \nabla\,\mathbf{v}_1 + \nabla\,\mathbf{v}_1^T, \\ \mathbf{B} &= \{\nabla\,\mathbf{A}\}\,\mathbf{v}_1 + \mathbf{A}\,\nabla\,\mathbf{v}_1 + \nabla\,\mathbf{v}_1^T\,\mathbf{A}, \\ \mathbf{C} &= \{\nabla\,\mathbf{B}\}\,\mathbf{v}_1 + \mathbf{B}\,\nabla\,\mathbf{v}_1 + \nabla\,\mathbf{v}_1^T\,\mathbf{B}, \\ \mathbf{W} &= \{\nabla\,\mathbf{C}\}\,\mathbf{v}_1 + \mathbf{C}\,\nabla\,\mathbf{v}_1 + \nabla\,\mathbf{v}_1^T\,\mathbf{C}.\end{aligned} \tag{41.5}$$

Thus, **A**, **B**, **C** and **W** are respectively the first four RIVLIN-ERICKSEN tensors

23. [1964 : 25].
24. LANGLOIS and RIVLIN [1963 : 6].

for the steady velocity field $\mathbf{v}_1$. Indeed, for the velocity field $\mathbf{v}$ given by

$$\mathbf{v} = \varepsilon\, \mathbf{v}_1 + \varepsilon^2\, \mathbf{v}_2 + \varepsilon^3\, \mathbf{v}_3 + \varepsilon^4\, \mathbf{v}_4 + o(\varepsilon^4), \qquad (41.1bis)$$

one has

$$\begin{aligned} \mathbf{A}_1 &= \varepsilon\, \mathbf{A} + \varepsilon^2\, \mathbf{D} + \varepsilon^3\, \mathbf{M} + \varepsilon^4\, \mathbf{N} + \mathrm{o}(\varepsilon^4), \\ \mathbf{A}_2 &= \varepsilon^2\, \mathbf{B} + \varepsilon^3\, \mathbf{E} + \varepsilon^4\, \mathbf{K} + \mathrm{o}(\varepsilon^4), \\ \mathbf{A}_3 &= \varepsilon^3\, \mathbf{C} + \varepsilon^4\, \mathbf{P} + \mathrm{o}(\varepsilon^4), \\ \mathbf{A}_4 &= \varepsilon^4\, \mathbf{W} + \mathrm{o}(\varepsilon^4). \end{aligned} \qquad (41.6)$$

Hence,

$$\mathbf{D} = \nabla\, \mathbf{v}_2 + \nabla\, \mathbf{v}_2^T, \qquad (41.7a)$$

$$\begin{aligned} \mathbf{E} = {} & \{\nabla \mathbf{D}\}\, \mathbf{v}_1 + \mathbf{D} \nabla \mathbf{v}_1 + \nabla\, \mathbf{v}_1^T\, \mathbf{D} + \{\nabla\, \mathbf{A}\}\, \mathbf{v}_2 + \\ & + \mathbf{A}\, \nabla\, \mathbf{v}_2 + \nabla\, \mathbf{v}_2^T\, \mathbf{A}, \end{aligned} \qquad (41.7b)$$

$$\mathbf{M} = \nabla\, \mathbf{v}_3 + \nabla\, \mathbf{v}_3^T, \qquad (41.7c)$$

$$\mathbf{N} = \nabla\, \mathbf{v}_4 + \nabla\, \mathbf{v}_4^T, \qquad (41.7d)$$

$$\begin{aligned} \mathbf{K} = {} & \{\nabla \mathbf{M}\}\, \mathbf{v}_1 + \mathbf{M} \nabla \mathbf{v}_1 + \nabla\, \mathbf{v}_1^T\, \mathbf{M} + \{\nabla \mathbf{D}\}\, \mathbf{v}_2 + \mathbf{D}\, \nabla\, \mathbf{v}_2 + \\ & + \nabla\, \mathbf{v}_2^T\, \mathbf{D} + \{\nabla \mathbf{A}\}\, \mathbf{v}_3 + \mathbf{A}\, \nabla\, \mathbf{v}_3 + \nabla\, \mathbf{v}_3^T\, \mathbf{A}, \end{aligned} \qquad (41.7e)$$

$$\begin{aligned} \mathbf{P} = {} & \{\nabla\, \mathbf{B}\}\, \mathbf{v}_2 + \mathbf{B}\, \nabla \mathbf{v}_2 + \nabla\, \mathbf{v}_2^T\, \mathbf{B} + \{\nabla \mathbf{C}\}\, \mathbf{v}_1 + \mathbf{C} \nabla \mathbf{v}_1 + \\ & + \nabla\, \mathbf{v}_1^T\, \mathbf{C}. \end{aligned} \qquad (41.7f)$$

Let it be assumed that the pressure field is given by

$$p = \varepsilon\, p_1 + \varepsilon^2\, p_2 + \varepsilon^3\, p_3 + \varepsilon^4\, p_4 + o(\varepsilon^4). \qquad (41.8)$$

At this juncture, two definitions are needed :

Slow Flow Constitutive Equation (*SFCE*). An r-th order SFCE (r=1, 2, 3, 4) is the sum $\sum_{i=1}^{r} \varepsilon^i \overset{(i)}{\mathbf{S}}$, where $\overset{(i)}{\mathbf{S}}$ are given by (41.2) – (41.5), and (41.6) – (41.7).

Order of Solution. A solution to the equations of motion by using an r-th order SFCE is called an r-th order solution, that is, the velocity field has the form $\mathbf{v} = \sum_{i=1}^{r} \varepsilon^i\, \mathbf{v}_i$.

In this section, we confine our attention to creeping flows, that is, the inertial terms are neglected. Generally speaking, in such problems the Newtonian fluid solution, which is also the first-order solution, is obtained, and this is used to derive the second-order solution. In this section, though we discuss the methodology of such an approach, it is not used. Instead, the conditions wherein the first- and second-order solutions are identical will be examined. In the next section, solutions upto the fourth order will be obtained for a problem.

Now, let $\mathbf{v}_1$ be the solution for the creeping flow of a Newtonian incompressible fluid. In other words, we put $\rho = 0$, tr $\mathbf{A} = 0$, and use (41.2c) and (41.8) to obtain

$$\varepsilon \nabla p_1 = \varepsilon \eta_0 \operatorname{div} \mathbf{A}, \tag{41.9}$$

or div $\mathbf{A}$ is the gradient of a scalar $P = p_{1/\eta_0}$. It $\varepsilon \mathbf{v}_1 + \varepsilon^2 \mathbf{v}_2$ is a possible solution for a fluid obeying the second-order SFCE, by using (41.2c), (41.8) and (41.9) one obtains the equation

$$-\varepsilon^2 \nabla p_2 + \varepsilon^2 \eta_0 \operatorname{div} \mathbf{D} = -\varepsilon^2(\beta \operatorname{div} \mathbf{A}^2 + \gamma \operatorname{div} \mathbf{B}). \tag{41.10}$$

If (41.9) and (41.10) both hold, then $\varepsilon \mathbf{v}_1 + \varepsilon^2 \mathbf{v}_2$ is possible in a fluid obeying a second-order SFCE and the total stress is

$$\mathbf{T} = -(\varepsilon p_1 + \varepsilon^2 p_2) + \varepsilon\eta_0\mathbf{A} + \varepsilon^2(\eta_0 \mathbf{D} + \beta\mathbf{A}^2 + \gamma\mathbf{B}). \tag{41.11}$$

The above scheme is self-explanatory but could be paraphrased as :

(i) calculate a creeping flow solution $(\mathbf{v}_1, p_1)$ to the equations of motion for an incompressible Newtonian fluid with appropriate boundary conditions ;

(ii) by using the equation for the second-order SFCE, derive the equations of motion to be satisfied by $(\mathbf{v}_2, p_2)$. Under null boundary conditions, if a solution $(\mathbf{v}_2, p_2)$ can be found, then $\varepsilon \mathbf{v}_1 + \varepsilon^2 \mathbf{v}_2$ is possible in such a fluid.

Let us examine (41.10) now in detail. In it $\mathbf{A}^2$ and $\mathbf{B}$ are obtained from the velocity field $\mathbf{v}_1$, while $\mathbf{D}$ is obtained from $\mathbf{v}_2$. If div $\mathbf{A}^2$ and div $\mathbf{B}$ are both gradients of scalars, then the right side of (41.10) is irrotational and one may take $\mathbf{D} \equiv 0$ as a possible solution. In other words, $\mathbf{v} = \varepsilon \mathbf{v}_1$ will also be the second-order solution. If this is true, then the second-order SFCE becomes identical to the constitutive equation of the second-order fluid (35.24), with the latter stresses being computed from $\varepsilon \mathbf{v}_1$. Or, the Newtonian and second-order fluids will then have the same velocity fields, though the stresses will differ. In this case, one may omit the parameter ε because it is no longer necessary. Following PIPKIN[25], the condition on $\mathbf{v}_1$ will now be found for $\mathbf{v}_1$ to be a second-order solution. For convenience put $\mathbf{v}_1 = \mathbf{u}$. Then in Cartesian coordinates :

$$\begin{aligned}
\operatorname{div} \mathbf{B} = B_{ij,j} &= \left(\frac{d}{dt} A_{ij}\right)_{,j} + (A_{ik}\, u_{k,j})_{,j} + \\
&\quad + (A_{jk}\, u_{k,i})_{,j} \\
&= \frac{d}{dt} (A_{ij,j}) + u_{k,j}\, A_{ik,j} + (A_{ik}\, u_{k,j})_{,j} + \\
&\quad + A_{jk,j}\, u_{k,i} + A_{jk}\, u_{k,ij}.
\end{aligned} \tag{41.12}$$

25. [1969 : 18], [1972 : 13].

But $u_{j,j} = 0$ and thus

$$(A_{ik} A_{kj})_{,j} = (A_{ik} u_{k,j})_{,j} + A_{ik,j} u_{j,k}, \tag{41.13}$$

or the second and third terms in (41.12) are equivalent to div $\mathbf{A}^2$. The last term in (41.12) is $\frac{1}{4}\nabla(\mathrm{tr}\,\mathbf{A}^2)$, since

$$\begin{aligned}(A_{jk} A_{kj})_{,i} &= 2A_{jk} A_{kj,i} = 2A_{jk}(u_{k,ji} + u_{j,ki}) \\ &= 4A_{jk} u_{k,ji}.\end{aligned} \tag{41.14}$$

Finally, $\nabla P = \nabla p_1/\eta_0 = \mathrm{div}\,\mathbf{A}$ and the steadiness of the flow implies $dP/dt = P_{,k} u_k$ and thus

$$\begin{aligned}(P_{,k} u_k)_{,i} &= P_{,ki} u_k + P_{,k} u_{k,i} \\ &= P_{,ik} u_k + P_{,k} u_{k,i} \\ &= \frac{d}{dt}(A_{ij,j}) + A_{jk,i} u_{k,i}.\end{aligned} \tag{41.15}$$

Using (41.12)–(41.15), we have

$$\mathrm{div}\,(\mathbf{B} - \mathbf{A}^2) = \nabla\left(\frac{dP}{dt} + \frac{1}{2}\zeta^2\right), \tag{41.16}$$

where $2\zeta^2 = \mathrm{tr}\,\mathbf{A}^2$. One may rewrite (41.10), by using the above results and assuming $\mathbf{D} \equiv \mathbf{0}$ as :

$$\nabla\left[p_2 - \gamma\left(\frac{dP}{dt} + \frac{1}{2}\zeta^2\right)\right] = (\beta + \gamma)\,\mathrm{div}\,\mathbf{A}^2. \tag{41.17}$$

Hence follows the GIESEKUS[26]-PIPKIN[27]

THEOREM 41.1 : *Let* $\mathbf{u}$ *be the velocity field in the creeping flow of an incompressible Newtonian fluid. Then the same velocity field is possible in a second-order fluid, subject to the original boundary conditions, if* div $\mathbf{A}^2$ *is irrotational.*

If $\mathbf{u} = \nabla\varphi$ or the flow is a potential flow[28], then div $\mathbf{u} = \nabla^2\varphi = 0$. Hence div $\mathbf{A}^2 = \nabla\zeta^2$, where $2\zeta^2 = \mathrm{tr}\,\mathbf{A}^2$. Or, the same velocity field occurs in second-order fluids, but the second-order pressure is given by $p_1 + p_2$, where

$$2p_2 = (2\beta + 3\gamma)\,\zeta^2. \tag{41.18}$$

In plane creeping flows, $\mathbf{A}$ is two-dimensional and since tr $\mathbf{A} = 0$, by CAYLEY-HAMILTON theorem,

$$\mathbf{A}^2 = -(\det \mathbf{A})\,\mathbf{1}, \tag{41.19}$$

whence

$$\mathrm{div}\,\mathbf{A}^2 = -\nabla(\det \mathbf{A}) = \nabla(\tfrac{1}{2}\,\mathrm{tr}\,\mathbf{A}^2) = \nabla\zeta^2. \tag{41.20}$$

Therefore, the Newtonian velocity field in a steady creeping flow is also possible in the steady plane flow of a second-order fluid[29], with the pressure

26. In [1963 : 3], GIESEKUS has shown that div $(\dot{\mathbf{A}} + \mathbf{AW} - \mathbf{WA})$ is the gradient of a scalar if div $\mathbf{A} = \nabla P$. I am grateful to Professor D. D. JOSEPH for bringing this to my attention.
27. Rediscovered independently by PIPKIN [1969 : 18], [1972 : 13].
28. In [1969 : 18], PIPKIN attributes this result to B. CASWELL.
29. TANNER [1966 : 18].

field p_2 given by[30]

$$p_2 = \frac{\gamma}{\eta_0}\frac{dp_1}{dt} + (2\beta + 3\gamma)\frac{\zeta^2}{2}. \tag{41.21}$$

The method outlined above implies that if div $\mathbf{A}^2$ is the gradient of a scalar, then the Newtonian solution is also a second-order solution. It does not answer whether it is the unique solution. To obtain such a result, one should start afresh with a given problem, e.g., plane flow, for the second-order fluid, and obtain the differential equation satisfied by the stream function and then show that the Newtonian stream-function and the second-order stream-function are identical. This we shall discuss next.

Following TANNER[31], let $\psi(x, y)$ be the stream function in a two-dimensional steady creeping flow of an incompressible fluid. Then,

$$u = \frac{\partial\psi}{\partial y}, \qquad v = -\frac{\partial\psi}{\partial x} \tag{41.22}$$

are the Cartesian components of the velocity field. If the stream function $\psi(x, y)$ is used to calculate the extra stresses in an incompressible Newtonian fluid and use is made of the equations of motion with conservative body forces and without acceleration, we obtain that ψ satisfies the biharmonic equation

$$\Delta^2\psi = 0, \tag{41.23}$$

when the fact that the mixed partial derivatives of the Newtonian pressure p_1 commute, that is,

$$\partial^2 p_1/\partial x\partial y = \partial^2 p_1/\partial y\partial x, \tag{41.24}$$

is used. If $\psi(x, y)$ is used in the second-order fluid constitutive equation (35.24) to compute the extra stresses, and we let $\mathbf{T}_E = \mathbf{T} + p\mathbf{1}$, then

$$\partial^2 p/\partial y\partial x = \partial^2 p/\partial x\partial y \tag{41.25}$$

implies that ψ satisfies

$$\Delta^2\psi + \delta\,\mathbf{v}\cdot\nabla(\Delta^2\psi) = 0, \tag{41.26}$$

where $\delta = \gamma/\eta_0$, $\mathbf{v} = u\mathbf{i} + v\mathbf{j}$ is the velocity field, defined in terms of ψ through (41.22). Clearly, if $\Delta^2\psi = 0$, ψ satisfies (41.26). Thus, we have proved the following

THEOREM 41.2 (TANNER)[32]. *Let ψ be the stream function and $\mathbf{v}$ the velocity field of the plane creeping flow of an incompressible Newtonian fluid. Then the same stream function ψ and velocity field $\mathbf{v}$ are a pair of solutions to the corresponding problem for an incompressible second-order fluid.*

The formula (41.21) says that the p_2 term in the pressure field $p = p_1 + p_2$ of the second-order fluid is given in terms of dp_1/dt and ζ^2.

30. TANNER and PIPKIN [1969 : 23], Eq. (6).
31. [1966 : 19].
32. [1966 : 19].

Before we examine an application of the preceding results, we call the attention of the reader to the following comments.

Let the plane creeping flow occur in a region Ω, bounded or unbounded; however, let Ω by simply- or doubly-connected with piecewise smooth boundaries. Then, given $\mathbf{f} = (\partial\psi/\partial y)\mathbf{i} - (\partial\psi/\partial x)\mathbf{j}$ on the boundary $\partial\Omega$[33], the stream function ψ is uniquely determined, to within a constant, by Eq. (41.23). We say that $(\psi, \mathbf{v})$ *is a Newtonian pair if* ψ *obeys* (41.23) *in* Ω, $\mathbf{v}$ *is derivable from* ψ *through* (41.22) *and* ψ *satisfies* $(\partial\psi/\partial y)\mathbf{i} - (\partial\psi/\partial x)\mathbf{j} = \mathbf{f}$ *on* $\partial\Omega$. We note that any two Newtonian pairs differ by constants. Thus, whenever we speak of a unique Newtonian pair, we mean it being unique to within a constant value for ψ, but to within a null difference for $\mathbf{v}$.

Next, write Eq. (41.26) in the form

$$w + \delta\,\mathbf{v}\cdot\nabla w = 0, \quad w = \Delta^2\psi. \tag{41.27}$$

Multiply throughout by w and integrate over Ω and recall that $\operatorname{div}\mathbf{v} = 0$. Then, we have

$$\int_\Omega w^2\,da + \delta\int_{\partial\Omega} w^2\,\mathbf{f}\cdot\mathbf{n}\,ds = 0, \tag{41.28}$$

by using the divergence theorem, with $\mathbf{n}$ representing the unit external normal to the boundary $\partial\Omega$.

Let us assume that $\delta = \gamma/\eta_0 < 0$. In Sec. 34, we have presented the theoretical evidence to prove this claim[34]. Then $w \equiv 0$ in Ω, wherever

$$\int_{\partial\Omega} w^2\,\mathbf{f}\cdot\mathbf{n}\,ds \leqq 0. \tag{41.29}$$

But $w \equiv 0$ in Ω, with ψ obeying the boundary condition on $\partial\Omega$, yields the Newtonian pair. Thus we have proved the following

THEOREM 41.3 (HUILGOL[35]) : *Let* $(\psi, \mathbf{v})$ *be a Newtonian pair, obeying the boundary condition* $(\partial\psi/\partial y)\,\mathbf{i} - (\partial\psi/\partial x)\,\mathbf{j} = \mathbf{f}$ *on* $\partial\Omega$. *Then it is the unique solution pair to Eq.* (41.26) *for the same boundary condition if* $w = \Delta^2\psi$ *and* w *obeys the inequality* (41.29).

It will be shown later in this section that in the plane flow across a deep slot, inequality (41.29) is satisfied and hence one can use the foregoing formulae for the pressure p_2, etc.

But here we turn to the situation when the pair $(\psi, \mathbf{v})$ of the second-order fluid may not be a Newtonian pair for a given boundary condition $\mathbf{f}$. Let (v_r, v_θ) be the radial and tangential components of the velocity field of an incompressible second-order fluid in creeping flow. Suppose that the domain Ω is the finite annulus $0 < R_1 \leqq r \leqq R_2$, $0 \leqq \theta \leqq 2\pi$ and that the velocity

33. In free surface flow, with $y =$ const. along the surface, ψ obeys $\partial\psi/\partial x = 0$ and $\partial^2\psi/\partial y^2 = 0$ along it. In such problems, the definition of the Newtonian pair is suitably altered.
34. For an experimental verification, see Chap. 11.
35. [1973 : 2].

field is radial at the boundary, that is,

$$\mathbf{f}\Big|_{R_\beta} = \frac{a}{R_\beta}\,\mathbf{e}_r, \quad \beta = 1, 2. \tag{41.30}$$

Then, assuming that v_r and v_θ are both functions of r only, we have

$$v_r = \frac{a}{r} \quad \text{in} \quad \Omega, \tag{41.31}$$

and $v_\theta = v$ obeys[36]

$$v''(r) + \left(\frac{r}{\lambda} + \frac{3}{r}\right) v' = \frac{b}{r^2}, \tag{41.32}$$

where $\lambda = a\delta = a\gamma/\eta_0$, and $b/r^2 = T_E \langle r\theta \rangle$, the shear stress in the fluid. Because of (41.30), $v(R_1) = v(R_2) = 0$. Then we can show immediately that :

1°. For the Newtonian fluid, $\gamma = 0$ and thus (41.32) reduces to

$$\eta_0\, r\, v'(r) = b/r^2\,; \tag{41.33}$$

thus the solution obeying the boundary condition is $v \equiv 0$; also $b = 0$.

2°. For the second-order fluid, if

(i) $b = 0$, then with A and C being two constants,

$$v(r) = A + \int_R^r \frac{C}{\xi^3} \exp\,(-\xi^2/2\lambda)\, d\xi. \tag{41.34}$$

Thus, $v = 0$ because $v(R_1) = v(R_2) = 0$.

(ii) if $b \neq 0$, then $v(r)$ is given by

$$v(r) = A - \frac{\lambda b}{r^2} + \int_R^r \frac{C}{\xi^3} \exp\,(-\xi^2/2\lambda)\, d\xi. \tag{41.35}$$

The constant A can be used with C to satisfy the boundary conditions on R_1 and R_2. Then we get a non-zero $v = v_\theta$ in the annulus.

Hence for the same boundary condition, we have exhibited two velocity fields in the annulus[37]. In other words, solutions to (41.26) may not be uniquely determined by prescribing **f** alone on the boundary. We emphasize that this is not the case for Newtonian fluids.

Indeed, one can say something more. Let us rewrite (41.27) as

$$w(\mathbf{X}, t) + \delta\,\frac{dw(\mathbf{X}, t)}{dt} = 0, \tag{41.36}$$

where **X** is the material position of a particle occupying **x** at time t and dw/dt is the material derivative. Integration leads to

$$w(\mathbf{X}, t) = A(\mathbf{X}) \exp\,(-t/\delta). \tag{41.37}$$

Suppose, given a velocity function **f** on $\partial\Omega$, the motion inside Ω is spatially

36. See Fosdick and Bernstein [1969 : 3]. Put $\rho = 0$ in Sec. 3, to receive (41.32).

37. Huilgol [1973 : 2] ; Fosdick and Bernstein [1969 : 3].

periodic, that is, streamlines are closed. Then as $t \to \infty$, $w(\mathbf{X}, t) \to \infty$, since $\delta < 0$. However, such a situation is physically impossible, since the motion is steady. Thus $w(\mathbf{X}, t) \equiv 0$ in Ω, which leads us to the Newtonian pair. But, now suppose that the domain Ω is such that motion is not spatially periodic, but that the fluid particle traverses it in a finite time. During this transit time, which is finite, w will be bounded, for each particle. Thus $w \neq 0$ may be a possible solution. In the example above, this situation arose and we were confronted with a 'non-Newtonian velocity' field[38].

Now, let us revert to the problem of the creeping flow of a second-order incompressible fluid across a deep slot (see Fig. 41.1)[39]. At distances $x = \pm R$,

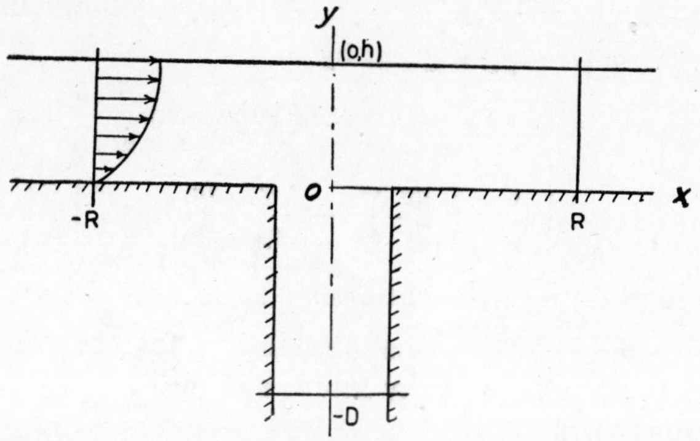

Fig. 41.1. Flow over a deep slot.

where R is fairly large, the velocity field is a function of y only. On the fixed parts of the boundary, the velocity is zero, while on $y = h$, it is normal to $\mathbf{n}$. Hence, in the limit as $R \to \infty$, $-R \to -\infty$ and $y = -D \to -\infty$,

$$\int_{\partial\Omega} w^2 \, \mathbf{f} \cdot \mathbf{n} \, ds = 0. \tag{41.38}$$

So the Theorem 41.3 tells us that the Newtonian solution is the unique solution to the problem of the flow over a deep slot of the second-order fluid, and this is true whether the surface at $y = h$ is free or is in contact with a boundary.

Now, let an undisturbed flow, far away from the slot be the shear flow $\dot{x} = u(y)$, $\dot{y} = 0$. Also, let h be large enough so that near the plate at $y = h$, the flow is still a shearing flow as the fluid moves over the slot. Then for the Newtonian fluid, the velocity field is symmetrical about the centerline, which means that close to $x = 0$, the shear stress $T\langle xy\rangle$ is an even function of x and thus $\left.\dfrac{\partial T\langle yx\rangle}{\partial x}\right|_{x=0} = 0$. So, along $x = 0$, the equations of motion imply that the stress $T\langle yy\rangle$ = constant. Now near $y = h$ on the centerline

38. The conclusion is the content of a theorem in [1973 : 2].
39. The remainder of this section follows TANNER and PIPKIN [1969 : 23].

$x = 0$, $T\langle yy\rangle = -p_1$, the hydrostatic pressure term, and since $\dot{y} = 0$ along this centerline, we have that $T\langle yy\rangle = -p_1$ along the centerline. For the second-order fluid, from (41.21) we have

$$T\langle yy\rangle\Big|_{\substack{y=h\\x=0}} = \Big[-p_1 - \frac{\gamma}{\eta_0}\frac{dp_1}{dx}u(y) - \frac{1}{2}\Big((2\beta + 3\gamma)\Big(\frac{du}{dy}\Big)^2 + (\beta + 2\gamma)\Big(\frac{du}{dy}\Big)^2\Big]_{\substack{y=h\\x=0}}$$

$$= \Big[-p_1 - \frac{\gamma}{\eta^0}\frac{dp_1}{dx}u(y) + \frac{1}{2}\gamma\Big(\frac{du}{dy}\Big)^2\Big]_{\substack{y=h\\x=0}}. \quad (41.39)$$

Since at $y = -\infty$, the stress $T\langle yy\rangle|_{x=0} = -p_1$, it follows that the difference, called *pressure error* p_e, is given by

$$p_e = T\langle yy\rangle\Big|_{\substack{y=h\\x=0}} - T\langle yy\rangle\Big|_{\substack{x=0\\y=-\infty}}$$

$$= \Big[\frac{1}{2}\gamma\Big(\frac{du}{dy}\Big)^2 - \frac{\gamma}{\eta_0}\frac{dp_1}{dx}u(y)\Big]_{\substack{x=0\\y=h}}. \quad (41.40)$$

The crucial steps in the argument are:

(i) the center-line $x = 0$ is a line of symmetry for the creeping flow, so that on this line the velocity field is still in the x-direction only and is a function of y;

(ii) for the Newtonian fluid this means that $T\langle yy\rangle = -p_1$ is a constant along the line $x = 0$;

(iii) for the second-order fluid, $T\langle yy\rangle$ is given by (41.39) at $y = h$ on $x = 0$ and by $-p_1$ at $y = -\infty$, on $x = 0$.

In simple shear, the far field is $u(y) = \varkappa y$, and $dp_1/dx = 0$. Hence, in the far field $T\langle yy\rangle\Big|_{y=0}$ is the same as $T\langle yy\rangle\Big|_{\substack{x=0\\y=h}}$. Thus the measurement of $T\langle yy\rangle\Big|_{x=0}$ at the bottom of the slot differs from the far field $T\langle yy\rangle\Big|_{\substack{x=0\\y=0}}$ by the amount

$$p_e = \frac{1}{2}\gamma\varkappa^2. \quad (41.41)$$

In a second-order fluid, the first normal stress difference $\bar{N}_1(\varkappa^2) = -2\gamma\varkappa^2$ (Sec. 35). Hence

$$p_e = -\frac{1}{4}\bar{N}_1(\varkappa^2), \quad (41.42)$$

in simple shear. This result is independent of the slot width[40].

In plane POISEUILLE flow across a channel of width $2h$, the Newtonian velocity field is

$$u(y) = U[1 - (y/h)^2]. \quad (41.43)$$

40. TANNER and PIPKIN [1969 : 23], p. 481.

If a slot, sufficiently narrow such that the flow at $y = h$ is not disturbed, is introduced at $y = -h$, then the stress $T\langle yy\rangle\Big|_{\substack{x=0\\y=h}}$ is given by (41.39). Hence

$$p_e = \frac{1}{2}\gamma\left(\frac{du}{dy}\right)^2\Big|_{\substack{y=h\\x=0}} = 2\gamma\left(\frac{U}{h}\right)^2 = -\frac{1}{4}N_1\left(\left(\frac{U}{h}\right)^2\right). \tag{41.44}$$

Though $dp_1/dx \neq 0$ for this flow, $u(y)|_{y=h} = 0$ and thus (41.44) is the exact result. Since the flow down an inclined plane can be thought of as one-half of a plane POISEUILLE flow, the results (41.42) or (41.44) hold there too.

The above theoretical results indicate that if one measures the pressure at the bottom of a slot, it is not equal to $T\langle yy\rangle$ on the base in the undisturbed flow. The consequences of these observations on experiments will be discussed in the next chapter.

42. Rectilinear Flow : Flow Along a Slot and in Pipes of Arbitrary Cross-Section

If the flow is truly rectilinear and steady, that is, $\dot{x}_1 = 0, \dot{x}_2 = 0, \dot{x}_3 = w(x_1, x_2)$, then the acceleration is zero. So the results of the previous section utilising the assumption of the neglect of inertia apply to rectilinear flows. Importantly, the rectilinear flow is viscometric and $w(x_1, x_2)$ is determined by the viscosity function alone in many fluids (*cf.* Sec. 30). In particular, the Newtonian (first-order) and second-order fluids have the same velocity distribution in rectilinear flow, since in the former $\bar{n}_2(\varkappa^2) = 0$ and in the latter, $\bar{n}_2(\varkappa^2) = \text{const}$, and the question of uniqueness does not arise here, for it has already been settled in Sec. 30. While the two velocity fields are identical, the pressure fields will differ. In this section, we shall follow KEARSLEY[41] and determine the pressure field in the second-order fluid[42] if the Newtonian pressure field p_1 is known.

Let $\mathbf{u} = w(x_1, x_2)\mathbf{k}$ be the velocity field. Then

$$u_{i,k}\, u_{k,i} = u_{j,k}\, u_{k,i} = 0. \tag{42.1}$$

If $\mathbf{u}$ is the Newtonian solution, then $\operatorname{div} \mathbf{u} = 0$ and

$$\eta_0 \Delta \mathbf{u} = \nabla p_1 = \text{const}. \tag{42.2}$$

For this velocity field $\mathbf{u}$, let us compute $\operatorname{div} \mathbf{A}^2$ with $\mathbf{A}$ being the first RIVLIN-ERICKSEN tensor computed by using $\mathbf{u}$. Then, on noting (42.1), we have

$$A_{ik}\, A_{kj} = u_{i,k}\, u_{j,k} + u_{k,i}\, u_{k,j} + u_{k,i}\, u_{j,k}. \tag{42.3}$$

On using $u_{i,i} = 0$, and $u_{i,kj}\, u_{k,j} = 0$ from (42.1), one obtains

$$(A_{ik}\, A_{kj})_{,j} = \frac{1}{2}(u_{k,j}\, u_{k,j})_{,i} + u_{k,i}\, u_{k,jj}. \tag{42.4}$$

41. [1970 : 15].
42. As remarked in Sec. 41, if $\nu_2 = 0$ then the stresses in a second-order SFCE fluid and a second-order fluid are identical, both being obtained from ν_1. For later use, we retain the parameter ε here.

Hence

$$\text{div } \mathbf{A}^2 = \frac{1}{2} \nabla[\text{tr}(\mathbf{L}\mathbf{L}^T)] + \frac{1}{\eta_0} \nabla(\mathbf{u} \cdot \nabla p_1), \quad \mathbf{L} = \nabla \mathbf{u}, \tag{42.5}$$

or div $\mathbf{A}^2$ is the gradient of a scalar. Substitution of (42.5) into (41.17) yields

$$p_2 = \frac{\gamma}{\eta_0} \frac{dp_1}{dt} + \frac{\gamma}{4} \text{tr } \mathbf{A}^2 + \\ + (\beta + \gamma)\left[\frac{1}{\eta_0} \mathbf{u} \cdot \nabla p_1 + \frac{1}{2} \text{tr}(\mathbf{L}\mathbf{L}^T)\right] + \text{const.} \tag{42.6}$$

Now consider the rectilinear flow of a Newtonian fluid along the z-axis,

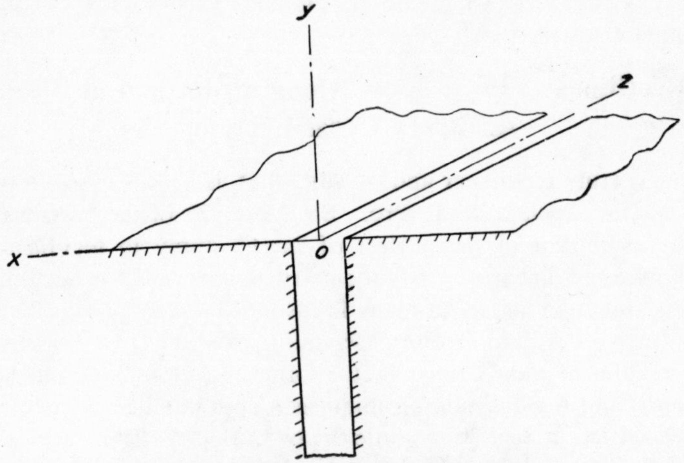

Fig. 42.1. Rectilinear flow along a slot.

with a slot along the flow direction (see Fig. 42.1). It is trivial to verify that for this problem of rectilinear flow,

$$(dp_1/dt) = (\partial p_1/\partial z)\, w = \mathbf{u} \cdot \nabla p_1. \tag{42.7}$$

Moreover, for the Newtonian fluid, the equations of motion imply that $T\langle yy \rangle = -\varepsilon\, p_1(z)$, a function of z everywhere, and this is also the stress that is measured at the bottom of the slot in the flow of a second-order fluid.

At the wall $y = 0$, for the second-order fluid,

$$T\langle yy \rangle\Big|_{y=0} - T\langle yy \rangle\Big|_{y=-\infty} = -\varepsilon^2 p_2\Big|_{y=0} + \varepsilon^2(\beta + 2\gamma)\left(\frac{\partial w}{\partial y}\right)^2\Big|_{y=0} \\ = -\varepsilon^2 \frac{(\beta + 2\gamma)}{2}\left[\left(\frac{\partial w}{\partial y}\right)^2 - \left(\frac{\partial w}{\partial x}\right)^2 + \frac{1}{\eta_0}\frac{dp}{dz}\, w\right]_{y=0}. \tag{42.8}$$

Since $w = 0$ at $y = 0$, the pressure error is

$$\varepsilon^2 p_e = -\varepsilon^2 \frac{\beta + 2\gamma}{2}\left[\left(\frac{\partial w}{\partial y}\right)^2 - \left(\frac{\partial w}{\partial x}\right)^2\right]_{y=0}. \tag{42.9}$$

Far away from the slot on $y = 0$, where the effects of the slot have vanished, p_e is given by

$$p_e = -\frac{\beta + 2\gamma}{2} \varkappa_w^2, \tag{42.10}$$

where $\varkappa_w$ is the wall rate of shear. From Sec. 35 we see that in second-order fluids

$$p_e = \frac{1}{2} \bar{N}_2(\varkappa^2). \tag{42.11}$$ [43]

Again, the above results indicate that the measurement of the pressure at the bottom of a slot does not give the normal stress, which would exist at the entrance to the slot, had the slot been absent. We shall need (42.11) in Sec. 48.

We now turn to the flow of a second-order fluid through a pipe of elliptic cross-section[44]. Introduce the dimensionless variables through

$$x = \frac{x_1}{L}, \quad y = \frac{x_2}{L}, \quad u = \frac{\eta_0}{L^2 G} w(x_1, x_2), \tag{42.12}$$

where L is a characteristic length and G is defined through

$$p_1 = -G x_3, \tag{42.13}$$

or G is the pressure drop per unit length of the pipe. Then (42.2) becomes

$$\Delta u = -1, \qquad \Delta = (\partial^2/\partial x^2) + (\partial^2/\partial y^2). \tag{42.14}$$

The boundary curve $\partial\Omega$ of an ellipse Ω with semi-axes L and hL can be written as

$$x^2 + (y/h)^2 = 1 \quad \text{on} \quad \partial\Omega. \tag{42.15}$$

Thus the solution of (42.14), obeying $u = 0$ on $\partial\Omega$ is

$$u = \frac{1 - x^2 - (y/h)^2}{2(1 + h^{-2})}. \tag{42.16}$$

The second-order fluid pressure $p = \varepsilon p_1 + \varepsilon^2 p_2$ is given by (42.13) and (42.6) as

$$p = -\varepsilon G x_3 + \varepsilon^2 \frac{\beta + 2\gamma}{2}\left[\left(\frac{\partial w}{\partial x_1}\right)^2 + \left(\frac{\partial w}{\partial x_2}\right)^2 + 2\frac{G}{\eta_0} w(x_1, x_2)\right], \tag{42.17}$$

and the extra stresses, over and above the Newtonian extra stresses $\varepsilon\eta_0 \mathbf{A}$, are

$$T_E\langle\mu\nu\rangle = \varepsilon^2(\beta + 2\gamma)\frac{\partial w}{\partial x_\mu}\frac{\partial w}{\partial x_\nu}, \quad \mu, \nu = 1, 2. \tag{42.18}$$

Combining the two equations above, the *non-dimensional total stresses* on a cross-section $x_3 =$ const. are given by

$$T\langle\mu\nu\rangle = \varepsilon^2\left(\frac{GL}{\eta_0}\right)^2 \frac{\beta + 2\gamma}{2}\left\{2u_{,\mu} u_{,\nu} - \left(u_{,\lambda} u_{,\lambda} - 2u\right)\delta_{\mu\nu}\right\}, \tag{42.19}$$

43. KEARSLEY [1970 : 15], Eq. (14).
44. PIPKIN and RIVLIN [1963 : 7].

where $\lambda, \mu, \nu = 1, 2$, and differentiation with respect to x is denoted by $\mu, \nu = 1$ and with respect to y by $\mu, \nu = 2$.

Now, if $\mathbf{n}$ is the unit external normal to $\partial\Omega$, then $T\langle\mu\nu\rangle\, n_\mu\, n_\nu = N$, the non-dimensional normal stress at that point. However, on $\partial\Omega$, $u,_\mu\, n_\mu = \partial u/\partial n$ is the normal derivative ; but $u = 0$ on $\partial\Omega$ implies that $|\,\partial u/\partial n\,| = |\nabla u|$, the magnitude of the gradient. Hence

$$N = \varepsilon^2 \left(\frac{GL}{\eta_0}\right)^2 \frac{\beta + 2\gamma}{2}\, \nabla u \cdot \nabla u. \tag{42.20}$$

For the elliptical pipe, using (42.16), one has that

$$\nabla u \cdot \nabla u = (x^2(h^2 - 1) + 1)/(h + h^{-1})^2. \tag{42.21}$$

Clearly, if we measure $N(\pm 1, 0)$ and $N(0, \pm h)$, we find the differences to be a maximum. Thus

$$\overline{\Delta N} = N(\pm 1, 0) - N(0, \pm h) \propto (h^2 - 1)/(h + h^{-1})^2. \tag{42.22}$$ [45]

Hence the measurement of the normal stresses at $(\pm 1, 0)$ and $(0, \pm h)$ yields a method of determining the coefficient $-(\beta + 2\gamma)$, which is the second normal stress difference for the second-order fluid.

PIPKIN and RIVLIN[46] have computed the velocity fields $u(x, y)$ for a tube of equilateral triangular cross-section and a rectangular cross-section. For the former, by representing the boundary $\partial\Omega$ through

$$y = 0, \quad y = \sqrt{3}\,x, \quad y = \sqrt{3}\,(1 - x), \tag{42.23}$$

they obtain

$$u(x, y) = \frac{\sqrt{3}}{6}\, y(y - \sqrt{3}\,x)\,[y + \sqrt{3}\,(x - 1)]. \tag{42.24}$$

For a rectangle with $x = \pm 1$, $y = \pm h$, they obtain

$$u(x, y) = \frac{1}{2}(1 - x^2) - \frac{2}{\pi^3} \sum_{n=0}^{\infty} \frac{(-1)^n}{\left(n + \frac{1}{2}\right)^3} \frac{\cosh\left(n + \frac{1}{2}\right)\pi y}{\cosh\left(n + \frac{1}{2}\right)\pi h} \times$$
$$\times \cos\left(n + \frac{1}{2}\right)\pi x. \tag{42.25}$$

From these, ∇u may be computed and shown to yield

$$|\nabla u|^2 = \frac{3}{4} x^2(1 - x^2) \quad \text{for the triangle on } y = 0. \tag{42.26}$$

This means that the maximum difference in N occurs between $N(0, 0)$ and

45. We draw the attention of the reader to the minor corrections to Eq. (4.4) in [1963 : 7].
46. [1963 : 7].

$N(\frac{1}{2}, 0)$ and this is proportional to 9/64. For the rectangle, on the boundary $y = h$,

$$|\nabla u|^2 = \frac{4}{\pi^4}\left\{\sum_{n=0}^{\infty} \frac{(-1)^n}{\left(n+\frac{1}{2}\right)^2} \tanh\left(n+\frac{1}{2}\right)\pi h \cos\left(n+\frac{1}{2}\right)\pi x\right\}^2. \tag{42.27}$$

Thus $\overline{\Delta N} = N(0, h) - N(\pm 1, h)$ is the maximum difference, and since $\cos(n + \frac{1}{2})\pi = 0$,

$$\overline{\Delta N}_{\max} \propto \frac{4}{\pi^4}\left\{\sum_{n=0}^{\infty} \frac{(-1)^n}{\left(n+\frac{1}{2}\right)^2} \tanh\left(n+\frac{1}{2}\right)\pi h\right\}. \tag{42.28}$$

If $h = 1$, that is, the rectangle is a square, then

$$\overline{\Delta N}_{\max} \propto 0.46. \tag{42.29}$$

Again this difference $\overline{\Delta N}$ yields $\overline{N}_2(\kappa^2)$ for the second-order fluid.

So far, we have concerned ourselves with fluids for which rectilinear flow is possible in a pipe of arbitrary cross-section Ω, and boundary $\partial\Omega$. Let us now examine whether a transverse flow component appears in a fluid obeying the third-order SFCE. For this fluid, the viscosity $\bar{\eta}(\kappa^2)$ varies with the rate of shear (*cf.* Sec. 35) and the second normal stress difference coefficient $\bar{n}_2(\kappa^2)$, which is a constant, is not proportional to $\bar{\eta}(\kappa^2)$. Thus, in the strict sense, rectilinear flow is not possible; however, we shall see that in the perturbation scheme we have developed, rectilinear flow is nevertheless possible.

Since the velocity field of a second-order solution in rectilinear flow is identical to the first-order solution, the velocity field of a third-order solution must have the form

$$\mathbf{v} = \varepsilon\mathbf{v}_1 + \varepsilon^3\mathbf{v}_3 = \varepsilon\mathbf{u} + \varepsilon^3\tilde{\mathbf{u}} \text{ (say)}, \tag{42.30}$$

where $\mathbf{u}$ obeys (42.2) and has the form $\mathbf{u} = w(x_1, x_2)\mathbf{k}$. Thus $\mathbf{u}$ is acceleration free, and though $\tilde{\mathbf{u}}$ may not turn out to be so, the acceleration field $\mathbf{L}\mathbf{v}$ is zero to $O(\varepsilon^3)$. Next, in the third-order SFCE (41.2)—(41.3), we note that

$$\mathbf{A} = \nabla\mathbf{u} + \nabla\mathbf{u}^T, \quad \mathbf{D} \equiv \mathbf{0}, \quad \mathbf{F} = \nabla\tilde{\mathbf{u}} + \nabla\tilde{\mathbf{u}}^T. \tag{42.31}$$

Hence $\mathbf{A}_1$ in (41.6) involves $\mathbf{u}$ and $\tilde{\mathbf{u}}$, but $\mathbf{A}_1^2$, $\mathbf{A}_2$, $\mathbf{A}_1\mathbf{A}_2$, and $\mathbf{A}_3$ do not involve $\tilde{\mathbf{u}}$ at all. Thus the equations of motion reduce to ($p = \varepsilon p_1 + \varepsilon^2 p_2 + \varepsilon^3 p_3$):

$$-\nabla p_3 + \eta_0 \Delta\tilde{\mathbf{u}} = \operatorname{div}\left[\alpha_0(\operatorname{tr}\mathbf{A}^2)\,\mathbf{A} + \alpha_1(\mathbf{AB} + \mathbf{BA}) + \alpha_2\,\mathbf{C}\right], \tag{42.32}$$

where $\mathbf{A}$, $\mathbf{B}$ and $\mathbf{C}$ are defined by (41.5), or they are the first three RIVLIN-ERICKSEN tensors based on $\mathbf{u}$, and since $\mathbf{u}$ is viscometric, $\mathbf{C} \equiv \mathbf{0}$. In deriving (42.32), it is noted that $\varepsilon p_1 + \varepsilon^2 p_2$ equilibrates the stress field arising from the second-order SFCE. Since the right side of (42.32) is a function of

(x_1, x_2) only, we shall assume

$$\tilde{\mathbf{u}} = \tilde{\mathbf{u}}(x_1, x_2), \quad p_3 = p_3(x_1, x_2). \tag{42.33}$$

Next, incompressibility demands that div $\tilde{u} = 0$, or $\tilde{u}_3 = \tilde{u}_3\ (x_1, x_2)$ and there exists a stream function $\tilde{\psi}(x_1, x_2)$ such that

$$\tilde{u}_1 = \partial\tilde{\psi}/\partial x_2, \quad \tilde{u}_2 = -\ \partial\tilde{\psi}/\partial x_1. \tag{42.34}$$

Using (42.33) and (42.34) in (42.32) leads to

$$\frac{\partial p_3}{\partial x_\alpha} = \eta_0\left(\frac{\partial^2 \tilde{u}_\alpha}{\partial x_1^2} + \frac{\partial^2 \tilde{u}_\alpha}{\partial x_2^2}\right), \quad \alpha = 1, 2, \tag{42.35}$$

$$\eta_0\left(\frac{\partial^2 \tilde{u}_3}{\partial x_1^2} + \frac{\partial^2 \tilde{u}_3}{\partial x_2^2}\right) = (\alpha_0 + \alpha_1)\left[-\left(\frac{G}{\eta_0}\right) \text{tr}\ \mathbf{A}^2 + \nabla(\text{tr}\ \mathbf{A}^2)\cdot\nabla w\right], \tag{42.36}$$

where we have used (42.13) to evaluate $\Delta\mathbf{u} = \Delta w(x_1, x_2)\ \mathbf{k}$. From $(42.35)_{1,2}$ it follows that the stream function $\tilde{\psi}(x_1, x_2)$ obeys $\Delta^2\ \tilde{\psi} = 0$ in Ω. However, $\tilde{u}_\alpha = 0$ on $\partial\Omega$ implies that $\tilde{\psi} =$ const. in Ω(*cf.* Sec. 41), and thus $\tilde{u}_\alpha = 0$ $(\alpha = 1, 2)$ in Ω. With this, $p_3 \equiv 0$ also. Thus the determination of the velocity field (42.30) is accomplished by solving the POISSON's equation (42.36) in Ω, with $\tilde{u}_3 = 0$ on $\partial\Omega$.

Using the above theory as the basis, RIVLIN[47] has shown that for the elliptical tube $b^2 x_1^2 + a^2 x_2^2 = a^2 b^2$, $\tilde{u}_3$ has the form

$$\tilde{u}_3 = R(a^2 b^2 - b^2 x_1^2 - a^2 x_2^2)\ (S x_1^2 + T x_2^2 + W), \tag{42.37}$$

where

$$\begin{aligned} R &= 16\ K^3(\alpha_0 + \alpha_1)/\eta_0, \\ S &= \frac{b^2}{12(a^4 + b^4 + 6a^2b^2)}\ [b^2(a^3 + 3b^2)\ (b^2 + 6a^2) - a^4(b^2 + 3a^2)], \\ T &= \frac{a^2}{12(a^4 + b^4 + 6a^2b^2)}\ [a^2(b^2 + 3a^2)\ (a^2 + 6b^2) - b^4(a^2 + 3b^2)], \\ W &= a^2b^3(S + T)/(a^2 + b^2), \end{aligned} \tag{42.38}$$

and we have used (42.16) in the *dimensional form* with the characteristic length $L = a$:

$$w(x_1, x_2) = K(a^2b^2 - b^2x_1^2 - a^2x_2^2),$$

$$K = \frac{G}{2\eta_0}\frac{1}{(a^2 + b^2)}. \tag{42.39}$$

We have seen therefore that not even a fluid obeying the third-order SFCE exhibits a transverse motion superposed on the axial motion. So we

47. [1962 : 9], Eqs. (3.13) — (3.14). See also LANGLOIS and RIVLIN [1963 : 6], Eq. (6.5).

now proceed to examine the velocity field for a fluid obeying a fourth-order SFCE. We take

$$\mathbf{v} = \varepsilon\, \mathbf{v}_1 + \varepsilon^3\, \mathbf{v}_3 + \varepsilon^4\, \mathbf{v}_4 = \varepsilon\, \mathbf{u} + \varepsilon^3\, \tilde{\mathbf{u}} + \varepsilon^4\, \bar{\mathbf{u}}. \tag{42.40}$$

Again, to $O(\varepsilon^4)$, the acceleration field is zero and thus the equations of motion do not involve any inertial terms. Using (42.40) in (41.5) and noting that terms involving ε^3 and less are equilibrated by the pressure field $\varepsilon p_1 + \varepsilon^2 p_2 + \varepsilon^3 p_3$, and assuming that the new velocity field $\bar{\mathbf{u}}$ and the extra pressure field p_4 obey

$$\bar{\mathbf{u}} = \bar{\mathbf{u}}(x_1, x_2), \quad p_4 = p_4(x_1, x_2), \tag{42.41}$$

one obtains that the z-component obeys

$$\Delta \bar{u}_3(x_1, x_2) = 0 \text{ in } \Omega, \quad \bar{u}_3 = 0 \text{ on } \partial\Omega, \tag{42.42}$$

which implies $\bar{u}_3 \equiv 0$ in Ω. Next, with the introduction of the stream function $\bar{\psi}$ for $\bar{u}_\alpha(x_1, x_2)$, $\alpha = 1, 2$, one has

$$\begin{aligned}
&\eta_0\, \Delta^2 \bar{\psi} + 2(\beta + 2\gamma) \left[\frac{\partial^2}{\partial x_1\, \partial x_2}\left(\frac{\partial w}{\partial x_1}\frac{\partial \tilde{u}_3}{\partial x_1} - \frac{\partial w}{\partial x_2}\frac{\partial \tilde{u}_3}{\partial x_2}\right)\right.\\
&\quad \left. + \frac{1}{2}\left(\frac{\partial^2}{\partial x_2^2} - \frac{\partial^2}{\partial x_1^2}\right)\left(\frac{\partial w}{\partial x_1}\frac{\partial \tilde{u}_3}{\partial x_2} + \frac{\partial w}{\partial x_2}\frac{\partial \tilde{u}_3}{\partial x_1}\right)\right]\\
&\quad + \bar{\alpha}\left[\frac{\partial^2}{\partial x_1\, \partial x_2}\left((\text{tr}\, \mathbf{A}^2)\left\{\left(\frac{\partial w}{\partial x_1}\right)^2 - \left(\frac{\partial w}{\partial x_2}\right)^2\right\}\right.\right.\\
&\quad \left.\left. + \left(\frac{\partial^2}{\partial x_2^2} - \frac{\partial^2}{\partial x_1^2}\right)\left((\text{tr}\, \mathbf{A}^2)\, \frac{\partial w}{\partial x_1}\frac{\partial w}{\partial x_2}\right)\right] = 0.
\end{aligned} \tag{42.43}$$

In (42.43)

$$\bar{\alpha} = \alpha_5 + 2\alpha_6 + 2\alpha_7 + 2\alpha_8. \tag{42.44}$$

Importantly, (42.43) is a non-homogeneous biharmonic equation for $\bar{\psi}$ with $(\partial\bar{\psi}/\partial x_\alpha) = 0$, $\alpha = 1, 2$, on the boundary.

For the elliptic tube $b^2 x_1^2 + a^2 x_2^2 = a^2 b^2$, it can be shown that (42.41) reduces to[48]

$$\eta_0\, \Delta^2\, \bar{\psi} = -\, 4\, H\, x_1\, x_2, \tag{42.45}$$

where H is a constant. The solution to (42.45), subject to the relevant boundary conditions, is

$$\bar{\psi} = -\, \frac{2H}{3\eta_0}\, \frac{(b^2 x_1^2 + a^2 x_2^2 - a^2 b^2)^2\, x_1 x_2}{(5a^4 + 6a^2\, b^2 + 5b^4)}, \tag{42.46}$$

with H explicitly given by

$$H = -\, \frac{3}{4}\left(\frac{G}{\eta_0}\right)^4 + \left[\frac{2(\alpha_0 + \alpha_1)}{\eta_0}\,(\beta + 2\gamma) - 4\,\bar{\alpha}\right]\frac{a^2 b^2 (a^2 - b^2)}{a^2 + b^2}\,. \tag{42.47}$$

48 LANGLOIS and RIVLIN [1963 : 6], Eq. 6.7.

The streamlines for the transverse flow field obtained through $\bar{\psi}$, when $H > 0$, are depicted in Fig. 42.2. If $H < 0$, the directions of the arrows are reversed.

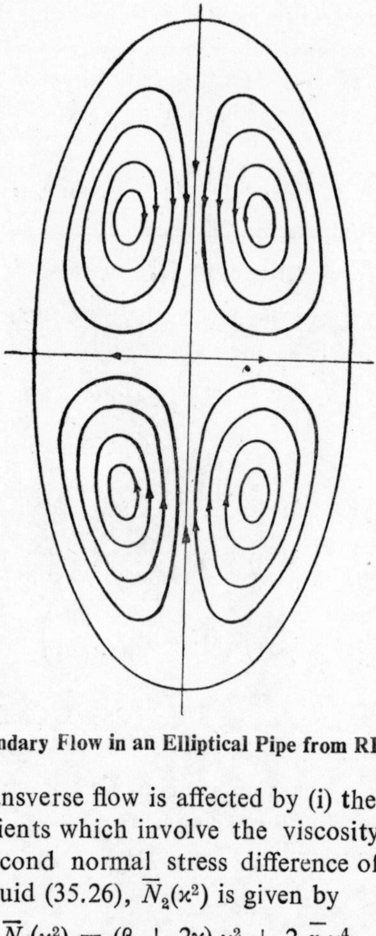

Fig. 42.2. Secondary Flow in an Elliptical Pipe from RIVLIN [1962 : 9].

We see that the transverse flow is affected by (i) the non-circularity of the tube and (ii) by coefficients which involve the viscosity parameters and the components of the second normal stress difference of the *fourth-order fluid.* For the fourth-order fluid (35.26), $\overline{N}_2(\varkappa^2)$ is given by

$$\overline{N}_2(\varkappa^2) = (\beta + 2\gamma)\,\varkappa^2 + 2\,\bar{\alpha}\,\varkappa^4, \tag{42.48}$$

while its viscosity is

$$\bar{\eta}(\varkappa^2) = \eta_0 + 2(\alpha_0 + \alpha_1)\,\varkappa^2\,. \tag{42.49}$$

We see from (42.47) that a fluid with a constant viscosity η_0 and $\overline{N}_2(\varkappa^2) = \bar{\alpha}\,\varkappa^4$ exhibits a secondary flow. However, this fluid has the same pipe efflux formula (28.36) as the Newtonian fluid and is therefore indistinguishable from it in that flow. We now offer a few additional comments:

(i) One can identify the perturbation parameter ε in the pipe flow problem with G, the pressure gradient, quite readily since each $\mathbf{v}_i$, $i = 1, 2, 3, 4$ is of order G^i.

(ii) For the *third-order fluid*, $\bar{n}_2(\varkappa^2)$ is given by

$$\bar{n}_2(\varkappa^2) = \alpha\,\bar{\eta}(\varkappa^2) - 2\,\alpha(\alpha_0 + \alpha_1)\,\varkappa^2, \qquad \alpha = (\beta + 2\gamma)/\eta_0, \tag{42.48}$$

while for the *fourth-order fluid*

$$\bar{n}_2(\varkappa^2) = \alpha\,\bar{\eta}(\varkappa^2) + 2[\,\bar{\alpha} - \alpha(\alpha_0 + \alpha_1)]\,\varkappa^2, \quad \alpha = (\beta + 2\gamma)/\eta_0. \tag{42.49}$$

So the two fluids have similar 'out of proportionality' factors as far as $\bar{n}_2(\cdot)$ and $\bar{\eta}(\cdot)$ are concerned. Yet in the perturbation scheme one exhibits transverse flow and not the other. While it is true that both fluids must exhibit transverse flow at the general level, the perturbation scheme conceals it. Therefore, one ought to be cautious in interpreting statements like 'it is the out of proportionality factor that causes transverse flows.'

43. Free Surface Flow. Convexity of the Free Surface

In this section, a study of the free surface flow of a viscoelastic fluid down an inclined trough will be made. The body forces due to gravity are included, but inertial forces turn out to be zero to the first order of approximation.

Let α be the angle of inclination of the trough with the horizontal (see Fig. 43.1), and let α be so small that $\sin\alpha \approx \alpha$. So, to ε we assign $\varepsilon = \sin\alpha$,

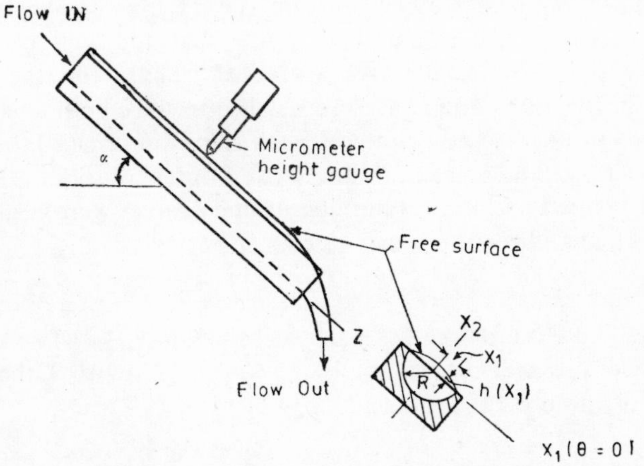

Fig. 43.1. Sketch of semicircular channel used in experiments.

and seek solutions to the equations of motion for the slow flow of a viscoelastic fluid, correct to $O(\varepsilon)$.

From the pioneering researches of STOKES it is known that if an incompressible Newtonian fluid flows down the trough, one can find the velocity field by assuming the trough to be replaced by a pipe with a cross-section which is composed of the given cross-section and its mirror image in the plane $x_2 = 0$. For a fluid obeying the first-order SFCE, the pressure field is given by

$$\varepsilon\,p_1 = -x_2\,\rho g \cos\alpha, \tag{43.1}$$

with the velocity field given by $\varepsilon\, \mathbf{v}_1 = w\mathbf{k}$ and

$$\eta_0\, \Delta\, \bar{w}(x_1, x_2) = -\rho\, g, \quad w = \bar{w} \sin \alpha. \tag{43.2}$$

In deriving (43.1) and (43.2), the body force is assumed to be given by $\rho\mathbf{b} = \{0, -\rho g \cos \alpha, \rho g \sin \alpha\}$. Now, $\bar{w}$ obeys the condition $\bar{w} = 0$ on the wall of the trough, and for the Newtonian fluid the free surface condition, demanding the absence of any surface traction on $x_2 = 0$, is satisfied if

$$\partial \bar{w}/\partial x_2 = 0 \quad \text{on} \quad x_2 = 0. \tag{43.3}$$

For, one may take $p_1 = 0$ on $x_2 = 0$, as indeed we have done in (43.1), and if (43.3) is valid, then $T\langle ij\rangle\, n_j = 0$ is satisfied on $x_2 = 0$, since the unit normal $\mathbf{n}$ has the components $\mathbf{n} = \{0, 1, 0\}$. Also, $\mathbf{v}_1 = \bar{w}\mathbf{k}$ obeys $\mathbf{v}_1 \cdot \mathbf{n} = 0$ trivially, which is the other free surface condition.

If ε is quite small, then so is the driving force $\rho g \sin \alpha = \varepsilon\rho g$. So, to $0(\varepsilon)$, (43.2) and (43.3) constitute the problem for the determination of the axial velocity for the slow flow of a viscoelastic fluid. From Sec. 42, we know that in rectilinear pipe flow, the first and second-order solutions are identical; also, the third-order solution has *no transverse component* in rectilinear pipe flow. This latter result is not true for free surface problems, since the transverse component stream function $\tilde{\psi}(x_1, x_2)$, while obeying $\Delta^2\, \tilde{\psi} = 0$ in the cross-section of the trough and $\nabla\, \tilde{\psi} = \mathbf{0}$ at the wall of the trough, does not obey $\nabla\, \tilde{\psi} = \mathbf{0}$ on the free surface.[49] Thus, for the third-order approximation the flow down a trough will introduce both axial and transverse additions to the velocity field with each addition of $O(\varepsilon^3)$.

But here, let us concentrate on the second-order solution. The formula (42.6) cannot be used because in deriving it, body forces were neglected. So, let us recall (41.9) and rewrite it as:

$$\Delta\psi_1 = \eta_0 \operatorname{div} \mathbf{A}, \tag{43.4}$$

where $\mathbf{A}$ is the first RIVLIN-ERICKSEN tensor based on $\mathbf{v}_1 = \bar{w}(x_1, x_2)\mathbf{k}$ and ψ_1 is the modified pressure, given by $\varepsilon\psi_1 = \varepsilon p_1 + \rho\chi$, with χ the body force potential. For the present problem,

$$\chi = g x_2 \cos \alpha - g\, x_3 \sin \alpha. \tag{43.5}$$

For the second-order approximation the equation analogous to (41.10) is

$$\Delta\psi_2 = \beta \operatorname{div} \mathbf{A}^2 + \gamma \operatorname{div} \mathbf{B}, \tag{43.6}$$

where $\mathbf{A}^2$ and $\mathbf{B}$ have the usual meaning. Since the body force is not altered, $\psi_2 = p_2$. Retracing the steps from (41.12) – (41.17), one can now see that $\mathbf{v}_1 = \bar{w}\mathbf{k}$ is also the second-order solution, if

$$\nabla \left[p_2 - \dot{\gamma}\left(\frac{1}{\eta_0} \frac{d\psi_1}{dt} + \frac{1}{2}\, \zeta^2 \right) \right] = (\beta + \gamma) \operatorname{div} \mathbf{A}^2 \,. \tag{43.7}$$

49. It does obey $\partial\tilde{\psi}/\partial x_1 = 0$ on $x_2 = 0$.

From (42.5), one has that div $\mathbf{A}^2$ is the gradient of a scalar:

$$\operatorname{div} \mathbf{A}^2 = \frac{1}{2} \nabla[\operatorname{tr} \mathbf{L}\,\mathbf{L}^T] + \frac{1}{\eta_0} \nabla(\mathbf{v}_1 \cdot \nabla\varphi_1). \tag{43.8}$$

Hence, on using $(d\psi_1/dt) = \mathbf{v}_1 \cdot \nabla\psi_1$, one derives that $\bar{w}\mathbf{k}$ is the solution and

$$p_2 = \frac{\beta + 2\gamma}{2\eta_0} \mathbf{v}_1 \cdot \nabla \varphi_1 + \frac{\beta + \gamma}{2} \operatorname{tr}(\mathbf{L}\,\mathbf{L}^T) + \frac{\gamma}{4} \operatorname{tr} \mathbf{A}^2 + C_0, \tag{43.9}$$

where C_0 is a constant. Using $\mathbf{v}_1 = \bar{w}\mathbf{k}$ in (43.9) yields

$$p_2 = C_0 + \frac{\beta + 2\gamma}{2}\left[\left(\frac{\partial \bar{w}}{\partial x_1}\right)^2 + \left(\frac{\partial \bar{w}}{\partial x_2}\right)^2 - \frac{1}{\eta_0} \rho g \bar{w}\right]. \tag{43.10}$$

If the free surface is to remain flat as $x_2 = 0$, then (43.3) is to be satisfied there. Though the second-order SFCE exhibits non-zero stresses for all the nine components of $T\langle ij \rangle$, there exists only the pressure $p_2|_{x_2=0}$ on $x_2 = 0$ if (43.3) is satisfied. Now, consider the consequence of not supplying this pressure. The free surface will not remain at $x_2 = 0$. It will be described by a function $x_2 = h(x_1)$. The new velocity field will still obey (43.2), but not the boundary condition (43.3). This will be replaced by the vanishing of the stress vector on $x_2 = h(x_1)$. As a first approximation, to find $h(x_1)$, one should put (this is another place where the domain perturbation technique to be presented in Sec. 44 can be applied and was indeed done by WINEMAN and PIPKIN though they did not explicitly use this terminology):

$$\rho g\, x_2 = \rho g h(x_1) = \varepsilon^2\, p_2 \Big|_{x_2=0}. \tag{43.11}$$

Then a simple computation shows that

$$\mathbf{v} = \varepsilon \bar{w}\mathbf{k} + \mathbf{O}(\varepsilon^3), \tag{43.12}$$

where $\bar{w}$ obeys (43.2) and (43.3), or $\varepsilon\bar{w}\mathbf{k}$ is the second-order solution once again[50]. In other words, corrections are of third order in ε. One could therefore interpret (43.11) to imply that the liquid acts as its own pressure gague[51], and changes its surface elevation to equilibrate it. In (43.12), it must also be noted that the third-order terms will include axial perturbations and transverse components, as remarked earlier.

As an example consider the case of a semi-elliptical trough with a cross-section described by $a^2b^2 - b^2x_1^2 - a^2x_2^2 = 0$, $x_2 < 0$. The velocity field $\bar{w}(x_1, x_2)$ is given by [*cf.* (42.39)]:

$$\bar{w}(x_1, x_2) = [\rho g/2\eta_0(a^2 + b^2)]\,(a^2b^2 - b^2x_1^2 - a^2x_2^2), \qquad x_2 < 0. \tag{43.13}$$

Substituting (43.13) into (43.10) and using (43.12) one gets

$$h(x_1) = \frac{(\beta + 2\gamma)\, b^2\sin^2\alpha}{\eta_0^2(a^2 + b^2)^2} [(2b^2 + a^2)\, x_1^2 - a^4 - a^2b^2] + C_1, \tag{43.14}$$

50. In [1966 : 20), WINEMAN and PIPKIN substantiate this in a detailed manner.
51. TANNER [1970: 22].

where C_1 is a constant, which may be fixed by assuming $h(\pm a) = 0$. In any case, one finds that

$$h(0) - h(\pm a) = - \frac{\beta + 2\gamma}{\eta_0^2} \sin^2\alpha \frac{(2b^2 + a^2)\, a^2 b^2}{(a^2 + b^2)^2}. \tag{43.15}$$

For a semi-circle, $h(0) - h(\pm a)$ is proportional to $3a^2/4$, while for a deep slot $(b \to \infty)$ it is proportional to 2. If the surface is convex, that is, $h(0) - (h \pm a) > 0$, then from (43.15) it follows that $(\beta + 2\gamma) < 0$. Now $(\beta + 2\gamma) = \bar{n}_2(\varkappa^2)$ for the second-order fluid. Hence the slow flow down a trough provides an excellent way of measuring the sign of $\bar{N}_2(\varkappa^2)$, the second normal stress difference, at low values of $\varkappa$.

TANNER[52] has presented an analysis for the flow of an incompressible simple fluid down a semi-circular trough under some restrictive assumptions, which are however well motivated experimentally. A summary of this method for the determination of $\bar{N}_2(\varkappa^2)$ will now be given. The assumptions used in this analysis based on a cylindrical polar coordinate system (r, θ, z), are (see Fig. 43.1) :

(i) the velocity field is given by $\mathbf{v} = w(r)\,\mathbf{k} + \mathbf{O}(\varepsilon)$, where

$$\varepsilon = N_2(\varkappa^2)/\rho g R \cos\alpha \ll 1 ;$$

(ii) to calculate $w(r)$, one solves the equations of motion for the POISEUILLE flow down a circular pipe under the gravitational force $\rho g \sin\alpha$ in the z-direction :

(iii) from this $w(r)$, the pressure field p is calculated ;

(iv) though this pressure is not available at the free surface, and the free surface will no longer be described by $x_2 = 0$, but by $x_2 = h(x_1)$, it will be assumed that the flow is rectilinear and that $p=0$ on $x_2=h(x_1)$.

To facilitate the analysis, let the constitutive equation for the fluid in the viscometric flow $w(r)\mathbf{k}$ be given by

$$\mathbf{T}_E = \bar{\eta}(\varkappa^2)\, \mathbf{A}_1 + [\bar{n}_1(\varkappa^2) + \bar{n}_2(\varkappa^2)]\, \mathbf{A}_1^2 - \frac{1}{2} \bar{n}_1(\varkappa^2)\, \mathbf{A}_2, \tag{43.16}$$

where $\varkappa = dw/dr$. The equations of motion now yield

$$T\langle rz \rangle = \bar{\eta}(\varkappa^2) \frac{dw}{dr} = - \frac{1}{2} \rho g r \sin\alpha, \tag{43.17}$$

from which $w(r)$ is determined. Also, one obtains

$$p = C - \rho g r \cos\alpha \sin\theta + \bar{N}_2(\varkappa^2) + \int_0^r \bar{N}_2[\varkappa(\xi)^2] \frac{d\xi}{\xi}, \tag{43.18}$$

where it is noted that (43.16) implies $T_E\langle\theta\theta\rangle = 0$, $T_E\langle rr \rangle = \bar{N}_2(\varkappa^2)$. To find the constant C, let it be assumed that $p = 0$ at $r = 0$. Then $C = 0$.

52. [1970 : 22].

Now, as per the assumption (iv), $p = 0$ on $h(x_1) = r \sin \theta$. Hence (43.18) becomes

$$\rho g h(x_1) \cos \alpha = \overline{N}_2(\varkappa^2) + \int_0^{x_1} \overline{N}_2 \left[\left(\frac{dw}{dx} \right)^2 \right] \frac{dx}{x}, \qquad (43.19)$$

where we have put $x_1 \doteq r$. We shall now solve (43.20) for $\overline{N}_2(\varkappa^2)$. From (43.17), the shear stress τ is given by

$$\tau = \frac{1}{2} \rho g x_1 \sin \alpha = \overline{\eta}(\varkappa^2) \varkappa. \qquad (43.20)$$

Let the liquid height $h(R) = h_w$. Then (43.20) may be written as

$$\frac{d}{d\tau_w} (\rho g\, h_w \cos \alpha) = \frac{d\tilde{N}_2}{d\tau_w} + \frac{\tilde{N}_2}{\tau_w}, \qquad (43.21)$$

where it has been assumed that the inversion $\varkappa = \varkappa(\tau_w)$ is possible and hence $N_2 = \tilde{N}_2(\tau_w)$ and that τ_w is the wall shear stress, obtained by replacing x_1 by R in (43.20). Intergrating (43.20), one obtains

$$\tilde{N}_2 = \rho g\, h_w \cos\alpha - \frac{1}{\tau_w} \int_0^{\tau_w} \rho g h_w \cos\alpha \, d\tau. \qquad (43.22)$$

Since $\tau_w = \frac{1}{2} \rho g R \sin \alpha$, one may keep ρ, g and R fixed, but vary α, and obtain a set of results of h_w vs α. Then

$$\tilde{N}_2\left(\frac{1}{2} \rho g\, R \sin\alpha \right) = \rho g \left\{ h_w(\alpha) \cos\alpha - \right.$$
$$\left. - (\sin\alpha)^{-1} \int_0^{\alpha} h_w(\varphi) \cos\varphi \, d(\sin\varphi) \right\}. \qquad (43.23)$$

The above method describes an approximate way of obtaining $\overline{N}_2(\varkappa^2)$; it becomes exact at low values of α, but then at low values of α, the more exact analysis of WINEMEN and PIPKIN should be used. However, it has been found that (43.23) yields experimentally acceptable results (see Sec. 48).

44. Free Surface in Couette Flow. The Domain Perturbation Method[53]

One of the intriguing problems in the flow of a non-Newtonian fluid between two concentric cylinders in relative rotation is the determination of the shape and the parameters which determine the shape of the free surface. For example, OLDROYD[54], ERICKSEN[55], LODGE[56] as well as COLEMAN,

53. This section is based on the work of JOSEPH and FOSDICK [1973 : 8] and JOSEPH, *et al.* [1973 : 7], who have extended the earlier method of JOSEPH [1967 : 15] to apply to this problem. See Chap. II, Sec. 12 of JOSEPH and FOSDICK ; and Secs. 15-16 of the next paper.
54. [1958 : 4]. See Sec. 37 in Chap. 8 above.
55. [1959 : 3].
56. [1964 : 14].

et al.[57] have argued that the rod climbing effect (WEISSENBERG effect) noticed in polymeric fluids is entirely due to the force in the axial direction in the fluid over and above that of the atmosphere. Indeed, SERRIN[58] employed this idea and computed the shape of the free surface, basing it on the parameters of the REINER-RIVLIN fluid. For general fluids, it is believed by these investigators that the rod climbing effect is qualitatively related to the normal stress functions of the fluid. It is therefore of interest to see if one could establish rigorously that this is indeed so, quantitatively as well.

Recently, JOSEPH and FOSDICK determined the shape of the free surface for a fourth-order fluid, and showed that the viscometric functions (not just the normal stresses) determine the free surface. They have investigated the effects of surface tension as well, found that these are vital and shown how to incorporate these into the calculations. In this section, we present this work in detail since it can be applied in the examination of the free surface shape in the cone-and-plate device as well as the die swell ratio in slow flows of Newtonian and non-Newtonian liquids. The die swell problem has received the attention of JOSEPH[59] (see Sec. 50) more recently, and in Sec. 50 we shall present this work in more detail.

For the flow between two cylinders in relative rotation, the equations of motion are given by $(A6.3)-(A6.4)$. The velocity field obeys div $\mathbf{v} = 0$ in $\mathcal{V}_\Omega$,

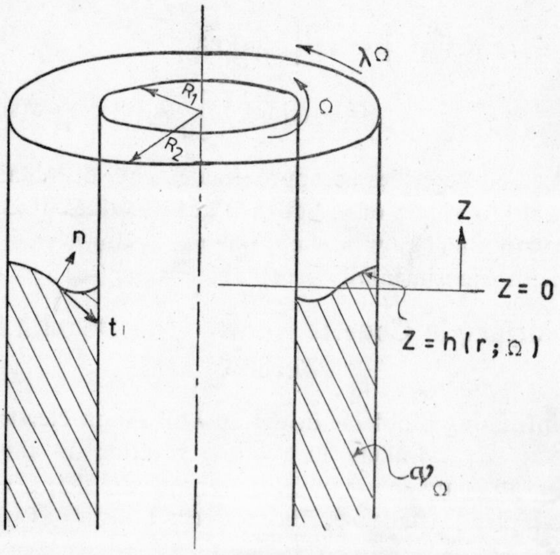

Fig. 44.1 The free surface between rotating cylinders.

the domain of the flow (see Fig. 44.1). It will be assumed that all quantities

57. [1966 : 4],
58. [1959 : 7].
59. [1973 : 6], [1974 : 2].

are independent of θ and hence the modified pressure φ is :

$$\varphi(r, z) = p(r, z) + \rho g z. \tag{44.1}$$

The velocity field obeys

$$\mathbf{v} = \begin{cases} \Omega\, R_1\, \mathbf{e}_\theta, & r = R_1, \\ \lambda\, \Omega\, R_2\, \mathbf{e}_\theta, & r = R_2. \end{cases} \tag{44.2}$$

On the free surface of the fluid, described by $z = h(r\,;\,\Omega)$, the following conditions hold :

$$\mathbf{v} \cdot \mathbf{n} = T\langle n\theta \rangle = T\langle nz \rangle = 0, \tag{44.3}$$

implying the vanishing of the normal component of the velocity as well as the shear stresses. Moreover, as $z \to -\infty$, the flow must become more like the COUETTE flow, or

$$\mathbf{v} \cdot \mathbf{e}_z,\ \ T\langle \theta z \rangle,\ \ T\langle rz \rangle \to 0 \qquad \text{as} \quad z \to -\infty. \tag{44.4}$$

Let σ be the surface tension and J the mean curvature of the free surface. Then the jump in the normal stress across the free surface is balanced by the surface tension. Or,

$$\mathbf{n} \cdot \| \mathbf{T} \| \, \mathbf{n} = \sigma\, J = \frac{\sigma}{r} (rh'/\sqrt{1 + h'^2})', \tag{44.5}$$

where $\| \mathbf{T} \|$ is the jump of the stress tensor across $h(r\,;\,\Omega)$, and $h' = dh/dr$. If the outside pressure is atmospheric ($= p_a$), then

$$\mathbf{n} \cdot \| \mathbf{T} \| \, \mathbf{n} = -\varphi + p_a + T_E\langle nn \rangle + \rho g h(r\,;\,\Omega). \tag{44.6}$$

The modified pressure φ being determined up to an additive constant by the equations of motion, we fix this constant by requiring that

$$\int_{R_1}^{R_2} h(r\,;\,\Omega)\, r\, dr = 0, \tag{44.7}$$

which expresses the condition that the total volume of the fluid below the free surface is conserved. In terms of the angle α between $\mathbf{e}_z$ and $\mathbf{n}[-(\pi/2) \leqq \alpha \leqq (\pi/2)]$, one has

$$\sin\alpha = h'/\sqrt{1 + h'^2}. \tag{44.8}$$

Let $\{u, v, w\}$ be the physical components of the velocity vector $\mathbf{v}$. Then on $z = h(r\,;\,\Omega)$,

$$\begin{aligned} \mathbf{v} \cdot \mathbf{n} &= w \cos\alpha - u \sin\alpha = 0, \\ T\langle n\theta \rangle &= T\langle \theta z \rangle \cos\alpha - T\langle r\theta \rangle \sin\alpha = 0, \\ T\langle nt \rangle &= \sin\alpha \cos\alpha\, (T_E\langle zz \rangle - T_E\langle rr \rangle) + (\cos^2\alpha - \sin^2\alpha)\, T\langle rz \rangle = 0, \\ T_E\langle nn \rangle &= T_E\langle zz \rangle \cos^2\alpha + T_E\langle rr \rangle \sin^2\alpha - \sin 2\alpha\, T\langle rz \rangle. \end{aligned} \tag{44.9}$$

Using $(44.9)_3$, one eliminates $T_E \langle rr \rangle$ to obtain

$$T_E \langle nn \rangle = T_E \langle zz \rangle - \tan\alpha\, T\langle rz \rangle, \quad \text{on} \quad z = h(r\,;\,\Omega). \tag{44.10}$$

We shall collect all the above equations together to state the problem as :

$$\left.\begin{aligned} \rho \mathbf{a} &= -\nabla\varphi + \operatorname{div} \mathbf{T}_E, \\ \operatorname{div} \mathbf{v} &= 0; \end{aligned}\right\} \text{ in } \mathcal{V}_\Omega, \qquad (44.11)_{1,2}$$

$$\mathbf{v} = \begin{cases} \Omega R_1 \mathbf{e}_\theta, & r = R_1, \\ \lambda\Omega R_2 \mathbf{e}_\theta, & r = R_2; \end{cases} \qquad (44.11)_{3,4}$$

$$\begin{aligned} w - uh' &= T\langle \theta z \rangle - h'\, T\langle r\theta \rangle = h'(T_E \langle zz \rangle - T_E \langle rr \rangle) + \\ &+ (1 - h'^2)\, T\langle rz \rangle = 0 \quad \text{on} \quad z = h(r\,;\,\Omega); \qquad (44.11)_{5,6,7} \end{aligned}$$

$$w = T\langle \theta z \rangle = T\langle rz \rangle = 0 \quad \text{as} \quad z \to -\infty\,; \qquad (44.11)_{8,9,10}$$

$$p_a - \varphi + T_E \langle zz \rangle - h'\, T\langle rz \rangle + \rho g h = \frac{\sigma}{r}\,(rh'/\sqrt{1 + h'^2})'$$

$$\text{on} \quad z = h(r\,;\,\Omega); \qquad (44.11)_{11}$$

$$\int_{R_1}^{R_2} h(r\,;\,\Omega) r\, dr = 0. \qquad (44.11)_{12}$$

44.1. Zero Surface Tension. Let us consider the solution of the system (44.11) with $\sigma = 0$, and λ given on the outer cylinder. We shall use Ω as the parameter which defines the perturbed domain $\mathcal{V}_\Omega$. By the latter, we mean that the domain $\mathcal{V}_\Omega$ is related to the domain

$$\mathcal{V}_0 = \{(r_0, \theta_0, z_0)\,;\; R_1 \leqq r_0 \leqq R_2\,;\; 0 \leqq \theta_0 \leqq 2\pi\,; -\infty < z_0 \leqq h^{[0]}(r_0\,;\,0)\}$$

through

$$r = r_0, \quad \theta = \theta_0, \quad z = f(r_0, z_0\,;\,\Omega), \tag{44.12}$$

where $(r_0, \theta_0, z_0) \in \mathcal{V}_0$, $(r, \theta, z) \in \mathcal{V}_\Omega$. We call $f(\cdot)$ the *domain deformation mapping*. We do not need to know $f(\,\cdot\,)$ in $\mathcal{V}_\Omega$ in detail ; however, we must impose on it the following obvious restrictions :

$$\begin{aligned} f(r_0, z_0\,;\,0) &= z_0, \\ &\qquad -\infty < f(r_0, z_0\,;\,\Omega) \leqq h(r\,;\,\Omega), \\ f(r_0, 0\,;\,\Omega) &= h(r\,;\,\Omega). \end{aligned} \tag{44.13}$$

Because of the domain deformation (44.12), the velocity of the deformation mapping $\mathbf{v}_f$ is given by

$$\mathbf{v}_f = v_\Omega\, \mathbf{e}_z \equiv \frac{df}{d\Omega}\, \mathbf{e}_z, \tag{44.14}$$

from which follows the *substantial derivative*

$$\frac{d(\,\cdot\,)}{d\Omega} = \frac{\partial}{\partial\Omega}(\,\cdot\,) + v_\Omega \frac{\partial(\,\cdot\,)}{\partial z}. \tag{44.15}$$

We shall assume that the velocity field $\mathbf{v}$, pressure field φ and the free surface elevation h are expressible as :

$$\begin{Bmatrix} \mathbf{v} \\ \varphi \\ h \end{Bmatrix} = \sum_{n=0}^{\infty} \frac{1}{n!} \begin{Bmatrix} \mathbf{v}^{[n]} \\ \varphi^{[n]} \\ h^{[n]} \end{Bmatrix} \Omega^n, \tag{44.16}$$ [69]

where, for example,

$$\mathbf{v}^{[n]}(r_0, z_0) = \left.\frac{d^n\mathbf{v}}{d\Omega^n}\right|_{\Omega=0}. \tag{44.17}$$

Since the final solution is sought in $\mathcal{V}_\Omega$, rather than $\mathcal{V}_0$, the variables (r_0, z_0) must be replaced by (r, z) via the inverse of (44.12) — (44.13). But this does not mean that the final answer will depend on $f(\cdot)$. Since the latter is not contained in the original problem, its introduction is one of convenience. So in the final summation of (44.16), the function $f(\cdot)$ must disappear ; however, a general proof of this and the convergence of the series is lacking at present.

In the system (44.11), all equations are identities in Ω, or they are true for all values of Ω. Hence they may be substantially differentiated with respect to Ω [*cf.* (44.15)] as many times as desired. There are also some equations, e.g., the equations of motion, the continuity equation and the velocity boundary conditions on $r = R_1$ and $r = R_2$, which are valid for all z also. Hence these can be partially differentiated with respect to z as many times as desired. For example, since div $\mathbf{v} = 0$, one can prove that

$$\frac{\partial}{\partial\Omega^n}\ (\text{div } \mathbf{v}) = 0 \quad \text{on} \quad \mathcal{V}_\Omega. \tag{44.18}$$

For, if

$$\frac{\partial^{n-1}}{\partial\Omega^{n-1}}\ \text{div } \mathbf{v} = \text{div } \frac{\partial^{n-1}\mathbf{v}}{\partial\Omega^{n-1}} = 0 \tag{44.19}$$

is true for $1, 2, \ldots, n-1$, then

$$\begin{aligned} \frac{d}{d\Omega}\ \text{div } \frac{\partial^{n-1}\mathbf{v}}{\partial\Omega^{n-1}} &= \text{div } \frac{\partial^n\mathbf{v}}{\partial\Omega^n} + \mathcal{V}_\Omega \frac{\partial}{\partial z}\left(\text{div } \frac{\partial^{n-1}\mathbf{v}}{\partial\Omega^{n-1}}\right) \\ &= \text{div } \frac{\partial^n\mathbf{v}}{\partial\Omega^n} = 0 \quad \text{in} \quad \mathcal{V}_\Omega. \end{aligned} \tag{44.20}$$

Since (44.19) is true for $n = 1$, it is thus true for all n.

However, the free surface equations in (44.11) are true for only $z \in h(r\,;\Omega)$ and hence they are not identities in z and cannot obey (44.18). But the substantial derivative operation may be used on $h(r\,;\Omega)$, and because of $(44.13)_2$ and $(44.16)_3$, $df/d\Omega = dh/d\Omega$ on $z = h(r\,;\Omega)$. To facilitate the

60. In this section, $\mathbf{v}^{[n]}$, $\varphi^{[n]}$, $h^{[n]}$ is the n-th order solution.

calculations, we introduce the notation:

$$(\cdot)^{(n)} = \frac{\partial^n}{\partial\Omega^n}(\cdot)\Big|_{\Omega=0}, \tag{44.21}$$

which is to be contrasted with (44.17).

Now, if $\Omega = 0$, then clearly the difference between $(\cdot)^{[0]}$ and $(\cdot)^{(0)}$ vanishes and

$$\mathbf{v}^{[0]} = \mathbf{0}, \quad h^{[0]} = 0, \quad \mathbf{T}_E^{[0]} = \mathbf{0}, \quad \varphi^{[0]} = p_a. \tag{44.22}$$

In other words, the fluid at rest has a flat surface in the absence of surface tension, and hence the region $\mathcal{V}_0 = \{(r_0, \theta_0, z_0)\,;\, R_1 \leqq r_0 \leqq R_2\,;\, 0 \leqq \theta_0 \leqq 2\pi\,;\, -\infty < z_0 \leqq 0\}$. From now on, we shall write (r, θ, z) for $(r_0, \theta_0, z_0) \in \mathcal{V}_0$ since all solutions will be found in $\mathcal{V}_0$ only.

To find the higher-order solutions, let us write

$$\mathbf{v} = v(r, z)\,\mathbf{e}_\theta + \mathbf{q}, \tag{44.23}$$

and with $\nabla_2 = (\partial/\partial r)\,\mathbf{e}_r + (\partial/\partial z)\,\mathbf{e}_z$,

$$\mathbf{q} = u(r, z)\,\mathbf{e}_r + w(r, z)\mathbf{e}_z = \mathbf{e}_\theta \times (1/r)\,\nabla_2\psi, \tag{44.24}$$

where ψ is the stream function for axisymmetric flows. It is easy to show that the acceleration field may be written as

$$\mathbf{a} = \left(\nabla_2 v \cdot \mathbf{q} + \frac{uv}{r}\right)\mathbf{e}_r - \frac{v^2}{r}\,\mathbf{e}_r + (\nabla_2\,\mathbf{q})\,\mathbf{q}. \tag{44.25}$$

Now, write the extra stress tensor $\mathbf{T}_E$ as

$$\mathbf{T}_E = S\,\mathbf{e}_\theta \otimes \mathbf{e}_\theta + \mathbf{t} \otimes \mathbf{e}_\theta + \mathbf{e}_\theta \otimes \mathbf{t} + \boldsymbol{\pi}, \tag{44.26}$$

where $\mathbf{t}$ is a vector normal to $\mathbf{e}_\theta$:

$$t\langle i\rangle = \{T\langle r\theta\rangle, T\langle \theta z\rangle\} = \{t_r, t_z\}, \tag{44.27}$$

and $\boldsymbol{\pi}$ is the two-dimensional stress tensor

$$[\pi\langle ij\rangle] = \begin{bmatrix} \pi\langle rr\rangle & \pi\langle rz\rangle \\ \cdot & \pi\langle zz\rangle \end{bmatrix} = \begin{bmatrix} T_E\langle rr\rangle & T\langle rz\rangle \\ \cdot & T_E\langle zz\rangle \end{bmatrix}. \tag{44.28}$$

Hence

$$\operatorname{div}\mathbf{T}_E = -\frac{S}{r}\,\mathbf{e}_r + \left(\nabla_2 \cdot \mathbf{t} + \frac{2}{r}\,t_r\right)\mathbf{e}_\theta + \operatorname{Div}\boldsymbol{\pi}, \tag{44.29}$$

where $\operatorname{Div}\boldsymbol{\pi}$ is normal to $\mathbf{e}_\theta$ and the equations of motion become

$$\nabla_2 \cdot \mathbf{t} + \frac{2}{r}\,t_r = \rho\left(\nabla_2 v \cdot \mathbf{q} + \frac{uv}{r}\right), \tag{44.30}$$

$$-\nabla_2\varphi - \frac{1}{r}\,S\,\mathbf{e}_r + \operatorname{Div}\boldsymbol{\pi} = \rho\left(-\frac{v^2}{r}\,\mathbf{e}_r + (\nabla_2\,\mathbf{q})\,\mathbf{q}\right).$$

The boundary conditions of (44.11) may be rewritten as

$$v = \begin{cases} \Omega R_1, & r = R_1, \\ \lambda\,\Omega R_2, & r = R_2\,; \end{cases} \qquad \psi = \frac{\partial \psi}{\partial r} = 0 \quad \text{at} \quad r = R_1, R_2\,;$$

$$\frac{\partial \psi}{\partial r} + h' \frac{\partial \psi}{\partial z} = t_z - h'\, t_r = h'(\pi\langle zz\rangle - \pi\langle rr\rangle) + (1 - h'^2)\, \pi\langle rz\rangle$$
$$= p_a - \varphi + \pi\langle zz\rangle - h'\, \pi\langle rz\rangle + \rho g h = 0$$
$$\text{on} \quad z = h(r\,;\Omega)\,; \qquad (44.31)$$

$$\partial\psi/\partial r = \pi\langle rz\rangle = t_z = 0 \quad \text{as} \quad z \to -\infty\,; \qquad \int_{R_1}^{R_2} r h(r\,;\Omega)\, dr = 0.$$

Assuming that $\mathbf{t}$, $\boldsymbol{\pi}$, $\mathbf{v}$, ψ, S, φ and $\mathbf{a}$ are all differentiable with respect to Ω and that they have TAYLOR series representations in Ω as in (44.16), we perform the following operations :

(i) those equations valid in $\mathcal{V}_\Omega$, because of (44.18), will be partially differentiated with respect to Ω, then evaluated at $\Omega = 0$ and written in terms of $\mathbf{t}^{(1)}$, $\mathbf{t}^{(2)}$, $\mathbf{t}^{(3)}$, $\mathbf{t}^{(4)}$, etc. ; because of this these equations are now valid in $\mathcal{V}_0$;

(ii) those conditions which apply on the free surface will be subjected to substantial differention, and use will be made of the fact (already mentioned) that on $z = h(r\,;\Omega)$, $v_\Omega = dh/d\Omega$; again the equations will be valid in $\mathcal{V}_0$;

For the convenience of the reader, we note that on the free surface $z = h(r\,;\Omega)$,

$$(\cdot)^{[0]} = (\cdot)^{(0)}, \qquad (44.32)$$
$$(\cdot)^{[1]} = (\cdot)^{(1)} + h^{[1]} \left\{\frac{\partial(\cdot)}{\partial z}\right\}^{(0)},$$
$$(\cdot)^{[2]} = (\cdot)^{(2)} + 2h^{[1]} \left\{\frac{\partial(\cdot)}{\partial z}\right\}^{(1)} + (h^{[1]})^2 \left\{\frac{\partial^2(\cdot)}{\partial z^2}\right\}^{(0)} + h^{[2]} \left\{\frac{\partial(\cdot)}{\partial z}\right\}^{(0)},$$

with similarly computed derivatives for higher orders.

To obtain the first-order solution, we compute $(44.30)_{1,2}{}^{(1)}$, and $(44.31)_{1,2,3,4}{}^{(1)}$. Secondly, because of (44.22) and $(44.32)_2$, the substantial derivatives of $(44.31)_{5-12}$ are cxpressed in terms of $(\cdot)^{(1)}$ except for $h^{[1]}$. Or, we have[61]

$$\nabla_2 \cdot \mathbf{t}^{(1)} + \frac{2}{r}\, t_r^{(1)} = \rho \left(\nabla_2 v \cdot \mathbf{q} + \frac{uv}{r}\right)^{(1)} = 0 \quad \text{in} \quad \mathcal{V}_0; \qquad (44.33)_1$$

61. Since $v^{(0)} = u^{(0)} = 0$, $\mathbf{q}^{(0)} = 0$, $[\nabla_2 v \cdot \mathbf{q} + (uv/r)]^{(1)} = 0$. Similarly, $(v^2)^{(1)} = 0$, and $[(\nabla_2\, \mathbf{q})\, \mathbf{q}]^{(1)} = 0$. Hence one obtains $(44.33)_{1-3}$.

$$-\nabla_2\varphi^{(1)} - \frac{1}{r} S^{(1)} \mathbf{e}_r + \operatorname{Div} \boldsymbol{\pi}^{(1)} = \rho\left(-\frac{v^2}{r}\mathbf{e}_r + (\nabla_2 \mathbf{q})\,\mathbf{q}\right)^{(1)} = \mathbf{0}$$

$$\text{in } \mathcal{V}_0; \qquad (44.33)_{2,3}$$

$$v^{(1)} = R_1 \text{ at } r = R_1; \qquad v^{(1)} = \lambda R_2 \text{ at } r = R_2; \qquad (44.33)_{4,5}$$

$$\psi^{(1)} = \frac{\partial \psi^{(1)}}{\partial r} = 0 \text{ at } r = R_1, R_2; \qquad (44.33)_{6,7,8,9}$$

$$\frac{\partial \psi^{(1)}}{\partial r} = t_z^{(1)} = \pi\langle rz\rangle^{(1)} = -\pi\langle zz\rangle^{(1)} + \varphi^{(1)} - \rho g h^{[1]} = 0$$

$$\text{at } z = 0; \qquad (44.33)_{10,11,12,13}$$

$$\frac{\partial \psi^{(1)}}{\partial r} = t_z^{(1)} = \pi\,\langle rz\rangle^{(1)} = 0 \text{ at } z = -\infty; \qquad (44.33)_{14,15,16}$$

$$\int_{R_1}^{R_2} r h^{[1]}\, dr = 0. \qquad (44.33)_{17}$$

One has to exercise care in computing the extra stresses. To avoid confusion with the methods of Secs. 41- 43, we shall write $\mathbf{T}_E$ as

$$\mathbf{T}_E = \mathbf{S}_1 + \mathbf{S}_2 + \mathbf{S}_3 + \mathbf{S}_4, \qquad (44.34)$$

with $\mathbf{S}_1$, $\mathbf{S}_2$, $\mathbf{S}_3$, $\mathbf{S}_4$ being given by the right sides of $(35.9)_{1,2,3,4}$ respectively. In other words, (44.34) means that the problem will always be considered for a fourth-order fluid. However, the process of approximation leads to the derivatives $\mathbf{A}_2^{(1)}$, $\mathbf{A}_3^{(1)}$ and $\mathbf{A}_4^{(1)}$ being zero for the first approximation, with similar results for higher approximations. This facilitates the computations in the sequel.

Now using the current notation,

$$\mathbf{T}_E^{(n)} = \mathbf{S}_1^{(n)} + \mathbf{S}_2^{(n)} + \mathbf{S}_3^{(n)} + \mathbf{S}_4^{(n)}, \qquad n = 1, 2, 3, 4, \qquad (44.35)$$

where we repeat that $(\cdot)^{(n)} = \left.\dfrac{\partial^n}{\partial\Omega^n}(\cdot)\right|_{\Omega=0}$. Next, because $\mathbf{v}^{(0)} \equiv \mathbf{0}$, only $\mathbf{A}_1^{(1)} \neq \mathbf{0}$, while $(\mathbf{A}_1^2)^{(1)}$, etc. are all zero. Thus $\mathbf{T}_E^{(1)} = \eta_0 \mathbf{A}_1^{(1)}$, with

$$\mathbf{A}_1^{(1)} = \nabla\mathbf{v}^{(1)} + \nabla\mathbf{v}^{(1)T},$$

or

$$S^{(1)} = \eta_0 \frac{2u^{(1)}}{r}, \qquad \mathbf{t}^{(1)} = \eta_0\left(\nabla_2 v^{(1)} - \frac{v^{(1)}}{r}\mathbf{e}_r\right), \qquad (44.36)$$
$$\boldsymbol{\pi}^{(1)} = \eta_0(\nabla_2\mathbf{q}^{(1)} + \nabla_2\mathbf{q}^{(1)T}), \qquad \mathbf{q}^{(1)} = \mathbf{e}_\theta \times \frac{1}{r}\nabla_2\psi^{(1)}.$$

Substitution of this set into (44.33) yields the field equation for $v^{(1)}$ as :

$$\nabla_2 \cdot \nabla_2 v^{(1)} + \frac{1}{r}\frac{\partial v^{(1)}}{\partial r} - \frac{v^{(1)}}{r^2} = 0 \text{ in } \mathcal{V}_0; \qquad (44.37)$$

$$v^{(1)} = \begin{cases} R_1 & \text{at} \quad r = R_1, \\ \lambda R_2 & \text{at} \quad r = R_2\,; \end{cases}$$

$$t_z^{(1)} = \eta_0 \frac{\partial v^{(1)}}{\partial z} = 0 \quad \text{at} \quad z = 0\,; \qquad t_z^{(1)} = \eta_0 \frac{\partial v^{(1)}}{\partial z} \to 0 \quad \text{as} \quad z \to -\infty.$$

Hence the solution is

$$v^{(1)} = Ar + \frac{B}{r} \quad \text{in} \quad \mathcal{V}_0,$$

$$A = \frac{\lambda R_2^2 - R_1^2}{R_2^2 - R_1^2}, \qquad B = \frac{R_1^2 R_2^2 (1-\lambda)}{R_2^2 - R_1^2}. \tag{44.38}$$

To obtain the differential equation for $\psi^{(1)}$, we note that a solution for $\psi^{(1)}$ will not exist unless $\text{Div}\, \boldsymbol{\pi}^{(1)} - \frac{1}{r} S^{(1)}\, \mathbf{e}_r$ is irrotational ; or, we must have

$$\nabla_2 \cdot \left\{ \mathbf{e}_\theta \times \left(\text{Div}\, \boldsymbol{\pi}^{(1)} - \frac{1}{r} S^{(1)}\, \mathbf{e}_r \right) \right\} = 0 \quad \text{in} \quad \mathcal{V}_0. \tag{44.39}$$

Using (44.36) in (44.39) yields

$$\frac{\eta_0}{r} \mathcal{L}^2 \psi^{(1)} = 0 \quad \text{in} \quad \mathcal{V}_0, \tag{44.40}$$

$$\mathcal{L}(\cdot) \equiv \left(\nabla_2 \cdot \nabla_2 - \frac{1}{r}\frac{\partial}{\partial r} \right)(\cdot) \equiv \left(\frac{\partial^2}{\partial r^2} + \frac{\partial^2}{\partial z^2} - \frac{1}{r}\frac{\partial}{\partial r} \right)(\cdot).$$

Using the boundary conditions on $\psi^{(1)}$, we have

$$\psi^{(1)} = 0, \quad u^{(1)} = w^{(1)} = 0, \qquad \mathbf{q}^{(1)} = \mathbf{0} \quad \text{in} \quad \mathcal{V}_0\,. \tag{44.41}$$

Hence $\psi^{(1)}$ is a constant :

$$\psi^{(1)} = C_1 \quad \text{in} \quad \mathcal{V}_0\,. \tag{44.42}$$

Since $\boldsymbol{\pi}^{(1)} = \mathbf{0}$, $\pi \langle zz \rangle^{(1)} = 0$ and thus from $(44.33)_{13}$

$$\rho\, g h^{[1]} = C_1 \quad \text{at} \quad z = 0, \tag{44.43}$$

which because of $(44.33)_{17}$ becomes

$$h^{[1]} = C_1 = 0. \tag{44.44}$$

Now (44.38), (44.41), (44.42) and (44.44) furnish the complete first-order solution.

For the second-order solution, the set (44.33) is unaltered except that

(i) all superscripts (1) are replaced by (2), with [1] changing to [2]; and (ii) the equation for $\varphi^{(2)}$ becomes

$$-\nabla\varphi^{(2)} - \frac{1}{r} S^{(2)}\, \mathbf{e}_r + \text{Div}\, \boldsymbol{\pi}^{(2)} = \rho \left(-\frac{v^2}{r} \mathbf{e}_r + (\nabla_2 \mathbf{q})\, \mathbf{q} \right)^{(2)}$$

$$= -2\rho \frac{\{v^{(1)}\}^2}{r} \mathbf{e}_r \quad \text{in} \quad \mathcal{V}_0. \tag{44.45}$$

The stress system, because of the differential operations involved, becomes $\mathbf{T}_E^{(2)} = \mathbf{S}_1^{(2)} + \mathbf{S}_2^{(2)}$, where

$$\mathbf{S}_1^{(2)} = \eta_0 \mathbf{A}_1^{(2)} = \eta_0(\nabla \mathbf{v}^{(2)} + \nabla \mathbf{v}^{(2)T}), \tag{44.46}$$

$$\mathbf{S}_2^{(2)} = \gamma \mathbf{A}_2^{(2)} + 2\beta(\mathbf{A}_1^{(1)})^2, \tag{44.47}$$

with

$$\mathbf{A}_2^{(2)} = 2[\{\nabla \mathbf{A}_1^{(1)}\}\, \mathbf{v}^{(1)} + \mathbf{A}_1^{(1)}\, \nabla \mathbf{v}^{(1)} + \nabla \mathbf{v}^{(1)T}\, \mathbf{A}_1^{(1)}]. \tag{44.48}$$

It can now be shown that

$$v^{(2)} = 0 \quad \text{in} \quad \mathcal{V}_0, \tag{44.49}$$

and that $\psi^{(2)}$ obeys

$$(\eta_0/r)\, \mathcal{L}^2 \psi^{(2)} = 0 \quad \text{in} \quad \mathcal{V}_0; \tag{44.50}$$

which yields $\psi^{(2)} = 0$ in $\mathcal{V}_0$ which means that $\mathbf{q}^{(2)} = 0$ in $\mathcal{V}_0$. Now, $\varphi^{(2)}$ obeys

$$\nabla \varphi^{(2)} + 16B^2\,(2\beta + 3\gamma)\, \frac{1}{r^5}\, \mathbf{e}_r = 2\rho\, \frac{(v^{(1)})^2}{r}\, \mathbf{e}_r \quad \text{in} \quad \mathcal{V}_0, \tag{44.51}$$

or

$$\varphi^{(2)} = \rho\left(A^2 r^2 + 4AB \log r - \frac{B^2}{r^2}\right) + \frac{4B^2(2\beta + 3\gamma)}{r^4} + C_2 \text{ in } \mathcal{V}_0. \tag{44.52}$$

This constant C_2 is fixed by

$$\int_{R_1}^{R_2} r\, \psi^{(2)}\, dr = 0. \tag{44.53}$$

Hence

$$h = \frac{1}{2}\, \Omega^2 h^{[2]} + O(\Omega^4), \tag{44.54}$$

$$\mathbf{v} = \Omega v^{(1)}\, \mathbf{e}_\theta + O(\Omega^3),$$

$$\varphi = p_a + \frac{1}{2}\, \Omega^2 \varphi^{[2]} + O(\Omega^4).$$

We note that $v^{[1]} = v^{(1)}$, and $\psi^{[2]} = \psi^{(2)}$ here; and when (44.53) is used, it is found that

$$h^{[2]} = \frac{1}{\rho g}\, \varphi^{[2]} = \frac{1}{g}\left\{A^2\left(r^2 - \frac{R_1^2 + R_2^2}{2}\right) + 4AB\left(\log r + \frac{1}{2} - \right.\right.$$
$$\left.\left. - \frac{R_2^2 \log R_2 - R_1^2 \log R_1}{R_2^2 - R_1^2}\right) - B^2\left(\frac{1}{r^2} - \frac{2 \log (R_2/R_1)}{R_2^2 - R_1^2}\right)\right\} +$$
$$+ \frac{4B^2}{\rho g}\,(2\beta + 3\gamma)\left(\frac{1}{r^4} - \frac{1}{R_2^2 R_1^2}\right). \tag{44.55}$$

For a purely Newtonian fluid, the above formula is valid, except that $\beta = \gamma = 0$. Thus the height discrepancy between a simple fluid (h_s) and a Newtonian fluid (h_N) is

$$\delta h = h_N - h_s = -\frac{2\Omega^2 B^2}{\rho g}(2\beta + 3\gamma)\left(\frac{1}{r^4} - \frac{1}{R_1^2 R_2^2}\right) + O(\Omega^4). \tag{44.56}$$

Thus if $2\beta + 3\gamma > 0$, the height discrepancy is negative at the inner cylinder, or the simple fluid has a tendency to climb the inner cylinder. Another interesting consequence is the slope of the free surface:

$$\frac{dh}{dr} = \frac{B^2\Omega^2}{\rho g r^5}\left\{\rho r^2\left(1 + \frac{Ar^2}{B}\right)^2 - 8(2\beta + 3\gamma)\right\}. \tag{44.57}$$ [62]

If R_1 is very small,

$$\left.\frac{dh}{dr}\right|_{r=R_1} = -\frac{8\Omega^2(1-\lambda)^2}{\rho g R_1}(2\beta + 3\gamma) \quad \text{as} \quad R_1 \to 0. \tag{44.58}$$

Hence for sufficiently small R_1, if $2\beta + 3\gamma > 0$, the fluid will slope sharply up the inner cylinder. Also, as $R_1 \to 0$, one has

$$\lim_{R_1\to 0} h(R_1; \Omega) = -\frac{\lambda^2 R_2^2 \Omega^2}{4g} + \frac{2\Omega^2(1-\lambda)^2}{\rho g}(2\beta + 3\gamma). \tag{44.59}$$

This remarkable result states that in the limit of vanishing radius of the inner cylinder, the height of the free surface is the sum of the standard paraboloidal depression together with a rise due to $2\beta + 3\gamma > 0$. If the outer cylinder is fixed, or $\lambda = 0$, the measured rise in height should yield the constant $2\beta + 3\gamma$ directly as cylinders of successively smaller radii are used.

In the manner described here, higher-order solutions may be obtained. However, we shall only list here the conclusions due to JOSEPH and FOSDICK:

(i) at fourth order, the azimuthal velocity is

$$\begin{aligned}
v &= \Omega v^{(1)} + \frac{1}{6}\Omega^3 v^{(3)} + O(\Omega^5),\\
v^{(3)} &= v_P^{(3)} + v_H^{(3)},\\
v_P^{(3)} &= -\frac{16B^3}{\eta_0}(\alpha_0 + \alpha_1)\left\{\frac{1}{r^5} + \frac{(R_1^2 + R_2^2)\,r}{R_1^4 R_2^4} - \frac{(R_1^2 + R_2^2)^2 - R_1^2 R_2^2}{R_1^2 R_2^2\, r}\right\},\\
v_H^{(3)} &= \sum_{n=1}^{\infty} B_n e^{\lambda_n z}\, C(\lambda_n r),
\end{aligned} \tag{44.60}$$

62. In [1959 : 7], SERRIN derived this result for the REINER-RIVLIN fluid where $\gamma = 0$.

where $\mathcal{C}(\lambda_n r)$ are the cylindrical functions :

$$\mathcal{C}(\lambda_n r) = J_1(\lambda_n R_1)\, Y_1(\lambda_n r) - J_1(\lambda_n r)\, Y_1(\lambda_n R_1), \tag{44.61}$$

λ_n are the positive roots of

$$\mathcal{C}(\lambda_n R_2) = 0, \tag{44.62}$$

and the constants B_n are chosen such that

$$v^{(3)} = 0 \quad \text{at} \quad r = R_1, R_2. \tag{44.63}$$

(ii) The stream function $\psi^{(4)}$ is determined from (note $\psi^{(3)} = 0$) :

$$\eta_0 \mathcal{L}^2 \psi^{(4)} = -8 \frac{\partial}{\partial z} \left\{ \frac{4B}{r^2} (\beta + \gamma) \left(\frac{\partial v^{(3)}}{\partial r} - \frac{v^{(3)}}{r} \right) + \rho v^{(1)} v^{(3)} \right\} \text{ in } \mathcal{V}_0, \tag{44.64}$$

$$\psi^{(4)} = \frac{\partial \psi^{(4)}}{\partial r} = 0 \quad \text{at} \quad r = R_1, R_2 :$$

$$\frac{\partial \psi^{(4)}}{\partial r} = \frac{\partial^2 \psi^{(4)}}{\partial r^2} - \frac{\partial^2 \psi^{(4)}}{\partial z^2} = 0 \quad \text{at} \quad z = 0, -\infty.$$

(iii) Since $v^{(1)} - v^{(4)}$ have been determined ($v^{(4)} = 0$) and $\psi^{(1)} - \psi^{(4)}$ are also known (note $\psi^{(1)} = \psi^{(3)} = 0$), one can find $\varphi^{(4)}$ from

$$-\nabla_2 \varphi^{(4)} - \frac{1}{r} S^{(4)} \mathbf{e}_r + \operatorname{Div} \boldsymbol{\pi}^{(4)} = -8\rho \frac{v^{(1)} v^{(3)}}{r} \mathbf{e}_r. \tag{44.65}$$

In deriving this, use has been made of the fact that $v^{(2)} = v^{(4)} = 0$ in $\mathcal{V}_0$. From symmetry we realize that $\psi^{(3)} = 0$, $\varphi^{(3)} = 0$, $h^{[3]} = 0$ also.

(iv) So, finally one can find $h^{[4]}$ from

$$\frac{2\eta_0}{r} \frac{\partial^2 \psi^{(4)}}{\partial r\, \partial z} + \varphi^{(4)} - \rho g h^{[4]} = 0 \quad \text{at} \quad z = 0, \tag{44.66}$$

with $\varphi^{(4)}$ obeying

$$\int_{R_1}^{R_2} r \varphi^{(4)} = 0. \tag{44.67}$$

Now, as (44.64) shows, the function $\psi^{(4)}$ depends on $(\beta + \gamma)$ and η_0, the second-order fluid constants, which are of course the viscometric functions for this fluid. A direct calculation shows that

$$S^{(4)} = \eta_0 \frac{2u^{(4)}}{r} - \frac{16}{r^2} B\beta \left(\frac{\partial v^{(3)}}{\partial r} + \frac{v^{(3)}}{r} \right) + \frac{768}{r^8} B^4 \alpha_5, \tag{44.68}$$

$$\begin{aligned} \boldsymbol{\pi}^{(4)} = {} & \eta_0 (\nabla_2 \mathbf{q}^{(4)} + \nabla_2 \mathbf{q}^{(4)T}) - \frac{8}{r^2} B(\beta + 2\gamma) \times \\ & \times \left(\mathbf{e}_r \otimes \nabla_2 v^{(3)} + \nabla_2 v^{(3)} \otimes \mathbf{e}_r - \frac{2v^{(3)}}{r} \mathbf{e}_r \otimes \mathbf{e}_r \right) + \\ & + \frac{1536 B^4}{r^8} \left(\frac{1}{2} \alpha_5 + \alpha_6 + \alpha_7 + \alpha_8 \right). \end{aligned} \tag{44.69}$$

Hence $\phi^{(4)}$ is determined by η_0, $(\beta + \gamma)$, $(\alpha_0 + \alpha_1)$, β, α_5, $(\beta + 2\gamma)$ as well as $\frac{1}{2}(\alpha_5 + 2\alpha_6 + 2\alpha_7 + 2\alpha_8)$. All these quantities occur in the viscometric functions of the fourth-order fluid (see Sec. 35). Hence $h^{[4]}$ is determined by viscometric functions only. In other words, for the fourth-order fluid, the *normal stress functions do not solely determine the free surface height.* Whether such is the case for the general simple fluid is not known however.

Finally, all of the above solutions have been expressed in terms of $\mathcal{V}_0$ and hence Eq. (44.12) must be used to invert $\mathcal{V}_0 \to \mathcal{V}_\Omega$. If one is interested in the free surface only, knowing $r = r_0$, $z = f(r_0, z_0; \Omega)$ can be written as $z = f(r, 0; \Omega) = h(r; \Omega)$ quite readily.

42.2. Effects of Surface Tension. Once the surface tension enters the picture, the problem becomes enormously complicated because the equations of motion and the free surface boundary conditions [*cf.* (44.11)] form a simultaneous set of differential equations governing $\mathbf{v}$ and $h(r; \Omega)$. In addition, the wetting angle at the solid-liquid-gas contact point under dynamic conditions is not well understood.[63] However, one may assume the wetting angle conditions relevant to the static situation and proceed.[64]

Let the tangent vector at the wall to the free surface be drawn such that it points into $\mathcal{V}_\Omega$. Let $\mathbf{n}$ be the unit normal to $h(r; \Omega)$ at this point, away from the fluid. The wetting angle χ is defined as the angle between the wall and the tangent vector, measured in the direction of the normal (see Fig. 44.1). We shall now assume that

$$h'(R_1; \Omega) = \cot \chi, \qquad h'(R_2; \Omega) = \cot \chi_2. \tag{44.70}$$

The relevant equation is [*cf.* (44.11)] :

$$p_a - \varphi + T_E\langle zz\rangle - h'\, T\langle rz\rangle + \rho g h = \frac{\sigma}{r}(rh'/\sqrt{1 + h'^2})'. \tag{44.71}$$

In the rest state, that is, with $\Omega = 0$, the stresses $T_E\langle zz\rangle$ and $T\langle rz\rangle$ vanish and $p_a - \varphi$ reduces to a constant C. The constant volume condition [*cf.* (44.11)] now shows that this constant C is given by

$$C = \sigma\left[rh_s'/\sqrt{1 + h_s'^2}\right]_{R_1}^{R_2}, \tag{44.72}$$

where h_s is the static rise.

Now suppose that either (i) $\sigma = 0$, or (ii) $h'(R_2; 0) = h'(R_1; 0) = 0$, the latter being called *neutral wetting conditions*, since $\chi_1 = \chi_2 = \pi/2$. Then $C = 0$. If $\sigma = 0$, of course $h \equiv 0$ trivially. If $\sigma \neq 0$, then (44.71) can be multiplied by rh_s to obtain

$$\rho g \int_{R_1}^{R_2} rh_s^2\, dr + \sigma \int_{R_1}^{R_2} \frac{rh_s'^2\, dr}{(1 + h_s'^2)^{1/2}} - \sigma\left[\frac{r\, h_s\, h_s'}{(1 + h_s'^2)^{1/2}}\right]_{R_1}^{R_2} = 0, \tag{44.73}$$

63. See HUH and SCRIVEN [1971 : 11] for a discussion of the contact angle and related references.

64. [1973 : 7], [1973 : 8].

which implies that $h_s \equiv 0$ under neutral wetting conditions.

Instead of considering the fluid to be contained between two cylinders, let us consider a rod in an infinite sea of liquid. Then, in this infinite domain we shall demand that

$$(\mathbf{v}, \mathbf{T}_E, h) \to 0 \qquad \text{as} \quad r \to \infty \tag{44.74}$$

along with their derivatives. These conditions replace $(44.11)_{4,12}$. Since we are now concerned with surface tension effects, assume that $(\mathbf{v}, \varphi, h)$ have a double TAYLOR series expension in Ω and ε, the latter being defined through

$$\varepsilon = -\tan\alpha = -h'(R_1), \tag{44.75}$$

where R_1 is the radius of the rod. The series are :

$$\begin{Bmatrix} \mathbf{v} \\ \varphi \\ h \end{Bmatrix} = \sum_{n=0}^{\infty} \sum_{m=0}^{\infty} \frac{\Omega^n}{n!} \frac{\varepsilon^m}{m!} \begin{Bmatrix} \mathbf{v}^{[n,m]} \\ \varphi^{[n,m]} \\ h^{[n,m]} \end{Bmatrix}, \tag{44.76}$$

where

$$\mathbf{v}^{[n,m]} = \frac{\partial^{n+m}}{\partial\Omega^n \partial\varepsilon^m} \mathbf{v}(r, z(\Omega, \varepsilon) ; \Omega, \varepsilon)\Big|_{\Omega=\varepsilon=0}, \tag{44.77}$$

with similar meanings for the other derivatives.

As in (44.13), let a domain mapping be defined through

$$\begin{aligned} z(\Omega, \varepsilon) &= f(r_0, z_0 ; \Omega, \varepsilon) \\ z_0 &= f(r_0, z_0 ; 0, 0), \end{aligned} \tag{44.78}$$

and

$$h(r ; \Omega, \varepsilon) = f(r_0, 0 ; \Omega, \varepsilon). \tag{44.79}$$

Since z depends on Ω and ε, the differentiation in (44.77) becomes algebraically more complex. For example.

$$\mathbf{v}^{[1,0]} = \mathbf{v}^{(1,0)} + f^{[1,0]}\partial_z \mathbf{v}^{(0,0)}, \tag{44.80}$$

where [*cf.* (44.21)] :

$$\mathbf{v}^{(n,m)} = \frac{\partial^{n+m}}{\partial\Omega^n \partial\varepsilon^m} \mathbf{v}(r, f^{[0,0]} ; \Omega, \varepsilon)\Big|_{\Omega=\varepsilon=0}. \tag{44.81}$$

Since we are interested in finding $h(r ; \Omega, \varepsilon)$, we have to find the coefficients $h^{[n,m]}$. This is facilitated by the following observations :

(i) $h^{[0,0]} = 0$, that is, in the absence of motion and surface tension, the free surface is flat ;

(ii) symmetry of the shape with respect to Ω suggests that

$$h^{[2n+1,m]} = 0, \qquad n, m \geqq 0 ; \tag{44.82}$$

(iii) if $\Omega = 0$, one notes that (44.71) leads one to the following equation in the modified infinite domain $\mathcal{V}_0$:

$$\left(\frac{rh_s'}{(1+h_s'^2)^{1/2}}\right)' - r\frac{\rho g}{\sigma}h_s = 0, \quad z = 0, \tag{44.83}$$

where h_s is the static free surface. Now h_s obeys

$$h_s'(R_1) = -\varepsilon, \quad (h_s, h_s') \to (0, 0) \text{ as } r \to \infty. \tag{44.84}$$

For low values of ε, one may linearize (44.83) and obtain a solution in terms of the modified BESSEL functions $K_0(\cdot)$ and $K_1(\cdot)$ as :

$$h_s(r\,;\varepsilon) = \varepsilon\, K_0(r\, S^{1/2})/S^{1/2}\, K_1(R_1 S^{1/2}), \tag{44.85}$$
$$S = \rho g/\sigma.$$

If $-30° < \alpha < 30°$, this expression is good to within an error of 10%. At any rate, h_s is an *odd* function of ε, that is, $h_s(r\,;\varepsilon) = \varepsilon\, H(r\,;\varepsilon^2)$. The above observations lead one to accept that $h(r\,;\Omega, \varepsilon)$ must be given by

$$h(r\,;\Omega, \varepsilon) = h_s(r\,;\varepsilon) + \frac{\Omega^2}{2}\, h^{[2,0]}(r) + \ldots \tag{44.86}$$

To find the equation satisfied by $h^{[2,0]}(r)$, neglect the omitted terms in (44.86), and substitute $h(r\,;\Omega,\varepsilon) = h_s(r\,;\varepsilon) + (\Omega^2/2)\, h^{[2,0]}(r)$ into (44.71). One obtains that $h^{[2,0]}$ obeys the equation

$$-\varphi^{(2)} + T_E\langle zs\rangle^{(2)} - (h'\, T\langle rz\rangle^{(2)} + \rho g h^{[2,0]} = \frac{\sigma}{r}\left(\frac{rh'}{(1+h'^2)^{1/2}}\right)^{(2)}, \tag{44.87}$$

or with $h^{[2,0]}(r) = R_1^4\, \bar{H}(r)$,

$$\left.\begin{aligned}\frac{1}{r}(R_1\,\bar{H}')' - SH &= -\frac{\varphi^{(2)}}{\sigma R_1^4},\\ \bar{H}'(R_1) = 0, \quad (\bar{H}, \bar{H}') &\to (0,0) \text{ as } r \to \infty.\end{aligned}\right\} \tag{44.88}$$

In deriving this equation, we have used the earlier results (44.45)—(44.55). In the solution for $\varphi^{(2)}$ in (44.52), let us assume that $A \to 0$, $B^2 \to R_1^4$. Then

$$\frac{\varphi^{(2)}}{R_1^4} = \frac{\rho}{r^2} - \frac{4(2\beta + 3\gamma)}{r^4}. \tag{44.89}$$

Using this in (44.88), one can show that

$$h^{[2,0]}(\gamma) \doteq \left\{\frac{\delta}{16-\mu^2}\left[\frac{4R_1^\mu}{\mu r^\mu} - \frac{R_1^4}{r^4}\right] + \frac{\nu R_1^2}{4-\mu^2}\left[\frac{R_1^2}{r^2} - \frac{2R_1^\mu}{\mu r^\mu}\right]\right\}, \tag{44.90}$$

where

$$\delta = 4\,R_1^2\,(2\beta + 3\gamma)/\sigma, \qquad \nu = \rho R_1^2\,/\,\sigma, \qquad \mu^2 = R_1^2\, S. \tag{44.91}$$

It has been shown by JOSEPH, *et al.* that (44.90) yields values in close agreement with experimental results[65] for low values of Ω.

For general simple fluids, TRUESDELL[65] notes that FOSDICK and SERRIN have proved that the rise of the free surface is confined to the annulus whose outer radius is

$$[(N_1''(0) + 5N_2''(0)]/\rho)^{1/2}, \qquad \text{where} \qquad N_i''(\varkappa) = d^2N_i(\varkappa)/d\varkappa^2, \qquad i = 1, 2.$$

In the work cited above, JOSEPH and FOSDICK have also studied the effects of the surface tension in a state of near rigid rotation, and the boundary layer due to capillarity in this state. Their conclusion is that this latter effect is important within a gap of width of $O(\sigma/\rho g)^{1/2}$. For instance, the liquid which climbs up the rod is in a state of near rigid rotation and it is the surface tension which maintains its balloon-like shape against the centrifugal forces. So we now have a quantitative idea of the boundary layer near the rod and the radial distance in which the rise occurs.

To conclude, the domain perturbation technique is a new tool which can be used to solve many free surface problems at slow speeds. We shall mention this technique again in connection with die swell in Sec. 50.

45. New Rheometrical Design. Inertia Effects and End Corrections

So far, the constitutive equations used in the solution of problems have been the 'order' fluids. Here we shall examine a technique for obtaining solutions to problems with the constitutive equation of finite linear viscoelasticity. Primarily, the emphasis is on obtaining an inertial correction to the flow in the MAXWELL rheometer[67] and for an end correction to the flow in the eccentric cylinder rotational viscometer[67].

Let the coordinate system be cylindrical polar (r, θ, z) and let the physical components of the velocity field be written in complex form as

$$\begin{aligned} v\langle r\rangle &= \alpha u(r, z)\, e^{i\theta}, \\ v\langle\theta\rangle &= \Omega r + \alpha v(r, z)\, e^{i\theta}, \\ v\langle z\rangle &= \alpha w(r, z)\, e^{i\theta}, \end{aligned} \tag{45.1}$$

where only the *real part* is implied. Eq. (45.1) can be regarded as a small perturbation (α is small) superposed on a steady rigid body rotation of the fluid with an angular velocity Ω about the z-axis. For the velocity field

65. KUNDU [1973 : 9] has also studied the problem of the rod climbing effect, and calculated the theoretical predictions of the second-order fluids and the OLDROYD fluid (A5.17). However his results are not as complete as those presented here. KAYE [1973 : B] also solved this problem for second-order fluids and BÖHME [1974 : C, D, E] developed the theory for these fluids between two solid surfaces in rotation. We have presented the JOSEPH-FOSDICK analysis for it is more comprehensive.

66. [1973 : 14].

67. See Sec. 34 ter for the solutions without inertia and edge effects.

(45.1), one uses the OLDROYD method of calculating path lines [*cf.* (2.33)] and obtains to $O(\alpha)$ the following:[68]

$$\xi = r + \frac{i\alpha u}{\Omega}\, e^{i\theta}\,(1 - e^{-i\Omega s}),$$

$$\eta = \theta - \Omega s + \frac{i\alpha v}{\Omega r}\, e^{i\theta}\,(1 - e^{-i\Omega s}), \tag{45.2}$$

$$\zeta = z + \frac{i\alpha w}{\Omega}\, e^{i\theta}\,(1 - e^{-i\Omega s}),$$

where (ξ, η, ζ) are the coordinates of the particle at time $t - s$, with the particle occupying (r, θ, z) at time t.

A computation of the strain history for (45.2) may be made as in (3.10)-(3.11), and terms to $O(\alpha)$ should be retained. When use is made of the constitutive relation (34.19) of finite linear viscoelasticity, the extra stresses are :

$$
\begin{aligned}
T_E\langle rr\rangle &= 2\eta^*(\Omega)\,\alpha e^{i\theta}\,\frac{\partial u}{\partial r},\\
T_E\langle \theta\theta\rangle &= 2\eta^*(\Omega)\,\alpha e^{i\theta}\,\frac{iv + u}{r},\\
T_E\langle zz\rangle &= 2\eta^*(\Omega)\,\alpha e^{i\theta}\,\frac{\partial w}{\partial z},\\
T\langle r\theta\rangle &= \eta^*(\Omega)\,\alpha e^{i\theta}\left[\frac{iu}{r} + r\frac{\partial}{\partial r}\left(\frac{v}{r}\right)\right],\\
T\langle rz\rangle &= \eta^*(\Omega)\,\alpha e^{i\theta}\left(\frac{\partial u}{\partial z} + \frac{\partial w}{\partial r}\right),\\
T\langle \theta z\rangle &= \eta^*(\Omega)\,\alpha e^{i\theta}\left(\frac{\partial v}{\partial z} + \frac{iw}{r}\right),
\end{aligned} \tag{45.3}
$$

where again the real parts have any meaning. In (45.3), $\eta^*(\Omega)$ is given by (34.*b*10).

If these stresses are used in the equations of motion (A6.3)-(A6.4) with an assumed pressure distribution p given by

$$p = p_0 - \rho g z + \rho\,\frac{\Omega^2 r^2}{2} + \alpha\bar{p}(r, z)\, e^{i\theta}, \tag{45.4}$$

one obtains the following equations, correct to $O(\alpha)$[69] :

$$
\begin{aligned}
\rho\Omega(iu - 2v) &= -\frac{\partial \bar{p}}{\partial r} + \eta^*\left(\frac{\partial^2 u}{\partial r^2} + \frac{\partial^2 u}{\partial z^2} - \frac{2iv}{r^2} - \frac{2u}{r^2} + \frac{1}{r}\frac{\partial u}{\partial r}\right),\\
\rho\Omega(iv + 2u) &= -\frac{i\bar{p}}{r} + \eta^*\left(\frac{\partial^2 v}{\partial r^2} + \frac{\partial^2 v}{\partial z^2} + \frac{2iu}{r^2} + \frac{1}{r}\frac{\partial v}{\partial r} - \frac{2v}{r^2}\right),\\
\rho\Omega iw &= -\frac{\partial \bar{p}}{\partial z} + \eta^*\left[\frac{\partial^2 w}{\partial z^2} + \frac{1}{r}\frac{\partial}{\partial r}\left(r\frac{\partial w}{\partial r}\right) - \frac{w}{r^2}\right].
\end{aligned} \tag{45.5}
$$

68. ABBOTT and WALTERS [1970 : 1], Eq. (33).
69. ABBOTT and WALTERS [1970 : 1], Eqs. (36)-(38).

These are precisely the NAVIER-STOKES equations[70] of an incompressible Newtonian fluid for velocity field (45.1), with a complex viscosity η^*.

The above result has been stated by ABBOTT, *et al.*, as[71]

THEOREM 45.1. *Let an incompressible viscoelastic liquid, obeying the constitutive equation of finite linear viscoelasticity, be subjected to a steady velocity field equivalent to a slight perturbation of a state of rigid-body rotation with an angular velocity Ω. If the perturbations are sinusoidal in the direction of primary rotation, the governing equations of motion are those of an incompressible Newtonian fluid with a complex viscosity $\eta^*(\Omega)$, where $\eta^*(\Omega)$ is the complex viscosity of the viscoelastic liquid computed on the basis of a small sinusoidal disturbance of frequency Ω*[72].

For the MAXWELL rheometer flow, the velocity field $\dot{x} = -\Omega y + \Omega\psi z$, $\dot{y} = \Omega x$, $\dot{z} = 0$ can be written in cylindrical polar coordinates as

$$\begin{aligned} v\langle r\rangle &= \Omega\psi z \cos\theta, \\ v\langle\theta\rangle &= \Omega r - \Omega\psi z \sin\theta, \\ v\langle z\rangle &= 0. \end{aligned} \tag{45.6}$$

Here (r, θ, z) is centered at the center of the bottom plate (see Fig. 7.1). Hence, if inertia is neglected, and pressure is taken to be constant (as has been done in Sec. 32), or equivalently, if one puts $u = u(z)$, $v = v(z)$, $w = 0$, $\rho = 0$ and $\bar{p} = 0$ in (45.5), one should obtain (45.6). It is fairly easy to see that under the above assumptions (45.5) reduce to

$$\frac{d^2u}{dz^2} = 0, \qquad \frac{d^2v}{dz^2} = 0. \tag{45.7}$$

The boundary conditions are :

$$\begin{aligned} u(0) &= 0, \qquad u(b) = \Omega, \\ v(0) &= 0, \qquad v(b) = i\Omega. \end{aligned} \tag{45.8}$$

We have also identified α with a, the distance between the axes of rotation. Then the solutions to (45.7), which are consistent with (45.6), are

$$n(z) = \Omega z/b, \qquad v(z) = i\Omega z/b. \tag{45.9}$$

To obtain the next approximation, we shall retain inertial terms, but keep the assumption that the plates are infinitely large, that is, put $w(z) = 0$. Then, substituting $u = u(z)$, $v = v(z)$, $\bar{p} = 0$ and the result from the continuity equation, viz., $u + iv = 0$ into (45.5), one obtains

$$u'' - \lambda^2 u = 0, \quad v'' = \lambda^2 v = 0, \quad \lambda^2 = -i\Omega\rho/\eta^*. \tag{45.10}$$

70. See e.g., LANGLOIS [1964 : 13], p. 89.
71. [1971 : 1], p. 191.
72. Additional comments on the above theorem may be found in JONES [1971 : 19].

The boundary conditions (45.8) imply that

$$u(z) = \frac{\Omega}{\sinh \lambda b} \sinh \lambda z,$$
$$v(z) = \frac{i\Omega}{\sinh \lambda b} \sinh \lambda z. \tag{45.11}$$

Using this in $(45.3)_{5,6}$, one can calculate the shear stresses $T\langle rz\rangle$ and $T\langle \theta z\rangle$ by retaining the real parts, this is,

$$T\langle rz\rangle = \mathrm{Re}\left\{\frac{a\eta^*(\Omega)\, e^{i\theta}\, \Omega\lambda \cosh \lambda z}{\sinh \lambda b}\right\}, \tag{45.12}$$

$$T\langle \theta z\rangle = \mathrm{Re}\left\{\frac{a\eta^*(\Omega)\, e^{i\theta}\, i\Omega\lambda \cosh \lambda z}{\sinh \lambda b}\right\}. \tag{45.13}$$

Let F_x and F_y be the forces in the x and y directions respectively on a plate of radius R. Thus

$$F_x = \int_0^R \int_0^{2\pi} (T\langle rz\rangle \cos\theta - T\langle \theta z\rangle \sin\theta)\, r d\theta\, dr, \tag{45.14}$$

$$F_y = \int_0^R \int_0^{2\pi} (T\langle rz\rangle \sin\theta + T\langle \theta z\rangle \cos\theta)\, r d\theta\, dr. \tag{45.15}$$

Hence on the plate at $z = b$,

$$F_x - iF_y = \frac{a\pi\Omega R^2}{2}\, \eta^*(\Omega)\, \lambda \coth \lambda b. \tag{45.16}$$[73]

As $\lambda \to 0$, by L'HÔPITAL's rule, one obtains that

$$F_x - iF_y = \frac{a\,\pi\, R^2\Omega}{2}\, \eta^*(\Omega), \tag{45.17}$$

which is inherent in $(34.C10)$–$(34.C11)$.

We now turn to the eccentric cylinder rotational viscometer. It has been shown that (see Sec. 34 ter) if one assumes the velocity field to be given by (45.1), and assumes further that the cylinders are infinitely long, that is, $w(z) = 0$, and neglects inertia ($\rho = 0$), then in this viscometer the forces F_x and F_y in the x and y directions respectively, are given by

$$F_x - iF_y = \frac{4\alpha\pi\, \Omega\eta^*(\Omega)\, L}{\ln\beta - (\beta^3 - 1)/(\beta^2 + 1)}, \qquad \beta = r_2/r_1. \tag{45.18}$$

Now, following BROADBENT and WALTERS[74], we shall consider the case of a

73. The above result differs from that given by ABBOTT and WALTERS [1970 : 1], Eq. (49). The reason is that their coordinate system does not coincide with the one chosen here. The forces they compute act on a circle of radius R in the top plate with its center at $\left(0, \frac{a}{2}, b\right)$ while the present calculations have their center at $(0, 0, b)$, but the upper plate rotates about $(0, a, b)$. However, as $\lambda \to 0$, these differences vanish, as they should.

74. [1971 : 6]. In [1971 : 13], JONES presents additional support to the theory of ABBOTT and WALTERS [1970 : 2] given in Sec. 34 ter.

finite inner cylinder of length L. It will be assumed moreover that the base of the inner cylinder is sufficiently far away from the base of the outer cylinder so that one may ignore the effects of the narrow gap at the bottom. We now let the velocity field be given by (45.1), the pressure field by

$$p = p_0 + \Omega\, \alpha\, \bar{p}(r, z)\, e^{i\theta}, \tag{45.19}$$

and take the boundary conditions to be

$$\begin{aligned} u(r_1, z) &= 0, \quad & v(r_1, z) &= 0, \quad & w(r_1, z) &= 0, \\ u(r_2, z) &= \Omega, \quad & v(r_2, z) &= i\Omega, \quad & w(r_2, z) &= 0. \end{aligned} \tag{45.20}$$

Let the gap be $h = r_2 - r_1$, and assumed to be very small. Let us denote the radial position r between the cylinders through

$$r = r_1 + h\xi, \qquad 0 \leqq \xi \leqq 1. \tag{45.21}$$

This is called the *lubrication approximation* since it appears in the theory of hydrodynamic lubrication of journal bearings. We shall seek solutions valid to $O(h/r_1)$. The continuity equation is given by

$$\frac{1}{r}\frac{\partial}{\partial r}(ru) + \frac{iv}{r} + \frac{\partial w}{\partial z} = 0. \tag{45.22}$$

Let us turn to the earlier solution (Sec. 34 ter) and note that if $r_2 - r_1 = h$ is very small then in $F(r)$ [see (34.C24)], $A\, r^2 = O(r^2/r_1^2)$, $C/r^2 = O(r^2/r_1^2)$, $B \ln r = O(\ln r)$, and $D = O(1)$. In other words, for a very narrow gap, one may take

$$F(r) = (C/r^2) + D, \tag{45.23}$$

with $F(r_1) = 0$ and $F(r_2) = 1$ determining C and D. With this assumption, from (34.C21) we see that $\partial\bar{p}/\partial r = 0$.

Using this in the current problem, one is led to

$$\partial\bar{p}/\partial\xi = 0, \tag{45.24}$$

as the equation for the r-direction in $(45.5)_1$. Next, Eq. $(45.5)_2$ for the θ-direction, because of the lubrication approximation (45.21), becomes

$$\bar{p} = -\, i\eta^* \frac{r_1}{h^2}\frac{\partial^2 v}{\partial \xi^2}, \tag{45.25}$$

which, on account of (45.24), can be integrated along with the boundary condition (45.20) to yield

$$v = \frac{iph^2}{2r_1\eta^*}\xi(\xi - 1) + i\Omega\xi. \tag{45.26}$$

The z-direction equation $(45.5)_3$ leads to

$$\frac{d\bar{p}}{dz} = \frac{\eta^*}{h^2}\frac{\partial^2 w}{\partial \xi^2}, \tag{45.27}$$

which can be integrated to yield, with (45.20), the following

$$w = \frac{h^2}{2\eta^*}\frac{d\bar{p}}{dz}\,\xi(\xi - 1). \tag{45.28}$$

Writing the continuity equation (45.22) as

$$\frac{1}{h}\frac{\partial u}{\partial \xi} + \frac{iv}{r_1} + \frac{\partial w}{\partial z} = 0, \tag{45.29}$$

and substituting (45.26) and (45.28) into it, we obtain

$$\frac{1}{h}\frac{\partial u}{\partial \xi} = \frac{h^2}{2r_1^2\eta^*}\bar{p}(z)\,\xi(\xi - 1) + \frac{\Omega\xi}{r_1} - \frac{h^2}{2\eta^*}\frac{d^2\bar{p}}{dz^2}\,\xi(\xi - 1). \tag{45.30}$$

On integrating this with respect to ξ between (0, 1), and using the boundary condition (45.20), one has

$$\frac{\Omega}{h} = \frac{h^2}{12\eta^*}\frac{d^2\bar{p}}{dz^2} - \frac{h^2}{2r_1^2\eta^*}\bar{p}(z) + \frac{\Omega}{2r_1}, \tag{45.31}$$

or

$$\frac{d^2\bar{p}}{dz^2} - \frac{\bar{p}}{r_1^2} = \frac{12\eta^*\Omega}{h^3} - \frac{6\,\Omega\,h^2\eta^*}{r_1}, \tag{45.32}$$

where we shall neglect the h^2/r_1 term. To find the end conditions on $\bar{p}(z)$ at $z = \pm L/2$, we note that the surface $z = L/2$ is a free surface. Hence $T\langle rz\rangle$, $T\langle \theta z\rangle$ and $T\langle zz\rangle$ must vanish there. Using the last one we find that $\bar{p} - 2\eta^*(\partial w/\partial z) = 0$ on $z = L/2$. But $\partial w/\partial z$ is of order r_1^2/h, while $\bar{p}$ is of order r_1^2/h^3. Thus we may take $\bar{p} = 0$ at $z = L/2$. Since body forces have been neglected, and the outer cylinder is assumed not to affect the flow at $z = -L/2$, we may take $\bar{p} = 0$ at $z = L/2$ also. Thus

$$\bar{p}(z) = -\frac{12r_1^2\eta^*\Omega}{h^3}\left[1 - \frac{\cosh(z/r_1)}{\cosh(L/2r_1)}\right]. \tag{45.33}$$

Now, to compute the forces in the x and y directions on the inner cylinder [*cf.* (34.C28)], we need p, $T_E\langle rr\rangle$ and $T\langle r\theta\rangle$ on $\xi = 0$. But $T_E\langle rr\rangle = 0$ on $\xi = 0$ and $T\langle r\theta\rangle$ is of order r_1/h^2, while p is of order r_1^2/h^3. Hence the forces F_x and F_y are

$$F_x - iF_y = \frac{12\pi\alpha\Omega\eta^* L r_1^3}{h^3}\left(1 - \frac{2r_1}{L}\tanh\frac{L}{2r_1}\right). \tag{45.34}$$

To obtain a solution to the second order in h/r_1, one notes that one may still take the pressure to be $p = \bar{p}(z) + \tilde{p}(z)$. Then (45.25) is not changed, but (45.27) is ; and so is (45.29). Going through the calculations one finds that[75]

$$\tilde{p}(z) = -\frac{12\eta^* r_1^2}{h^3}\left(\frac{h}{r_1}\right)\left\{1 - \frac{\cosh(z/r_1)}{\cosh(L/2r_1)} + \frac{1}{\cosh(L/2r_1)} \times \right.$$
$$\left. \times \left(\frac{z}{2r_1}\sinh\frac{z}{r_1} - \frac{L}{4r_1}\tanh\frac{L}{2r_1}\cosh\frac{z}{r_1}\right)\right\}, \tag{45.35}$$

and

$$F_x - iF_y = \frac{12\alpha\Omega\pi\eta^* r_1^3}{h^3}\left\{1 - \frac{2r_1}{L}\tanh\frac{L}{2r_1} + \frac{2h}{r_1} \times \right.$$
$$\left. \times \left(1 - \frac{2r_1}{L}\tanh\frac{L}{2r_1} - \frac{1}{4}\tanh^2\frac{L}{2r_1}\right)\right\}. \tag{45.36}$$

The above material has been presented here from an instructional point of view, for the above formulae for the determination of the forces are too cumbersome to be useful in practice. Moreover, as the authors have noted, the inner cylinder lags visibly behind the outer one in rotation, and this slip negates the entire analysis. However, by suitable modifications of the apparatus one could force the inner cylinder not to lag behind the other, and use the analytical results given here (see Sec. 47).

46. Unsteady Flows: Initial Value Problems and Periodic Motions

It has been proved in the Appendix to Chap. 8 that second-order fluids have physically unacceptable responses in initial value problems ; at this stage it is not known whether higher-order fluids have similar characteristics or not. This lack of knowledge is not only due to the complexity of models, but also to the fact that viscometric flows do not yield sufficient restrictions on all the constants appearing in these constitutive relations. So, to solve initial-value problems one is forced, on account of the desire to obtain physically reasonable answers as well as the fact that in finite linear viscoelasticity the only unknown quantity is the relaxation function, to consider integral models. As an illustration, let us consider the so-called 'start up' problem. Here the fluid which is initially at rest is set in motion by a suddenly applied pressure gradient, or by a sudden movement of one of the bounding surfaces. The transitional velocity field as well as the final velocity field are of interest in the analysis of this flow.

Next, the periodic motion of a viscoelastic liquid under a periodic pressure gradient will also be studied. For the second-order fluid[76], and the n-th-order fluid[77], such oscillatory motions have been shown to exist, but here

75. Broadbent and Walters [1971 : 6], Eqs. (24) and (25).
76. Markovitz and Coleman [1964 : 15].
77. Pipkin [1964 : 19], [1964 : 20].

we shall concentrate on the results obtained for the theory of finite linear viscoelasticity[78] and briefly mention the work of BARNES, *et al.*,[79] on the flow due to a pulsating pressure gradient superposed on a mean value.

To begin, consider the axial flow of a finitely linear viscoelastic liquid. The flow will be assumed to be down a pipe of arbitrary cross-section. Taking the velocity field to be given in Cartesian coordinates by

$$\mathbf{v} = w(x_1, x_2, t)\,\mathbf{k}, \tag{46.1}$$

we shall seek the qualitative behaviour[80] of the velocity field for large t. For the velocity field (46.1), the path lines are given by [*cf.* (2.4)—(2.5)]:

$$\xi = x_1, \quad \eta = x_2, \quad \zeta = x_3 - \int_0^s w(x_1, x_2, t - \sigma)\, d\sigma. \tag{46.2}$$

Let us assume that w is so small that the products $w_{,\alpha} w_{,\beta}$ $(\alpha, \beta = 1, 2)$ may be neglected. Then the matrix of $\mathbf{C}_t(t - s)$ has the components

$$[C_t(t-s)\langle ij\rangle] = \begin{bmatrix} 1 & 0 & -\int_0^s w_{,1}\, d\sigma, \\ \cdot & 1 & -\int_0^s w_{,2}\, d\sigma \\ \cdot & \cdot & 1 \end{bmatrix}. \tag{46.3}$$

Writing the constitutive equation of finite linear viscoelasticity as (34.*b*5), that is, as

$$\mathbf{T}_E = -\int_0^\infty G(s)\, \frac{d\,\mathbf{G}(s)}{ds}\, ds, \tag{46.4}$$

and using (46.3), one obtains

$$T_E\langle 3\alpha\rangle = \int_0^\infty G(s)\, \frac{\partial w}{\partial x_\alpha}(x_1, x_2, t - s)\, ds, \qquad \alpha = 1, 2. \tag{46.4}$$

If the pressure gradient in the x_3 direction, that is, $\partial\psi/\partial x_3 = -C$, is a constant, the equations of motion become

$$\rho\, \frac{\partial w}{\partial t} = C + \Delta \int_0^\infty G(s)\, w(x_1, x_2, t - s)\, ds, \quad \Delta = \frac{\partial^2}{\partial x_1^2} + \frac{\partial^2}{\partial x_2^2}. \tag{46.5}$$

When the flow becomes steady, Eq. (46.5) reads

$$0 = C + \eta_0\, \Delta w_\infty, \tag{46.6}$$

where η_0 is the steady-shearing viscosity (34.*C*5), and w_∞ the steady velocity. To consider the start up problem, let us introduce a displacement function

$$W(x_1, x_2, t) = \int_0^t w(x_1, x_2, \tau)\, d\tau, \tag{46.7}$$

78. JONES and WALTERS [1967 : 14].
79. BARNES, TOWNSEND and WALTERS [1971 : 2].
80. NORMAN and PIPKIN [1967 : 19]:

so that in (46.5), $s \in [0, t)$ only. Integration of (46.5) with respect to t yields

$$\rho w(x_1, x_2, t) = Ct + \Delta \int_0^t \left\{ \int_0^\tau G(s)\, w(x_1, x_2, \tau - s)\, ds \right\} d\tau \tag{46.8}$$

$$= Ct + \Delta \int_0^t G(s)\, W(x_1, x_2, t - s)\, ds, \tag{46.9}$$

where we have used the fact that $W(x_1, x_2, \sigma) = 0$ for $\sigma < 0$.

Write the displacement $W(x_1, x_2, t)$ as

$$W = w_\infty t + W_\infty, \tag{46.10}$$

where w_∞ is defined by (46.6) and is thus the steady state Newtonian velocity, and W_∞ is a constant (as $t \to \infty$) which must incorporate the elasticity of the fluid as well as the inertial decrement in displacement. Substitution of (46.10) into (46.9) yields

$$\rho w = Ct + t(\Delta w_\infty) \int_0^t G(s)\, ds - \Delta w_\infty \int_0^t s\, G(s)\, ds + $$
$$+ \int_0^t G(s)\, \Delta W_\infty(x_1, x_2, t - s)\, ds. \tag{46.11}$$

Now from (46.6)

$$\lim_{t\to\infty} \left[Ct + t\Delta w_\infty \int_0^t G(s)\, ds \right] = \frac{C}{\eta_0} \lim_{t\to\infty} \left[\left(\eta_0 - \int_0^t G(s)\, ds \right) (1/t)^{-1} \right]$$

$$= \frac{C}{\eta_0} \lim_{t\to\infty} t^2 G(t) = 0, \tag{46.12}$$

because in (34.b9) we have assumed that this is so.

Introducing a relaxation time λ through

$$\eta_0 \lambda = \int_0^\infty s\, G(s)\, ds, \tag{46.13}$$

one can write the limit of (46.11) as $t \to \infty$ to be:

$$\rho w_\infty = \eta_0 \lambda + \mu \Delta W_\infty(x, y, \infty), \tag{46.14}$$

where one again recalls that W_∞ is independent of time for large t. Let W_∞ be decomposed into an elastic displacement W_E and an inertial decrement W_I through (at large t) :

$$W_\infty = W_E - W_I. \tag{46.15}$$

Suppose that $W_E = 0$, $W_I = 0$ on the boundary of the pipe. Then $W_\infty = 0$ on it. Now W_E and W_I obey

$$\Delta W_E = - C\lambda/\eta_0, \qquad \Delta W_I = - \rho w_\infty/\eta_0. \tag{46.16}$$

We note that the elastic displacement $W_E = \lambda w_\infty$ because of (46.6). Hence the elasticity adds an extra displacement to the fluid particle, while the inertial decrement W_I is the same as in the Newtonian fluid.

Unlike in the steady flow problem of Sec. 42, the non-circularity of the cross-section and the presence of the second normal stress difference mean that the transverse flow disturbances are of the same order as the perturbation in the axial velocity[81]. So, one should consider, as a next approximation, the velocity field

$$\mathbf{v} = w\mathbf{k} + \varepsilon \mathbf{v}_1, \quad 0 < \varepsilon \ll 1. \tag{46.17}$$

However, this does not affect the calculations listed above[82].

Let us return to (46.5) once more, but this time let the cross-section be a circular pipe of radius R. One can write (46.6) as

$$\rho \frac{\partial w}{\partial t} = C + \Delta \int_0^t G(\lambda)\, w(r, t - \lambda)\, d\lambda, \tag{46.18}$$

where the axial velocity is assumed to be $w = w(r, t)$, obeying

$$w(R, t) = 0; \quad w(r, 0) = 0, \quad 0 \leqq r \leqq R; \quad \frac{\partial w\,(0, t)}{\partial r} = 0, \tag{46.19}$$

and $\Delta = \dfrac{\partial^2}{\partial r^2} + \dfrac{1}{r}\dfrac{\partial}{\partial r}$. Now, the analytical difficulty in solving (46.18) rests on the relaxation function $G(\lambda)$. If one can find a relaxation function $G(\lambda)$ which is amenable to analysis, it can be used in (46.18) to compute $w(r, t)$.

Let us now examine one such choice of $G(s)$. A fluid with such a relaxation function has been studied by WALTERS[83] in connection with liquids called A' and B'. He assumes that

$$G(\lambda) = \int_0^\infty \frac{N(\tau)}{\tau}\, e^{-\lambda/\tau}\, d\tau, \tag{46.20}$$

where $N(\tau)$ is a distribution function of relaxation times τ. Let $\bar{w}$ denote the LAPLACE transform of w, that is,

$$\bar{w}\,(r, s) = \int_0^\infty w(r, t)\, e^{-st}\, dt. \tag{46.21}$$

Then, on using (46.20), (46.18) becomes[84]

$$\frac{\partial^2 \bar{w}}{\partial r^2} + \frac{1}{r}\frac{\partial \bar{w}}{\partial r} - q^2 \bar{w} = \frac{q^2 c}{\rho s^2}, \tag{46.22}$$

$$q^2 = \rho s \bigg/ \int_0^\infty \frac{N(\tau)}{1 + s\tau}\, d\tau. \tag{46.23}$$

Let the distribution function $N(\tau)$ be assumed to be

$$N(\tau) = \eta_0 \frac{\lambda_2}{\lambda_1} \delta(\tau) + \eta_0 \frac{\lambda_1 - \lambda_2}{\lambda_1} \delta(\tau - \lambda_1), \tag{46.24}$$

81. PIPKIN [1966 : 15].

82. NORMAN and PIPKIN [1967 : 19]. See this paper for a constant flux problem generated by an impulsive pressure gradient.

83. WALTERS [1962 : 12]. See Eqs. (A5.20) and (A5.21). See also (37.38).

84. WATERS and KING [1971 : 29], Eqs. (12)-(13).

where $\delta(\cdot)$ is the usual DIRAC delta function. It is assumed that $\lambda_1 > \lambda_2 > 0$; λ_1 is called the *relaxation time* and λ_2 the *retardation time*. Using (46.24) in (A5.20) leads to the liquid A and in (A5.21) to liquid B of OLDROYD [see (A5.17)].

The solution to (46.22) is given by

$$\frac{w(r_1, t_1)}{w_0} = (1 - r_1^2) - 8 \sum_{Z_n} \frac{J_0(r_1 Z_n)}{J_1(Z_n)\, Z_n^3} \exp\left(-\frac{\alpha_n t_1}{2S_1}\right) G_n(t_1), \tag{46.25}$$

where

$$G_n(t_1) = \cosh\left(\frac{\beta_n t_1}{2S_1}\right) + \frac{\{1 + Z_n^2 (S_2 - 2S_1)\}}{\beta_n} \sinh\left(\frac{\beta_n t_1}{2S_1}\right). \tag{46.26}$$

Also,

$$\left.\begin{aligned} &r_1 = \frac{r}{R}, & &t_1 = \frac{\eta_0 t}{\rho R^2}, & &S_\alpha = \frac{\eta_0 \lambda_\alpha}{\rho R^2}, \quad \alpha = 1, 2, \\ &w_0 = -\frac{CR^2}{4\eta_0}, & &\alpha_n = 1 + S_2 Z_n^2, & &Z_n = i q_1, \\ &q_1 = \frac{\sigma(1 + S_1\sigma)}{(1 + S_2\sigma)}, & &\sigma = \frac{\rho R^2}{\eta_0} s, & &\beta_n = \{(1 + S_2 Z_n^2) - 4 S_1 Z_n^2\}^{1/2}. \end{aligned}\right\} \tag{46.27}$$

For a Newtonian fluid, $S_1 = S_2 = 0$ and thus

$$\frac{w(r_1, t_1)}{w_0} = (1 - r_1^2) - 8 \sum_{Z_n} \frac{J_0(r_1 Z_n)}{J_1(Z_n)\, Z_n^3} \exp(-Z_n^2 t_1). \tag{46.28}$$

Note that $J_0(\cdot)$ and $J_1(\cdot)$ are the usual BESSEL functions of the first kind and order 0 and 1 respectively. The Z_n in (46.27) determine the singularities of the following inversion integral:

$$w(r_1, t_1) = \frac{1}{2\pi i} \int_{\gamma - i\infty}^{\gamma + i\infty} \bar{w}_1(r_1, \sigma) \exp(\sigma t_1)\, d\sigma, \tag{46.29}$$

where

$$\bar{w}_1(r_1, \sigma) = \frac{4w_0}{\sigma^2}\left(1 - \frac{J_0(i q_1 r_1)}{J_0(i q_1)}\right). \tag{46.30}$$

Straightforward but tedious algebraical manipulations will recast (46.25) and (46.28) in the form (46.10).

WATERS and KING have computed the effects of S_1 and S_2 on the velocity at the center of the pipe. It is found (see Fig. 46.1) that the velocity at the center oscillates about the Newtonian value before reaching it, as $t \to \infty$.

To examine the behaviour of the flow under a periodic pressure gradient, let us assume that the pressure and velocity fields are given by[85]

$$-p = \operatorname{Re}\{x_3 A e^{i\omega t}\}, \qquad w(r, t) = \operatorname{Re}\{f(r)\, e^{i\omega t}\}, \tag{46.31}$$

85. JONES and WALTERS [1967 : 14] ; PIPKIN [1964 : 20].

where A is a real constant, and $f(r)$ may be complex.

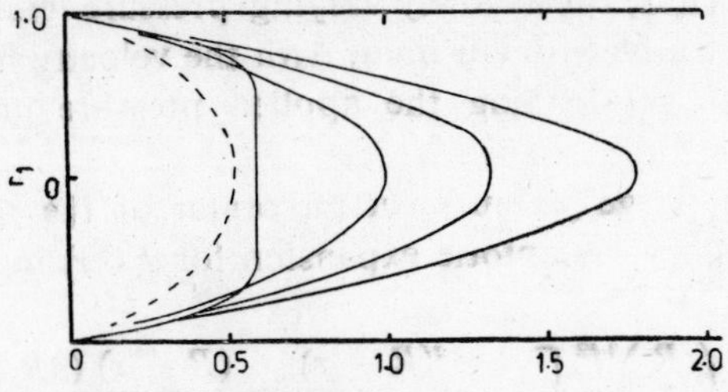

A. Velocity profiles for the generation of Poiseuille flow when $S_1=0.4$ and $S_2=0.04$ for $t_1=0.15$, ∞, 0.5, 0.85 (from left to right). The broken curve is the Newtonian profile for $t_1=0.15$.

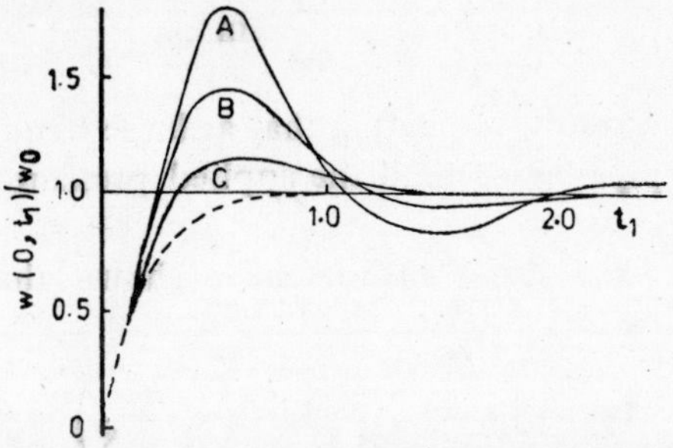

B. The velocity at the center of the pipe against t_1 for $S_1=0.4$ when A, $S_2=0.04$; B, $S_2=0.12$; C, $S_2=0.25$. The broken curve is the Newtonian case.

Fig. 46.1. Poiseuille flow generation.

Now, the equations of motion for the axial flow of an incompressible liquid become [cf. (46.22)]:

$$\frac{d^2 f}{dr^2} + \frac{1}{r}\frac{df}{dr} + k^2 f = \frac{Ak^2}{i\rho\omega}, \tag{46.32}$$

where

$$k^2 = -i\rho\omega\left(\int_0^\infty \frac{N(\tau)}{1+i\omega\tau}\,d\tau\right)^{-1}. \tag{46.33}$$

The solution to (46.32), obeying the conditions $f(R) = 0$, $|f(0)| < \infty$ is given by[86]

$$f(r) = \frac{A}{i\rho\omega}[1 - J_0(kr)/J_0(kR)], \tag{46.34}$$

where $J_0(\cdot)$ is the BESSEL function of the first kind of order zero.

For small frequencies ω, that is, small k, one can approximate $f(r)$ in (46.34) and obtain $w(r, t)$ as

$$w(r, t) = \frac{A}{|\eta^*|}(R^2 - r^2)\cos(\omega t + \delta), \tag{46.35}$$ [87]

86. JONES and WALTERS [1967 : 14], PIPKIN [1964 : 20], Eq. (3.4).

87. This is the more exact result of PIPKIN [1964 : 20], Eq. (6.2). JONES and WALTERS [1967 : 14] consider the real viscosity only.

where $\eta^*(\omega)$ is the zero shear rate complex viscosity, written as $\eta^*(\omega) = |\eta^*| \exp(-i\delta)$. Thus, for a slowly varying pressure gradient, the flow is equivalent to that of a Newtonian fluid, with the velocity field being parabolic at each instant, but leading the applied pressure gradient by a phase angle δ.

When $|kr|$ is very large (except near the center of the pipe $r = 0$), JONES and WALTERS obtain an asymptotic expansion for $J_0(kr)$ in powers of $(kr)^{-1}$, whence

$$f(r) \sim \frac{A}{i\rho\omega} \left\{1 - \left(\frac{R}{r}\right)^{1/2} \left[1 - \frac{i(R-r)}{8kRr} - \frac{(R-r)(9R+7r)}{128k^2R^2r^2}\right] e^{-ik(R-r)}\right\}, \tag{46.36}$$

while near the center, for large values of kR, (46.34) yields

$$w(r, t) = \frac{A}{\omega\rho} \sin \omega t. \tag{46.37}$$

This is an interesting result, indicating that at large frequencies, the velocity at the center of the pipe lags behind the applied pressure gradient by $\pi/2$ in phase.

Also, for large $|kr|$, these authors have shown that the mean square velocity $\overline{w^2}$ is given by

$$\overline{w^2} = \frac{1}{2}[(\operatorname{Re} w)^2 + (\operatorname{Im} w)^2] = \frac{A^2}{2\omega^2\rho^2} \left| 1 - \left(\frac{B}{B-S}\right)^{1/2} \times \right.$$
$$\left. \times \left\{1 - \frac{iS}{8B(B-S)K} - \frac{S(16B-7S)}{128B^2(B-S)^2K^2}\right\} e^{-iKs} \right|^2, \tag{46.38}$$

where

$$K = \left(\frac{\omega}{\nu_0}\right)^{1/2} k, \qquad \nu_0 = \eta_0/\rho, \qquad B = \left(\frac{\omega}{\nu_0}\right) R, \qquad S = \left(\frac{\omega}{\nu_0}\right)^{1/2} (R-r). \tag{46.39}$$

For a fluid with the distribution function $N(\tau)$ given by (46.24), the graph of $2\rho^2\omega^2\,\overline{w^2}/A^2$ vs S shows that (see Fig. 46.2):

(i) the mean square velocity has a higher value than the Newtonian value;
(ii) this higher value occurs closer to the wall than in a Newtonian fluid;[88]
(iii) the thickness of the boundary layer in which the rapid velocity fluctuations occur is decreased from that for the Newtonian fluid.

For the flow under a small periodic pressure gradient superposed on a constant mean value, the above problem becomes considerably more complicated. If the pressure gradient dp/dz is given by

$$\frac{dp}{dz} = -\bar{p}\,(1 + \varepsilon\, e^{i\omega t}), \tag{46.40}$$

88. PIPKIN [1964 : 20] calls this the *annular effect.*

then the axial velocity field may be assumed to be given by[88]

$$w(r, t) = w_0(r) + \varepsilon w_1(r) e^{i\omega t} + \varepsilon^2 [w_2(r) + w_2^{(1)}(r) e^{i\omega t} + w_2^{(2)}(r) e^{2i\omega t}]. \tag{46.41}$$

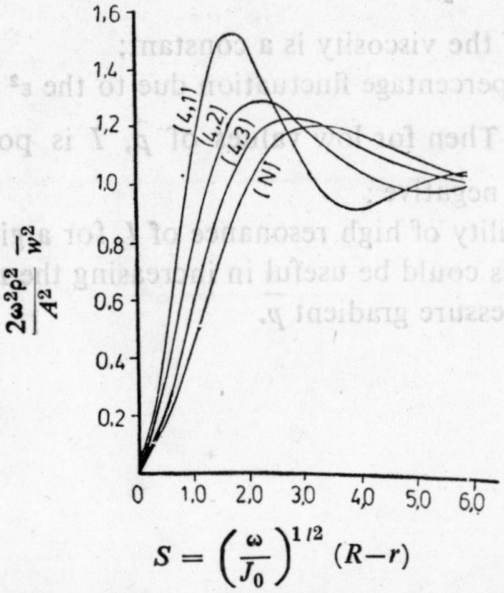

Curves showing the variation of mean square velocity profiles with the parameters ($\omega\lambda_1$, $\omega\lambda_2$) of spectrum ω_j; the profile (N) corresponds to that in the case of a Newtonian liquid

Fig. 46.2. The annular effect.

Taking the shear stress $T\langle rz\rangle$ to be[90]

$$T\langle rz\rangle = \eta(\varkappa_0)\varkappa_0 + \int_0^\infty \varphi_1(\varkappa_0, s)\,\gamma(s)\,ds + \int_0^\infty\int_0^\infty \varphi_2(\varkappa_0, s_1, s_2) \times \gamma(s_1)\,\gamma(s_2)\,ds_1 ds_2, \tag{46.42}$$

where $\varkappa_0 = dw_0/dr$, we have put

$$\gamma(s) = \frac{\partial}{\partial r}\int_0^s w(r, t-\sigma)\,d\sigma - \varkappa_0 s \tag{46.43}$$

as the oscillatory shear part, correct to $O(\varepsilon^2)$, and $\varphi_1(\cdot\,,\cdot)$ and $\varphi_2(\cdot\,,\cdot\,,\cdot)$ are the memory functions. Note that these kernels obey the consistency relations

$$\int_0^\infty \varphi_1(\varkappa_0, s)\,s\,ds = \frac{d}{d\varkappa}(\eta(\varkappa)\varkappa)\Big|_{\varkappa=\varkappa_0},$$

$$\int_0^\infty\int_0^\infty \varphi_2(\varkappa_0, s_1, s_2)\,s_1 s_2\,ds_1\,ds_2 = \frac{1}{2}\frac{d^2}{d\varkappa^2}(\eta(\varkappa)\varkappa)\Big|_{\varkappa=\varkappa_0}. \tag{46.44}$$

89. BARNES, TOWNSEND and WALTERS [1971 : 2].
90. PIPKIN [1966 : 15].

BARNES, *et al.* have used the above relations along with certain approximate forms for $\varphi_1(\cdot\,,\cdot)$ and $\varphi_2(\cdot\,,\cdot\,,\cdot)$ to arrive at the following conclusions:

(i) the mean flow rate

$$Q = \int_0^R (w_0 + \varepsilon^2 w_2)\, 2\pi r\, dr \tag{46.45}$$

is not affected if the viscosity is a constant;

(ii) let I denote the percentage fluctuation due to the ε^2 term in (45.45) over $\int_0^R w_0\, 2\pi r\, dr$. Then for low values of $\bar{p}$, I is positive and for high values of $\bar{p}$, I is negative;

(iii) there is a possibility of high resonance of I, for a given ω, at a certain value of $\bar{p}$. This could be useful in increasing the mean flow rate Q for a given mean pressure gradient $\bar{p}$.

CHAPTER 11

EXPERIMENTAL MEASUREMENTS—THEORETICAL BASIS

It is not the purpose of this monograph to instruct the reader in the experimental techniques of measuring the fluid properties. So, we shall not dwell on the preparation of the solution, the stability of the sample[1] and other equally important matters. However, the main objective of this chapter is to acquaint the reader with the major techniques used by experimentalists to measure the viscosity, normal stress differences, extensional viscosity and die swell ; and to show the theoretical basis underlying each experiment. Also, we mention areas in which innovative experiments are needed.

In Sec. 47, we list the four experimental methods by which the zero shear viscosity, the dynamic viscosity, the viscometric viscosity and the dynamic viscosity in small motions on large can be measured. Detailed discussion about the best apparatus available and the limitations of the theoretical analyses are given.

In Sec. 48, the theory for the measurements of the normal stress differences is discussed. Here one learns the pressure distribution as well as the total thrust techniques for measuring these viscometric material functions. Consideration is given to edge effects, free as well as of the bath type, inertial effects and pressure hole errors.

In Sec. 49, the experimental results on extensional viscosity are given and the theoretical attempts to describe such data are listed.

In Sec. 50, the flow of a viscoelastic jet as it emerges from a die is shown to yield the primary normal stress difference at high shear rates. However, at low shear rates the die swell which occurs in such a flow is shown to be measurable by an analysis of the elastic properties of the solidified extrudate or the flow properties of the molten polymer. We shall also inquire into the effects of surface tension.

Even in the areas which may be considered to be closed, there is a need for a lot of careful study, both from an experimental and a theoretical point of view. For example, to understand the flow of a fluid through a pipe, one would have to consider entrance effects as well as die swell at the exit. In the entrance region, it would seem that a boundary layer theory[2] as well as

1. See e.g. [1966 : 4] or [1968 : 14].

2. Since BEARD and WALTERS [1964 : 2] proposed such a theory for a finitely linear viscoelastic fluid, there have been arguments against it by FRATER [1969 : 4], [1970 : 10] and for it by WALTERS [1970 : 28]. So this matter needs additional investigation. For some qualitative discussions of boundary layers in viscoelastic fluids, see METZNER and ASTARITA [1967 : 18].

the extensional flow theory[3] of a body of revolution will be needed. In the exit region, one would have an extensional flow which may be considered as a reversal of the type in the entrance region. Also, one would have to consider the decay of the stresses in the sample[4], and surface tension effects (see Sec. 50).

As a second example, one may consider the method of reduced variables.[5] Is this method applicable to nonlinear continuum mechanics? WALTERS[6] has suggested that for a special linearly finite theory it is not. Then, what is the rightful place of the method of reduced variables in nonlinear continuum mechanics; how can one incorporate molecular weight into a continuum mechanical approach to derive such a theory? Such questions are wide open.

Since a student of rational mechanics can benefit by a careful study of the theoretical bases of experiments, we shall present them here.

47. Measurement of the Viscosity

By the term viscosity, we mean the ratio of the shear stress to the rate of shear in viscometric flows. This definition is unambiguous in such motions; the term viscosity is also clear when defining the quantity 'dynamic viscosity', if this dynamic viscosity is computed by measuring the shear stress in small oscillations, superposed on steady simple shearing or not. We note, because of CASEWELL's results in Sec. 3, that the flow of such a fluid past a sphere must be viscometric in the boundary layer adjacent to the sphere. Thus the quantity η, which appears in the equation describing the drag experienced by a sphere moving at very low speeds in a viscoelastic liquid, can be described as the viscosity at zero shear rate. However, the term viscosity of the fluid with memory is difficult to interpret in MWCSH-II or MWCSH-III, though we shall follow the conventional usage and call the ratio of axial stress to strain rate as *extensional viscosity* in Sec. 49.

So then, one has the following values of the 'ordinary' viscosity to measure:

(i) zero shear viscosity;
(ii) dynamic viscosity about a state of rest;
(iii) viscometric flow viscosity or steady shear viscosity;
(iv) dynamic viscosity in superposed motions.

We now present some experimental techniques and results for the above four types.

47.1. Zero shear viscosity. In a series of papers, CASWELL[7] has studied the motion of a sphere falling freely in a viscoelastic liquid. As is well

3. A calculation of such a flow was made by METZNER and WHITE [1965 : 12] for a rate type fluid.
4. A beginning has been made by POWELL and MIDDLEMAN [1969 : 19].
5. See FERRY [1970 : 9] for its application to linear viscoelasticity.
6. [1962 : 13].
7. [1970 : 6], [1971 : 7], [1972 : 3].

known, if the motion occurs in an infinite sea of an incompressible Newtonian liquid, the drag D on the sphere is given by

$$D = 6\pi a \eta_0 U_s , \tag{47.1}$$

where a is the sphere radius and η_0 is the Newtonian viscosity. The motion is assumed to be very slow (REYNOLDS numbers of order 0.5 or less) so that inertia may be neglected. We shall call U_s the *Stokes velocity*, since this drag was first computed by STOKES[8].

For the third-order fluid (35.25) it may be shown that[9]

$$U_\infty(D) = U_s \,[1 - (\lambda/\eta_0)^2 (D/6\pi a^2)^2 + O(D/6\pi a^2)^4], \tag{47.2}$$

where λ is a characteristic time depending on the third-order fluid constants. However, it is not necessary to know λ to compute η_0. Hence we omit defining it in terms of the third-order fluid constants.

We call $U_\infty(D)$ the *terminal velocity of the sphere* in an infinite sea of the liquid. We call the quantity η obtained from

$$\frac{1}{\eta} = \frac{6\pi a U_\infty}{D} = \frac{1}{\eta_0} - \frac{(\lambda\tau)^2}{\eta_0^3} + O(\tau^4) \qquad (47.2)^{10}$$

the *apparent viscosity*. Here τ is some characteristic shear stress, given by $\tau = D/6\pi a^2$. Hence a plot of $1/\eta$ vs τ^2 should become linear with an intercept $1/\eta_0$ as $\tau \to 0$.

Of course in practice, one cannot find $U_\infty(D)$, but $U(D)$, where $U(D)$ is the terminal velocity attained by the sphere falling in a confined space, such as a vertical tube. In such cases, CASEWELL[11] has shown that

$$U(D) = U_\infty(D) - (D/6\pi \eta_0 R)\, W(a/R) + O(D/R)^3, \tag{47.4}$$

where R is the radius of the tube and $W(a/R)$ is a complicated function given by

$$W(a/R) = 2.1044 - 2.0888\,(a/R)^2 + \ldots \tag{47.5}$$

In addition to the above correction due to the proximity of the boundary, one has to include a minor correction due to the eccentricity of the sphere in the tube. If this eccentricity be b, then the measured velocity, denoted by

8. STOKES published this result in 1851. For a modern treatment, see HAPPEL and BRENNER [1965 : 9], or BATCHELOR [1967 : 3].

9. For a second-order fluid, the drag is not altered, though the velocity field is. See CASWELL and SCHWARZ [1962 : 1], for example. These authors present the third-order fluid calculations also. Similar results were obtained by LESLIE [1961 : 3] for a 6-constant OLDROYD fluid and GIESEKUS [1963 : 3] for the third-order fluid.

10. SUBBARAMAN, *et al.* [1971 : 27] present additional experimental evidence to confirm the above equation, and state that the theory of CASWELL [1970 : 6] is most appropriate for falling sphere viscometry. See also MASHELKAR, *et al.* [1972 : 8].

11. [1970 : 6].

$U(b/R)$, is related to $U(D)$ through[12]

$$U(b/R) = U(D) + 0.6977\,(b/R)^2\,(D/6\pi\,\eta_0\,R) + O(b^2/R^4). \qquad (47.6)$$

Thus given a sphere of radius a, falling in a tube of radius R with a *weight D in the fluid* and eccentricity b relative to the center line of the tube, one measures $U(b/R)$, computes $U(D)$ from (47.6) and $U_\infty(D)$ from (47.4). Using the definition $(47.3)_1$, one is then led to the determination of the apparent viscosity.

We have presented the results of CYGAN and CASWELL[13] (see Fig. 47.1). It is interesting that the theory predicts that $1/\eta$ is proportional to τ^2 while

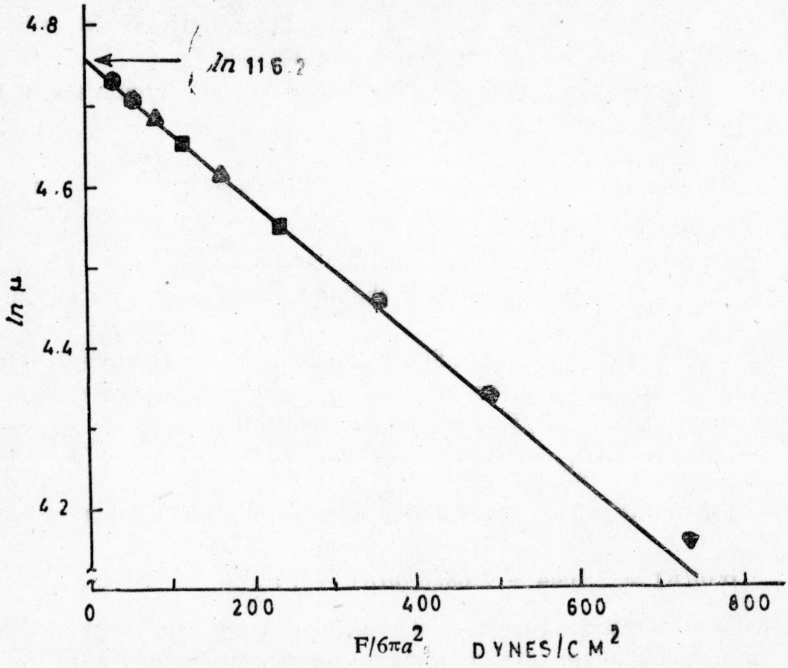

Fig. 47.1. **Log of the apparent viscosity versus shear stress—▼ carbide; ● : nylon, ▲ : ruby; ■ : steel.**

experiments indicate that $1/\eta$ is proportional[14] to τ. Indeed, according to PETERLIN[15] the habit of plotting the viscosity η vs log τ seems to indicate that $1/\eta$ has no term linear in τ as $\tau \to 0$, but if one were to plot η vs $1/\tau$, then at low values of τ one would observe that η was proportional to $1/\tau$. It seems to us that a rational explanation of this disagreement between theory and ex-

12. CYGAN and CASWELL [1971 : 7]. Eqs. (8)–(9).
13. CYGAN and CASWELL [1971 : 7].
14. This had been observed earlier by TURIAN [1964 : 26].
15. [1968 : 19], pp. 234-236.

periment lies in that the experiments did not reach the Newtonian regime rather than that the analytical predictions are in error.

We now comment upon other attempts to determine the zero-shear viscosity η_0. In Newtonian fluids, if a part of a rheometer instrument (e.g. the upper plate of the rheogoniometer in a parallel plate configuration) is turned through a small angle and released, then the damped oscillations provide the viscosity[16] η_0. However, attempts[16,17] to determine this viscosity for viscoelastic liquids by observing the damping of free oscillations have so far not made available easy to work with formulae.

47.2. Dynamic Viscosity about a State of Rest[18]. By the expression dynamic viscosity about a state of rest, we mean the measurement of the shear stresses in a liquid specimen being subjected to forced oscillations, and the computation of the complex viscosity $\eta^*(\omega)$ from the shear stress $\tau(\omega)$, given by

$$\eta^*(\omega) = \tau(\omega)/\omega. \tag{47.7}$$

There are many basic apparatus systems which yield the desired information, of which we list a few.

47.21. THE CONCENTRIC CYLINDER APPARATUS. Here the outer cylinder of radius $R_2(> R_1)$ is subjected to forced harmonic oscillations while the inner cylinder of radius R_1 is constrained by a torsional wire. The liquid in-between the cylinders is assumed to execute periodic oscillations.

Let θ_1 and θ_2 be the angular amplitudes of the inner and outer cylinder respectively; let $\Theta = \theta_1/\theta_2$, and δ be the phase lag of the inner cylinder behind the outer cylinder. Then it can be shown that[19]

$$\frac{\exp(i\delta)}{\Theta} = 1 + \alpha^2\left[\frac{\delta R_1^2}{2R_2 R_1^2}(4 R_2R_1 - 3R_1^2 - R_2^2) - \frac{R_1(R_2 - R_1)^2}{2 R_2}\right], \tag{47.8}$$

from which α is to be computed, with

$$S = \frac{1}{2\pi \rho R_1^3 L}\left(\frac{K}{\omega^2} - I\right). \tag{47.9}$$

Here ρ and L are respectively the density and height of the liquid column, K the restoring constant of the torsion wire, and I the polar moment of inertia of the inner cylinder.

Let us define a new quantity T through

$$T = \rho\left[\frac{SR_1^2}{2 R_2R_1^2}(4 R_2R_1 - 3 R_1^2 - R_2^2) - \frac{R_1(R_2 - R_1)^2}{2 R^2}\right]. \tag{47.10}$$

16. WALTERS and KEMP [1968 : 27].
17. JONES and WALTERS [1968 : 11].
18. For experimental data, see Figs. 39.1 and 39.2.
19. A summary of results needed here can also be found in WALTERS [1968 : 25]. See also WALTERS [1961 : 6].

Then the dynamic viscosity $\eta'(\omega)$ and dynamic rigidity $G'(\omega)$ are given by

$$\eta'(\omega) = -\frac{\omega\, T\, \Theta \sin \delta}{\Theta^2 - 2\Theta \cos \delta + 1}, \tag{47.11}$$

$$G'(\omega) = \frac{\omega^2 T\, \Theta\, (\cos \delta - \Theta)}{\Theta^2 - 2\, \Theta \cos \delta + 1}. \tag{47.12}$$

Measuring Θ and δ and computing T leads to the desired results immediately.

47.22. THE CONE-AND-PLATE SYSTEM. In this configuration, the bottom plate is excited externally, while the cone is used to record the transmitted forces. It can be shown[20] that if I represents the polar moment of inertia of the cone, R the wetted cone radius, and S is given by

$$S = \frac{3}{2\pi\rho\, R^S}\left(\frac{K}{\omega^2} - I\right), \tag{47.13}$$

with

$$T = S\, \alpha_0\, R^2 \rho, \tag{47.14}$$

where α_0 is the gap angle, then $\eta'(\omega)$ and $G'(\omega)$ are again given by (47.11) and (47.12) respectively, provided inertia is neglected.

47.23. THE PARALLEL-PLATE SYSTEM. In this system, one keeps the bottom plate as the driving member, and measures the amplitude ratio and phase lag of the upper plate to obtain $\eta'(\omega)$ and $G'(\omega)$. In this set up, these dynamic properties are again given by[21] (47.11) and (47.12) respectively, but

$$T = Sh^2\rho, \qquad S = \frac{2}{\pi\rho R^4 h}\left(\frac{K}{\omega^2} - I\right), \tag{47.15}$$

where h is the gap between the parallel plates oscillating about a common axis and each plate has a radius R; again K is the stiffness of the torsion wire and I the polar moment of inertia of the top plate.

In each of the above three instruments, one assumes a periodic velocity field, uses the constitutive relation of finite linear viscoelasticity to compute the torque on the part held by the torsion wire and uses this in the equation of motion of that part to determine $\eta'(\omega)$ and $G'(\omega)$.

To illustrate the calculations involved, let us consider the parallel-plate set up. Here the liquid, held between the plates, undergoes azimuthal oscillations. Let the velocity field be given by[21]

$$\dot{r} = \dot{z} = 0, \qquad \dot{\theta} = Re(F(z)\, e^{i\omega t}), \tag{47.16}$$

where $F(z)$ is complex. For this velocity field, the only non-vanishing shear stress is $T\langle \theta z \rangle$, when one uses the constitutive equation (34.b5). This is

20. WALTERS and KEMP [1968 : 26].
21. WALTERS and KEMP [1968 : 27].

given by

$$T\langle\theta z\rangle = r\ \frac{dF}{dz}\ e^{i\omega t} \int_0^\infty G(s)\, e^{-i\omega s}\ ds. \tag{47.17}$$

In the θ-direction, this stress appears in the equations of motion (*A*6.3)—(*A*6.4) to yield

$$\partial T\langle\theta z\rangle/\partial z = \rho\gamma\ F(z)\ e^{i\omega t}\ i\omega. \tag{47.18}$$

Now (47.17) and (47.18) imply that $F(z)$ obeys

$$(d^2F/dz^2) + \alpha^2 F = 0, \tag{47.19}$$

$$\alpha^2 = -\left(i\omega\rho \Big/ \int_0^\infty G(s)\, e^{-i\omega s}\ ds\right) = -\left\{i\omega\rho/[\eta'(\omega) - i(G'(\omega)/\omega)]\right\} \tag{47.20}$$

where we have used (34.*b*10). The boundary conditions on $F(z)$ are obtained by assuming that the bottom plate ($z = 0$) executes forced oscillations and that it leads the upper plate ($z = h$) by δ in phase. Then

$$F(0) = i\omega\theta_2\, e^{i\delta}, \qquad F(h) = i\omega\theta_1. \tag{47.21}$$

The solution of (47.19), subject to (47.21), is

$$\left.\begin{aligned} F(z) &= A \sin \alpha z + B \cos \alpha z, \qquad B = i\omega\theta_2\, e^{i\delta}, \\ A &= i\omega\,(\theta_1 \operatorname{cosec} \alpha h - \theta_2\, e^{i\delta} \cot \alpha h). \end{aligned}\right\} \tag{47.22}$$

The couple C on the upper plate is given by

$$C = \int_0^R T\langle\theta z\rangle\, 2\pi r^2 dr = \frac{\pi\,\omega^2 R^4}{2\alpha}\ (-\theta_1 \cot \alpha h + \theta_2\, e^{i\delta} \operatorname{cosec}\ \alpha h)\, e^{i\omega t}. \tag{47.23}$$

But the equation of motion for the upper plate is

$$C = (K - I\,\omega^2)\,\theta_1\, e^{i\omega t}, \tag{47.24}$$

and hence

$$\frac{e^{i\delta}}{\Theta} = \cos \alpha h + S\,\alpha h \sin \alpha h, \tag{47.25}$$

where S obeys (47.15). For small values of αh, (47.25) reads

$$\frac{e^{i\delta}}{\Theta} = 1 + \alpha^2 h^2\left(S - \frac{1}{2}\right) + \alpha^4 h^4\left(\frac{1}{24} - \frac{S}{6}\right) + O(\alpha^6 h^6). \tag{47.26}$$

If inertia can be ignored, then

$$\frac{e^{i\delta}}{\Theta} = 1 + S\,\alpha^2 h^2. \tag{47.27}$$

Measuring Θ and δ, and solving for α from (47.27), or the exact transcendental equation (47.25), lead to (47.11) and (47.12) for the determination of $\eta'(\omega)$ and $G'(\omega)$.

But a word of caution. If $\omega = \omega_0$, where ω_0 is the natural frequency of the instrument, that is, ω_0 obeys

$$K - I\,\omega_0^2 = 0, \tag{47.28}$$

then (47.11) and (47.12) have no meaning, because in this case $\omega = \omega_0$, $\Theta = 1$, $\delta = 0$. Thus measurements near ω_0 require the utmost care ; otherwise small errors tend to be magnified considerably.[22] So, one should take readings far from ω_0 and assume continuous curves for $\eta'(\omega)$ and $G'(\omega)$ through this frequency.

47.24. OTHER ROTATIONAL RHEOMETERS. The measurement of $\eta^*(\omega)$ at very low frequencies is not easily performed in the above three rheometers. Indeed, there is an almost total lack of data for $\eta^*(\omega)$ at extremely low values of ω. One bonus of the determination of $\eta^*(\omega)$ as $\omega \to 0$ is the test which determines whether the fluid being sheared is a simple fluid with fading memory or not.[23] It is therefore of interest to consider a new series of rheometers in which the fluid is subjected to a motion which is almost equivalent to a rigid rotation ; the small periodic strain that is introduced onto the fluid element occurs because the axes of rotation of the two elements containing the fluid are not collinear. As illustrations, recall the MAXWELL orthogonal rheometer, the KEPES apparatus and the eccentric rotational viscometer. It is hard to recommend any one of these as being superior to the others, since in each instrument the free member (on which the forces or the torque are measured) does not rotate at the same speed as the driven member. In other words, there is a slip between the two rotating parts despite the use of air bearings. For example, in the MAXWELL rheometer flow, this means that the angular velocity Ω varies across the fluid gap and the motion is no longer a MWCSH. Let us examine the consequences of this. Let Ω_1 be the angular velocity of the driven member and let $\Delta\Omega$ be the slip. Then, assuming a linear angular velocity distribution, the angular velocity is given by (see Sec. 7 for notation):

$$\Omega(z) = \Omega_1\left(1 - \frac{\alpha}{b}\,z\right), \qquad \alpha = \Delta\Omega/\Omega_1. \tag{47.29}$$

The velocity field is given by [*cf.* (7.30)] :

$$\begin{aligned} \dot{x} &= -\Omega(z)\,y + \Omega(z)\,\psi\,z, \\ \dot{y} &= \Omega(z)\,x, \\ \dot{z} &= 0. \end{aligned} \tag{47.30}$$

22. JONES and WALTERS [1971 : 15] offer this as the explanation for the anomalous behaviour of the rheogoniometer near ω_0.

23. See Eq. (34.C39) and the ensuing discussion.

The path lines corresponding to (47.30) are :

$$\begin{aligned} \xi &= x \cos \Omega(z)\, s + (y - \psi z) \sin \Omega(z)\, s, \\ \eta &= -x \sin \Omega(z)\, s + (y - \psi z) \cos \Omega(z)\, s + \psi z, \\ \zeta &= z. \end{aligned} \tag{47.31}$$

The strain history for this flow is :

$$[C_t(t-s)] = \begin{bmatrix} 1 & 0 & -\psi \sin \Omega(z)\, s - \dfrac{\alpha s \Omega_1}{b}(y - \psi z) \\ . & 1 & -\psi(1 - \cos \Omega(z)\, s) \\ . & . & 1 + 2\psi^2(1 - \cos \Omega\, s) + 2(y - \psi z)\dfrac{\psi \alpha \Omega_1}{b}\, s + \\ & & + \dfrac{\alpha^2 \Omega_1^2}{b^2}(y - \psi z)^2 s^2 \end{bmatrix} \tag{47.32}$$

Comparison of (47.32) with (7.45) shows that the $\langle 31 \rangle$ and $\langle 33 \rangle$ components have been altered. Hence, the constitutive equation of finite linear visco-elasticity (if it applies) would yield [*cf.* (34.C10) — (34.C11)] :

$$T\langle xz \rangle = \psi\, G''(\Omega) + \frac{\alpha(y - \psi z)\, \Omega_1}{b}\, \eta_0, \tag{47.33}$$

$$T\langle yz \rangle = \psi\, G'(\Omega). \tag{47.34}$$

On the upper plate, where $\psi z = a$, the force normal to the line of centers (that is, the x-axis) is altered by the slip,[24] while the force along the line of centers (the x-axis) is not—these two statements depend on the constitutive equation used, of course.

An ingenious method, which is useful in the MAXWELL rheometer, to eliminate the slip is to blow air over vanes attached to the free member so as to increase its speed.

The foregoing comments indicate the reasons for the remarks following the elaborate analysis of BROADBENT and WALTLRS for the eccentric cylinder viscometer presented in Sec. 45.

47.3. Viscometric Flow or Steady Shear Viscosity $\bar{\eta}(\varkappa^2)$[25]. The role of the viscometric flow viscosity $\bar{\eta}(\varkappa^2)$ in determining the velocity profiles of viscometric flows was examined in detail in Secs. 28-30. From the POISEUILLE flow, the COUETTE flow and the cone-and-plate analyses, we shall now record the methods for determining $\bar{\eta}(\varkappa^2)$. The basic reference in this field is the book by VAN WAZER, *et al.*[26], where many other methods are discussed.

---

24. These comments have been included because Professor TANNER showed me that such is the case experimentally.

25. See Fig. 39.2 or 48.5.

26. [1963 : 8].

47.31. POISEUILLE FLOW METHOD OR CAPILLARY VISCOMETER. In POISEUILLE flow [see (28.34)], the flow rate Q was shown to be given by

$$Q = \frac{8\pi}{c^3} \int_0^{cR/2} \tau^2 \varkappa(\tau) \, d\tau, \tag{47.35}$$

where R is the radius of the pipe, $\varkappa(\tau)$ is the shear rate corresponding to a shear stress τ, and c is the pressure drop per unit length. Since the wall shear stress τ_w is [*cf.* (28.9)]

$$\tau_w = cR/2, \tag{47.36}$$ [27]

(47.35) may be rewritten as

$$\frac{Q}{\pi R^3} = \frac{1}{\tau_w^3} \int_0^{\tau_w} \tau^2 \varkappa(\tau) \, d\tau. \tag{47.37}$$

Let $F = -4Q/\pi R^3$. Then (47.37) may be differentiated with respect to τ_w to yield

$$\varkappa(\tau_w) = \frac{3}{4} F(\tau_w) + \frac{\tau_w}{4} \frac{dF}{d\tau_w}. \tag{47.38}$$

This is often called the *Weissenberg-Rabinowitsch-Mooney equation,* though MARKOVITZ[28] submits evidence to show that the sole credit for the derivation of the above equation goes to WEISSENBERG, who obtained this around 1928-9.

These historical comments aside, measuring Q and τ_w through

$$\tau_w = \frac{\Delta P}{L} \frac{R}{2}, \tag{47.39}$$

where ΔP is the pressure drop over a length L, leads one to compute the shear rate at the wall from (47.38). But as TANNER and WILLIAMS[29] have remarked, the experimental scatter makes the task of computing $dF/d\tau_w$ very difficult. So, these authors have developed an iterative numerical technique for inverting integral equations of the type :

$$f(x) = \int_a^b K(x, s) \, \varphi(s) \, ds. \tag{47.40}$$

In other words, given $f(x)$ at a set of discrete points, and given $K(x, s)$, their method can be used to determine $\varphi(s)$. The results obtained from their method applied to (47.37) agree well with those obtained by using (47.38).

In actual practice, if one uses the capillary to generate a POISEUILLE flow and wishes to use (47.38), then a host of corrections become necessary. First of all, the fluid will travel a certain length of the tube before the flow is fully developed ; these are called *entrance effects.* There are losses at the exit

27. Putting τ_w to be $-R/2$ will also yield (47.37).
28. [1968 : 15].
29. [1970: 24]. The computer program may be obtained from these authors.

because the velocity profile is affected by the presence of the exit hole. There are other losses such as the kinetic energy loss due to the fluid leaving with this energy; and there may even be some elastic stored energy loss due to the fluid acquiring some stored energy in the deformation and not releasing it in the capillary. Corrections to such effects are discussed by VAN WAZER, *et al.*[30] We shall not go into these in detail here.

47.32. COUETTE FLOW METHOD OR THE COAXIAL CYLINDER VISCOMETER. The theory for the flow between two concentric cylinders with the inner one rotating and the outer one fixed was given in (28.24)—(28.31). Let us consider the opposite situation here, that is, let $\omega(R_1) = 0$, $\omega(R_2) = \Omega$, $R_2 > R_1$; moreover, let τ_1 be the shear stress on the inner cylinder. Then the MOONEY equation (28.26) becomes[31]

$$2\tau_1(d\Omega/d\tau_1) = \varkappa(\tau_1) - \varkappa(\beta^{-2}\tau_1), \tag{47.41}$$

where $\varkappa(\tau_1)$ is the shear rate-shear stress function and $\beta = R_2/R_1 > 1$. Since $\varkappa(0) = 0$, by letting

$$h(\tau_1) = 2(d\Omega/d \ln \tau_1), \tag{47.42}$$

one can recast (47.41) as

$$\sum_{n=0}^{\infty} h(\beta^{-2n}\tau_1) = \varkappa(\tau_1). \tag{47.43}$$

KRIEGER has shown that $\varkappa(\tau_1)$ may be approximated as

$$\varkappa(\tau_1) = \frac{2\,\Omega\,N}{1-\beta^{-2N}}\left\{1 + \frac{N'}{N^2} f(2N \ln \beta)\right\}, \qquad (47.44)^{32}$$

where

$$N = \frac{1}{\Omega}\frac{d \ln \Omega}{d \ln \tau_1}, \qquad \frac{N'}{N^2} = -\frac{d}{d \ln \tau_1}(1/N),$$

$$f(t) = \{t[e^t(t-2) + (t+2)]\}/2(e^t - 1)^2. \tag{47.45}$$

For power law fluids, $\varkappa(\tau) = a\tau^N$ and hence[33]

$$\varkappa(\tau_1) = 2\,\Omega\,N/(1 - \beta^{-2N}). \tag{47.46}$$

Thus, Eq. (47.44) presents the correction to the power law expression if it is needed and moreover, since $f(t) \leqq 0.11$, this correction is usually small.

To use (47.44), one measures the total torque T on the inner cylinder, divides it by $2\pi R_1^2 L$ to obtain τ_1. Then a graph of log Ω vs. log τ_1 is prepared. At each value of τ_1, one obtains the slope N, permitting the calculation of the power law shear rate $\varkappa(\tau_1) = 2 N \Omega/(1 - \beta^{-2N})$. When needed,

30. [1963 : 8], see pp. 211-215.
31. We follow KRIEGER's [1968 : 13] presentation.
32. [1968 : 13]. For the exact result, see [1969 : 13].
33. FARROW, *et al.* [1928 : 1].

one can obtain the correction term by plotting $1/N$ against $\log \tau_1$, which yields $(-N^{-2}N')$.

There are two major sources of error in this instrument. The first is the so-called end correction,[34] which arises because the inner cylinder usually has a flat base and its motion relative to the base of the outer tube exerts an extra torque on the inner cylinder; the second is due to the free surface. VAN WAZER, *et al.*[35] state that if l is the clearance between the base of the cylinder and that of the tube and if $l \gg R_1$, and $R_1/R_2 \doteq 1$, then the correction becomes negligible. However, ROSCOE[36] has shown that reducing the annular gap does not eliminate the end correction. For example, if the gap is 1/10 of the vessel radius, the correction is 12%. An exceptionally good instrument which eliminates the need for end correction by keeping the base of the outer (rotating) cylinder also stationary is described in detail by KAYE and SAUNDERS[37]. But it is still necessary to keep the free surface at a height above the rim of the cylinder to reduce free surface effects.

47.33. CONE-AND-PLATE SYSTEM. The idea of shearing a viscoelastic liquid, held between a rotating plate and a cone restrained by a torsion wire to measure the viscosity is the basis of the WEISSENBERG rheogoniometer. The theory for the flow was presented in (27.28)—(27.33).

The rate of shear in the gap is $\varkappa = \Omega/\alpha_0$, where Ω is the angular velocity of the bottom plate and α_0 is the gap, and $\varkappa$ is thus a constant. The shear stress on the cone being given by $\tau(\varkappa)$, the torque M on it is

$$M = \tau(\varkappa) \cdot 2\pi R^3/3, \tag{47.47}$$

and thus

$$\bar{\eta}(\varkappa^2) = 3\alpha_0 M/2\pi R^3 \Omega, \tag{47.48}$$

where R is the wetted radius of the cone.

The simple expression relating the viscosity and the rate of shear is the reason why the cone-and-plate instrument has become extremely popular for measuring $\bar{\eta}(\varkappa^2)$. However, as with the other two methods, there are some errors associated with the use of this instrument. Of course, the machining of the cone is especially important, and the errors caused in the measurement of $\bar{\eta}(\varkappa^2)$ by an imperfect cone have been discussed by CHENG and DAVIS.[38] We shall not elaborate this point, though it is quite important in its own right.

However, we shall comment on the effect of the flow not persisting as a viscometric flow up to the free surface, on secondary flows induced by

34. HIGHGATE and WHORLOW [1969: 7] report values up to 36% for non-Newtonian fluids.

35. [1963 : 8], p. 70.

36. [1962 : 10].

37. [1964 : 11].

38. [1968 : 5].

the neglect of inertia, and on the importance of keeping the angular gap $\alpha_0(\leqq 4°)$ small.

For the flow to remain viscometric throughout the entire gap and at the free boundary, the free surface must be spherical. Let us examine this point next. If $\mathbf{n}$ is the unit normal at a point in the free surface with $n\langle r\rangle$, $n\langle\theta\rangle$, $n\langle\varphi\rangle$ being its physical components, then the components of the stress vector $\mathbf{t}$ on the boundary are

$$\mathbf{t} = \{T\langle rr\rangle\, n\langle r\rangle, \quad T\langle\theta\theta\rangle\, n\langle\theta\rangle + T\langle\theta\phi\rangle\, n\langle\phi\rangle, \\ T\langle\theta\phi\rangle\, n\langle\theta\rangle + T\langle\phi\phi\rangle\, n\langle\phi\rangle\}. \quad (47.49)$$

But the atmospheric pressure has the form $\{p_a\, n\langle r\rangle, 0, 0\}$. Hence this force will not be compatible with $\mathbf{t}$ unless $n\langle\theta\rangle = 0$, and $n\langle\varphi\rangle = 0$. Even if we take $n\langle\varphi\rangle = 0$ by assuming symmetry of the surface about the axis of rotation, the situation does not improve appreciably, since

$$\mathbf{t} = \{T\langle rr\rangle\, n\langle r\rangle, \qquad T\langle\theta\theta\rangle\, n\langle\theta\rangle, \qquad T\langle\theta\phi\rangle\, n\langle\theta\rangle\}. \quad (47.50)$$

For the Newtonian fluid, the pressure field is a constant p_0 [see (27.32)], and this is the value of $-T\langle rr\rangle$. Also, $T\langle\theta\theta\rangle = -p_0$. So unless $n\langle\theta\rangle = 0$, or the surface is spherical, the flow will not be viscometric up to the boundary even in Newtonian fluids, and secondary flows may occur near the edge. At present no detailed analysis of this is available[39], though the domain perturbation theory of Sec. 44 could be used to derive the answers.

The effects of secondary flows arising out of the neglect of inertia and maintaining a large angular gap have been examined by WALTERS and WATERS.[40] That the existence of secondary flows is not an academic exercise has been known ever since such flows were computed to exist and experimentally verified to occur around rotating spheres and in cone-and-plate devices, etc.[41]

For the purely viscous fluid, the secondary flow in an infinite cone-and-plate system with a large cone angle ($\alpha_0 \doteqdot 30°$) consists of a radially inward migration along the flat plate and an outward one along the cone (if the cone is rotating). For a viscoelastic liquid, under the same set up, this secondary flow pattern exists far away from the axis of rotation, while closer to the axis there is a closed cell like structure (see Figs. 47.2-47.3). These effects were

---

39. TURIAN [1972 : 15] has considered this problem for Newtonian fluids, by using perturbation techniques in terms of a 'REYNOLDS number' and a function of the cone angle. See the paper by JOSEPH [1974: 2], where such an analysis is developed for the torsional flow. This paper apppeared after this book was finished.

40. [1968 : 28].

41. See LANGLOIS [1963 : 5] for calculations of the flow between rotating spheres. GIESEKUS has demonstrated, both by theory and experiment, the existence of secondary flows in a variety of geometries [1965 : 7], [1967 : 8], [1967 : 9], [1968 : 9] such as in a cone-and-plate device, flow around a sphere, flow through an orifice, etc.

studied by WALTERS and WATERS[42] and they concluded that if $\alpha_0 \leqq 4°$, the neglect of secondary flows and the inertia has less than 0.25% effect on the

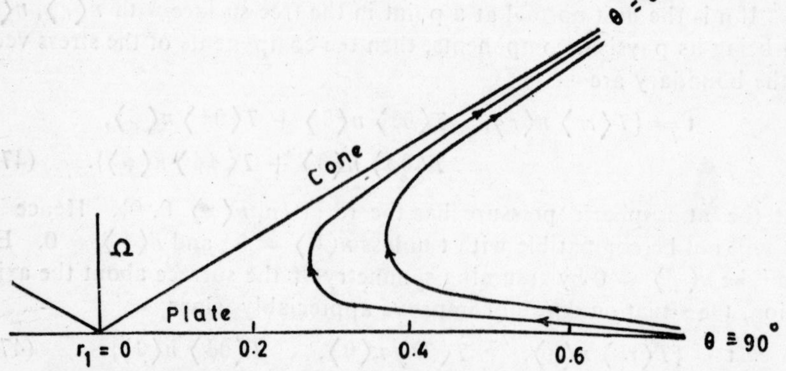

Fig. 47.2. Typical streamlines of the secondary flow for a Newtonian liquid between a cone (of angle 60°) and a plate.

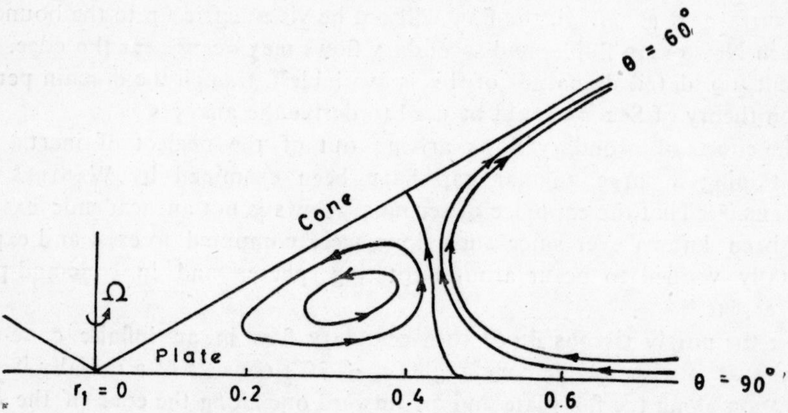

Fig. 47.3. Typical streamlines of the secondary flow for a second-order liquid between a cone (of angle 60°) and a plate with $\beta + \gamma = 0.5$. From WALTERS and WATERS [1968 : 28].

couple measured. A more elaborate analysis of secondary flows by KING and WATERS[43] supports this conclusion, indicating that corrections become necessary for $\alpha_0 > 4°$.

However, we emphasize that so far no discussion of the effect of the secondary flows near the free surface exists.

It is of course possible to immerse the cone in a bath and to avoid the free surface effects completely. Such a set up is more useful in the measure-

42. [1968 : 28].
43. [1970 : 16].

ment of the first normal stress difference and will be considered in Sec. 48. In passing, we note that GRIFFITHS and WALTERS[44] have discussed the immersed cone-and-plate and parallel-plate systems with second-order fluids and found that if in the former the cone angle $\alpha_0 \leqq 4°$ and in the latter if the ratio of the gap to the radius $h/R < 0.07$, then the error in the torque measurements is less than 2%.

47.4. Dynamic Viscosity in Superposed Motions[45]. In Sec. 39.2, we have discussed the conceptual techniques for the determination of the dynamic viscosities $\eta'_{\|}(\varkappa, \omega)$ or $\eta'_{\perp}(\varkappa, \omega)$. Here we shall re-examine these techniques briefly in the light of the theory behind the measurements of $\eta'(\omega)$ and $G'(\omega)$ in various instruments presented above.

In a cone-and-plate instrument, if an oscillatory motion is superposed on the steady rotation of the bottom plate and the amplitude ratio Θ of the oscillations and phase lag δ of the cone, held by a torsion wire, are measured, one can again compute the dynamic moduli from the formulae[46] [*cf.* (47.11)—(47.12)] :

$$\eta'_{\|}(\varkappa, \omega) = -\omega T \Theta \sin \delta/(\Theta^2 - 2\Theta \cos \delta + 1),$$

$$G'_{\|}(\varkappa, \omega) = \omega^2 T\, \Theta(\cos \delta - \Theta/(\Theta^2 - 2\Theta \cos \delta + 1), \tag{47.51}$$

with T given by [*cf.* (47.14)]

$$T = \frac{3\alpha_0}{2\pi R^3}\left(\frac{K}{\omega^2} - I\right). \tag{47.52}$$

Of course, the influence of $\varkappa$ on $\eta'_{\|}$ and $\eta'_{\perp}$ will be apparent through Θ and δ. Various authors have reported experimental results for the above dynamic moduli as well as the plase lag δ.[47]

To obtain the dynamic moduli $\eta'_{\perp}(\varkappa, \omega)$ and $G'_{\perp}(\varkappa, \omega)$, when the oscillations are orthogonal to steady shearing, is not possible in a commercially available instrument. SIMMONS[48] constructed his own apparatus and reported his conclusions[49], which we have already mentioned in Sec. 39.2. For his apparatus, SIMNONS has recorded the technique used in measuring these dynamic moduli and we refer the reader to his paper. More recently, TANNER and WILILAMS[50] have repeated SIMMONS' experiments with a modified apparatus and have obtained improved results.

This section will not be complete without a mention of the relations between these dynamic moduli $\eta'(\omega)$ and $G'(\omega)$ and the relaxation spectrum,

44. [1970 : 12]. This conclusion supersedes that in [1969 : 6].
45. See Figs. (39.1) and (39.2) for data.
46. JONES and WALTERS [1971 : 14].
47. See the references in Sec. 39.2.
48. [1966 : 18].
49. [1968 : 22].
50. [1971 : 28].

as well as the meaning of relaxation and retardation times. This is undertaken in what follows.

47.5. Relaxation Functions. Consider the constitutive relation (A5.17) as an example. If, for all instants $t \geqq t_0$, the velocity field vanishes, we obtain

$$\left(1 + \lambda_1 \frac{\partial}{\partial t}\right) T^E_{ik} = 0, \tag{47.53}$$

or,

$$T^E_{ik}(t) = T^E_{ik}(t_0) \exp\left[-(t - t_0)/\lambda_1\right]; \tag{47.54}$$

or the stresses decay from the original value of $T^E_{ik}(t_0)$ to zero. At $t - t_0 = \lambda_1$, $T^E_{ik}(t_0 + \lambda_1) = \frac{1}{e} T^E_{ik}(t_0)$, and this is used in the definition of the *relaxation time* λ_1, as the time interval during which the stresses decay to $1/e$ times the value they had at the cessation of motion.

For the same constitutive relation consider the form, obtained when $T^E_{ik} = 0$ for all $t \geqq t_0$. This reads :

$$\left(1 + \lambda_2 \frac{\delta}{\delta t}\right) d_{ik} = 0. \tag{47.55}$$

Omitting the product terms in the operator $\delta/\delta t$ [see (A5.11)], one derives

$$\left(1 + \lambda_2 \frac{\partial}{\partial t}\right) d_{ik} = 0, \tag{47.56}$$

or,

$$d_{ik}(t) = d_{ik}(t_0) \exp\left[-(t - t_0)/\lambda_2\right], \tag{47.57}$$

which leads to the definition of the *retardation time* λ_2. It is in this context that λ_1 and λ_2 are respectively called the *relaxation* and *retardation times* in (36.24).

Turning to the more general case, let the memory function $G(s)$ in (34.*b*5) be represented as

$$G(s) = \sum_{n=1}^{N} G_n \exp\left(-s/\lambda_n\right). \tag{47.58}$$

The G_n and λ_n are constants, and the λ_n are called *relaxation times*. If one uses (47.58) or a more general function in the finite linear theory to describe a relaxation experiment one ends up with a mean relaxation time T. In many instances, one could visualize T as the major memory of a fluid [*cf*. Sec. 35)].

In the literature[51], many methods are available to determine the λ_n experimentally. These can be obtained from dynamic viscosity measurements, or stress relaxation experiments, or creep tests. Rheologists have found that it is easier to find the relaxation spectra, to be defined below, rather than $G(s)$.

51. See e.g. FERRY [1970 : 9].

In Sec. 34 bis, we have shown that the dynamic moduli are related to the shear relaxation modulus $G(t)$ by

$$\eta'(\omega) - i\,\frac{G'(\omega)}{\omega} = \int_0^\infty G(s)\,e^{-i\omega s}\,ds. \tag{47.59}$$

It is usually assumed that $G(t)$ can be written as the LAPLACE transform of a function[52] $M(s)$, that is,

$$G(t) = \int_0^\infty e^{-st}\,M(s)\,ds. \tag{47.60}$$

There are other ways of writing $G(t)$:

$$G(t) = \int_0^\infty e^{-t/\tau}\,F(\tau)\,d\tau = \int_{-\infty}^\infty e^{-t/\tau}\,H(\tau)\,d\ln\tau, \qquad (47.61)^{53}$$

where $F(s^{-1}) = s^2 M(s) = sH(s^{-1})$. Both $F(\tau)$ and $H(\tau)$ are called *relaxation spectra.* In terms of these new functions,

$$G'(\omega) = \int_0^\infty \frac{\omega^2 M(s)}{s^2+\omega^2}\,ds = \int_0^\infty \frac{F(\tau)\,\omega^2\tau^2}{1+\omega^2\tau^2}\,d\tau$$
$$= \int_{-\infty}^\infty \frac{H(\tau)\,\omega^2\tau^2}{1+\omega^2\tau^2}\,d\ln\tau\,; \tag{47.62}$$

$$G''(\omega) = \int_0^\infty \frac{M(s)\,\omega s}{s^2+\omega^2}\,ds = \int_0^\infty \frac{F(\tau)\,\omega\tau}{1+\omega^2\tau^2}\,d\tau =$$
$$= \int_{-\infty}^\infty \frac{H(\tau)\,\omega\tau}{1+\omega^2\tau^2}\,d\ln\tau. \tag{47.63}$$

There are many iterative schemes to find $H(\tau)$, given $G''(\omega)$. Previously, in discussing the capillary viscometer theory, we have mentioned the work of TANNER and WILLIAMS[54], which could be used here. Perhaps, one should follow SHROFF[55] and conclude that the best technique to compute $H(\tau)$ is the one due to FERRY and WILLIAMS[56]. Many other iterative procedures are available in the literature to which the reader is referred.

Leaving the linear theory aside, can one perform similar experiments and determine the memory functions for nonlinear constitutive relations? An answer to this was given by YAMAMOTO[57], who defined a relaxation spectrum for the BIRD-CARREAU equation (36.31) and showed that the stress relaxation experiment, the experimental measurements in the initiation of simple shearing flow and dynamic viscosity measurements can be used to predict this

---

52. See e.g. FERRY [1970 : 9]. If (47.59) holds, the spectrum is said to be *continuous.* If (47.58) holds, the spectrum is said to be *discrete.*
53. For the models of finite linear viscoelasticity discussed in Sec. 45, $F(\tau) = N(\tau)/\tau$.
54. [1970 : 24]. See also TANNER [1968 : 23].
55. [1971 : 23].
56. [1952 : 1]. See also [1970 : 9].
57. [1971 : 30].

spectrum. It would be interesting to see these ideas extended to other integral models.

Finally, we also note the interesting work of KING and WATERS[58], who have shown that the function $N(\tau)$ in (46.24) can be determined by the unsteady motion of a sphere moving from rest under the action of a constant force in a viscoelastic liquid. The limitations and extensions of this procedure should also be explored.

48. Measurement of the Normal Stress Differences

There are quite a number of systems which measure N_1 and N_2. Here we shall examine COUETTE flow (gives N_1), parallel-plate flow (combination of N_1 and N_2), cone-and-plate flow (both N_1 and N_2), as well as the flow between a body of revolution and a plate. Many excellent papers[59] on the measurement of normal stress differences of recent origin are recommended to the reader. However, our presentation looks at all the literature critically.

We shall outline the theory for the experimental measurement of N_1 and N_2 first and at the end of the section, we shall recommend the best techniques and present the results for some polymeric fluids.

48.1 Couette Flow Apparatus. The theory of the viscometric flow between two cylinders in relative rotation has been presented in Secs. 27 and 28. From (27.*b*8), the difference in the radial stresses on the outer and inner cylinders is given by

$$T\langle rr\rangle(R_2) - T\langle rr\rangle(R_1) = \int_{R_1}^{R_2}\left\{\frac{1}{r}\,\overline{N}_1(\varkappa^2(r_1)) - \rho r\omega^2(r)\right\}dr. \qquad (48.1)$$

To measure the radial stress, one can use

(i) slots parallel to the cylinder axis; or
(ii) slots which are circumferential; or
(iii) radial holes.

Note that flush mounted transducers cannot be placed on the curved surfaces and hence the above three exhaust the alternatives; of course, transducers may be used inside the slots or holes.

The problem that arises immediately is to ascribe a value to the hole error in each case. In Sec. 41, we have shown that for second-order fluids undergoing a creeping flow across a slot, (41.42) holds. One may assume that the set up (i) above approximates this type of flow and assume that for the simple fluid,

$$p_e(\varkappa_{R_i}) = -\frac{1}{4}\,\overline{N}_1(\varkappa^2_{R_i}), \qquad i = 1, 2. \qquad (48.2)$$

58. [1968 : 1].
59. PRITCHARD [1971 : 20]; PIPKIN and TANNER [1973 : 11]; MILLER and CHRISTIANSEN [1972 : 10]; OLABISI and WILLIAMS [1972 : 12].

Hence for parallel slots,

$$p_g(R_1) - p_g(R_2) = \frac{1}{4}\left[\overline{N}_1(\varkappa^2_{R_2}) - \overline{N}_1(\varkappa^2_{R_1})\right] + \\ + \int_{R_1}^{R_2}\left\{\frac{1}{r}\overline{N}_1(\varkappa^2(r)) - \rho r\omega^2(r)\right\} dr, \qquad (48.3)$$

where $p_g(R_i) = [T\langle rr\rangle(R_i)]_g$, with the subscript g referring to the gauge pressure. If p_u is the undisturbed normal pressure, then $p_e = p_g - p_u$.

For the case of the circumferential slot, we may assume that the flow is locally two-dimensional and rectilinear, and that $\overline{n}_2(\varkappa^2) = \alpha\overline{\eta}(\varkappa^2)$. In other words, we assume that the flow corresponds to that over a deep slot as in Sec. 42. Then the results of Sec. 30 on rectilinear flow are applicable and the formula for the pressure (30.21) becomes

$$p = (2a + \rho g)\, z + \frac{2a\,\overline{n}_2(\varkappa^2)}{\overline{\eta}(\varkappa^2)}\, w(x, y) + I_2(\varkappa^2), \qquad (48.4)$$

where we have used one of the following moments[60] of n_α:

$$I_\alpha(\varkappa^2) = \int_0^{\varkappa} \overline{n}_\alpha(\xi^2)\,\xi\, d\xi, \qquad \alpha = 1, 2. \qquad (48.5)$$

On $y = 0$, which is the wall (see Fig. 42.1), the undisturbed normal stress is

$$-T\langle yy\rangle\Big|_{y=0} = p_u = (2a + \rho g)z + I_2(\varkappa^2_W) - T_E\langle yy\rangle\Big|_{y=0} \qquad (48.6)$$

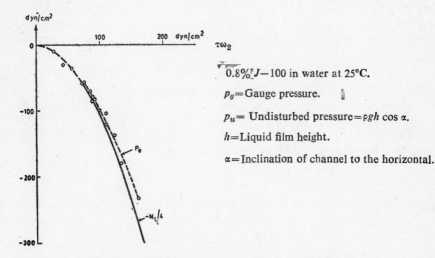

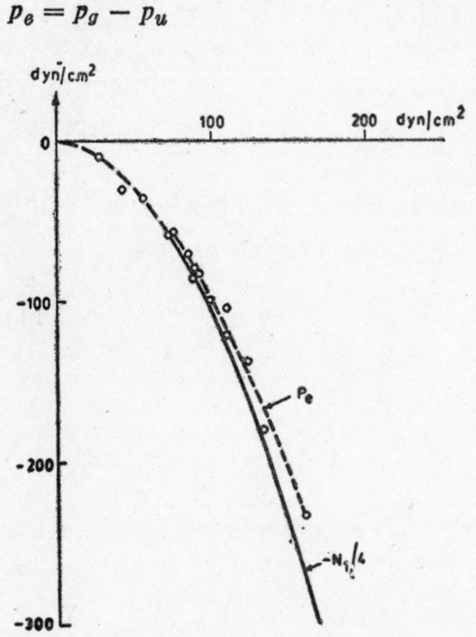

0.8% J—100 in water at 25°C.

p_g=Gauge pressure.

p_u= Undisturbed pressure=$\rho gh \cos\alpha$.

h=Liquid film height.

α=Inclination of channel to the horizontal.

Fig. 48.1. Pressure error compared with $-N_1/4$ [1969 : 23].

because $w \equiv 0$ on the wall. By using the constitutive relation (26.66) which

60. These moments occur in PIPKIN and TANNER [1973 : 11], Eq. (3.5).

corresponds to (30.21), one has that

$$T_E \langle yy \rangle \Big|_{y=0} = \overline{N}_2 (\varkappa_W^2). \tag{48.7}$$

In (48.6)—(48.7), $\varkappa_W$ is the wall shear rate $\left.\frac{\partial w}{\partial y}\right|_{y=0}$. Deep in the slot, that is, as $y \to -\infty$, $T \langle yy \rangle$ reduces to $(2a + \rho g)z$. Hence the pressure error for a rectilinear flow along a slot is[61]

$$p_e(\varkappa_W) = p_g - p_u = \overline{N}_2 (\varkappa_W^2) - I_2 (\varkappa_W^2). \tag{48.8}$$

If $N_2 = (\beta + 2\gamma)\,\varkappa^2$, that is, if it corresponds to the second-order fluid, then from (48.5) and (48.8), one recovers (42.11), the result due to KEARSLEY[62].

Pressure hole errors have also been deduced by HIGASHITANI and PRITCHARD[63]. They derive the value

$$p_e = -\frac{1}{2}\int_0^\tau \frac{\hat{N}_1(\lambda)}{\lambda}\, d\lambda \tag{48.9}$$

for the flow across a slot [*cf.* (48.2)], while the value for flow along a slot is

$$p_e = \int_0^\tau \frac{\hat{N}_2(\lambda)}{\lambda}\, d\lambda. \tag{48.10}$$

Now, in the derivation of (48.9) and (48.10), these authors assumed that $N_\alpha \propto \tau^{m_\alpha}$, $\alpha = 1, 2$, $m_\alpha \equiv$ const.

While it is not easy to choose between (48.2) and (48.9) theoretically[64], it is quite straightforward to note that (48.8) is better than (48.10). For, while both of these latter results depend on rectilinear flow along a slot, (48.8) is exact since it depends on $\overline{n}_2(\varkappa^2) \propto \overline{\eta}(\varkappa^2)$, a prerequisite for rectilinear flow. This proportionality is not used by HIGASHITANI and PRITCHARD in deriving (48.10).

These authors have deduced corrections for *circular holes* also. According to them,

$$p_e = -\frac{1}{3}\int_0^\tau \frac{\hat{N}_1(\lambda) - \hat{N}_2(\lambda)}{\lambda}\, d\lambda \tag{48.11}$$

in this case. For the second-order fluid, (48.11) yields

$$p_e = -\frac{1}{6}(N_1 - N_2). \tag{48.12}$$

It has been suggested by PIPKIN and TANNER[65] that an acceptable formula for

61. PIPKIN and TANNER [1973 : 11], Eq. (16.6).
62. [1970 : 15].
63. [1972 : 6], [1971 : 20].
64. Experimentally (48.2) is excellent. See Fig. 48.1. More recent results of NOVOTNY and ECKERT [1973 : 10] suggest that for a polyox solution, p_e is negative upto $\varkappa = 100$ sec^{-1}.
65. [1973: 11], Eq. (15.10).

circular holes may be taken as (for general fluids):

$$p_e = C_1 N_1 + C_2 N_2, \tag{48.13}$$

where $-C_1 = C_2 \doteq 1/6$ may be used. Since N_2 is small, p_e will be insensitive to the value of C_2.

If we accept (48.13) to be fairly reasonable, then the measured difference (48.3) is altered to

$$p_g(R_1) - p_g(R_2) = \frac{1}{6}\left[\bar{N}_1(\varkappa^2_{R_2}) - \bar{N}_1(\varkappa^2_{R_1})\right] + \int_{R_1}^{R_2}\left\{\frac{1}{r}N_1 - \rho r\omega^2\right\}dr, \tag{48.14}$$

when the contribution from N_2 is neglected. So KEARSLEY's observation[66] that the measured pressure difference was *lower* for the *parallel slot configuration* than the *circular hole configuration* is accounted for.

Proceeding in this way, we can see that if circumferential slots are employed, the measured and theoretical values should not differ widely since the contribution by N_2 to the hole error [*cf.* (48.8)] is very small. This has also been observed by KEARSLEY.

Suppose that one can make allowances for hole errors in this instrument, and find N_1. This may not be the true value because the rotation of the circular hole on the inner cylinder makes the motion spatially unsteady.

It follows therefore that the COUETTE flow, while theoretically leading to an elegant formula for N_1, is not the best method to determine N_1. So one has to discover other techniques which yield N_1.

Finally, we note that the elegant analysis of JOSEPH and FOSDICK presented in Sec. 44 casts doubt on the qualitative results of COLEMAN, *et al.*[67] and LODGE[68], who have claimed that the WEISSENBERG effect (or the rod climbing effect) is solely due to N_1, with rod climbing being implied by a positive N_1 and vice versa. Therefore, rod climbing cannot yield N_1.

48.2. Generalized Torsional Flow. Rotation of an Almost Flat Solid Body of Revolution over a Flat Plate. As we have proved in Sec. 27, the flows in the cone-and-plate and parallel plate apparatus are compatible with the laws of motion if and only if inertia is neglected. So the flow[69] between an almost flat plate rotating over a flat plate will be possible under the neglect of inertia. Let the height of the rotating plate over the fixed flat plate be $h(r)$, assumed small ; moreover, $h(r)$ varies very slowly with r, the radial distance along the fixed plate. Use of a cylindrical polar coordinate system will be made in this subsection.

---

66. Paper presented at the 41st Annual Meeting of the Society of Rheology, Princeton, 1970.
67. [1966 : 4], p. 48.
68. [1964 : 14], p. 193.
69. This will be called a *generalized torsional flow*. Some of the results here are derived in a manner different to that of PIPKIN and TANNER [1973 : 11] who obtained them first. See [1973 : 4], where the case of a spherical surface is discussed.

Let the almost flat body rotate with a steady angular velocity Ω. Then the shear rate is approximately given by

$$\varkappa \doteq \Omega r/h(r), \tag{48.15}$$

or it is a function of r. Since we are interested in the force and stress distribution on the fixed plate in the axial (z) direction, we shall examine this first.

Let θ be the flow direction, ζ the direction of the velocity gradient and ρ the direction normal to both (see Fig. 48.2). Actually, ρ will be tangential

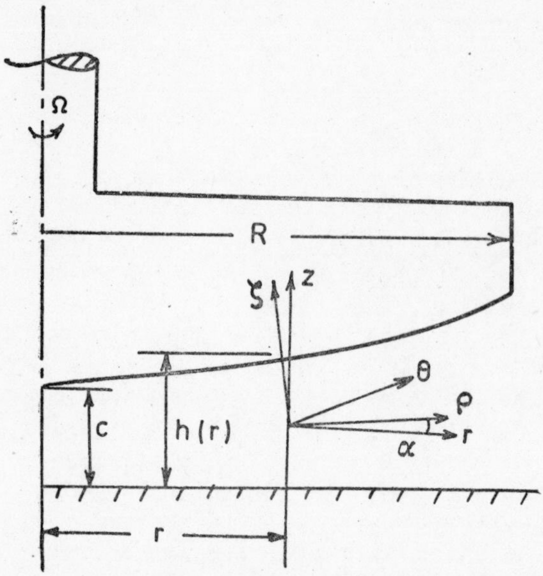

Fig. 48.2. Generalized torsional flow [1973 : 4].

to surface of constant angular velocity. Let α be the angle between the ρ and r directions. Then, one can compute the stresses $T\langle rr\rangle$, $T\langle\theta\theta\rangle$, $T\langle zz\rangle$, $T\langle r\theta\rangle$, $T\langle\theta z\rangle$ and $T\langle r\theta\rangle$ in terms of the *viscometric stresses* $T\langle\rho\rho\rangle$, $T\langle\theta\theta\rangle$, $T\langle\zeta\zeta\rangle$ and $T\langle\theta\zeta\rangle$. Since the viscometric stresses depend on r through $\varkappa$, the equation of motion in the radial direction becomes

$$\frac{\partial T\langle rr\rangle}{\partial r} = \frac{1}{r}\{T\langle\theta\theta\rangle - T\langle rr\rangle\} + (T\langle\zeta\zeta\rangle - T\langle\rho\rho\rangle)\frac{\partial}{\partial z}(\cos\alpha\sin\alpha). \tag{48.16}$$ [70]

Now, the slope of a surface of costant angular velocity changes from 0 at $z = 0$ to $h'(r)$ at $z = h(r)$. Thus to a first approximation, the slope inbetween is $\tan\alpha \doteq h'(r)\, z/h(r)$. Since α is very small, take $\tan\alpha \doteq \alpha$. Then

70. One notes that $T\langle rr\rangle = T\langle\rho\rho\rangle\cos^2\alpha + T\langle\zeta\zeta\rangle\sin^2\alpha$, and $T\langle rz\rangle = (T\langle\rho\rho\rangle - T\langle\zeta\zeta\rangle)\sin 2\alpha = -N_2 \sin 2\alpha/2$.

we have

$$\frac{dT\langle rr\rangle}{dr} = \frac{1}{r}(N_1 + N_2) + \frac{N_2 h'(r)}{h(r)}, \qquad z = 0. \quad (48.17)$$

Thus

$$T\langle rr\rangle = -p_R + \int_R^r \left\{\frac{1}{\xi}\left(N_1 + N_2\right) + \frac{N_2 h'(\xi)}{h(\xi)}\right\} d\xi, \quad z = 0, \quad (48.18)$$

whence the axial stress is $T\langle rr\rangle + (T\langle zz\rangle - T\langle rr\rangle)$, or

$$T\langle zz\rangle = -p_R + \int_R^r \left\{\frac{1}{\xi}\left(N_1 + N_2\right) + \frac{N_2 h'(\xi)}{h(\xi)}\right\} d\xi + \\ + \cos 2\alpha(T\langle\zeta\zeta\rangle - T\langle\rho\rho\rangle), \quad z = 0. \qquad (48.19)$$

Thus, on the flat plate

$$T\langle zz\rangle = -p_R + N_2 + \int_R^r \left\{\frac{1}{\xi}\left(N_1 + N_2\right) + \frac{N_2 h'}{h}\right\} d\xi, \ z = 0. \quad (48.20)$$

We specialize this result to a few flow systems.

(i) *Parallel-plate.* There $h(r) = h =$ const., and $h'(r) = 0$. Thus $\varkappa = \Omega r/h$, and

$$T\langle zz\rangle = -p_R + N_2 + \int_R^r \frac{1}{\xi}(N_1 + N_2)\, d\xi, \qquad z = 0. \quad (48.21)$$

(ii) *Cone-and-plate.* Here $h(r) = r \tan \alpha_0 \doteq r\, \alpha_0$, wherc α_0 is the cone angle. Hence $\varkappa = \Omega/\alpha_0$, a constant, and

$$T\langle zz\rangle = -p_R + N_2 + (N_1 + 2N_2) \ln \frac{r}{R}, \qquad z = 0. \quad (48.22)$$ [71]

(iii) *Cone-and-plate with an axial separation.* In this system first proposed by JACKSON and KAYE[72] and extended by MARSH and PEARSON[73],

$$h(r) = c + r \tan \alpha_0. \qquad (48.23)$$

Hence

$$T\langle zz\rangle = -p_R + N_2 + \int_R^r \left\{\frac{1}{\xi}\left(N_1 + N_2\right) + \frac{N_2 \tan \alpha_0}{c + r \tan \alpha_0}\right\} d\xi, \\ z = 0. \quad (48.24)$$

The flat plate provides an excellent opportunity to mount flush transducers to measure the axial stress. If the free surface is stress free and surface tension is neglected, (48.21) and (48.22) provide an opportunity to measure N_1 and N_2. Such an attempt has been made by MILLER and CHRISTIANSEN[74] to measure N_1 and N_2. They used a cone-and-plate device with a free surface. We shall present their data later on.

---

71. These results agree with those in LODGE [1964 : 14], for example.
72. [1966 : 10].
73. [1968 : 16]. There is a misprint in their Eq. (16).
74. [1972 : 10].

Also, in the cone-and-plate system with a free surface, the *gauge stress* $T\langle zz \rangle$ yields N_2 directly at the outer rim,[75] $r = R$. However, this is not a very accurate method for determining N_2. One can also deduce formulae for N_1 and N_2 by differentiating $T\langle zz \rangle$ with respect to r, but we do not record these here.

We recall (48.20) and compute the axial *thrust* next:

$$F = \pi R^2 p_R - 2\pi \int_0^R r\left[N_2 + \int_R^r \left\{ \frac{1}{\xi}\left(N_1 + N_2\right) + \frac{N_2\, h'(\xi)}{h(\xi)} \right\} d\xi \right] dr. \tag{48.25}$$

But from (48.15),

$$\frac{h'(r)}{h(r)} = \frac{1}{r} - \frac{1}{\varkappa}\,\frac{d\varkappa}{dr}. \tag{48.26}$$

Hence, integration by parts yields

$$F = \pi R^2 p_R + \int_0^R \pi(r\, N_1 - r^2 n_2(\varkappa^2)\, \varkappa\, \varkappa')\, dr. \tag{48.27}$$

In the parallel-plate and the cone-and-plate system with free surfaces, one may assume $p_R = 0$. Then one can derive the following formulae for the total thrust.

Parallel-Plate: $$F = \int_0^R \pi(N_1 - N_2)\, r dr, \qquad \varkappa = \Omega r/h. \tag{48.28}$$

Cone-and-Plate: $$F = \frac{\pi R^2}{2} N_1, \qquad \varkappa = \Omega/\alpha_0. \tag{48.29}$$

Thus the total thrust measurement in a cone-and-plate device yields N_1 directly; using this in (48.28), one may find N_2.

To investigate the effect of a vertical separation at $r = 0$, whether or not this separation is intentional, one may set

$$h(r) = h^*(r) + c, \tag{48.30}$$

and let c be a variable.

Assuming p_R to be independent of c, one may differentiate F in (48.27) with respect to c to obtain

$$\frac{dF}{dc} = \frac{\pi}{\Omega} \int_0^R \left[\frac{dN_1}{d\left(\frac{1}{\varkappa}\right)} - \varkappa N_2 + r\,\frac{d}{dr}\,(N_2\,\varkappa) \right] dr, \tag{48.31}$$

where $\varkappa = \Omega r/[c + h^*(r)]$. For a cone-and-plate system, $c = 0$, and $\varkappa$ is independent of r also. Hence

$$\left.\frac{dF}{dc}\right|_{c=0} = \frac{\pi R}{\Omega}\left(\frac{dN_1}{d\left(\frac{1}{\varkappa}\right)} - \varkappa N_2 \right), \qquad \varkappa = \Omega/\alpha_0. \tag{48.32}$$

75. Lodge [1964 : 14].

This was first derived by JACKSON and KAYE[76], who suggested that it be used to compute N_2, if $dN_1/d(1/\varkappa)$ be known. However, as remarked by COWSLEY[77], the next derivative d^2F/dc^2 is infinite at $c = 0$, since

$$d\varkappa/dc = -(\varkappa^2/\Omega r), \tag{48.33}$$

and this makes the integrand of d^2F/dc^2 infinite at $r = 0$. Hence (48.32) does not provide a good method for determining N_2.

If on the other hand, the rotating element is a cone with a flat tip[78] of radius $R_f \ll R$ (see Fig. 48.3), then d^2F/dc^2 has a term of order $\log (R/R_f)$ in

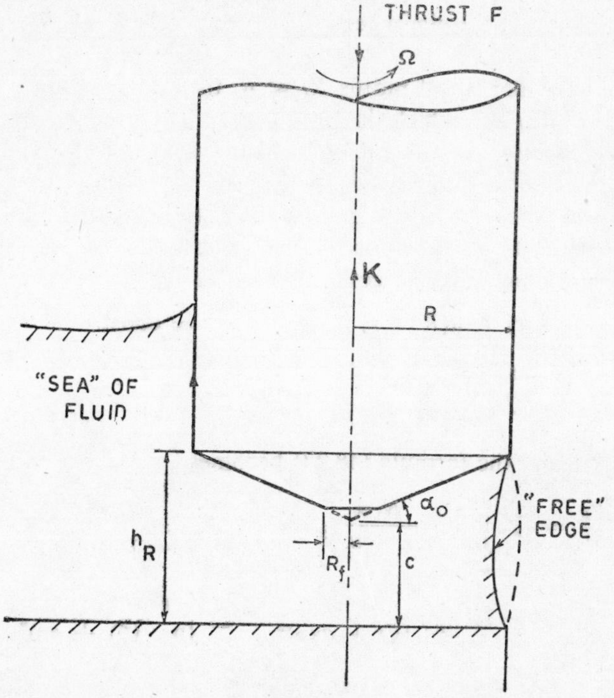

Fig. 48.3. Free surface and bath type flow [1970 : 22].

i, which will be quite large. Thus attempts to find N_2 from dF/dc will not be very successful. This conclusion is due to PIPKIN and TANNER[79], who also deduced that the moment required to hold the plate in generalized torsional flow, viz.

$$M = 2\pi \int_0^R \varkappa \bar{\eta}(\varkappa^2) r \, dr \tag{48.34}$$

76 [1966 : 10].

77 [1970 : 8].

78 LODGE [1971 : 17] reports some normal stress calculations for using such a device.

79 [1973 : 11].

depends on c through

$$\frac{dM}{dc} = \frac{2\pi}{\Omega}\int_0^R \frac{d(\eta\varkappa)}{d\left(\frac{1}{\varkappa}\right)}\, r\, dr. \tag{48.35}$$

Hence for a cone with $c = 0$, the integrand is a constant multiple of r, or the relative error dM/M near $c = 0$ is found to be

$$\frac{dM}{M} = -\frac{3}{2}\left.\frac{d\log(\eta\varkappa)}{d\log\varkappa}\right|_{c=0}\cdot\frac{dc}{h_R}, \tag{48.36}$$

where h_R is the gap at the rim when $c = 0$, and thus $h_R \doteq R\alpha$, α being the cone angle.

If the cone has a flat tip of radius $R_f \ll R$, then dM/dc differs from (48.35) by a factor of $O(R_f^2 / R^2)$ which is negligible. Thus the moment measured with a small separation at the tip of the cone does not introduce a serious error, but such is not the case for F.

We close this subsection by pointing out two recent innovations for measuring normal stress differences. The first is due to COWSLEY[80], who suggested the use of an inverted cone, that is, let

$$h(r) = a(R - r) + c, \qquad a, c = \text{const.} \tag{48.37}$$

Solving for r in terms of $\varkappa$, one has[81]

$$r = \varkappa(aR + c)/(\Omega + a\varkappa). \tag{48.38}$$

Then $\varkappa_R = aR/c$ and the formula (48.27) becomes

$$\frac{F - \pi R^2 p_R}{\pi(aR + c)^2} = \Omega\int_0^{\varkappa_R} \frac{N_1\varkappa}{(\Omega + a\varkappa)^3}\, d\varkappa - \int_0^{\varkappa_R} \frac{N_2\varkappa}{(\Omega + a\varkappa)^2}\, d\varkappa \tag{48.39}$$

$$= -I_2(\varkappa_R^2)\frac{\varkappa_R^2}{(\Omega + a\varkappa_R)^2} + \Omega\int_0^{\varkappa_R} \frac{\varkappa(N_1 + 2I_2)}{(\Omega + a\varkappa)^3}\, d\varkappa. \tag{48.40}$$

Now as $c \to 0$, $\varkappa_R \to \infty$ and it is anticipated that $I_2(\varkappa_R^2)$ dominates the integral, since the integrand has a denominator of order $\varkappa^2$ as $\varkappa$ becomes large while in the definition of $I_2(\cdot)$ [see (48.5)], the denominator is $\varkappa$.

A more direct way of obtaining $I_2(\varkappa_0^2)$ for any $\varkappa_0$ is to let $\Omega \to 0$, $c \to 0$, but let $\Omega R/c \to \varkappa_0$. Then (48.40) becomes

$$F - \pi R^2 p_R = -I_2(\varkappa_0^2)\,\pi R^2. \tag{48.41}$$

It can be shown that in the limit as Ω approaches zero linearly in c, the derivative $(\partial/\partial c)\,[(F - \pi R^2 p_R)/\pi(aR + c)^2]_{c=0}$ is bounded, and hence this suggestion does not have the drawback associated with (48.32). By letting $\Omega \to 0$

80. COWSLEY [1970 : 8]. See also PIPKIN and TANNER [1973 : 9].
81. HUILGOL [1973 : 4].

linearly in c, but at various rates of $\varkappa_0$, one can obtain $I_2(\varkappa_0^2)$ for many values of $\varkappa_0$.

A second idea, due to VAN ES[82], is to use a cylindrical well in the flat plate (see Fig. 48.4). Because of the geometry, the shearing occurs between $R_1 \leqq r \leqq R_2$ only and in this range (48.22) applies; moreover,

$$T\langle zz\rangle(r) = N_2 + (N_1 + 2N_2) \ln \frac{r}{R_2}, \tag{48.42}$$

where we have assumed $p_{R_2} = 0$, the atmospheric pressure. The normal thrust contribution F' from this stress distribution is given by [*cf.* (48.27)] :

$$F' = \frac{\pi}{2} N_1 (R_2^2 - R_1^2) + \pi R_1^2 (N_1 + 2N_2) \ln \frac{R_1}{R_2}. \tag{48.43}$$

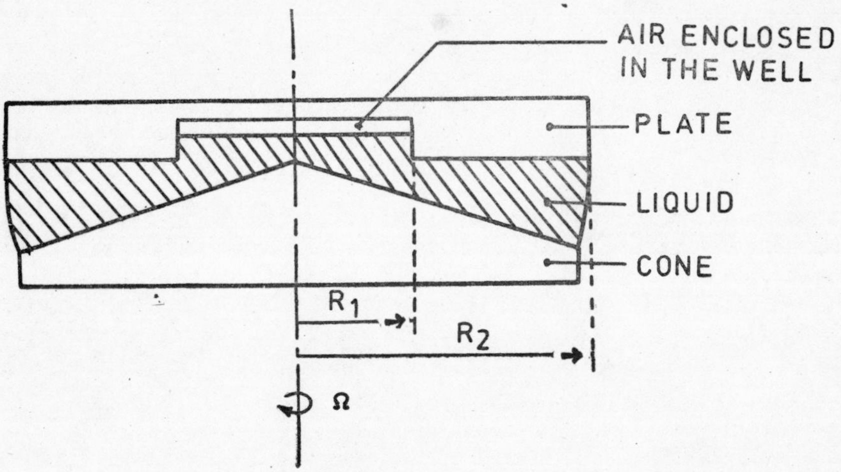

Fig. 48.4. Cone-and-ring flow system. From VAN ES [1972 : 16].

In the central region of the cone-and-cup, one may assume that the pressure p_c is equivalent to the radial normal stress (radial in the sense of spherical coordinates) at the inner rim R_1. In the context of generalized torsional flow, this stress is $T\langle \rho\rho\rangle$, and we have

$$p_c = -T\langle \rho\rho\rangle\Big|_{\rho=R_1} = -(N_1 + 2N_2) \ln \frac{R_1}{R_2}, \tag{48.44}$$

which is $(27.33)_1$ with a suitable modification. The contribution F'' to the thrust on the flat plate can be computed from (48.44) and added to (48.43) to yield the total thrust F as

$$F = \frac{\pi}{2} N_1(R_2^2 - R_1^2). \tag{48.45}$$

82. [1972 : 16]. I am indebted to Professor A. S. LODGE for this reference.

This measurement gives N_1. Using (48.44) and (48.45), we obtain

$$N_2 = -\frac{p_c}{2\ln(R_1/R_2)} - \frac{F}{\pi(R_2^2 - R_1^2)}. \tag{48.46}$$

VAN Es has obtained excellent results by using (48.45) and (48.46).

We now turn to various corrections that are necessary to account for inertial effects and others.

In what follows, it must be noted that in generalized torsional flows α is very small. Thus the formulae (48.20) and (48.27) are valid throughout the flow domain to within $O(\alpha)$. This will be used wherever necessary without notice.

48.3. Inertial Corrections, Surface Tension and Bath Effects. As we have mentioned previously, the flow in the parallel-plate and the cone-and-plate apparatus is theoretically possible under the neglect of inertia. Now, in practice, inertia is always present, and while it may be kept small by using very low angular velocities, it is not ignorable. Experimentally, the thrust due to inertia may be measured by using a Newtonian liquid of comparable density, and this may be subtracted from the measured thrust to obtain the true thrust. This technique is feasible whether or not the surface is free or immersed in a bath. We now report some theoretical attempts to account for inertia.

The first attempt was made by GREENSMITH and RIVLIN[83] for the torsional flow. The velocity varies from 0 to Ω over an axial length $z = h$. Thus, at a distance r from the axis, the *centripetal force* is given by $-\rho\,\Omega^2 r\,z^2/h^2$, and the equation of motion in the radial direction is

$$\frac{dT\langle rr\rangle}{dr} + \frac{T\langle rr\rangle - T\langle\theta\theta\rangle}{r} = -\rho\,\frac{\Omega^2 z^2 r}{h^2}, \tag{48.47}$$

and this is incompatible with the inertial term. The solution proposed by GREENSMITH and RIVLIN is to replace the right side by its average value over $z = 0$ to h, that is, by $-\frac{1}{3}\rho\Omega^2 r$, which contributes an extra stress of amount $-1/6\,\rho\,\Omega^2 r^2$ to $T\langle rr\rangle$. On assuming the fluid surface to be stress free at $r = R$, one has

$$T\langle zz\rangle = \frac{1}{6}\,\rho\,\Omega^2(R^2 - r^2) + \int_R^r \frac{N_1 + N_2}{\xi}\,d\xi + N_2, \quad z = 0, \tag{48.48}$$

which should be compared with (48.21). The extra stress means that the

83. [1953 : 1]. After this book was completed, the paper by JOSEPH [1974 : 2] appeared. This discusses the inertial effects in detail. It is argued that inertial effects are more important than previously thought, and a crucial experiment is suggested to measure the importance of these effects, driven by surface tension and inertia. See footnote 141 on p. 329 also.

formula for the thrust is

$$F = \int_0^R \pi(N_1 - N_2)r\,dr - c\,\rho\,\Omega^2 R^2\,\frac{\pi R^2}{2}, \qquad c = \frac{1}{6}. \quad (48.49)$$

For the cone-and-plate flow, if the cone angle is small ($\alpha_0 \leqq 4°$), the above correction is again valid ; so, instead of (48.29), one has

$$F = \frac{\pi R^2}{2}\,(N_1 - c\,\rho\,\pi^2 R^2), \qquad c = \frac{1}{6}. \quad (48.50)$$

Some experiments have shown[84,85] that the value $c = 1/6$ is in excess of the true value. A more comprehensive theoretical analysis of WALTERS and WATERS[86] has shown that $c = 0.15$. These authors also claim that this value fits the experimental results of ADAMS and LODGE more closely than the value $c = 1/6$. On the other hand, OLABISI and WILLIAMS[87] use $c = 1/6$ and claim that this is adequate. In any event, either value seems to be acceptable within experimental error.

Turning to the measurement of N_1, we note that many workers[88] have used the thrust in a cone-and-plate device with an inertial correction to measure N_1. In these cases, the surface is open to the atmosphere. As we have discussed in Sec. 47, the free surface will not be spherical and will exhibit a tendency to bulge outwards from the rotating member, drawing in on the fixed plate. At present no theoretical calculations are available to indicate the magnitude of the correction to the total thrust, though it should be possible to make a start on this by using the domain perturbation technique presented in Sec. 44.

KAYE, *et al.*[89] have stated that their experiments indicate that the effect of having a concave boundary (spherical = convex) was to lower the thrust measurement by upto 6%. They also note that change in the thrust measurement due to surface tension occurs if this surface tension changes during the motion. Otherwise it may be taken to be a constant and provided the shape of the surface does not change too much, its contribution to F may be ignored.

While we have dwelt upon some of the disadvantages of the free surface configuration, it tends to minimize secondary flows at low speeds and small cone angles ($\leqq 4°$).

On the other hand, because of these corrections it was felt by some investigators that the free surface be replaced by immersing it in a bath (see Fig. (48.3). It was believed that this did not alter the total thrust measurements

84. ADAMS and LODGE [1964 : 1]. See Fig. 6.
85. MILLER and CHRISTIANSEN [1972 : 10], Eqs. (27)—(28).
86. [1968 : 28].
87. [1972 : 12].
88. GINN and METZNER [1969 : 5] ; ADAMS and JACKON [1967 : 1] ; PRITCHARD [1971 : 20].
89. [1968 : 12], p. 373.

though secondary flows due to inertia would now be more common. An elaborate numerical analysis by GRIFFITHS and WALTERS[90] on a second-order fluid with $N_2 = 0$ shows that if the cone angle $\alpha_0 \leqq 4°$, and the cone has a spherical cap, the bath correction to the thrust is a fraction of one percent. They obtained similar results for the parallel-plate set up when $h/R \leqq 0.07$ (approx). In this study, these authors assumed that the primary flow persisted upto the edge and that the secondary flow was superposed on it. The persistence of the primary flow upto the edge has been challenged by TANNER[91], who performed theoretical calculations to determine the thrust and the stress distribution on the flat plate, when the upper cone (or plate) rotates in an immersed bath. He assumes that the primary motion persists upto a radial distance $R - h_R$ ($h_R \ll R$), and between $R - h_R$ to $R + h_R$ the motion is again locally two-dimensional in character, but the angular velocity ω depends nonlinearly on both r and z (for, a cone read $h_R = R\alpha_0$). The shear rate decreases from the primary value at $r = R - h_R$ to almost zero at $r = R + h_R$.

It was shown by TANNER that for a cone, the thrust (corrected for hydrostatic effects) is given by

$$F = \frac{\pi R^2}{2} (N_1 + N_2) + \text{inertial correction}, \tag{48.51}$$

and that the axial stress on the bottom plate was given by (modulo the hydrostatic pressure)

$$T\langle zz\rangle(R) = \frac{N_2}{2} = \frac{\beta + 2\gamma}{2} \varkappa_R^2, \qquad \varkappa_R = \Omega\, R/h_R. \tag{48.52}$$

Using this in (48.22) yields the stress distribution on the bottom plate for the *second-order fluid* as

$$T\langle zz\rangle = \frac{N_2}{2} + (N_1 + 2N_2) \ln \frac{r}{R}. \tag{48.53}$$

So, if $-N_2 > 0$, then $-T\langle zz\rangle$ is positive at $r = R$.

A more elaborate analysis may be made by assuming the flow to be a generalized torsional flow in the region $R \leqq r \leqq R + h_R$[92], $0 \leqq z \leqq h_R$, and that $T\langle zz\rangle(R + h_R) = -p_\infty = -\rho g H$, the hydrostatic pressure. Since the shear rate ($R + h_R$) is assumed to be zero, we have from (48.20), (48.5) and (48.26) that

$$p_R = \rho g H + \int_R^{R+h_R} \frac{1}{\xi} (N_1 + 2N_2)\, d\xi + I_2(\varkappa_R^2), \tag{48.54}$$

where we have used $N_2(0) = 0$. Then

$$T\langle zz\rangle(r) = -\rho g H + \int_{R+h}^{r} \frac{1}{\xi} (N_1 + 2N_2)\, d\xi + \overline{N}_2[\varkappa^2(r)] - I_2(\varkappa_R^2), \tag{48.55}$$

90. [1970 : 12].
91. [1970 : 22].
92. Here we assume the primary motion to persist upto the edge.

which differs essentially from that in ((48.20) in the $I_2(\cdot)$ term ; or

$$T\langle zz\rangle(r) = -\rho g H + \overline{N}_2[\varkappa^2(r)] - I_2(\varkappa_R^2) + N_1 \cdot O(h_R/R), \quad 0 \leqq r \leqq R + h. \tag{48.56}$$

If $h_R/R \ll 1$, then we may derive

$$T\langle zz\rangle(R) = -\rho g H + \overline{N}_2(\varkappa_R^2) - I_2(\varkappa_R^2). \qquad (48.57)^{93}$$

So the pressure at $r = R$ is not the hydrostatic pressure, depending only on N_2. Also, (48.55) reduces to (48.53) under appropriate simplifications.

An elaborate experimental analysis of the pressure distribution along the bottom plate in a cone-and-plate device, when the cone is immersed in a sea of liquid, has been made by OLABISI and WILLIAMS[94]. They find that $T\langle zz\rangle(r)$ is different from $\rho g H$ and close to the value predicted by (48.57).

It is fairly straightforward to use (48.55) to derive the total thrust, modulo the hydrostatic contribution. This is :

$$F = \int_0^R \pi(rN_1 - r^2 n_2\, \varkappa\varkappa')\, dr + \int_0^R 2\pi r[N_2(\varkappa^2) - I_2(\varkappa_R^2)]\, dr + \\ + N_1(\varkappa_R^2) \cdot O(h/R). \tag{48.58}$$

From this, one can deduce the formula (48.51) as well as the one given by TANNER[95] for a cone-and-plate device.

48.4. Birefringence Measurements. A method which yields N_1 directly is the birefringence measurement technique, which we shall now describe briefly[96]. If in a plane viscometric flow, e.g. COUETTE flow, the stresses $T\langle\theta\theta\rangle$, $T\langle rr\rangle$ and $\tau = T\langle r\theta\rangle$ are known, then the acute angle χ between τ and one of the principal stresses is given by

$$N_1 = T\langle\theta\theta\rangle - T\langle rr\rangle = 2\tau \cot 2\chi. \tag{48.59}$$

If a beam of plane polarized light is passed up the column of liquid being sheared between the two cylinders[96] and the locations of the two principal directions of the optic tensor in the $z =$ const. plane are determined by finding the extinction positions, and the birefringence Δn is determined, then by postulating that the principal directions of the optic tensor coincide with

93. PIPKIN and TANNER [1973 : 11], Eq. (17.9).
94. [1972 : 12].
95. [1970 : 22], Eq. (45).
96. A fairly elementary introduction is given by HARRIS [1970 : 13]. For a comprehensive review of theory and experiment, see JANESCHITZ-KRIEGEL [1969 : 9].

those of the stress tensor,[97] one has that

$$\Delta n \sin 2\chi = 2\, C\tau, \qquad \Delta n \cos 2\chi = CN_1, \tag{48.60}$$

where C is a constant, called the *stress-optical coefficient*. Thus, if one can measure the birefringence Δn and the shear stress τ, one obtains the coefficient C. Then N_1 is given by (48.60), where χ is now the extinction angle, which is also measured. It has been reported by PRITCHARD[98] that the values of N_1 obtained by the total thrust measurement in a cone-and-plate device agree with those obtained by KAYE through the birefringence technique on the KAYE-SAUNDERS instrument.[99]

In COUETTE flow, one finds the birefringence through

$$\Delta n = (\Delta/2\pi)\,(\lambda/l), \tag{48.61}$$

where λ is the wave length (generally equal to 550 mμ), l the liquid height between the cylinders undergoing COUETTE flow, and Δ the phase difference between the orthogonal light components travelling along the principal directions of the test material.[100]

Various techniques have been developed to measure the extinction angle χ, the phase difference Δ and to adduce the length l of the fluid undergoing the COUETTE flow; the latter correction is necessary since the entire fluid column will not be subjected to the assumed motion. We refer the reader to JANESCHITZ-KRIEGL for more details.

If one can find Δn and χ, then the stress optical coefficient C can be calculated and N_1 can be found from (48.60). Again, JANESCHITZ-KRIEGL has presented extensive experimental data to confirm the validity of (48.60) and (48.61).

We note in passing that the conclusion by PHILIPPOFF[101] that $N_2 = 0$ from birefringence measurements was due to an inherent inadequacy of the assumed stress-optic law rather than actual fact, for WALES and PHILIPPOFF[102] have recently obtained birefringence data which proves that $N_2 \neq 0$. In addition, one could deduce that $N_2 \neq 0$, at low shear rates at least, if the free surface in a slow flow down an inclined channel is not flat. The theory behind this statement and experimental data to confirm that N_2 is not zero will be presented next.

---

97. This is tantamount to saying that the optic tensor is an isotropic function of the stress tensor. Phenomenological theories, where it is assumed that these two tensors are given by two different functionals of the strain history, see e.g. COLEMAN, *et al.* [1970 : 7], RIVLIN and SMITH [1971 : 22], do not lead to such a conclusion. However, the validity of (48.59) stems from the experimental evidence available. See JANESCHITZ-KRIEGL [1969 : 9], or KAYE, *et al.* [1968 : 12].

98. [1971 : 20].

99. [1964 : 11].

100. It can also be defined as that created between the linearly polarized components, which oscillate in the two directions of extinction.

101. [1961 : 5].

102. [1973 : *C*]. I wish to thank Professor R. I. TANNER for bringing this work to my attention.

48.5. The Second Normal Stress Difference. The methods of Secs. 48.2 and 48.3 yield many (mechanical) measuring situations whereby N_1 and then N_2 may be found, though (48.41) and (48.46) yield N_2 directly. While (48.41) may be difficult to achieve experimentally, no such restriction applies to (48.46). Despite this it would be of interest to discover a flow situation (i) which would demonstrate unequivocally the sign of N_2 (which was of some dispute in the recent past), and (ii) from which one may find N_2 without subtracting two large numbers. This drawback of subtraction exists in (48.46), for example.

Now, from the simple fluid theory, it is not possible to obtain a restriction on the sign of N_2. It is therefore very interesting to note that by using a theory of deformable viscoelastic spheres suspended in a Newtonian liquid, ROSCOE[103] proved that at small shear rates (that is, $\varkappa \to 0$),

$$N_2 = -2/7\, N_1. \tag{48.62}$$

This is surprisingly close to the value found experimentally and seems to have been generally overlooked in the literature. We now return to the continuum mechanical approach.

As has been shown in Sec. 43, the shape of the liquid surface as it flows slowly down a semi-circular channel yields the sign of N_2 immediately. The experiments performed by TANNER[104] prove beyond doubt that the free surface is convex, leading to the conclusion that $N_2 < 0$. Surface tension effects have to be ignored here and may be shown to be fairly small. We shall now indicate how N_2 can be determined by measuring the central rise (in the middle of the semi-circular channel).

Let us integrate (43.24) by parts and write the resulting equation as

$$\frac{\hat{N}_2\left(\frac{1}{2}\rho g R \sin\alpha\right)}{\rho g R} = \frac{1}{\sin\alpha}\int_0^\alpha \sin\varphi\, d\left(\frac{h_w(\varphi)\cos\varphi}{R}\right). \tag{48.63}$$

In a semi-circular channel, if the channel is full so that $h_w = 0$, then (48.63) seems to lose its meaning. But it is experimentally acceptable, in the present situation, to take h_w to be the central rise h_c. Then for various values of α, one can measure h_c and draw a curve of $(h_c(\varphi)\cos\varphi)/R$ vs $\sin\varphi$. Integration of the data (see Fig. 48.5) yields $\hat{N}_2(\tau_w)$ directly. Since we have already determined the relation between τ and $\varkappa$, (e.g. see Fig. 28.1), one can express $\hat{N}_2(\tau_w)$ as the function $\bar{N}_2(\varkappa^2)$.

Since the above method is useful at low rates of shear only, we examine the

103. [1967 : 22]. I wish to thank Professor R. I. TANNER for bringing this to my attention, as well as for pointing out that the same result is implicit in Eq. (2.29) of GODDARD and MILLER [1967 : 10].

104. [1970 : 22]. Recently, JONES, DAVIS and THOMAS [1973 : A] have come to the conclusion that $N_2 < 0$ in a 250 p.p.m. aqueous solution of polyacrylamide, from a study of TAYLOR vortices.

set up used by TANNER[105] to measure N_2 at high rates of shear. Consider the axial flow of an incompressible simple fluid down an annulus between two concentric cylinders of radii R_1 and R_2 ($R_2 > R_1$) respectively. This is a

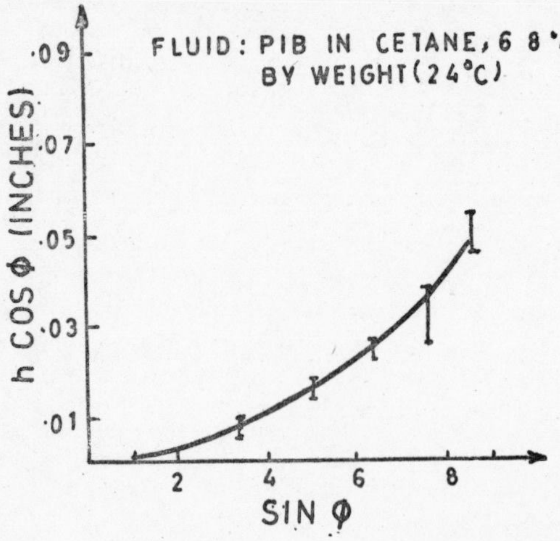

Fig. 48.5. Central rise × cos φ vs. sin φ for 6.8% polyisobutylene in cetane solution at 24°C.

special case of helical flow [*cf.* (28.3)], with $\omega(r) = 0$, $u = u(r)$ and

$$\varkappa = du/dr. \tag{48.64}$$

From (27.*b*8) it now follows that the pressure difference ΔP is given by

$$\Delta P = T\langle rr\rangle\Big|_{r=R_1} - T\langle rr\rangle\Big|_{r=R_2} = \int_{R_1}^{R_2} \frac{1}{r}\,\overline{N}_2(\varkappa^2(r))\,dr. \tag{48.65}$$

To invert this, we shall utilize the approach given below.

From (28.9), we have the following formula for the shear stress τ:

$$\tau = \frac{a}{2}\,(r - r_0^2\,/\,r), \tag{48.66}$$

where we have used $c = -a$, $2b = -cr_0^2$, and r_0 is called the *radius of zero shear rate*. Since (28.18) now reads

$$0 = \int_{R_1}^{R_2} \varkappa(r)\,dr, \tag{48.67}$$

there does exist one point r_0, $R_1 < r_0 < R_2$ such that $\varkappa(r_0) = 0$. If the gap $R_2 - R_1$ is small compared to R_1, then the mean radius $\bar{r} = (R_1 + R_2)/2$ may be assumed equal to r_0. Then

$$a\Delta P = \frac{2}{\bar{r}}\int_0^{\tau_W} \hat{N}_2(\tau)\,d\tau. \tag{48.68}$$

105. [1967 : 23]. This is an extension of the work by HAYES and TANNER [1965 : 10].

Since $\tau \doteq a(r - \bar{r})$,

$$\tau_W = a(R_2 - R_1)/2 = ah/2 \quad \text{(say)}. \tag{48.69}$$

Differentiating (48.68) with respect to a gives

$$\hat{N}_2\left(\frac{ah}{2}\right) = \frac{\bar{r}}{h}\left(a\frac{d\Delta P}{da} + \Delta P\right) + O\left(\frac{h^2}{\bar{r}^2}\right), \tag{48.70}$$

whence $\bar{N}_2(\varkappa^2)$ may be deduced, to within an error.

TANNER[106] has shown that this error term is less than 5% for second-order fluids if $R_1/R_2 = 2/3$, but the error analysis is difficult for general fluids; on the other hand, if $N_2 = \alpha \mid \tau \mid^m$, $m = 1$ or 2, an exact inversion of (48.65) is possible.

However, this method of determining the normal stress difference is open to the criticism that the pressure hole errors have been neglected. Also alignment of the inner tube to obtain concentricity with the outer tube appears to be very difficult. Moreover, if one desires to employ flush mounted transducers, the curvature of the tubes makes it a delicate operation. Since flat surfaces exist in parallel-plate, and cone-and-plate devices, these instruments permit the application of such transducers. However, as we have emphasized before, one needs N_1 to find N_2 by these methods, but this is not required in VAN Es' formula (48.46).

The only other flow which we have not yet mentioned is the axial motion of fanned planes. Here one plate is held at rest and another plate, tilted at an angle θ_0 to it, moves with a constant speed U, parallel to the line where the plates would intersect. The velocity field is $\mathbf{v} = (U\theta/\theta_0)\,\mathbf{e}_z$. It can be shown, by the use of the radial equation of motion and $T\langle rr\rangle(R) = -p_R$, that $\varkappa(r) = U/r\theta_0$ and

$$T\langle\theta\theta\rangle(r) = -p_R + \bar{N}_2(\varkappa^2(r)) + I_2(\varkappa_R^2) - I_2(\varkappa^2(r)). \tag{48.71}$$

One could again use flush mounted transducers on the fixed plate to find N_2. This method needs investigation.[107]

48.6. Experimental Results. From the literature, we reproduce the following figures : (i) Fig. 48.6 : The functions η, n_1 and n_2 for a 6.8% polyisobutylene in cetane at 24°C[108]; (ii) Fig. 48.7: The functions τ, N_1 and N_2 for a 3.54% solution of polyethylene oxide in water (Polyox grade WSR-301) at 25.2°C[109].

We believe that these represent a good set of results available at present.

In conclusion, the determination of the normal stress differences has involv-

106. [1967 : 23].
107. PIPKIN [1968 : 21]. In addition, see [1974 : *B*] for measuring N_1.
108. TANNER [1970 : 22]; PIPKIN and TANNER [1973 : 11].
109. MILLER and CHRISTIANSEN [1972 : 10].

ed many workers over the last 20 years. It is now possible to recommend

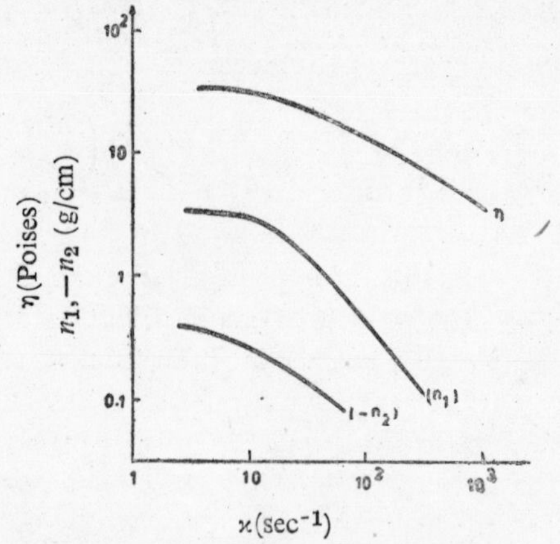

Fig. 48.6. The viscometric Functions η, n_1, n_2.

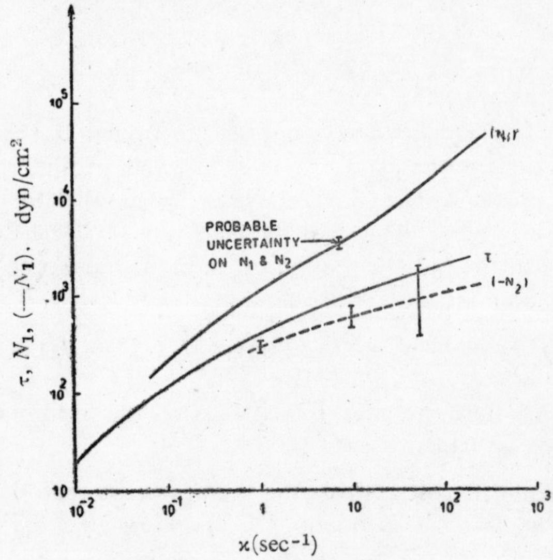

Fig. 48.7. Viscometric functions.

that

(i) The total thrust measurement in a cone-and-plate device with a free surface[110] provides an excellent method to find N_1;

(ii) While COWSLEY and VAN ES' methods and the fanned motion remain to be explored widely, N_2 can be found at low rates of shear by the flow

110. In other words, it is anticipated that the errors due to the free surface condition are more tolerable than those due to bath (and secondary flow) effects.

down a channel and at high rates of shear by flush mounted transducers, placed on the flat surface in a generalized torsional flow.

Based on the available experimental data, TANNER[111] has found interesting correlations between N_1, N_2, τ, the concentration C and the molecular weight M for polyisobutylene solutions. He finds that

$$N_1 C^2/\tau^2 = 1.2 \times 10^{-5} (CM/10^6)^{0.13} \tau^{-0.1},$$
$$N_2 = -0.15 N_1. \tag{48.72}$$

He also finds that the solvent has no effect on this correlation. It would be interesting to see if such a correlation holds for other polymers as well.

48.7. Comparison between Theory and Experiment. A Short Historical Survey. The late 1940's saw the use of the RIENER-RIVLIN fluid described by

$$\mathbf{T}_E = \alpha_1 \mathbf{A}_1 + \alpha_2 \mathbf{A}_1^2, \tag{48.73}$$

where α_1 and α_2 are functions of $\operatorname{tr} \mathbf{A}_1^2$ and $\operatorname{tr} \mathbf{A}_1^3$, to explain the flow behavior of polymeric fluids, especially the normal stress effects, in COUETTE flow, etc. The attack on the problems was aided by the remarkable discovery of RIVLIN that incompressibility and the undetermined pressure were a boon with the bonus that these could be exploited in the equations of motion. Now, as is well known, (48.73) predicted that $N_1 = 0$, $N_2 = \alpha_2 \kappa^2$; however, when it was found that experimental results did not predict that $N_1 = 0$, this constitutive equation was abandoned.

Though in 1950 OLDROYD had laid the foundation for forming general constitutive equations, the models he chose to illustrate his theory were too simple to describe the normal stress effects adequately. Hence the explicit constitutive equations (35.10), derived by RIVLIN in 1955, mark the beginning of a comprehensive theory. This was utilized by RIVLIN himself in 1956 and put into the form (26.65) by ERICKSEN in 1958, and the latter was shown to apply to simple fluids by COLEMAN and NOLL in 1959. They introduced the normal stress difference functions σ_1 and σ_2 (see Sec. 26) and exploited the viscosity function in an impressive manner to solve for the velocity fields in viscometric flows. Thus, by 1959 adequate theoretical predictions were available for viscometric flows. On the other side of the Atlantic, OLDROYD had published a paper in 1958 (Sec. 37) devoted to similar problems; however, the model he chose was again an approximate one, a drawback removed in his 1965 paper.

By the late 1950's, it was experimentally apparent that two normal stress differences existed in viscometric flows. A comprehensive series of experiments to determine these differences was undertaken by MARKOVITZ and his collaborators, the results of which are collected together in his book[112] with

111. [1973 : 13].
112. [1966 : 4].

COLEMAN and NOLL. His results imply that *both* N_1 *and* $N_2 > 0$. The matter could have rested here but for that the measurements obtained from the cone-and-plate apparatus did not agree with those from the parallel-plate apparatus[113]. The way out of the dilemma, which seemed rather strange at that time, was to ascribe systematic hole pressure errors[114] to measurements wherever holes had been employed to gather data. The theoretical results of TANNER and PIPKIN (Sec. 41), KEARSLEY (Sec. 42), and HIGASHITANI and PRITCHARD (Sec. 48), and SALATHE and SAWERS[115] have shown that such errors must exist, and be independent of the slot or hole size[116]. If such errors are accepted, then all the discrepancies in measurements disappear[117] and one finds that $N_1 > 0$, $N_2 < 0$, a result confirmed independently by TANNER's trough flows tests (Sec. 48.5).

This preoccupation to correlate viscometric theory with experiment has formed the basis of many studies, and the empirical single integral models have become more complex to do just this. Similarly, when the second-order fluid was proposed as a constitutive equation, it was noted that it yields unequal N_1 and N_2. So do the third- and higher-order fluids, and the OLDROYD equation (37.1) is far better here than these order fluids. Later on additional theoretical investigation was done (Chap. 8, Appendix) and it was found that the second-order fluids yielded crude predictions; consequently, its use declined, though it has not gone completely out of vogue.

In conclusion, there exists a complete agreement between the predictions of the simple fluid theory and viscometric experiments. It should again be emphasized that the simple integral models in Sec. 36 also offer an excellent agreement between theory and practice. However, as we have shown in Sec. 39, the crucial tests are the ultrasonic viscosity tests and here the additive functional theory emerges as the best model available so far to approximate the simple fluid behaviour[118].

49. Measurement of Elongational Viscosity

Let us consider the simple extensional flow[119] of an incompressible New-

113. ADAMS and LODGE [1964 : 1].

114. BROADBENT, *et al.* [1968 : 4].

115. [1973 : 12]. This article appeared after the book was completed. It confirms the analyses in Secs. 41 and 42, and shows them to be valid under more general conditions.

116. GREENSMITH and RIVLIN [1953 : 1] had shown that varying the hole sizes introduces no *additional errors* into measurements. It was not known at that time that errors exist, independent of the hole size.

117. PRITCHARD [1970 : 18]; PIPKIN and TANNER [1973 : 11].

118. As a postscript, we may mention the papers by VAN ES and CHRISTENSEN [1973 : 15], and SAKAI, KAJIURA and NAGASAWA [1974 : 4], which show that the empirical integra model (36.31) predicts results in disagreement with measurements of transient shear and normal stresses in the initiation of a viscometric flow. It would be interesting to see how the additive functional theory fares in this test.

119. Despite the use of extensional viscometry in the title, the three papers of BIERMAN [1971 : 5] are not concerned with the measurement of elongational viscosity.

tonian fluid. If the velocity field is given by

$$\dot{x} = -\frac{1}{2}\dot{\varepsilon}x, \quad \dot{y} = -\frac{1}{2}\dot{\varepsilon}y, \quad \dot{z} = \dot{\varepsilon}z, \tag{49.1}$$

with $\dot{\varepsilon}$ a constant, then the Newtonian stresses are :

$$T\langle xx\rangle = T\langle yy\rangle = -p - \eta_0\dot{\varepsilon}, \qquad T\langle zz\rangle = -p + 2\eta_0\dot{\varepsilon}. \tag{49.2}$$

Comparison of these equations with those in Sec. 33 shows that in (33.8), $\alpha_0 = 0$, $l = 0$, $h = \eta_0 =$ const. To derive the elongational viscosity for Newtonian liquids, let it be assumed that the lateral surface is stress free, that is, $T\langle xx\rangle = T\langle yy\rangle = 0$ on the surface of the cylinder of the fluid. Now p is a function of x, y and z as in (33.7). However, if we neglect inertia, then p may be taken to be a constant. Hence $T\langle xx\rangle = 0$ on the surface implies that it is zero everywhere ; or,

$$p = -\eta_0\dot{\varepsilon}, \tag{49.3}$$

from which follows the result

$$T\langle zz\rangle = 3\eta_0\dot{\varepsilon}. \tag{49.4}$$

Thus the *elongational viscocity* η_e becomes

$$\eta_e \overset{\text{def}}{=\!=} \frac{T\langle zz\rangle}{\dot{\varepsilon}} = 3\eta_0. \tag{49.5}$$

This is also called the *Trouton viscosity*[120].

Beginning with BALLMAN[121] in 1965, COGSWELL[122], MEISSNER[123], VINOGRADOV, *et al.*[124] and STEVENSON[125] have made measurements of the extensional viscosity for viscoelastic liquids. The main conclusions which can be drawn are that[125] :

(i) For isobutylene-isoprene copolymer at 100°C, the elongational stress increases monotonically to a steady value as a function of time. The elongational viscocity is almost a constant, being nearly three times the zero shear viscosity.

(ii) for a cis-1, 4-polyisoprene at 80°C, the elongational stress-strain curves are *S*-shaped and hence a steady state elongational viscosity does not exist.

An excellent summary of the results is given in a table by STEVENSON, which is reproduced here (see Table 49.1).[126]

---

120. On account of the discovery by TROUTON [1906 : 1].
121. [1965 : 1].
122. [1968 : 6], [1969 : 2], [1972 : 4].
123. [1971 : 18], [1972 : 9].
124. [1970 : 25], [1970 : 26].
125. [1972 : 14].
126. In the table, some earlier work of VINOGRADOV, *et al.* in 1968 is included. However, we have mentioned the 1970 work only.

TABLE 49.1—SUMMARY OF ELONGATIONAL VISCOSITY DATA

Investigator and references	*Material/ molecular weight*	*Temperature,* °C	*Zero-shear-rate viscosity* (poise)	$\bar{\eta}_0/\eta_0$	*Behavior of elongational viscosity*			*Behavior of recoverable strain*		
					SVC	SVD	SVPL	SVC	SVD	SVPL
BALLMAN [1965 : 1]	Atactic polystyrene $M_v = 3.2 \times 10^5$ $M_n = 1.3 \times 10^5$	149	—	—	—	C[1]	—	—	—	—
VINOGRADOV, *et al.* [1970 : 25, 26]	Polyisobutylene $M_v = 7.0 \times 10^4$	22	1.0×10^7	3.16	C	C	I	S	S	—
		40	2.8×10^6	2.76	C	C	—	—	—	—
		60	$\sim 7.3 \times 10^5$	3.03	C	C	—	—	—	—
	Atactic polystyrene $M_v = 3.0 \times 10^5$	130	$\sim 8 \times 10^7$	~ 3	C	I-U	U	S	S-U	—
MEISSNER [1969 : 15] [1971 : 18]	Low density polyethylene, $M_w = 4.8 \times 10^5$ $M_n = 1.7 \times 10^4$	150	5.0×10^5	—	[2]	U	U[3]	U	U	[3,4]
COGSWELL [1968 : 6] [1969 : 2]	Polymethyl-methacrylate	190	1.0×10^6	3.2	C	[5]	—	—	—	—
	Low density polyethylene	130	5.0×10^6	2.6	—	I	5	—	—	—
STEVENSON [1972 : 14]	Polyisobutylene-isoprene copolymer $M_v = 3.5 \times 10^5$	80	3.8×10^7	2.84	C	C	—	—	—	—
		100	8.0×10^6	2.91	C	C	—	—	—	—
	cis-1.4-polyisoprene	80	$\sim 10^8$	—	U (no shear viscosity data)			—	—	—

[1] Since the reported strain rates for elongational viscosity data and shear viscosity data did not overlap, this result is based on a reasonable extrapolation of the shear viscosity data.

[2] At the lowest experimental elongation rate, the elongational viscosity apparently attained a time-independent value, but the recoverable elongational strain was still time-dependent.

[3] Since a complete shear viscosity curve was not reported, it is not possible to be certain that the highest elongation rates are in the SVPL group.

[4] At the highest experimental elongation rate, the recoverable strain approached a time-independent value, but the elongational stress was time-dependent and ultimately began to decrease with increasing time (or strain).

[5] A convergent flow experiment which required many questionable assumptions in the data analysis showed that the elongational viscosity is independent of elongation rate in this elongation rate group.

This table gives a qualitative comparison elongational flow behavior for several materials by referring the behavior of the material in elongational flow to its behavior in shear flow.

For the purposes of this comparison, shear rates are divided into three groups characterized by the behavior of the shear viscosity. These groups are designated : SVC (shear viscosity constant), shear rates at which the shear viscosity is constant; SVD (shear viscosity decreasing), shear rates at which the shear viscosity is decreasing but not yet in the power-law region ; and SVPL (power-law shear viscosity), shear rates in the power-law region.

The behavior of the elongational viscosity vs elongation rate curve is determined for three groups of elongation rates which correspond in numerical value to the three shear rate groups. The behavior of the elongational viscosity curve for each group of elongation rates can be approximately characterized as : *C* (constant), elongational viscosity independent of elongation rate ; *I* (increasing), elongational viscosity increases with increasing elongation rate; *D* (decreasing), elongational viscosity decreases with increasing elongation rate ; or *U* (unsteady), elongational viscosity during stress growth is not approaching steady state at the maximum experimental strain.

The recoverable elongational strain $\bar{\gamma}_r$ versus total elongational strain $\bar{\gamma}$ curve for a group of elongation rates can generally be characterized as : *S* (steady), recoverable elongational strain attains a steady state value ; or *U* (unsteady), recoverable elongational strain does not attain a steady state value. Here $\bar{\gamma} = \dot{\varepsilon}t$. If $\bar{\gamma}_m$ is the max. value of $\bar{\gamma}$, and at the end of the experiment $\bar{\gamma}_i$ is the irrecoverable strain, $\bar{\gamma}_n = \bar{\gamma}_m - \bar{\gamma}_i$.

The basic principle behind an extensional flow experiment is to pull a specimen and subject it to a deformation at a constant extensional rate. The specimen may float on a bath or be submerged in it, or be left free to the atmosphere. As shown by STEVENSON (for a particular case, no doubt), the effects of the neglect of inertia, that of not supplying the requisite radial stress along the specimen (*cf.* Sec. 33), or of ignoring the interfacial tension between the specimen and the bath liquid are very small. So these will be omitted in the ensuing brief discussion of the theory behind such experiments.

Since $\dot{z} = \dot{\varepsilon} z$, it means that

$$l(t) = z(t) = l_0 \exp \dot{\varepsilon}, \tag{49.6}$$

where l_0 is the length of the cylinder at time $t = 0$. Hence

$$\dot{\varepsilon} = \frac{d}{dt} \ln (l/l_0) ; \tag{49.7}$$

or, in any experiment for simple extensional flow, the length of the rod must increase exponentially with time. For example, in the setup used by BALLMAN, extension rates varying from 7.8×10^{-4} sec^{-1} to 2.2×10^{-2} were used, and the constant strain rate was obtained by using a function generator on the testing machine. He computed, for each constant extension rate, the actual stress in the specimen at time t and compared it with the *true strain*, that is, $\varepsilon_T = (l - l_0)/l_0$.

We reproduce his curves of η_e [resp. $\bar{\eta}(\varkappa^2)$] against the strain rate $\dot{\varepsilon}(\varkappa$, the rate of shear) in Fig. 49.1. It is seen that the elongational viscosity is almost a constant, and is nearly 160 times the steady shear viscosity at a strain rate of about 3×10^{-2} sec^{-1}. In Fig. 49.2, we have reproduced the curves of STEVENSON for two materials, one for which the axial stress attains an equilibrium and the other for which it has an S-shape.

Since the above experiments are somewhat difficult to perform, METZNER and METZNER[127] have suggested that the thrust exerted by a fluid jet as it emerges from a sharp orifice be used to calculate the extensional viscosity. However, we believe that the approach is based on a large number of assumptions to be adaptable readily.

Another method, used by KAYE and VALE[128], is to measure the diameter of a free falling jet and to attempt to relate it to extensional viscosity. Some success has been claimed for Newtonian liquids by these authors, but no information about non-Newtonian liquids is available.

Turning to theoretical attempts to predict the elongational viscosity, we note that if $T\langle rr \rangle \equiv 0$ in the cylinder of radius $R(t)$ (that is, we neglect iner-

127. [1970 : 17]. In [1971 : 19], METZNER further explores this idea of approximating certain flows as extensional flows.

128. [1969 : 12].

tia again), then (33.9)—(33.10) imply that (put $h \equiv \alpha_1$, $l \equiv \alpha_2$, $\dot{\varepsilon} = \alpha$ there):

$$T\langle zz\rangle(t) = 3(\alpha_1 + \dot{\varepsilon}\alpha_2)\,\dot{\varepsilon}, \tag{49.8}$$

whence

$$\eta_e = 3(\alpha_1 + \dot{\varepsilon}\alpha_2), \tag{49.9}$$

where α_1 and α_2 are functions of $\dot{\varepsilon}^2$ and $\dot{\varepsilon}^3$. This is what the simple fluid theory predicts. No attempt seems to have been made to determine $\alpha_1(\cdot)$ and $\alpha_2(\cdot)$.

However, most of the research has centered around the fact that some of the simpler models predict that $\eta_e \to \infty$ as $\dot{\varepsilon}$ tends to some value $\dot{\varepsilon}_\infty$ (say). For example, LODGE's theory predicts[129] that if $\dot{\varepsilon} > 1/\lambda_1$, where λ_1 is a relaxation time, then $\eta_e \to \infty$. To overcome such difficulties, METZNER[130] and his co-workers have suggested that there exists a maximum value of $\dot{\varepsilon}$ to which a material may be subjected. The reasoning behind this suggestion will now be explored. By using Eq. (34.C30), one has that

$$T\langle zz\rangle = \int_0^\infty \mu(s)\,(e^{-2\dot{\varepsilon}s} - e^{\dot{\varepsilon}s})\,ds. \tag{49.10}$$

It is clear from this that the stress $T\langle zz\rangle$ does not remain bounded unless $\lim_{s\to\infty} \mu(s)\,e^{\dot{\varepsilon}s} = 0$, and since usually one assumes that $\lim_{s\to\infty} s^3\mu(s) = 0$ [*cf.* (34.*b*9), the above comments demanding the finiteness of $\dot{\varepsilon}$ may be expected to be valid. However, we believe that such limitations are the consequences of the poor quality of the constitutive equations chosen rather than any such restriction on $\dot{\varepsilon}$. This comment is borne out by the analysis, both theoretical and experimental, of DENN and MARRUCCI[131], as well as the theoretical analysis of TANNER and BALLMAN[132]. We shall discuss the latter briefly.

These authors considered a constitutive equation of the form[133]

$$\mathbf{T}_E = \int_0^{\tau_R} \mu(s)(\mathbf{C}_t(t-s)^{-1} - \mathbf{1})\,ds, \tag{49.11}$$

where τ_R is some characteristic time, which varies from one flow to another. TANNER and BALLMAN assumed that $\mu(s)$ is given by

$$\mu(s) = \sum_{n=1}^{13} \frac{a_n}{\lambda_n^2} \exp(-s/\lambda_n), \tag{49.12}$$

129. [1964 : 14], p. 116.
130. [1967 : 17), [1967 : 18]. In [1969 : 21], this limiting stretch rate is applied to drag reduction in turbulent flows.
131. [1971 : 9].
132. [1969 : 22].
133. TANNER and SIMMONS [1967 : 24] proposed this model to explain SIMMONS' [1968 : 22] experimental results. It is actually a special case of the additive functional theory.

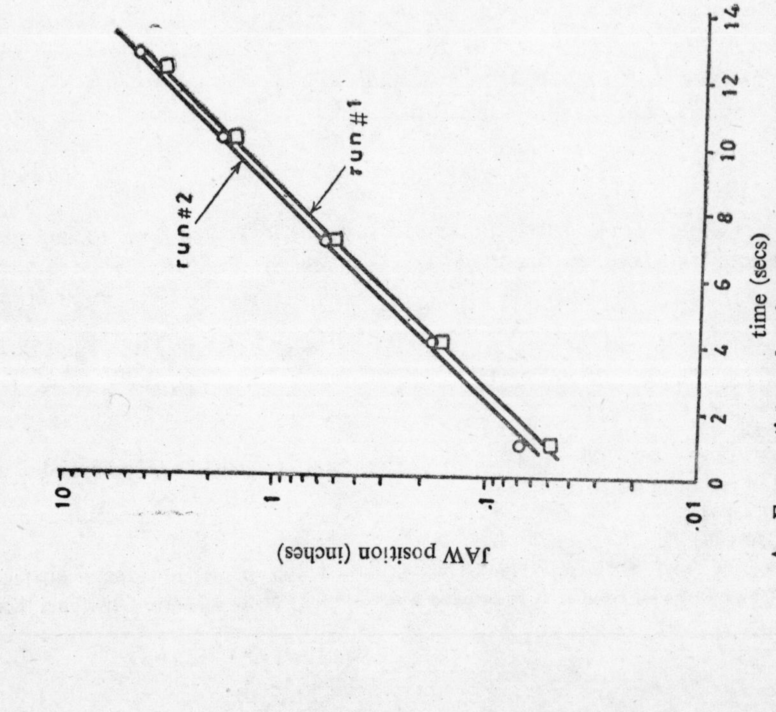

A. Exponential acceleration of tester crosshead, giving constant true strain rate.

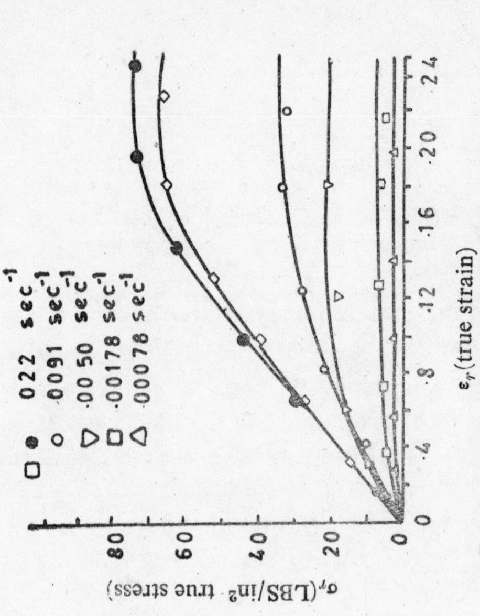

B. Polystyrene stress-strain curves at 300°F.

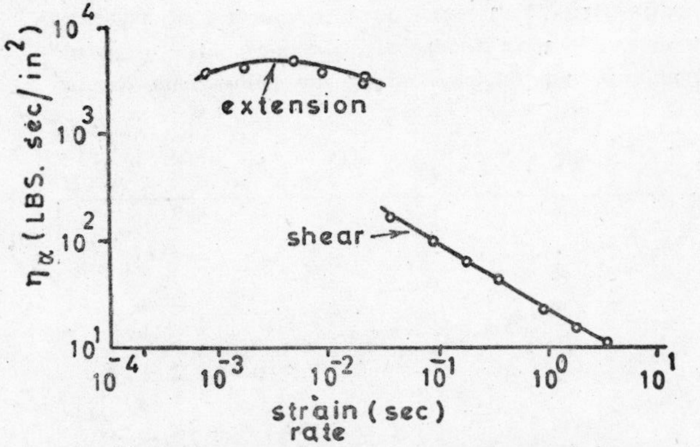

C. Comparison of extensional and shear viscosities vs. strain rate for polystyrene at 300°F.

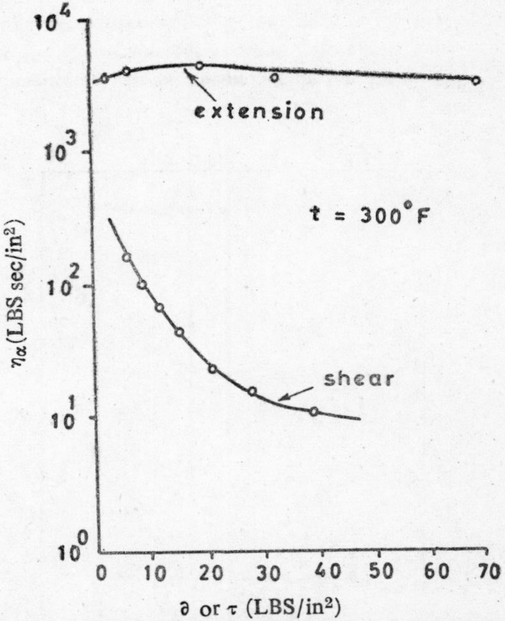

D. Comparision of stress dependence of tensile and shear viscosities of polystyrene melt.

Fig. 49.1. Elongational viscosity data (see BALLMAN [1965, 1]).

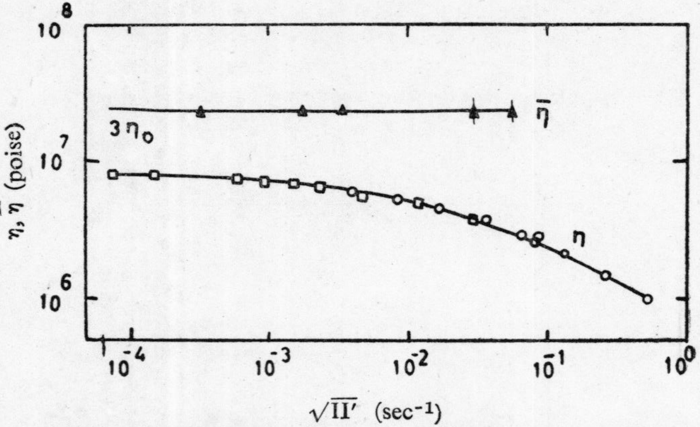

A. Steady elongational and shear viscosity data Butyl 035 at 100°C. The ordinate is the square root of the second invariant of the rate-of-strain tensor. △ Elongational viscosity; △ Elongational viscosity (extrapolated elongational stress); □ Shear viscosity (steady shear experiment); ○ Shear viscosity (oscillatory experiment).

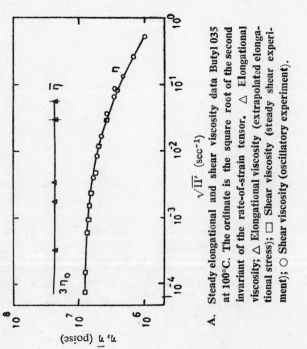

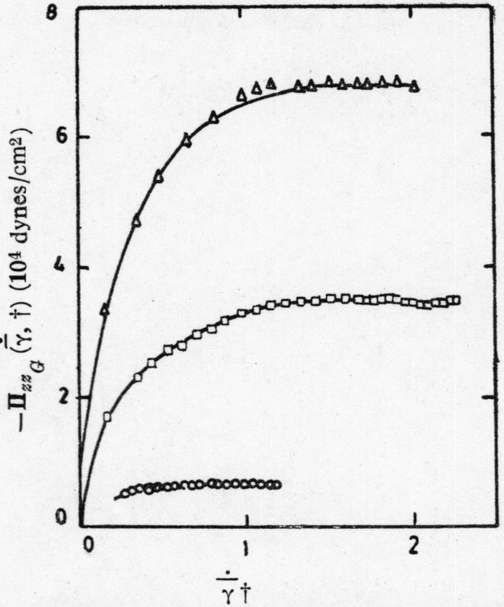

B. Elongational stress versus strain for Butyl 035 at 100°C (low elongation rates). Symbols represent elongation rates as indicated: ○ $2.85 \times 10^{-4}\ s^{-1}$; □ $1.43 \times 10^{-3}\ s^{-1}$; △ $2.82 \times 10^{-3}\ s^{-1}$. Note that the elongational stress attains steady state.

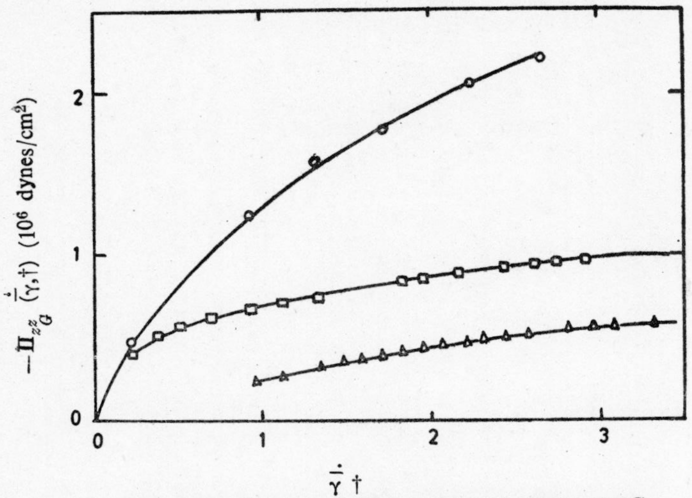

C. Elongational stress versus strain for Butyl 035 at 100°C (high elongation rates). Symbols represent elongation rates as indicated: △ $2.48 \times 10^{-2}\ s^{-1}$; □ $4.51 \times 10^{-2}\ s^{-1}$; ○ $1.35 \times 10^{-1}\ s^{-1}$. Note that the elongational stress is approaching steady state.

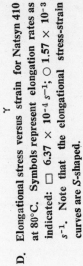

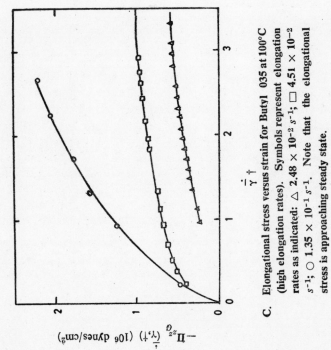

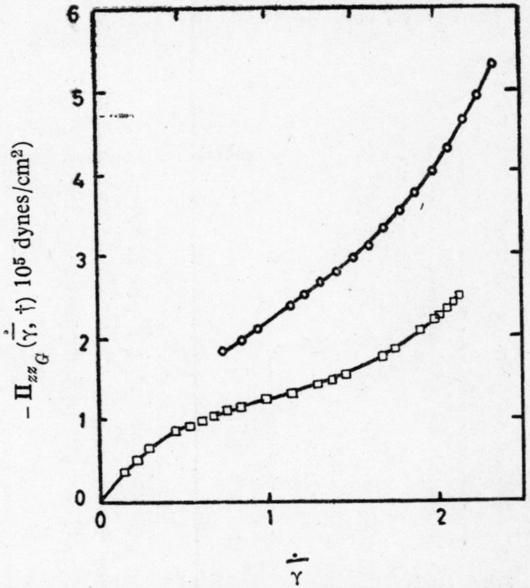

D. Elongational stress versus strain for Natsyn 410 at 80°C. Symbols represent elongation rates as indicated: □ $6.37 \times 10^{-4}\ s^{-1}$; ○ $1.57 \times 10^{-3}\ s^{-1}$. Note that the elongational stress-strain curves are *S*-shaped.

Fig. 49.2. Elongational viscosity data (See STEVENSON [1972 : 14]).

and used the available experimental data for $\eta'(\omega)$ to compute λ_n and a_n, with $\lambda_1 = 505$ seconds, $\lambda_{n+1} = \lambda_n/\sqrt{10}$, $n = 1, \ldots, 12$. Then this form of $\mu(s)$ was used to compute the steady shear viscosity $\bar{\eta}(\kappa^2)$ as well as the extensional viscosity η_e. It was found that the agreement between the theoretical predictions and experimental observations was of the right order of magnitude. This theory also predicts that

$$\ln \eta_e \to b \ln \dot{\varepsilon}, \qquad b = \text{const.} \tag{49.13}$$

From the experimental data available, it is difficult to say whether or not (49.13) is a reasonable prediction. The above prediction should be compared with that derived from a molecular theory[134] with LANGEVIN springs, which anticipates that η_e tends to a finite limit as $\dot{\varepsilon}$ becomes large.

Finally, the large values for extensional viscosity mean that the fluid will try to avoid such flows wherever possible.[135] METZNER, *et al.*[136] have reported that if one examines the velocity field of the Newtonian liquid upstream of a sudden contraction, it is found that the fluid particles approach the hole from all directions. In case of a viscoelastic liquid there is a narrow region of flow where simple extension seems to take place, while outside of it, there are secondary flows of a shearing nature. We believe that the rational explanation for such a narrow range of extensional flow lies in the greater resistance met in such motions, and the intuitive feeling that processes in nature follow the path of least resistance.

50. Viscoelastic Jets : First Normal Stress Difference and Die Swell

It has been known that a viscoelastic liquid on emerging from a capillary exhibits an increase in its radius[137]. The main cause for this expansion is the tension in the fluid along the flow direction created by the motion in the tube. When the bounding surfaces vanish, the fluid element contracts, leading to a *die swell*. We shall show how the issuance of a jet of viscoelastic fluid from an orifice can be used to predict the first normal stress difference N_1 (by assuming $N_2 = 0$) at very high speeds (or large shear rates), by ignoring surface tension. At the other end of the scale, we shall consider the slow flow of a special fluid, again with $N_2 = 0$, to predict the die swell ratio (diameter of extrudate/diameter of tube), again neglecting surface tension. Both ana-

134. See STEVENSON and BIRD [1971 : 25].
135. PIPKIN [1969 : 18].
136. [1969 : 17].
137. See METZNER [1969 : 16] for historical comments on the discovery of the radial expansion. He states that BARUS (c. 1893) did not record the radial expansion, while others preceded MERRINGTON (c. 1943). So, we will not call this *Barus effect* or the *Merrington phenomenon*.

lyses are based on other approximations[138], for in the case of the fast jet, one ignores die swell, while in the latter one ignores inertia. These assumptions as well as those made to compute the die swell in the latter case are not rigorous, and are motivated by experimental observations.

However, we wish to dispel the notion that analytical approaches have neither been formulated nor tried. First of all, we shall summarize the more recent work of JOSEPH[139], who has developed the correct equations governing the problem of die swell, and who has rephrased this problem along the lines of the domain perturbation technique presented in Sec. 44. Because of this work, we will not present the analytical work of RICHARDSON[140], who solved the two-dimensional creeping flow problem for a Newtonian fluid, issuing out of a channel, ignoring surface tension. His results while incomplete, were the first analytical results for this extremely difficult problem. Now, the analytical and experimental work on the free surface in COUETTE flow (see Sec. 44) we have presented means that surface tension should not be ignored at low speeds. So we have not presented a detailed summary of RICHARDSON's paper. However, it is in order to remark that his solution calls to attention the fact that the fluid can sense the exit, while it is still in the channel (remniscent of subsonic gas dynamics but for incompressibility), and thus one should not assume a fully developed flow to exist upto the exit. Hence, our summary on fast and slow viscoelastic jets presented later is not quantitatively correct.

Now consider a horizontal jet issuing out of a circular pipe (see Fig. 50.1).

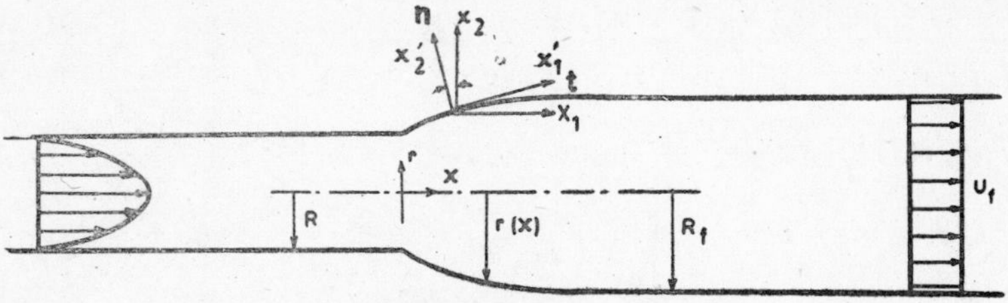

Fig. 50.1. The capillary jet.

Let the radius of the pipe be R and that of the jet for downstream be R_f. It will be assumed that gravitational forces are absent and that the final velocity U_f is uniform, that is, independent of the radial distance from the axis of the jet and if $r(x)$ is the radius of the jet in $x > 0$, then with U being the axial velocity,

$$U_f = \lim_{x\to\infty} \mathbf{v}(r, x) \cdot \mathbf{e}_x \; ; \quad R_f = \lim_{x\to\infty} r(x), \tag{50.1}$$

138. In addition, all effects such as length to diameter ratio, entrance losses, exit losses, etc. are ignored.

139. [1973 : 6]. He extends substantially the theory of SLATTERY and SCHOWALTER [1964 : 23]. See also [1974 : 2] and HUILGOL [1975 : A].

140. [1970 : 19], [1970 : 20].

where $\mathbf{v}(r, x)$ is the velocity field and $\mathbf{e}_x$ the unit vector in the x-direction. Consider a control volume bounded by $x = \pm L$. The momentum balance states that in the axial direction $(x = x^1)$,

$$\iint_S (T_{1j} - \rho U v_j)\, n_j\, da = 0, \tag{50.2}$$

where Cartesian tensor notation is employed. Now (50.2) can be expanded as

$$\int_0^{r(L)} [T_{11}(r, L) - \rho U^2(r, L)] r\, dr - \int_0^R [T_{11}(r, -L) - \rho U^2(r, -L) r\, dr + \\ + \int_0^L T_{1j}\, n_j\, r(x)\, ds + \int_{-L}^0 T_{1j}\, n_j\, R\, dx = 0, \tag{50.3}$$

where $ds = (dr^2 + dx^2)^{1/2}$ is the arc length along the free surface, with $\mathbf{n}$ the unit normal on it. Now in terms of the unit normal

$$T_{1j}\, n_j = T_{11} n_1 + T_{12} n_2 = -\sin\alpha\, T_{nn}, \tag{50.4}$$

where T_{nn} is the stress in the normal direction, and α is the angle of inclination. Moreover, this normal stress T_{nn} is given by

$$T_{nn} = -p_a + \sigma J, \tag{50.5}$$

where p_a is the atmospheric pressure, σ the surface tension and J is the mean curvature of the surface (in $x > 0$) given by

$$J = -\frac{1}{r\varphi} + \frac{h''}{\varphi^3}, \\ \varphi(r') = (1 + r'^2)^{1/2}. \tag{50.6}$$

Hence, by using $ds/dx = \sin\alpha$, we have

$$\int_0^L T_{1j} n_j\, r(x) \frac{ds}{dx}\, dx = -\int_0^L \tan\alpha\, T_{nn}\, r dx \\ = -\int_0^L r' r\, T_{nn}\, dx \\ = p_a \frac{r^2(x)}{2}\Big|_0^L + \frac{\sigma r(x)}{(1 + r'^2)^{1/2}}\Big|_0^L; \tag{50.7}$$

and on the surface $r = R$,

$$\int_{-L}^0 T_{1j}\, n_j\, R_0\, dx = R \int_{-L}^0 T_{12}\, dx. \tag{50.8}$$

Combining all of the above, we can show that (50.3) becomes

$$\int_0^{r(L)} [T_{11}(r, L) - \rho U^2(r, L)]\, r dr + \int_0^R [p(r, -L) - T^E_{11}(r, -L) + \\ + \rho U^2(r, -L)]\, r dr + p_a \frac{r^2(x)}{2}\Big|_0^L + \frac{\sigma r(x)}{(1 + r'^2)^{1/2}}\Big|_0^L + \\ + R \int_{-L}^0 T_{12}\, dx = 0. \tag{50.9}$$

Let us note in passing that in a state of rest, the pressure in the liquid is given by (in $x > 0$) :

$$p_0 = p_a + \frac{\sigma}{R}, \tag{50.10}$$

because $J = -1/R$ in this case. We shall now postulate a *Poiseuille flow hypothesis* (PFH)[141]:

All flow quantities are almost fully developed when $x < -\overline{L}$.

Now, we let $L \to \infty$ in (50.9) and obtain

$$\int_0^R [T_{11}^E(r, -\infty) - \rho U^2(r, -\infty)]\, rdr - R \int_{-\infty}^0 T_{12}\, dx$$
$$= \frac{T_{11}(R_f, \infty) - \rho U_f^2}{2} R_f^2 + \sigma \left[R_f - \frac{R}{(1 + R'^2)^{1/2}} \right] +$$
$$+ \frac{p_a}{2}(R_f^2 - R^2) + \int_0^R p\,(r, -\infty)\, rdr. \tag{50.11}$$

We have used the assumption that as $x \to \infty$, T_{11} and U are independent of x. Also, in (50.11), R' should be interpreted as $\lim\limits_{x \to 0+} \dfrac{dr(x)}{dx}$. It can be seen that, given PFH,

$$\int_0^R p(r, -\infty)\, rdr + R \int_{-\infty}^{-\overline{L}} T_{12}\, dx = \int_0^R p(r, -\overline{L})\, rdr. \tag{50.12}$$

Finally, as $L \to \infty$, $T_{11}(R_f, \infty) = -p(R_f, \infty)$, and this is given by

$$p_f = p_a + \frac{\sigma}{R_f}\cdot \tag{50.13}$$

Hence (50.9) becomes, on taking $p_a = 0$ for convenience,

$$\int_0^R [T_{11}(r, -\overline{L}) - \rho U^2(r, -\overline{L})]\, rdr - R \int_{-\overline{L}}^0 T_{12}\, dx$$
$$= \frac{\sigma}{2}\left[R_f - \frac{2R}{(1 + R'^2)^{1/2}} \right] - \frac{\rho U_f^2 R_f^2}{2}. \tag{50.14}$$

This is the correct momentum balance equation for the jet flow problem.

---

141. Eqs. (50.11) — (50.14) are based on HUILGOL [1975 : *A*], and differ from those in JOSEPH [1974 : 2] because of PFH. It is shown by the author that

$$\int_0^R [2\rho U^2 - 2N_1 - N_2]\, rdr - R^2 T_{rr}(r, -\overline{L}) - \frac{2CR\tau(\varkappa_w)R}{1-n}\left(\frac{\overline{L}}{R}\right)^{1-n} \leqq$$
$$\leqq \rho U_f^2 R_f^2 \leqq \int_0^R [2\rho U^2 - 2N_1 - N_2]\, rdr - R^2 T_{rr}(r, -\overline{L}) - 2R\tau(\varkappa_w)\overline{L}, \qquad (*)$$

where $C = 1.0$, $n = 0.5$, $\overline{L} = 0.8R$, $U = U(r, -\overline{L})$ and $\tau(\varkappa_w)$ is the wall shear stress in the POISEUILLE flow.

For a Newtonian jet, T^E_{11} $(r, -\infty) = 0$, and on the pipe surface n $-\overline{L} \leq x < 0$,

$$T_{12}(R) = \eta_0 \left.\frac{\partial U}{\partial r}\right|_{r=R} . \tag{50.15}$$

Also,

$$U(r, -\overline{L}) = \frac{4Q}{\pi R^2}\left(1 - \frac{r^2}{R^2}\right), \tag{50.16}$$

where Q is the flow rate. Assuming that the integral of T_{12} is proportional to

$$K_1(R_e)\, R_e^n , \tag{50.17}$$

where K_1 is a weak function of R_e, which is the REYNOLDS number, JOSEPH showed that if one chooses $n = 1/3$ according to experimental results of GOREN and WRONSKI[142], then the ratio $\chi = R_f/R$ obeys

$$\frac{4}{3} - \frac{1}{\chi^2} - \frac{K_1}{R_e{}^{2/3}} = 0, \tag{50.18}$$

when it is also assumed that surface tension effects are negligible. As $R_e \to \infty$, $\chi^2 \to 3/4$, which is HARMON's result[143].

The reader is referred to JOSEPH's paper for the following:

(i) the excellent agreement between experiments and theory, obtained by choosing $\chi = 1$ when $R_e = 14.4$, and using $K_1 = (14.4)^{2/3}/3 = 1.973$ from (50.18) (see Fig. 50.2);

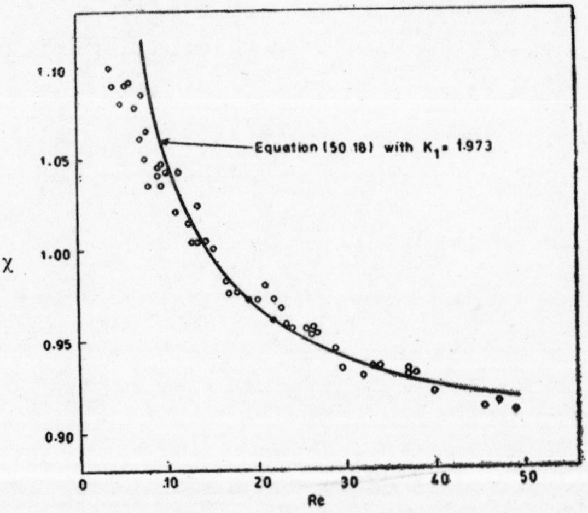

Fig. 50.2. Contraction ratio experiments of GOREN and WRONSKI [1966 : 9].

(ii) the setting up of a domain perturbation analysis based on the ideas presented in Sec. 44, and pivoted about the rest solution. The question

142. [1966 : 9].

143. [1955 : 1], See also DUDA and VRENTAS [1967 : 7], Eq. (94).

of the existence of higher-order solutions is at present unresolved, however.

One can see that by neglecting surface tension, and assuming that $\bar{L} \to 0$ at large REYNOLDS numbers, (50.14) becomes

$$2\int_0^R [\rho U^2(r, -\bar{L}) - T_{11}(r, -\bar{L})]\, rdr = \rho U_f^2 R_f^2 \tag{50.19}$$

From this, one can deduce that $\chi^2 \to 3/4$ for a Newtonian circular jet and $\chi \to 5/6$ for the plane (channel flow) jet,[144] on changing the analysis slightly to take care of the geometry of the flow domain.

We shall now neglect surface tension, let $\bar{L} = 0$ in the next two subsections and present analyses which indicate that the first normal stress difference may be found at high rates of shear by measuring the radial stress at the wall, and that an upper bound for die swell may be found from a knowledge of viscometric functions.

50.1. Fast Moving Jets.[145] Consider a fast moving jet issuing out of a circular capillary. Let the flow be laminar and be the fully developed POISEUILLE flow even at the exit, denoted by $x = 0$, with x being the flow axis. If $T\langle xx\rangle$ is the stress in the fluid at exit, then the thrust F exerted by the fluid is

$$F = \int_0^R \rho U^2 \cdot 2\pi\, rdr - \int_0^R T\langle xx\rangle\, 2\pi\, rdr. \tag{50.20}$$

In other words, the net thrust is the excess of the momentum efflux over the force exerted by the axial stress.

Following LODGE,[146] let the second integral be subjected to integration by parts. Then

$$F = \int_0^R 2\pi\rho\, rU^2 dr - \pi R^2\, T\langle xx\rangle\,(R, L) + \pi \int_0^R r^2 \frac{\partial T\langle xx\rangle}{\partial r}\, dr. \tag{50.21}$$

However,

$$\begin{aligned}\frac{\partial T\langle xx\rangle}{\partial r} &= \frac{\partial T\langle rr\rangle}{\partial r} + \frac{\partial}{\partial r}\left(T\langle xx\rangle - T\langle rr\rangle\right)\\ &= \frac{T\langle\theta\theta\rangle - T\langle rr\rangle}{r} + \frac{\partial}{\partial r}\left(T\langle xx\rangle - T\langle rr\rangle\right),\end{aligned} \tag{50.22}$$

144. RICHARDSON [1970 : 20]; TILLETT [1968 : 24].

145. The reader is referred to the review article on the subject by SHERTZER and METZNER [1965 : 16], where references to earlier papers may be found.

146. [1964 : 14], p. 219.

where we have used the r-component of the equations of motion and noted that radial acceleration is zero. Substituting (50.22) in (50.21) and integrating by parts once more, one is led to

$$F = 2\pi\rho \int_0^R U^2 r dr - \pi R^2 T\langle rr\rangle (R, L) - \pi \int_0^R r(2N_1 + N_2)\, dr. \tag{50.23}$$ [147]

Since, in a POISEUILLE flow, the shear stress τ is directly proportional to the radius, or

$$\tau/\tau_w = r/R, \tag{50.24}$$

where τ_w is the wall shear stress, one can write the integral in (50.23) in terms of τ as

$$F = \frac{2\pi R^2\rho}{\tau_w^2} \int_0^{\tau_w} \tau U^2 d\tau - \pi R^2 T\langle rr\rangle (R, L) - \frac{\pi R^2}{\tau_w^2} \int_0^{\tau_w} \tau(2N_1 + 2N_2)\, d\tau. \tag{50.25}$$

Hence

$$(2N_1 + N_2)\Big|_{\varkappa=\varkappa_w} = \frac{1}{\pi R^2 \tau_w} \frac{\partial}{\partial \tau_w}\left[\tau_w^2 (F + \pi R^2 T\langle rr\rangle)\right], \tag{50.26}$$ [148]

since

$$\frac{d}{d\tau_w} \int_0^{\tau_w} \tau U^2 d\tau = \tau_w U^2(\tau_w) = 0, \tag{50.27}$$

for the value of U on the wall of the pipe, where $\tau = \tau_w$, is zero. Measuring F, and the radial stress $T\langle rr\rangle (R, L)$ at exit on the wall, yields $2N_1 + N_2$. Omitting N_2, one obtains the first normal stress difference. However, the above method is open to the criticism that measurement of $T\langle rr\rangle$ at the exit may not only be difficult, but may introduce severe errors into that of the thrust F, because of pressure hole errors.

50.2. Creeping Jet. Here the fluid is assumed to move through the capillary (with a fully developed POISEUILLE flow up to the exit) very slowly so that inertial effects are unimportant. The process, while slow on one time scale, is fast enough for the viscoelastic fluid (on its own internal time clock) to register a sudden elastic response with a noticeable expansion in diameter. Later on, due to the decay of stresses by relaxation,[149] a further swelling occurs. This additional increase may or may not be negligible. If one cuts off a piece of the extrudate and anneals it completely so that full recovery occurs, then one designates this ratio of the fully recovered diameter D_r to the tube diameter D as the die swell ratio. In this case, one attempts to

147. N_1 and N_2 are the usual normal stress differences.

148. Our results correct a misprint in [1964 : 14], p. 211.

149. POWELL and MIDDLEMAN [1969 ; 19] have attempted an analysis of this relaxation for a special fluid.

relate D_r/D to the energy needed to pull the annealed extrudate till its diameter becomes D again. Such a method was originally proposed by NAKAJIMA and SHIDA,[150] and has been put into a more useful form by GRAESSLEY, *et al.*,[151] as well as by BAGLEY and DUFFEY.[152]

On the other hand, if one wishes to predict the die swell just as the fluid emerges from the jet, then one has to use a constitutive relation applicable to a fluid. Such a relation proposed and used by TANNER[153] is the following special case of the additive functional theory:

$$\mathbf{T}_E = \int_0^\infty f(I_1, s)\, \mathbf{C}_t^{-1}(t-s)\, ds, \tag{50.28}$$

where $\qquad I_1 = \operatorname{tr} \mathbf{C}_t^{-1}(t-s).$

Since we shall ignore the effect of length to diameter ratio and residence time in the tube and so on, we shall present the GRAESSLEY, *et al.*'s analysis for the fully annealed solid first. Let this solid be considered to be hyperelastic with a stored energy function (per unit volume) W of the neo-Hookean type, viz.

$$W = C_1(\lambda_1^2 + \lambda_2^2 + \lambda_3^2 - 3), \tag{50.29}$$

where the λ_i are the eigenvalues of the strain tensor $\mathbf{B}$, which is the left strain tensor (1.11) computed by using the swollen annealed configuration as the natural state, and C_1 is a constant. Let the piece of extrudate have a length L_r, and when its diameter D_r is reduced to D, let its length be L, and $\alpha = (D_r/D)^2$. Then, by assuming a uniform elongation along the axis (x_3-axis), the extensions λ_i are

$$\lambda_1^2 = \lambda_2^2 = \frac{1}{\alpha}, \qquad \lambda_3^2 = \alpha^2, \tag{50.30}$$

where the incompressibility of the solid has been employed. Using (50.30) in (50.29), one obtains

$$W = C_1\left(\alpha^2 + \frac{2}{\alpha} - 3\right). \tag{50.31}$$

For this neo-Hookean solid, $C_1 = \frac{1}{2}G$, where G is the shear modulus.[154] Assuming that $\langle e \rangle$ is the energy needed to deform the solid per unit volume, we can write (50.31) as

$$\langle e \rangle = \frac{1}{2}G\left(\alpha^2 + \frac{2}{\alpha} - 3\right). \tag{50.32}$$

150. [1966 : 14].
151. [1970 : 11].
152. [1970 : 3].
153. [1970 : 23].
154. See RIVLIN [1948 : 1].

Now, following GRAESSLEY, *et al.*, let us postulate that in the viscometric flow in the capillary, the energy e stored in the fluid is

$$e(r) = \frac{1}{2} J_0 \tau^2(r), \tag{50.33}$$

where $\tau(r)$ is the shear stress at the radial position and J_0 is the steady state compliance of the linear theory, defined by

$$J_0 = \frac{1}{\eta_0^2} \int_0^\infty s\, G(s)\, ds. \tag{50.34}$$

Equation (50.33) is an expression for the work done ($= \frac{1}{2}$ tr $\mathbf{T}\,\mathbf{A}_1$), and is exact in the linear theory, but is an assumption here. The stored energy $\langle e \rangle$ per unit volume is given by

$$\langle e \rangle = \frac{\int_0^R 2\pi r\, U(r)\, e(r)\, dr}{\int_0^R 2\pi r\, U(r)\, dr} = \frac{J_0\, \tau_w^2}{4} \frac{\int_0^1 \xi^5/\eta\, d\xi}{\int_0^1 \xi^3/\eta d\xi}, \tag{50.35}$$

where η is the steady shear viscosity, and use is made of the relation $\tau_r = \tau_w\, r/R = \bar{\eta}(\varkappa^2)\, dU/dr$. Combining (50.32) and (50.35) one has

$$\alpha^2 + \frac{2}{\alpha} - 3 = \frac{J_0\, \tau_w^2}{2G} \frac{\int_0^1 \xi^5/\eta\, d\xi}{\int_0^1 \xi^3/\eta\, d\xi}, \tag{50.36}$$

from which α may be calculated if τ_w, J_0, G and $\bar{\eta}(\varkappa^2)$ are known.

The above analysis assumes that J_0 and G are constants, while at the same time it uses a variable viscosity. Such a computation may yield some kind of an upper bound on the die swell ratio, but it is uncertain as to what α really represents in this analysis. A more accurate calculation to obtain this bound $\bar{\alpha}$ is to assume that the energy stored in fluid per unit volume is immediately turned into elastic energy, that is, we have

$$e(r) = \frac{1}{2} \bar{\eta}(\varkappa)\, \varkappa^2, \qquad \varkappa = dU/dr, \tag{50.37}$$

$$\langle e \rangle = \int_0^R 2\pi r\, U(r)\, e(r)\, dr \Big/ \int_0^R 2\pi r\, U(r)\, dr, \tag{50.38}$$

and the stored energy per unit volume W is given by

$$W = W(I_1 - 3, I_2 - 3), \tag{50.39}$$

where $I_1 = \bar{\alpha}^2 + (2/\bar{\alpha})$, $I_2 = 2\bar{\alpha} + (1/\bar{\alpha}^2)$. Then one solves the equation

$$W(I_1 - 3, I_2 - 3) = \langle e \rangle \tag{50.40}$$

for $\bar{\alpha}$. For example, from the empirical equations used in the rubber theory of elasticity, one has[155]

$$W(I_1 - 3, I_2 - 3) = \Sigma_s \mu_s \varphi(\alpha_s), \qquad (50.41)$$

where the μ_s are constants, $\varphi(\alpha_s)$ is defined by

$$\varphi(\alpha_s) = (a_1^{\alpha_s} + a_2^{\alpha_s} + a_3^{\alpha_s} - 3)/\alpha_s, \qquad (50.42)$$

a_i^2 $(i = 1, 2, 3)$ are the eigenvalues of the strain tensor **B** and the α_s are real numbers. Note that $\frac{1}{2}(I_1 - 3) = \varphi(2)$ and $-\frac{1}{2}(I_2 - 3) = \varphi(-3)$. OGDEN found that the function

$$6.3\ \varphi(1.3) + 0.012\ \varphi(5.0) - 0.1\ \varphi(-2) \qquad (50.43)$$

gave the best fit for experimental data obtained from simple tension, pure shear and biaxial tension.

It would therefore be of interest to obtain such a strain energy function for the solidified polymer by using (50.41), and then computing $\bar{\alpha}$ through (50.40). Since one has to perform two sets of experiments, one in the solid and the other in the liquid phase and since one may not deal with a solution having a solid phase at manageable temperatures, it would be desirable to obtain such a bound $\bar{\alpha}$ purely in terms of fluid properties. So we shall pursue TANNER's method[156] to achieve this goal, and will compute the ratio D_e/D where D_e is the extrudate diameter as it emerges from the die.

Let (r', θ', z') be the coordinates of a fluid particle inside the capillary at the instant t', such that up to t' the particle has experienced the viscometric flow history of a POISEUILLE flow. At this instant, the fluid particle is pushed out and suddenly it expands to occupy the position (r, θ, z) and from then on it is subject to a pure translation. This translation means that the fluid particle is held in a state of rest; or, the relative deformation gradient is given by

$$\mathbf{F}_t(t-s) = \begin{cases} \mathbf{1}, & 0 \leqq s < t' \\ \mathbf{F}_{t'}(t-s)\,\mathbf{F}_t(t'), & t' \leqq s < \infty, \end{cases} \qquad (50.44)$$

where $t > t'$ is the position currently occupied by the particle at time t [*cf.* (6.5 - (6.6)]. So, the stresses at the particle are given by [on using (50.28)]:

$$\lim_{t \to t'-} \mathbf{T}_E = \int_0^\infty f(I_1, s)\,[\mathbf{C}^0_{t'}(t'-s)]^{-1}\,ds, \qquad (50.45)$$

$$\lim_{t \to t'+} \mathbf{T}_E = \int_0^\infty f(I_1, s)\,[\mathbf{F}_t(t')^T\,\mathbf{C}^0_{t'}(t'-s)\,\mathbf{F}_t(t')]^{-1}\,ds\,; \qquad (50.46)$$

that is, just before the sudden deformation, the stress is viscometric and is given by (50.45), where $\mathbf{C}^0_{t'}(t'-s)$ is the POISEUILLE strain history, while just

155. OGDEN [1972 : 11] has discovered the best strain-energy function for rubber-like solids.
156. [1970 : 23].

after, it is given by (50.46). In (50.46) it is assumed that

$$\text{tr}\,\{[\mathbf{F}_t\,(t')^T\ \mathbf{C}^0_{t'}\,(t'-s)\ \mathbf{F}_t(t')]^{-1}\} = \text{tr}\,\{[\mathbf{C}^0_{t'}\,(t'-s)]^{-1}\}, \tag{50.47}$$

or the 'magnitude' of $\mathbf{F}_t(t')$ is small. Since $\mathbf{F}_t(t')$ is a volume-preserving deformation gradient, (r, θ, z) must be related to (r', θ', z') in an axially symmetric elongational type deformation by

$$r' = \lambda r, \quad \theta' = \theta, \quad z' = (z/\lambda^2) + g(r), \tag{50.48}$$

where $g(r)$ is an unknown function and λ is related to α of (50.30) by $\lambda^2 = 1/\alpha$. Hence the matrix of $\mathbf{F}_t(t')\ \mathbf{F}_t(t')^T$ is given by

$$[\mathbf{F}_t(t')\ \mathbf{F}_t(t')^T] = \begin{bmatrix} \lambda^2 & 0 & \lambda g' \\ \cdot & \lambda^2 & 0 \\ \cdot & \cdot & g'^2+1/\lambda^4 \end{bmatrix}. \tag{50.49}$$

We now assume that *the stress tensor at time t* $(t > t')$ *has only one non-zero component* $T\langle zz\rangle$ *and this obeys*

$$2\pi \int_0^{D_e/2} T\langle zz\rangle\, r dr = 0, \tag{50.50}$$

which is the condition of static equilibrium.

Let us now offer some justifications for the assumptions about the state of stress:

(i) since $T\langle r\theta\rangle$ and $T\langle \theta z\rangle$ do not exist in the tube, they are not expected to exist outside ;

(ii) if $\mathbf{T}$ is independent of z and θ (and note this means that stress relaxation is not being considered), then by the z-component of the equations of equilibrium ($\rho = 0$), $T\langle rz\rangle = 0$ since it is zero on the surface;

(iii) the assumed constitutive relation (50.46) yields that $T_E\langle rr\rangle = T_E\langle \theta\theta\rangle$. Hence $T\langle rr\rangle = 0$ because it is zero on the surface;

(iv) thereupon the only non-zero component is $T\langle zz\rangle$. Then the constitutive relation (50.28) yields

$$-p\mathbf{1} + \int_0^\infty f(I_1, s)\,[\mathbf{F}_t(t')^T\ \mathbf{C}^0_{t'}\,(t'-s)\ \mathbf{F}_t(t')]^{-1}\, ds = T\langle zz\rangle\, \mathbf{e}_z \otimes \mathbf{e}_z. \tag{50.51}$$

The pressure p is unknown; however, *the linearity of the constitutive relation* as well as the assumed deformation (50.48) permit us to determine p as follows. Rewrite (50.51) as

$$-p\ \mathbf{F}_t(t')\ \mathbf{F}_t(t')^T + \int_0^\infty f(\mathbf{T}_1, s)\,[\mathbf{C}^0_{t'}\,(t'-s)]^{-1}\ ds = T\langle zz\rangle\ \mathbf{F}_t(t')\ \mathbf{e}_z \otimes \mathbf{e}_z\ \mathbf{F}_t(t')^T. \tag{50.52}$$

The integral in (50.52) is just the viscometric stress and is given in the r', θ', z' coordinate system by

$$\begin{bmatrix} G & 0 & \tau \\ \cdot & G & 0 \\ \cdot & \cdot & G + N_1 \end{bmatrix}, \tag{50.53}$$

where

$$G, \tau, N_1 = \varkappa^n \int_0^\infty f(I_1, s)\, s^n ds, \quad n = 0, 1, 2 \tag{50.54}$$

respectively and $\varkappa = dU(r')/dr'$. Here G is some elastic modulus, while τ and N_1 are the usual viscometric functions. Using (50.53) in (50.52) we have

$$\begin{aligned} p\lambda^2 &= G, \\ p\lambda g' &= \tau, \\ N_1 + G - T\langle zz\rangle/\lambda^4 &= p\,(1/\lambda^4 + g'^2). \end{aligned} \tag{50.55}$$ [157]

Hence, eliminating p and g' one has

$$N_1 + G - T\langle zz\rangle/\lambda^4 = G/\lambda^6 + \tau^2/G. \tag{50.56}$$

Using (50.46) in (50.56), it is found that

$$\frac{1}{\lambda} = \frac{D_e}{D} = \left\{ \frac{\int_0^{D_e/2} (N_1 + G - \tau^2/G)\, r dr}{\int_0^{D_e/2} Gr\, dr} \right\}^{1/6}. \tag{50.57}$$

Since N_1, τ and G are given in terms of r', it is preferable to change the variable to $\xi = 2r'/D$, $r' = \lambda r$ and obtain

$$\frac{D_e}{D} = \left[\int_0^1 (N_1 + G - \tau^2/G)\, \xi\, d\xi \Big/ \int_0^1 G\, \xi d\xi \right]^{1/6}, \tag{50.58}$$

from which D_e/D may be calculated. Note that in (50.58), G, N_1 and τ are functions of $\varkappa(\xi) = (2/\lambda D)\, dU/d\xi$. Let us call

$$\bar{G} = 2 \int_0^1 G\,(\varkappa(\xi))\, \xi\, d\xi \tag{50.59}$$

an average modulus, and consider a special fluid for which

$$f(I_1, s) = \frac{\eta_0}{\lambda_1^2} \exp(-s/\lambda_1), \tag{50.60}$$ [158]

where λ_1 is a relaxation time. We are dealing now with a very special linear

157. The linearity permits one to go directly from (50.51) to (50.55) without using (50.52).

158. The λ which appears in Eq. (32) of TANNER [1970 : 23] is not related to the λ in his Eq. (23).

theory, of course. For this fluid,

$$N_1/2\tau = \tau/G, \qquad G = \bar{G} = \eta_0. \tag{50.61}$$

Hence (50.58) now becomes

$$\frac{D_e}{D} = \left[2 \int_0^1 \left\{ 1 + (N_1/2\tau)^2 \right\} \xi \, d\xi \right]^{1/6}. \tag{50.62}$$

Since, for this fluid,

$$\left(\frac{N_1}{2\tau}\right)^2 = \frac{\tau^2}{\bar{G}^2} = \frac{\tau_w^2}{\eta_0^2} \xi^2, \tag{50.63}$$

where τ_w is the wall shear stress and $\xi = 2r'/D$, one has

$$\frac{D_e}{D} = \left(1 + \tau_w^2 / 2\eta_0^2\right)^{1/6} = \left(1 + N_{1w}^2 / 8\tau_w^2\right)^{1/6}. \tag{50.64}$$

In principle, it is prerequisite to know the wall shear stress, the zero shear viscosity η_0, and λ_1, to compute the die swell ratio for this fluid. Now (50.64) represents an upper bound for the ratio D_e/D. A different upper bound may be obtained by noting that τ varies from 0 to τ_w as D_e varies from [0, 1], but $G(\varkappa(\xi)) \neq 0$. Thus another upper bound is

$$\overline{\left(\frac{D_e}{D}\right)} = \left[2 \int_0^1 (1 + N_1/\bar{G}) \, \xi \, d\xi \right]^{1/6}, \tag{50.65}$$

which is independent of any assumption on $f(I_1, s)$. For a liquid obeying (50.61), (50.65) becomes

$$\overline{\left(\frac{D_e}{D}\right)} = \left[2 \int_0^1 (1 + N_1^2 / 2\tau^2) \, \xi \, d\xi \right]^{1/6}, \tag{50.66}$$

instead of (50.62). We can also derive, instead of (50.65), the following:

$$\overline{\left(\frac{D_e}{D}\right)} = \left[2 \int_0^1 (1 + 3N_1 \varkappa_w/2\tau_w) \, \xi \, d\xi \right]^{1/6}, \tag{50.67}$$

since $\bar{G} \geq 2 \int_0^1 \tau\xi \, d\xi/\varkappa \geq 2\tau_w/3\varkappa_w$.

We now record (see Fig. 50.3) some experimental results of TANNER, which show that[159] the agreement between theory and experiment is good. Note that the 10% increase in die swell is an empirical addition to account for such an increase exhibited by Newtonian liquids,[160] and this seems to be in agreement with the results in Fig. 50.3.

At this stage, we may remark that a calculation similar to that for the capillary jet may be made for a two-dimensional jet issuing out of a channel.

159. [1970 : 23]. See also VLACHOPOULOS, *et al.* [1972 : 17].
160. *Cf.* GAVIS and MODAN [1967 : 8].

If the flow is in the x-direction, with $u = u(y)$, and if one assumes

$$x' = x/\lambda^2 + g(y), \quad y' = \lambda y, \quad z' = \lambda z,$$

where (x', y', z') and (x, y, z) are the coordinates of a particle just before and just after die swell, one can derive that for the ratio h_e/h, where h_e is the

+ 6.8% PIB/CETANE [1970 : 24].
△ GRAESSLEY D-A (TABLE 31[1970 : 13)]

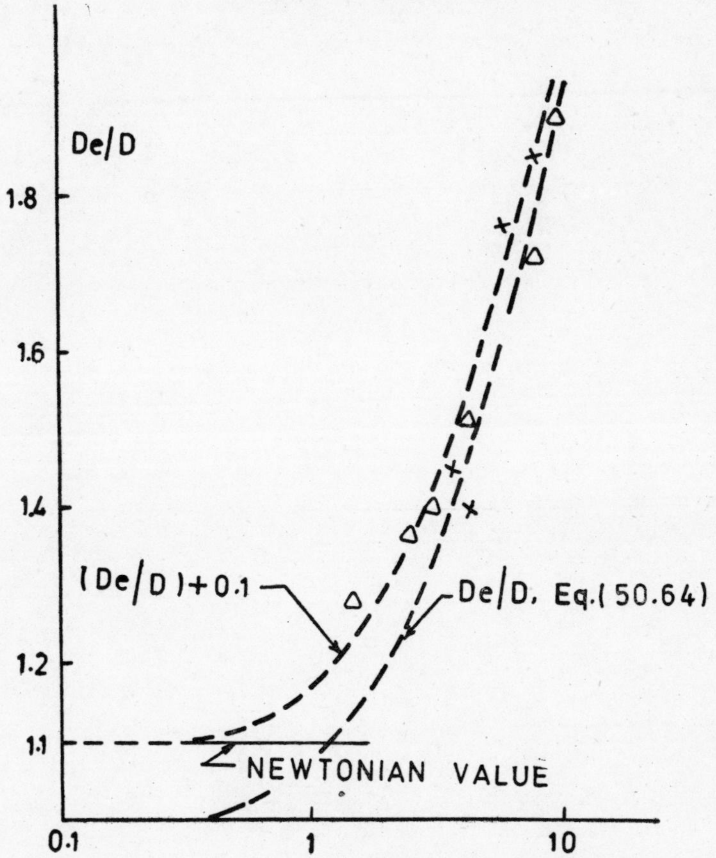

$f = N_{1w}/2\tau_w$

Fig. 50.3. Comparison between theory and experiment.

half-thickness of the sheet outside and h the semi-channel width, the following holds :

$$h_e/h = \left[\int_0^1 (N_1 + G - \tau^2/G)\, d\xi \Big/ \int_0^1 G\, d\xi \right]^{1/6}, \tag{50.68}$$

when it is assumed that

$$\int_0^{h_e} T \langle xx \rangle \, dy = 0. \tag{50.69}$$

As we remarked at the beginning of this section, the methods for determining N_1 and the die swell ratio are not exact. Indeed, it is known[161] that the values of N_1 thus computed are in serious disagreement with those obtained differently. However, the die swell ratios computed by the approximate method yield fairly good predictions. It is thus of interest to see whether in the near future both sets of approximations can be placed on a firmer footing by using the exact theory presented at the beginning of the section[162].

161. GRAESSLEY, *et al.* [1970 : 11].
162. See footnote 141 on p. 329.

REFERENCES

The references are arranged by the year and in alphabetical order under each year. A number in the parenthesis at the end of each reference denotes the section where the work has been cited, except that the symbol * denotes the introduction to a chapter and the letter A or B to the appendix following a chapter.

REFERENCES

[1906]

1. TROUTON, F. T. : Coefficient of viscous traction and its relation to viscosily. *Proc. Roy. Soc. Lond.* A **77**, 426-440. (49)

[1924]

1. WEATHERBURN, C. E. : *Advanced Vector Analysis.* Bell & Sons : London. (3)

[1928]

1. FARROW, F. D., LOWE, G. M. and NEALE, S. M. : The flow of starch pastes at high and low rates of shear. *J. Textile Inst.* **19**, T18-T31. (47)

[1931]

1. DUPONT, Y. : Quelques contributions a la theorie invariantive de l'élasticité. *Bull. Sci. Acad. Belg.* (5)**17**, 1152-1157. (4)

[1934]

1. FINZI, B. : Integrazione delle equazioni indefinite delle meccanica dei sistemi continui. *Rend. Lincei.* (6)**19**, 578-584 ; 620-623. (21)

[1946]

1. WEYL, H. : *Classical Groups.* Princeton Univ. Press : Princeton. (35)

[1948]

1. RIVLIN, R. S. : Large elastic deformations of isotropic materials. I. Fundamental concepts. *Phil. Trans. Roy. Soc. Lond.* A **240**, 459-490. (50)

[1950]

1. OLDROYD, J. G. : On the formulation of rheological equations of state. *Proc. Roy. Soc. Lond.* A **200**, 523-541. (2, 4, 22, 5A, 5B)

[1951]

1. BERKER, R. : Sur certaines propriétés de l'effort qui s'exerce sur une paroi en contact avec un fluide visqueux. *C.R. Acad. Sci. Paris*, **232**, 148-149. (3)
2. LODGE, A. S. : On the use of convected coordinate systems in the mechanics of continuous media. *Proc. Camb. Phil. Soc.* **47**, 575-584. (5A, 5B)

[1952]

1. FERRY, J. D. and WILLIAMS, M. L. : Second approximation methods for determining the relaxation time specturm of a viscoelastic material. J. *Colloid Sci.* **7**, 347-353. (47)

[1953]

1. GREENSMITH, H. W. and RIVLIN, R. S.: The hydrodynamics of non-Newtonian fluids. III. The normal stress effect in high-polymer solutions. *Phil. Trans. Roy. Soc. Lond.* A **245**, 399-428. (48)
2. KRIEGER, I. M. and ELROD, H.: Direct determination of the flow curves of non-Newtonian fluids. II. Shearing rate in the concentric cylinder viscometer. *J. Appl. Phys.* **24**, 134-136. (28)
3. RIVLIN, R. S.: The solution of problems in second order elasticity theory. *J. Ratl. Mech. Anal.* **2**, 53-81. (10*)

[1954]

1. TRUESDELL, C.: *The Kinematics of Vorticity*. Indiana Univ. Press: Bloomington. (3)

[1955]

1. HARMON, D. B. (JR.): Drop sizes from low speed jets. *J. Franklin Inst.* **259**, 519-522. (50)
2. NOLL, W.: On the continuity of the solid and fluid states. *J. Ratl. Mech. Anal.* **4**, 3-81. (4, 12, 14, 5B)
3. RIVLIN, R. S.: Further remarks on the stress-deformation relation for isotropic materials. *J. Ratl. Mech. Anal.* **4**, 681-712. (35)
4. RIVLIN, R. S. and ERICKSEN, J. L.: Stress-deformation relaxations for isotropic matetials. *J. Ratl. Mech. Anal.* **4**, 323-425. (4, 12, 5B, 35)

[1956]

1. ERICKSEN, J. L.: Overdetermination of the speed in rectilinear motion of non-Newtonian fluids. *Quart. Appl. Math.* **14**, 318-321. (30)
2. GREEN, A. E. and RIVLIN, R. S.: Steady flow of non-Newtonian fluids through tubes. *Quart. Appl. Math.* **14**, 299-308. (30)
3. LODGE, A. S.: A network theory of flow birefringence and stress in concentrated polymer solutions. *Trans. Faraday Soc.* **52**, 120-130. (8*)
4. RIVLIN, R. S.: Solution of some problems in the exact theory of visco-elasticity. *J. Ratl. Mech. Anal.* **5**, 179-188. (26, 28)

[1957]

1. GREEN, A. E. and RIVLIN, R. S.: The mechanics of non-linear materials with memory, Part I. *Arch. Ratl. Mech. Anal.* **1**, 1-21. (8*)

[1958]

1. CRIMINALE, W. O. (JR.), ERICKSEN, J. L. and FILBEY, G. L. (JR.): Steady shear flow of non-Newtonian fluids. *Arch. Ratl. Mech. Anal.* **1**, 410-417. (26, 30, 39)
2. ERICKSEN, J. L.: The behavior of certain viscoelastic materials in laminar shearing motions. In: *Viscoelasticity: Phenomenological Aspects*, pp. 77-91, (Ed.) J. T. BERGEN. Academic Press: New York, 1960. (26)
3. NOLL, W.: A mathematical theory of the mechanical behavior of continuous media. *Arch. Ratl. Mech. Anal.* **2**, 197-226. (4, 14, 22, 23, 24, 5B)
4. OLDROYD, J. G.: Non-Newtonian effects in steady motion of some idealized elastico-viscous liquids. *Proc. Roy. Soc. Lond.* A **245**, 278-297. (22, 5A, 8*, 37, 44)
5. SMITH, G. F. and RIVLIN, R. S.: The strain energy function for anisotropic elastic materials. *Trans. Amer. Math. Soc.* **88**, 175-193. (5B)

[1959]

1. COLEMAN, B. D. and NOLL, W.: On certain steady flows of general fluids. *Arch. Ratl. Mech. Anal.* 3, 289-303. (26, 28)
2. COLEMAN, B. D. and NOLL, W.: Helical flow of general fluids. *J. Appl. Phys.* **30**, 1508-1512. Erratum: *op. cit.* **35**, 2276 (1964). (26, 28)
3. ERICKSEN, J. L.: Secondary flow phenomena in non-linear fiuids. *TAPPI*, **42**, 773-775. (44)
4. LANGLOIS, W. E. and RIVLIN, R. S.: Steady flow of slightly viscoelastic fluids. Brown Univ. Div. Appl. Math. *Tech. Rept.* No 3. (8*, 10*)
5. PIPKIN, A. C. and RIVLIN, R. S.: The formulation of constitutive equations in continuum physics, I. *Arch. Ratl. Mech. Anal.* **4**, 129-144. (35)
6. SERRIN. J.: Mathematical principles of classical fluid mechanics. **In**: *Encyl. Phys.*, Vol. VIII, Pt. 1. (Ed.) S. FLÜGGE, pp. 125-263. (33, 7A)
7. SERRIN, J.: POISEUILLE and COUETTE flow of non-Newtonian fluids. *Zeit. angew. Math. Mech.* **39**, 295-299. (44)
8. SPENCER, A. J. M. and RIVLIN, R. S.: The theory of matrix polynomials and its application to the mechanics of isotropic continua. *Arch. Ratl. Mech. Anal.* **2**, 309-336. (35)
9. SPENCER, A. J. M. and RIVLIN, R. S.: Finite integrity bases for five or fewer symmetric 3×3 matrices. *Arch. Ratl. Mech. Anal.* **2**, 435-446. (35)

[1960]

1. COLEMAN, B. D. and NOLL, W.: An approximation theorem for functionals, with applications in continuum mechanics. *Arch. Ratl. Mech. Anal.* **6**, 355-370. (8*, 34, 35)
2. GREEN, A. E. and RIVLIN, R. S.: The mechanics of non-linear materials with memory, Part III. *Arch. Ratl. Mech. Anal.* **4**, 387-404. (34)
3. SMITH, G. F.: On the minimality of integrity bases for symmetric 3×3 matrices. *Arch. Ratl. Mech. Anal.* **5**, 382-389. (35)
4. SPENCER, A. J. M. and RIVLIN, R. S.: Further results in the theory of matrix polynominals. *Arch. Ratl. Mech. Anal.* **4**, 214-230. (35)

[1961]

1. COLEMAN, B. D. and NOLL, W.: Foundations of linear viscoelasticity. *Rev. Mod. Phys.* **33**, 239-249. Erratum: *op. cit.* 36, 1103 (1964). (5, 8*, 34, 34 bis)
2. COLEMAN, B. D. and NOLL, W.: Recent results in the continuum theory of viscoelastic fluids. *Ann. N.Y. Acad. Sci.* **89**, 672-714. (26, 8*)
3. LESLIE, F. M.: The slow flow of a visco-elastic liquid past a sphere, with an appendix by R. I. TANNER. *Quart. J. Mech. Appl. Math.* **14**, 36-48. (47)
4. OLDROYD, J. G.: The hydrodynamics of materials whose rheological properties are complicated. *Rheol. Acta* **1**, 337-344. (37)
5. PHILIPPOFF, W.: Experimental tests of symmetry conditions in laminar flow. *Trans. Soc. Rheol.* **5**, 149-162. (48)
6. WALTERS, K.: The motion of an elastico-viscous liquid contained between coaxial cylinders, III. *Quart. J. Mech. Appl. Math.* **14**, 431-436. (47)

[1962]

1. CASWELL, B. and SCHWARZ, W. H.: The creeping motion of a non-Newtonian fluid past a sphere. *J. Fluid Mech.* **13**, 417-426. (47)

2. COLEMAN, B. D. : Kinematical concepts with applications in the mechanics and thermodynamics of incompressible viscoelastic fluids. *Arch. Ratl. Mech. Anal.* **9**, 273-300. (8, 11)

3. COLEMAN, B. D. and NOLL, W. : Steady extension of incompressible simple fluids. *Phys. Fluids* **5**, 840-843. (7, 10, 33)

4. COLEMAN, B. D. and NOLL, W. : Simple fluids with fading memory. **In :** *Proc. Intl. Sympos. Second-Order Effects*, Haifa, (Ed.) M. REINER and D. ABIR, pp. 530-552. Pergamon Press : New York, 1964. (8*, 34)

5. COURANT, R. and HILBERT, D. : *Methods of Mathematical Physics*, Vol. 2. Interscience : New York. (30)

6. GIESEKUS, H. : Die rheologische Zustandsgleichung elasto-viskosen Flüssigkeiten—insbesondere von Weissenberg-Flüssigkeiten—fur allgemeine und stationäre Fliessvorgänge. *Zeit. angew Math. Mech.* **42**, 32-61. (7A)

7. NOLL, W. : Motions with constant stretch history. *Arch. Ratl. Mech. Anal.* **11**, 97-105. (7, 8, 2A)

8. RIVLIN, R. S. : Constraints on flow invariants due to incompressibility. *Zeit. angew. Math. Phys.* **13**, 589-591. (35)

9. RIVLIN, R. S.,: Second and higher-order theories for the flow of a viscoelastic fluid in a non-circular pipe. **In :** *Proc. Intl. Sympos. Second-Order Effects*, Haifa, (Ed.) M. REINER and D. ABIR, pp. 668-677. Pergamon Press : New York, 1964. (10*, 42)

10. ROSCOE, B. : The end correction for rotation viscometers. *Brit. J. Appl. Phys.* **13**, 362-366. (47)

11. TRUESDELL, C. : Second-order effects in the mechanics of materials. **In :** *Proc. Intl. Sympos. Second-Order Effects*, Haifa, (Ed.) M. REINER and D. ABIR, pp. 1-47. Pergamon Press : New York, 1964. (28)

12. WALTERS, K. : Non-Newtonian effects in some general elastico-viscous liquids. **In :** *Proc. Intl. Sympos. Second-Order Effects*, Haifa, (Ed.) M. REINER and D. ABIR, pp. 507-519. Pergamon Press : New York, 1964. (46)

13. WALTERS, K. : A test of the validity of FERRY's method of reduced variables in the case of certain dilute polymer solutions. *J. Polymer Sci.*, **56**, 449-454. (11*)

[1963]

1. BERNSTEIN, B., KEARSLEY, E. A. and ZAPAS, L. J. : A study of stress relaxation with finite strain. *Trans. Soc. Rheol.*, **7**, 391-410. (8*, 36)

2. CHURCHILL, R. V. : *Fourier Series and Boundary Value Problems.* 2nd Edition. McGraw-Hill ; New York. (34 ter, 39)

3. GIESEKUS, H. : Die simultane Translations—und Rotations—bewegung einer Kugel in einer elastoviskosen Flüssigkeit. *Rheol. Acta* **3**, 59-71. (41, 47)

4. LANGLOIS, W. E. : A recursive approach to the theory of slow, steady-state viscoelastic flow. *Trans. Soc. Rheol.* **7**, 75-99. (10*)

5. LANGLOIS, W. E. : Steady flow of a slightly viscoelastic fluid between rotating spheres. *Quart. Appl. Math.* **21**, 61-71. (10*, 47)

6. LANGLOIS, W. E. and RIVLIN, R. S. : Slow steady state flow of visco-elastic fluids through non-circular tubes. *Rend. Mat. Appl. Univ. Roma. Inst. Naz. Alta Mat.* **22**, 169-185. (10*, 41, 42)

7. PIPKIN, A. C. and RIVLIN, R. S. : Normal stresses in flow through tubes of non-circular cross-section. *Zeit. angew. Math. Phys.* **14**, 738-742. (42)

8. VAN WAZER, J. R., LYONS, J. W., KIM, K. Y. and COLWELL, R. E. : *Viscosity and Flow Measurements : A Laboratory Handbook of Rheology.* Wiley and Sons ; New York. (47)

[1964]

1. ADAMS, N. and LODGE, A. S.: Rheological properties of concentrated polymer solutions. II. A cone-and-plate and parallel-plate pressure distribution apparatus for determining normal stress differences in steady shear flow. *Phil. Trans. Roy. Soc. Lond.* **A256**, 149-184. (48)
2. BEARD, D. W. and WALTERS, K.: Elastico-viscous boundary layer flows I. Two-dimensional flow near a stagnation point. *Proc. Camb. Phil. Soc.*, **60**, 667-674. (11*)
3. BERNSTEIN, B., KEARSLEY, E. A. and ZAPAS, L. J.: Thermodynamics of perfect elastic fluids. *J. Res. Natl. Bur. Stds.* **68B**, 103-113. (36)
4. COLEMAN, B. D.: Thermodynamics of materials with memory. *Arch. Ratl. Mech. Anal.* **17**, 1-46. (6, 36)
5. COLEMAN, B. D.: On thermodynamics, strain impulses and viscoelasticity. *Arch. Ratl. Mech. Anal.* **17**, 230-254. (6)
6. COLEMAN, B. D. and MARKOVITZ, H.: Normal stress effects in second-order fluids. *J. Appl. Phys.* **35**, 1-9. (34 ter, 35)
7. GREEN, A. E.: A continuum theory of anisotropic fluids. *Proc. Camb. Phil. Soc.* **60**, 123-128. (5B)
8. GREEN, A. E.: Anisotropic simple fluids. *Proc. Roy. Soc. Lond.* A **279**, 437-445. (5B).
9. GREEN, A. E. and RIVLIN, R. S.: On CAUCHY's equations of motion. *Zeit. angew. Math. Phys.* **15**, 290-292. (19)
10. HARDY, G. H., LITTLEWOOD, J. E. and POLYA, G.: *Inequalities.* 2nd Edition. Cambridge Univ. Press: Cambridge. (10)
11. KAYE, A. and SAUNDERS, D. W.: A concentric cylinder viscometer for the measurement of flow birefringence and viscosity in concentrated polymer solutions. *J. Sci. Instrum.* **41**, 139-144. (47, 48)
12. LANGLOIS, W. E.: The recursive theory of slow visco-elastic flow applied to three basic problems of hydrodynamics. *Trans. Soc. Rheol.* **8**, 33-60. (10*)
13. LANGLOIS, W. E.: *Slow Viscous Flow.* Macmillan and Co.: New York. (45)
14. LODGE, A. S.: *Elastic Liquids.* Academic Press: New York. (5B, 44, 48, 49, 50)
15. MARKOVITZ, H. and COLEMAN, B. D.: Nonsteady helical flows of second-order fluids. *Phys. Fluids* **7**, 844-851. (46)
16. MARTIN, A. D. and MIZEL, V. J.: A representation theorem of certain nonlinear functionals. *Arch. Ratl. Mech. Anal.* **15**, 353-367. (36)
17. NIILER, P. P. and PIPKIN, A. C.: Finite amplitude shear waves in some non-Newtonian fluids. *Int. J. Engng. Sci.* **2**, 305-315. (37)
18. PIPKIN, A. C.: Small finite deformations of visco-elastic solids. *Rev. Mod. Phys.* **36**, 1034-1041. (34)
19. PIPKIN, A. C.: Alternating flow of non-Newtonian fluids in tubes of arbitrary cross-section. *Arch. Ratl. Mech. Anal.* **15**, 1-13. (46)
20. PIPKIN, A. C.: Annular effect in viscoelastic fluids. *Phys. Fluids* **7**, 1143-1146. (46)
21. REINER, M.: The DEBORAH number. *Physics Today*, **17**(1), 62. (10*)
22. SLATTERY, J. C.: Unsteady relative extension of incompressible simple fluids. *Phys. Fluids* **7**, 1913-1914. (27, 33)
23. SLATTERY, J. C. and SCHOWALTER, W. R.: Effect of surface tension in the measurement of the average normal stress at the exit of a capillary tube through an analysis of the capillary jet. *J. Appl. Polymer Sci.* **8**, 1941-1947. (50)
24. TADJBAKHSH, I. G. and TOUPIN, R. A.: On the equations of finite elastic deformations in deformed coordinates. IBM *Research Paper* RC-1111, January 30. (5A, 5B)
25. TRUESDELL, C.: The natural time of a viscoelastic fluid: its significance and measurement. *Phys. Fluids* **7**, 1134-1142. (10*)

26. TURIAN, R. M.: Thermal Phenomena and non-Newtonian Viscometry. *Ph.D. Thesis*, Dept. of Chemical Engineering, Univ. of Wisconsin. (47)

27. WINEMAN, A. S. and PIPKIN, A. C.: Material symmetry restrictions on constitutive equations. *Arch. Ratl. Mech. Anal.* **17**, 184-214. (8*, 34, 35)

[1965]

1. BALLMAN, R. L.: Extensional flow of polystyrene melt. *Rheol. Acta* **4**, 137-140. (49)
2. BERNSTEIN, B., KEARSLEY, E. and ZAPAS, L.: Elastic stress-strain relations in perfect elastic fluids. *Trans. Soc. Rheol.* **9**, 27-39. (36)
3. CHACON, R. V. and FRIEDMAN, N.: Additive functionals. *Arch. Ratl. Mech. Anal.* **18**, 230-240. (36)
4. COLEMAN, B. D.: Simple liquid crystals. *Arch. Ratl. Mech. Anal.* **20**, 41-58. (5B)
5. COLEMAN, B. D., DUFFIN, R. J. and MIZEL, V. J.: Instability, uniqueness, and nonexistence theorems for the equation $u_t = u_{xx} - u_{xtx}$ on a strip. *Arch. Ratl. Mech. Anal.* **19**, 100-116. (8A)
6. COLEMAN, B. D. and TRUESDELL, C.: Homogeneous motions of incompressible materials. *Zeit. angew. Math. Mech.* **45**, 547-551. (27)
7. GIESEKUS, H.: Sekundärströmungen in viskoelastischen Flüssigkeiten bei stationärer und periodischer Bewegung. *Rheol. Acta.* **4**, 85-101. (47)
8. GIESEKUS, H.: Some secondary flow phenomena in general viscoelastic fluids. **In**: *Proc. Fourth Int. Congr. Rheol.*, Part 1, (Ed.) E. H. LEE, pp. 249-266. Interscience: New York. (47)
9. HAPPEL, J. and BRENNER, H.: *Low Reynolds Number Hydrodynamics.* Prentice-Hall (Englewood Cliffs). (47)
10. HAYES, J. W. and TANNER, R. I.: Measurements of the second normal stress difference in polymer solutions. **In**: *Proc. Fourth. Intl. Congr. Rheol.*, Part **3**, (Ed.) E. H. LEE, pp. 389-399. Interscience: New York. (48)
11. MAXWELL, B. and CHARTOFF, R. P.: Studies of a polymer melt in an orthogonal rheometer. *Trans. Soc. Rheol.* **9**, 41-52. (7, 32)
12. METZNER, A. B. and WHITE, J. L.: Flow behavior of viscoelastic fluids in the inlet region of a channel. *A. I. Ch. E. Journal* **11**, 989-995. (11*)
13. OLDROYD, J. G.: Some steady flows of the general elastico-viscous liquid. *Proc. Roy. Soc. Lond.* A **283**, 115-133. (8, 30, 31, 34)
14. OSAKI, K., TAMURA, M., KURATA, M. and KOTAKA, T.: Complex modulus of concentrated polymer solutions in steady shear. *J. Phys. Chem.* **69**, 4183-4191. (39)
15. PIPKIN, A. C. and RIVLIN, R. S.: Mechanics of rate independent materials. *Zeit. angew. Math. Phys.* **16**, 313-327. (10*)
16. SHERTZER, C. R. and METZNER, A. B.: Measurement of normal stresses in viscoelastic materials at high shear rates. **In**: *Proc. Fourth. Intl. Congr. Rheol.*, Part 2, (Ed.) E. H. LEE, pp. 603-618. Interscience: New York. (50)
17. WANG, C.-C.: Stress relaxation and the principle of fading memory. *Arch. Ratl. Mech. Anal.* **18**, 117-126. (8*, 34)
18. WANG, C.-C.: The principle of fading memory. *Arch. Ratl. Mech. Anal.* **18**, 343-366. (8*, 34, 35)
19. WANG, C.-C.: A general theory of subfluids. *Arch. Ratl. Mech. Anal.* **20**, 1-40. (5B)
20. WANG, C.-C.: A representation theorem for the constitutive equation of a simple material in motions with constant stretch history. *Arch. Ratl. Mech. Anal.* **20**, 329-340. (8, 9, 10, 11, 32, 33)

[1966]

1. AKHEIZER, N. I. and GLAZMAN, I. M.: *Theory of Linear Operators in Hilbert Space*, Vol. I, Trans. by M. NESTELL. Ungar: New York. (34)

2. BERNSTEIN, B. : Time-dependent behavior of an incompressible elastic fluid. Some homogeneous deformation histories. *Acta Mechanica* **2**, 329-354. (36)
3. BOOIJ, H. C. : Influence of superimposed steady shear flow on the dynamic properties of non-Newtonian fluids, I. Measurements on non-Newtonian solutions. *Rheol. Acta* **5**, 215-221. (39)
4. COLEMAN, B. D., MARKOVITZ, H. and NOLL, W. : *Viscometric Flows of Non-Newtonian Fluids.* Springer-Verlag : New York. (26, 44, 11*, 48)
5. COLEMAN, B. D. and MIZEL, V. J. : Breakdown of laminar shearing flows for second-order fluids in channels of critical width. *Zeit. angew. Math. Mech.* **46**, 445-448. (8A)
6. COLEMAN, B. D. and MIZEL, V. J. : Norms and semi-groups in the theory of fading memory. *Arch. Ratl. Mech. Anal.* **23**, 87-123. (35)
7. FRIEDMAN, N. and KATZ, M. : Additive functions on L_p spaces. *Canad. J. Math.* **18**, 1264-1271. (36)
8. GODDARD, J. D. and MILLER, C. : An inverse for the JAUMANN derivative and some applications to the rheology of viscoelastic fluids. *Rheol. Acta* **5**, 177-184. (37)
9. GOREN, S. L. and WRONSKI, S. : The shape of low-speed capillary jets of Newtonian liquids. *J. Fluid Mech.* **25**, 185-198. (50)
10. JACKSON, R. and KAYE, A. : The measurement of the normal stress differences in a liquid undergoing simple shear flow using a cone-and-plate total thrust apparatus only. *Brit. J. Appl. Phys.* **17**, 1355-1360. (48)
11. MACDONALD, I. F. and BIRD, R. B. : Complex modulus of concentrated polymer solutions in steady shear. *J. Phys. Chem.* **70**, 2068-2069. (39)
12. MARTIN, A. D. and MIZEL, V. J. : *Introduction to Linear Algebra.* McGraw-Hill : New York. (1, 8)
13. METZNER, A. B., WHITE, J. L. and DENN, M. M. : Constitutive equations for viscoelastic fluids for short deformation periods and for rapidly changing flows : Significance of the DEBORAH number. *A. I. Ch. E. Journal* **12**, 863-866. (10*)
14. NAKAJIMA, N. and SHIDA, M. : Viscoelastic behavior of polyethylene in capillary flow expressed with three material functions. *Trans. Soc. Rheol.* **10**, 299-316. (50)
15. PIPKIN, A. C. : Approximate constitutive equations. In : *Modern Developments in the Mechanics of Continua,* (Ed.) S. ESKINAZI, pp. 89-108. Academic-Press : New York. (10*, 34, 46)
16. RIVLIN, R. S. : The fundamental equations of nonlinear continuum mechanics. **In** : *Dynamics of Fluids and Plasmas,* (Ed.) S. I. PAI, pp. 83-126. Academic Press : New York. (5B)
17. RUDIN, W. : *Real and Complex Analysis.* McGraw-Hill : New York. (34)
18. SIMMONS, J. M. : A servo-controlled rheometer for measurement of the dynamic modulus of viscoelastic liquids. *J. Sci. Instrum.* **43**, 887-892. (47)
19. TANNER. R. I. : Plane creeping flows of incompressible second-order fluids. *Phys. Fluids* **9**, 1246-1247. (35, 41)
20. WINEMAN, A. S. and PIPKIN, A. C. : Slow viscoelastic flow in tilted troughs. *Acta Mechanica* **2**, 104-115. (10*, 43)

[1967]

1. ADAMS, N. and JACKSON, R. A trifilar-suspension rheogoniometer. *J. Sci. Instrum.* **44**, 461-464. (48)
2. ASTARITA, G. : Two dimensionless groups relevant in the analysis of steady flows of viscoelastic materials. *Ind. Eng. Chem. Fund.* **6**, 257-262. (7, 8, 10*)
3. BATCHELOR, G. K. : *A Introduction to Fluid Dynamics.* Cambridge Univ. Press : Cambridge. (47)
4. BEATTY, M. F. : On the foundation principles of general classical mechanics. *Arch. Ratl. Mech. Anal.* **24**, 264-273. (19)

5. CARROLL, M. M.: Controllable deformations of incompressible simple materials. *Int. J. Engng. Sci.* **5**, 515-525. (34)

6. CASWELL, B.: Kinematics and stress on a surface of rest. *Arch. Ratl. Mech. Anal.* **26**, 385-399. (3, 26 bis)

7. DUDA, J. L. and VRENTAS, J. S.: Fluid mechanics of laminar liquid jets. *Chem. Engng. Sci.* **22**, 855-869. (50)

8. GAVIS, J. and MODAN, M.: Expansion and contraction of jets of Newtonian liquids in air: Effect of tube length. *Phys. Fluids* **10**, 487-497. (50)

9. GIESEKUS, H.: Die Sekundärstromung in einer Kegel-Platte-Anordnung: Abhängigkeit von der Rotationsgeschwindigkeit beiverschiedenen Polymersystemen. *Rheol. Acta* **6**, 339-353. (47)

10. GODDARD, J. D. and MILLER, C.: Nonlinear effects in the rheology of dilute suspensions. *J. Fluid Mech.* **28**, 657-673. (48)

11. GREEN, A. E. and LAWS, N.: On the formulation of constitutive equations in thermodynamical theories of continua. *Quart. J. Mech. Appl. Math.* **29**, 265-275. (8A)

12. GREENBERG, J. M.: The existence of steady shock waves in nonlinear materials with memory. *Arch. Ratl. Mech. Anal.* **24**, 1-21. (13)

13. HILLS, R. N.: On uniqueness of flows for a class of second order fluids. *Mathematika* **17**, 333-337. (8A)

14. JONES. J. R. and WALTERS, T. S.: Flow of elastico-viscous liquids in channels under the influence of a periodic pressure gradient. *Rheol. Acta* **6**, Part 1—240-245; Part 2—330-338. (46)

15. JOSEPH, D. D.: Parameter and domain dependence of eigenvalues of elliptic partial differential equations. *Arch. Ratl. Mech. Anal.* **24**, 325-351. (10*, 44)

16. KUROIWA, S. and NAKAMURA, M.: Influence of superimposed steady shear flow on the dynamic viscosity in polyelectrolyte solutions. *Kobunshi Kagaku* **24**, 807-808. (39)

17. METZNER, A. B.: Behavior of suspended matter in rapidly accelerating viscoelastic fluids; the UEBLER effect. *A. I. Ch. E. Journal* **13**, 316-318. (49)

18. METZNER, A. B. and ASTARITA, G.: External flows of visco-elastic materials: Fluid property restrictions on the use of velocity-sensitive probes. *A. I. Ch. E. Journal* **13**, 550-555. (11*, 49)

19. NORMAN, C. R. and PIPKIN, A. C.: Two problems of transient viscoelastic flow in tubes. *Trans. Soc. Rheol.* **11**, 335-345. (10*, 46)

20. PIPKIN, A. C. and OWEN, D. R.: Nearly viscometric flows, *Phys. Fluids* **10**, 836-843. (9*, 38)

21. PROTTER, M. H. and WEINBERGER, H. F.: *Maximum Principles in Differential Equations.* Prentice-Hall: Englewood Cliffs. (30)

22. ROSCOE, R.: On the rheology of a suspension of viscoelastic spheres in a viscous liquid. *J. Fluid Mech.* **28**, 273-293. (48)

23. TANNER, R. I.: Annular flows for measuring the second normal stress difference. *Trans. Soc. Rheol.* **11**, 347-360. (48)

24. TANNER, K. I. and SIMMONS, J. M.: An instability in some rate-type viscoelastic constitutive equations. *Chem. Eng. Sci.* **22**, 1079-1082. (37)

25. TANNER, R. I. and SIMMONS, J. M.: Combined simple and sinusoidal shearing in elastic liquids. *Chem. Eng. Sci.* **22**, 1803-1815. (49)

[1968]

1. BIRD, R. B. and CARREAU, P. J.: A nonlinear viscoelastic model for polymer solutions and melts—I. *Chem. Eng. Sci.* **23**, 427-434. (36, 47)

2. BIRD, R. B. and HARRIS, E. K. (JR.): Analysis of steady state shearing and stress relaxation in the MAXWELL Orthogonal Rheometer. *A. I. Ch. E. Journal* **14**, 758-761. (7)

3. BOOIJ, H. C.: Influence of superimosed steady shear flow on the dynamic properties of non-Newtonian fluids. III—Measurements on oscillatory normal stress components. *Rheol. Acta* **7**, 202-209. (39)

4. BROADBENT, J. M., KAYE, A., LODGE, A. S. and VALE, D. G.: Possible systematic error in the measurement of normal stress differences in polymer solutions in steady shear flow. *Nature* **217**, 55-56. (48)

5. CHENG, D. C.-H. and DAVIS, J. B.: The geometry of the "cones" used in cone-and-plate viscometers. *Rheol. Acta* **7**, 85-87. (47)

6. COGSWELL, F. N.: The rheology of polymer melts under tension. *Plastics & Polymers* **36**, 109-111. (49)

7. COLEMAN, B. D. and MIZEL, V. J.: On the general theory of fading memory. *Arch. Ratl. Mech. Anal.* **29**, 18-31. (35)

8. FOSDICK, R. L.: Dynamically possible motions of incompressible, isotropic, simple materials. *Arch. Ratl. Mech. Anal* **29**, 272-288. (34)

9. GIESEKUS, H.: Nicht-lineare Effekte beim Stromen viskoelastischer Flussigkeiten durch Schlitz—und Lochdusen. *Rheol. Acta* **7**, 127-138. (47)

10. GURTIN, M. E., MIZEL, V. J. and WILLIAMS, W. O.: A note on CAUCHY's stress theorem. *J. Math. Anal. Appl.* **22**, 398-401. (17, 19)

11. JONES, J. R. and WALTERS, T. S.: A vibrating-rod elasto-viscometer. *Rheol. Acta* **7**, 360-363. (47)

12. KAYE, A., LODGE, A. S. and VALE, D. G.: Determination of normal stress differences in steady shear flow. II-Flow birefringence, viscosity, and normal stress data for a polyisobutene liquid. *Rheol. Acta* **7**, 368-379. (48)

13. KRIEGER, I. M.: Shear rate in the COUETTE viscometer. *Trans. Soc. Rheol.* **12**, 5-11. (47)

14. LIPSON, J. M. and LODGE, A. S.: Determination of normal stress differences in steady shear flow. I—Stability of a polyisobutene liquid. *Rheol. Acta* **7**, 364-368. (11*)

15. MARKOVITZ, H.: Letters to the Editor. *Physics Today* **21** (8), 13-14. (28, 47)

16. MARSH, B. D. and PEARSON, J. R. A.: The measurement of normal-stress differences using a cone-and-plate total thrust appartus. *Rheol. Acta* **7**, 326-331. (48)

17. MAYNE, G.: Etude de l' ecoulement permanent rectiligne des fluids visqueux non-Newtoniens incompressibles. *Bull. Acad. Roy. Be lg. Classe des Sciences, 5e Serie*, **54**, 90-104. (30)

18. OWEN, D. R. and WILLIAMS, W. O.: On the concept of rate independence. *Quart. Appl. Math.* **26**, 321-329. (10*)

19. PETERLIN, A.: Non-Newtonian viscosity and the macromolecule. In: *Advances in Macromolecular Chemistry*, Vol. 1, (Ed.) W. M. PASIKA, pp. 225-281. Academic Press: New York. (47)

20. PIPKIN, A. C.: Small displacements superposed on viscometric flow. *Trans. Soc. Rheol.* **12**, 397-408. (5, 9*,39)

21. PIPKIN, A. C.: Controllable viscometric flows. *Quart. Appl. Math.* **26**, 87-100. (8, 28, 29, 30, 48)

22. SIMMONS, J. M.: Dynamic modulus of polyisobutylene solutions in superposed steady shear flow. *Rheol. Acta* **7**, 184-188. (39, 47, 49)

23. TANNER, R. I.: Note on the iterative calculation of relaxation spectra. *J. Appl. Polymer Sci.* **12**, 1649-1652. (47)

24. TILLETT, J. P. K.: On the laminar flow in a free jet of liquid at high REYNOLDS numbers. *J. Fluid Mech.* **32**, 273-292. (50)

25. WALTERS, K.: *Basic Concepts and Formulae for the Rheogoniometer*. Sangamo Controls Ltd., Bognor Regis: England. (47)

26. WALTERS, K. and KEMP, R. A.: On the use of a rheogoniometer. Part II—Oscilla-

tory shear. In: *Polymer Systems—Deformation and Flow*, (Eds.) R. E. WETTON and R. W. WHORLOW, pp. 237-250. Macmillan: London. (47)

27. WALTERS, K. and KEMP, R. A.: On the use of a rheogoniometer. Part III—Oscillatory shear between parallel-plates. *Rheol. Acta* **7**, 1-8. (47)
28. WALTERS, K. and WATERS, N. D.: On the use of a rheogoniometer. Part I—Steady shear. In: *Polymer Systems—Deformation and Flow*, (Eds.) R. E. WETTON and R. W. WHORLOW, pp. 211-236. Macmillan: London. (47, 48)

[1969]

1. BERNSTEIN, B.: Small shearing oscillations superposed on large steady shear of the BKZ Fluid. *Int. J. Nonlinear Mech.* **4**, 183-199. (39)
2. COGSWELL, F. N.: Tensile deformations in molten polymers. *Rheol. Acta* **8**, 187-194. (49)
3. FOSDICK, R. L. and BERNSTEIN, B.: Nonuniqueness of second-order fluids under steady radial flow in annuli. *Int. J. Engng. Sci.* **7**, 555-569. (41)
4. FRATER, K.: A boundary-layer in an elastico-viscous fluid. *Zeit. angew. Math. Phys.* **20**, 712-721. (11*)
5. GINN, R. F. and METZNER, A. B.: Measurement of stresses developed in steady laminar shearing flows of viscoelastic media. *Trans. Soc. Rheol.* **13**, 429-453. (48)
6. GRIFFITHS, D. F., JONES, D. T. and WALTERS, K.: A flow reversal due to edge effects. *J. Fluid Mech.* **36**, 161-175. (37, 47)
7. HIGHGATE, D. J. and WHORLOW, R. W.: End effects and particle migration effects in concentric cylinder rheometry. *Rheol. Acta* **8**, 142-151. (47).
8. HUILGOL, R R.: On the properties of the motion with constant stretch history occuring in the MAXWELL Rheometer. *Trans. Soc. Rheol.* **13**, 513-526. (7, 32, 34 ter)
9. JANESCHITZ-KRIEGL, H.: Flow birefringence of elastico-viscous polymer systems. *Adv. Polymer Sci.* **6**, 170-318. (48)
10. JONES, T. E. R. and WALTERS, K.: A theory for the balance rheometer. *Brit. J. Appl. Phys.* (*J. Phys. D*) *Ser. 2*, **2**, 815-819. (2, 3, 34 bis, 34 ter)
11. KATAOKA, T. and UEDA, S.: Influence of superimposed steady shear flow on the dynamic properties of polyethylene melts. *J. Polymer Sci. Pt. A-2*, **7**, 475-481. (39)
12. KAYE, A. and VALE, D G.: The shape of a vertically falling stream of a Newtonian liquid. *Rheol. Acta* **8**, 1-5. (49)
13. KRIEGER, I, M.: Computation of shear rate in the COUETTE viscometer. *Proc. 5th Int. Cong. Rheol.*, (Ed.) S. ONOGI. **1**, 511-516. Univ. Park Press: Baltimore. (47)
14. MARKOVITZ, H.: Small deformations superimposed on steady viscometric flows. *Proc. 5th Int. Congr. Rheol.*, (Ed.) S. ONOGI. **1**, 499-510. Univ. Park Press (Baltimore). (39)
15. MEISSNER, J: Rheometer zur Untersuchung der deformations-mechanischen Eigenschaften von Kunststoff-Schmelzen unter definierter Zugbeanspruchung. *Rheol. Acta* **8**, 78-88. (49)
16. METZNER, A. B.: Historical comments on stress relaxation following steady flows through a duct or orifice. *Trans. Soc. Rheol.* **13**, 467-470. (50)
17. METZNER, A. B., UEBLER. E. A. and CHAN MAN FONG, C. F.: Converging flows of viscoelastic materials. *A. I. Ch. E. Journal* **15**, 750-758. (49)
18. PIPKIN, A. C.: Non-linear phenomena in continua. *Tech. Rept. No. 2.* Div. Appl. Math. Brown. Univ. (30, 32, 7A, 34 ter, 10*, 41, 49)
19. POWELL, R. L. and MIDDLEMAN, S.: Decay of normal stresses developed in capillary flows of viscoelastic fluids. *Trans. Soc. Rheol.* **13**, 241-254. (11*, 50)
20. RIVLIN, R. S. and SMITH, G. F.: Orthogonal integrity basis for *N* symmetric matrices. In: *Contributions to Mechanics, Markus Reiner Eightieth Anniversary Volume*, (Ed.) D. ABIR, pp. 121-141. Pergamon Press: Oxford. (35)

21. SEYER, F. A. and METZNER, A. B.: Turbulence phenomena in drag reducing systems. *A. I. Ch. E. Journal* **15**, 426-434. (49)
22. TANNER, R. I. and BALLMAN, R. L.: Prediction of polystyrene melt tensile behavior. *Ind. Eng. Chem. Fund.* **8**, 588-589. (49)
23. TANNER, R. I. and PIPKIN, A. C.: Intrinsic errors in pressure-hole measurements. *Trans. Soc. Rheol.* **13**, 471-484. (10*, 41)
24. YIN, W.-L.: Kinematics of Viscometric Flows. *Ph. D. Thesis.* Div. Appl. Math. Brown University (Providence). (29)

[1970]

1. ABBOTT, T. N. G. and WALTERS, K.: Rheometrical flow systems. Part 2. Theory for the orthogonal rheometer, including an exact solution of the NAVIER-STOKES equations. *J. Fluid Mech.* **40**, 205-213. (32, 8*, 34 ter, 10*, 45)
2. ABBOTT, T. N. G. and WALTERS, K.: Rheometrical flow systems. Part 3. Flow between rotating eccentric cylinders. *J. Fluid Mech.* **43**, 257-267. (2, 3, 34 bis, 34 ter, 45)
3. BAGLEY, E. B. and DUFFEY, H. J.: Recoverable shear strain and the BARUS effect in polymer extrusion. *Trans. Soc. Rheol.* **14**, 545-553. (50)
4. BOGUE, D. C. and WHITE, J. L.: *Engineering Analysis of Non-Newtonian Fluids.* AGARDograph No. 144. Available from Natl. Aero. Space Admn., July. (36)
5. BOOIJ, H. C.: Effect of Superimposed Steady Shear Flow on Dynamic Properties of Polymeric Fluids. *Ph. D. thesis.* University of Leiden, Netherlands. (39)
6. CASWELL, B.: The effect of finite boundaries on the motion of particles in non-Newtonian fluids. *Chem. Eng. Sci.* **25**, 1167-1176. (47)
7. COLEMAN, B. D., DILL, E. H. and TOUPIN, R. A.: A phenomenological theory of streaming birefringence. *Arch. Ratl. Mech. Anal.* **39**, 358-399. (48)
8. COWSLEY, C. W.: Improvements to total thrust methods for the measurement of sceond normal stress differences. Univ. Cambridge Polymer Processing Research Centre *Rept. No. 6.* (48)
9. FERRY, J. D.: *Viscoelastic Properties of Polymers.* 2nd Ed. Wiley and Sons: New York. (11*, 47)
10. FRATER, K. R.: On the solution of some boundary-value problems arising in elastico-viscous fluid mechanics. *Zeit. angew. Math. Phys.* **21**, 134-317. (11*)
11. GRAESSLEY, W. W., GLASSCOCK, S. D. and CRAWLEY, R. L.: Die swell in molten polymers. *Trans. Soc. Rheol.* **14**, 519-544. (50)
12. GRIFFITHS, D. F. and WALTERS, K.: On edge effects in rheometry. *J. Fluid Mech.* **42**, 379-399. (47, 48)
13. HARRIS, J.: Dynamic streaming birefringence. *Rheol. Acta* **9**, 467-473. (48)
14. HUILGOL, R. R.: Relations between certain non-viscometric and viscometric material functions. *Trans. Soc. Rheol.* **14**, 425-437 (9*, 40)
15. KEARSLEY, E. A.: Intrinsic errors for pressure measurements in a slot along a flow. *Trans. Soc. Rheol.* **14**, 419-424. (10*, 42, 48)
16. KING, M. J. and WATERS, N. D.: The effect of secondary flows in the use of a rheogoniometer. *Rheol. Acta* **9**, 164-170. (47)
17. METZNER, A. B. and METZNER, A. P.: Stress levels in rapid extensional flows of polymeric fluids. *Rheol. Acta* **9**, 174-181. (49)
18. PRITCHARD, W. G.: The measurement of normal stresses by means of liquid-filled holes in a surface. *Rheol. Acta* **9**, 200-207. (48)
19. RICHARDSON, S.: A 'stick-slip' problem related to the motion of a free jet at low REYNOLDS numbers. *Proc. Camb. Phil. Soc.* **67**, 477-489. (50)
20. RICHARDSON, S.: The die swell phenomenon. *Rheol. Acta* **9**, 193-199. (50)
21. RIVLIN, R. S.: Red herrings and sundry unidentified fish in nonlinear continuum

mechanics. In: *Inelastic Behavior of Soilds*, (Ed.) M. F. KANNINEN, *et al.*, pp. 117-134. McGraw-Hill: New York. (5B)

22. TANNER, R. I.: Some methods for estimating the normal stress functions in viscometric flows. *Trans. Soc. Rheol.* **14**, 483-570. (30, 10*, 43, 48, 50)
23. TANNBR, R. I.: A theory of die swell, *J. Polymer Sci. Pt. A-2*, **8**, 2067-2078. (52)
24. TANNER, R. I. and WILLIAMS, G.: Iterative numerical method for some integral equations arising in rheology. *Trans. Soc. Rheol.* **14**, 19-38. (47)
25. VINOGRADOV, G. V., FIKHMAN, V D., RADUSHKEVICH, B. V. and MALKIN, A. YA.: Viscoelastic and relaxation properties of a polystyrene melt in axial extension. *J. Polymer Sci. Pt. A-2*, **8**, 657-678. (49)
26. VINDOGRADOV, G. V., RADUSHKEVICH, B. V. and FIKHMAN, V. D.: Extension of elastis liquids. Polyisobutylene. *J. Polymer Sci. Pt. A-2*, **8**, 1-17. (49)
27. WALTERS, K.: Rheometrical flow systems. Part 1. Flow between concentric spheres rotating about different axes. *J. Fluid Mech.* **40**, 191-203. (34 bis)
28. WALTERS, K.: On a boundry-layer controversy. *Zeit. angew. Math. Phys.* **21**, 276-281. (11*)
29. WALTERS, K. Relation between COLEMAN-NOLL, RIVLIN-ERICKEN, GREEN-RIVLIN and OLDROYD fluids. *Zeit. angew. Math. Phys.* **21**, 592-600. (37)
30. WALTERS, K. and JONES, T. E. R.: Further studies on the usefulness of the WEISSENBERG rheogoniometer. *Proc. 5th Int. Congr. Rheol.*, (Ed.) S. ONOGI. **4**, 337-350. Univ. Park Press: Baltimore. (30)
31. YIN, W. L. and PIPKIN, A. C.: Kinematics of viscometric flow. *Arch. Ratl. Mech. Anal.* **37**, 111-135. (8, 27, 29, 30)

[1971]

1. ABBOTT, T. N. G., BOWEN, G. W. and WALTERS, K.: Some suggestions for new rheometer design. 1. Theory. *J. Phys. D: Appl Phys.* **4**, 190-203. (8*, 34 ter, 45)
2. BARNES, H. A., TOWNSEND, P. and WALTERS, K.: On pulsatile flow of non-Newtonian liquids. *Rheol. Acta* **10**, 517-527. (46)
3. BERNSTEIN. B.: Steady flows in history format. *Illinois Inst. Tech. Rept.* (13)
4. BERNSTEIN, B. and HUILGOL, R. R.: On the ultrasonic dynamic viscosities in superposed oscillatory shear. *Trans. Soc. Rheol.* **15**, 731-739. (9*, 39)
5. BIERMANN, M.: Foundations of extensional viscometry. Part I: Prolegomena on motion groups encompassing viscometric motions. *Acta Mechanica* **11**, 293-298; Part II: Essentials as to design and analysis of hose viscometers. *Ibid.* **12**, 155-174; Part III: Analysis of combined extensional and squeezing motions in the open-end type hose viscometer. *Ibid.* **12**, 209-222. (1, 49)
6. BROADBENT, J. M. and WALTERS, K.: Some suggestions for new rheometer designs. II Interpretation of experimental results. *J. Phys. D.: Appl. Phys.* **4**, 1863-1879. (8*, 10*, 45)
7. CYGAN, D. A. and CASWELL, B.: Precision falling sphere viscometry. *Trans. Soc. Rheol.* **15**, 663-683. (47)
8. COLEMAN, B. D: On retardation theorems. *Arch. Ratl. Mech. Anal.* **43**, 1-23 (35)
9. DENN, M. M. and MARRUCCI, G.: Stretching of viscoelastic liquids. *A. I. Ch. E. Journal* **17**, 101-103. (49)
10. HUANG, C. R.: Determination of the shear rates of non-Newtonian fluids from rotational viscoelastic data. I. Concentric cylinder viscometer. *Trans. Soc. Rheol.* **15**, 25-30; see also: II. Cone-and-plate viscometer. *Ibid.* **15**, 31-37. (28)
11. HUH, C. and SCRIVEN, L. E.: Hydrodynamic model of steady movement of a solid/liquid/fluid contact line. *J. Colloid and Interface Science* **35**, 85-101. (44)
12. HUILGOL, R. R.: A class of motions with constant stretch history. *Quart. Appl. Math.* **29**, 1-15. (7, 8, 10, 13, 31, 32)

13. JONES, J. R.: Remarks on near-rigid rotations of elastico-viscous liquids. *Rheol. Acta* **10**, 451-456. (45)
14. JONES, T. E. R. and WALTERS, K.: The behaviour of materials under combined steady and oscillatory shear. *J. Phys. A : Gen. Phys.* **4**, 85-100. (39, 47)
15. JONES, T. E. R. and WALTERS, K.: An interpretation of discontinuities occurring in dynamic testing near the natural frequency. *Rheol. Acta* **10**, 365-367. (47)
16. LANGLOIS, W. E.: An elementary proof that the undetermined stress in an incompressible fluid is of the form —*p*1. *Amer. J. Phys.* **39**, 641-642. (25)
17. LODGE, A. S.: Determination of normal stress differences and hole pressure errors in shear flow using a plate- and-truncated-cone apparatus. *Rheol. Acta* **10**, 554-556. (48)
18. MEISSNER, J.: Dehnungsverhalten von Polyathylen-Schmelzen. *Rheol. Acta* **10**, 230-242. (49)
19. METZNER, A. B.: Extensional primary field approximations for viscoelastic media. *Rheol. Acta* **10**, 434-444. (49)
20. PRITCHARD, W. G.: Measurements of the viscometric functions for a fluid in steady shear flows. *Phil. Trans. Roy. Soc. Lond.*, A **270**, 507-556. (48)
21. RIVLIN, R. S. and SAWYERS, K. N.: Nonlinear continuum mechanics of viscoelastic fluids. *Annual Rev. of Fluid Mechanics* **3**, 117-146. (5B, 8*, 36)
22. RIVLIN, R. S. and SMITH, G. F.: Birefringence in viscoelastic materials. *Zeit. angew. Math. Phys.* **22**, 325-339. (48)
23. SHROFF, R. N.: Dynamic mechanical properties of polyethylene melts: calculation of relaxation spectrum from loss modulus. *Trans. Soc. Rheol.* **15**, 163-175. (47)
24. SPENCER, A. J. M.: Theory of invariants. In: *Continuum Physics*, Vol. I, (Ed.) A. C. ERINGEN, pp. 239-353. Academic Press: New York. (35)
25. STEVENSON, J. F. and BIRD, R. B.: Elongational viscosity of nonlinear elastic dumbbell suspensions. *Trans. Soc. Rheol.* **15**, 135-145. (49)
26. STIPPES, M.: Flux functions and balance laws. *J. Elasticity* **1**, 175-177. (17)
27. SUBBARAMAN, V., MASHELKAR, R. A. and ULBRECHT, J.: Extrapolation procedures for zero shear viscosity with a falling sphere viscometer. *Rheol. Acta* **10**, 429-433. (47)
28. TANNER, R. I. and WILLIAMS, G.: On the orthogonal superposition of simple shearing and small-strain oscillatory motions. *Rheol. Acta* **10**, 528-538. (39, 47)
29. WATERS, N. D. and KING, M. J.: The unsteady flow of an elastico-viscous liquid in a straight pipe of circular cross section. *J. Phys. D : Appl. Phys.* **4**, 204-211. (10*, 46)
30. YAMAMOTO, M.: Rate-dependent relaxation spectra and their determination. *Trans. Soc. Rheol.* **15**, 331-344. (47).
31. ZAHORSKI, S.: Flows with constant stretch history and extensional viscosity. *Arch. Mech. Stos.* **23**, 433-445. (33)

[1972]

1. BERNSTEIN, B.: A rheological relation between parallel and transverse superposed complex dynamic shear moduli. *Rheol. Acta* **11**, 210-215. (39)
2. BERNSTEIN, B., HUILGOL, R. R. and TANNER, R. I.: Certain asymptotic relations for the dynamic moduli in superposed oscillatory shear. *Int. J. Engng. Sci.* **10**, 263-272. (9*, 39)
3. CASWELL, B.: The stability of particle motion near a wall in Newtonian and non-Newtonian fluids. *Chem. Eng. Sci.* **27**, 373-389. (47)
4. COGSWELL, F. N.: Measuring the extensional rheology of polymer melts. *Trans. Soc, Rheol.* **16**, 383-403. (49)
5. GURTIN, M. E.: The Linear Theory of Elasticity. In: *Encyclopedia of Physics*. Vol. VI a/2, (Ed.) C. TRUESDELL, pp. 1-295. Springer-Verlag; New York. (17)
6. HIGASHITANI, K. and PRITCHARD, W. G.: A kinematic calculation of intrinsic errors in pressure measurements made with holes. *Trans. Soc. Rheol.* **16**, 687-696. (48)

7. KING, M. J. and WATERS, N. D.: The unsteady motion of a sphere in an elastico-viscous liquid. *J. Phys. D: Appl. Phys.* **5**, 141-510. (47)
8. MASHELKAR, R. A., KALE, D. D., KELKAR, J. V. and ULBRECHT, J.: Determination of material parameters of viscoelastic fluids by rotational non-viscometric flows. *Chem. Eng. Sci.* **27**, 973-985. (47)
9. MEISSNER, J.: Development of a universal extensional rheometer for the uniaxial extension of polymer melts. *Trans. Soc. Rheol.* **16**, 405-420. (49)
10. MILLER, M. J. and CHRISTIANSEN, E. B.: The stress state of elastic fluids in viscometric flow. *A. I. Ch. E. Journal*, **18**, 600-608. (48)
11. OGDEN, R. W.: Large deformation isotropic elasticity—on the correlation of theory and experiment for incompressible rubberlike solids. *Proc. Roy. Soc. Lond.* A **326**, 565-584. (50)
12. OLABISI, O. and WILLIAMS, M. C.: Secondary and primary normal stresses, hole error, and reservoir edge effects in cone-and-plate flow of polymer solutions. *Trans. Soc. Rheol.* **16**, 727-759. (48)
13. PIPKIN, A. C.: *Lectures on Viscoelasticity Theory.* Springer-Verlag: New York. (10*, 41)
14. STEVENSON, J. F.: Elongational flow of polymer melts. *A. I. Ch. E. Journal* **18**, 540-547. (49)
15. TURIAN, R. M.: Perturbation solution of the steady Newtonian flow in the cone-and-plate and parallel plate systems. *Ind. Eng. Chem. Fund.* **11**, 361-368. (47)
16. VAN ES, H. E.: A new method for determining the second normal stress difference in viscoelastic fluids. Koninklijke/Shell-Laboratorium, Amsterdam, *Report.*
17. VLACHOPOULOS, J., HORIE, M. and LIDORIKIS, S.: An evaluation of expressions predicting die swell. *Trans. Soc. Rheol.* **16**, 669-685. (50)

[1973]

1. FOSDICK, R. L. and SERRIN, J.: Rectilinear steady flows of simple fluids. *Proc. Roy. Soc. Lond.* A **332**, 331-333. (30)
2. HUILGOL, R. R.: On uniqueness and nonuniqueness in the plane creeping flows of second order fluids. *SIAM J. Appl. Math.* **24**, 226-233. (41)
3. HUILGOL, R. R.: On a characterization of simple extensional flows. *Rheol. Acta* **14** (1975). (11)
4. HUILGOL, R. R.: On generalized torsional flow. *Trans. Soc. Rheol.* **18**, 191-198 (1974). (48)
5. HUILGOL, R. R.: On the concept of the DEBORAH number, *Trans. Soc. Rheol.* **19** (1975). (10*)
6. JOSEPH, D. D.: Die Swell I: The final diameter of a capillary jet. Dept. of Aerospace Engrg. and Engrg. Mech. Univ. Minnesota, *Report.* Feb. 1. (50)
7. JOSEPH, D. D., BEAVERS, G. S. and FOSDICK, R. L.: The free surface on a liquid between cylinders rotating at different speeds. Part II. *Arch. Ratl. Mech. Anal.* **49**, 381-401. (10*, 44)
8. JOSEPH, D. D. and FOSDICK, R. L.: The free surface on a liquid between cylinders rotating at different speeds. Part I. *Arch. Ratl. Mech. Anal.* **49**, 321-380. (10*, 44).
9. KUNDU, P. J.: Normal stresses and WEISSENBERG effect. *Trans. Soc. Rheol.* **17**, 343-349 (1973). (44)
10. NOVOTNY, E. J., (JR.) and ECKERT, R. E.: Direct measurement of hole error for viscoelastic fluids in flow between infinite parallel plates. *Trans. Soc. Rheol.* **17**, 227-241 (1973). (48)
11. PIPKIN, A. C. and TANNER, R. I.: A survey of theory and experiment in viscometric flows of viscoelastic liquids. In: *Mechanics Today*, Vol. 1 (1972), (Ed.) S. NEMAT-NASSER, pp. 262-321. Pergamon: Oxford, 1974. (48).

12. SALATHE, E. P. and SAWYERS, K. N.: On intrinsic pressure hole errors for viscoelastic fluids. *Trans. Soc. Rheol.* **17**, 243-253. (48)
13. TANNER, R. I.: A correlation of normal stress data for polyisobutylene solutions. *Trans. Soc. Rheol.* **17**, 365-373. (48)
14. TRUESDELL, C.: The meaning of viscometry in fluid dynamics. *Annual Review of Fluid Mechanics* **6**, 111-146 (1974). (44)
15. VAN ES, H. E. and CHRISTENSEN, R. M.: A critical test for a class of nonlinear constitutive equations. *Trans. Soc. Rheol.* **17**, 325-330. (48)

[1974]

1. BERNSTEIN, B. and HUILGOL, R. R.: A note on ultrasonic dynamic moduli. *Trans. Soc. Rheol.* **18**, 583-590 (1974). (39)
2. JOSEPH, D. D.: Slow motion and viscometric motion; stability and bifurcation of the rest state of simple fluid. *Arch. Ratl. Mech. Anal.* **56**, 99-157. (48, 50)
3. MACDONALD, I. F.: On high-frequency behavior in superposed flow. *Trans. Soc. Rheol.* **18**, 299-312. (39)
4. SAKAI, M., KAJIURA, H. and NAGASAWA, M.: On the rate-dependent memory function. *Trans. Soc. Rheol.* **18**, 323-327. (48)

ADDITIONAL REFERENCES

1968 A. LEIGH, D. C.: Asymptotic constitutive approximations for rapid deformations of viscoelastic materials. *Acta Mechanica* **5**, 274-288. (34)

1972 A. LODGE, A. S.: On the description of rheological properties of viscoelastic continua I. Body-, space-, and Cartesian-space-tensor fields. *Rheol. Acta* **11**, 106-118. (58)

B. LODGE, A. S. and STARK, J. H,: On the description of rheological properties of viscoelastic continua II. Proof that OLDROYD'S 1950 formalism includes all 'simple fluids'. *Rheol. Acta* **11**, 119-126. (58)

1973 A. JONES, W. M., DAVIES, D. M. and THOMAS, M. C.: TAYLOR vortices and the evaluation of material constants: a critical assessment. *J. Fluid Mech.* **60**, 19-41. (43)

B. KAYE, A.: The shape of a liquid surface between rotating concentric cylinders. *Rheol. Acta* **12**, 206-211. (44)

C. WALES, J. L. S. and PHILIPPOFF, W.: The anisotropy of simple shearing flow. *Rheol. Acta* **12**, 25-34. (48)

1974 A. HUILGOL, R. R.: On objectivity, symmetry and extension principles. *Proc. Confce. Symmetry, Similarity and Group Theoretic Methods in Mechanics,* Calgary, (Ed.) P. G. GLOCKNER and M. C. SINGH, pp. 101-106. (5B)

B. LOBO, P. F. and OSMERS, H. R.: The First Normal Stress Function at High Shear Rates Using Pressure-driven Angular Annular Flow. Dept. of Chemical Engrg., Univ. Rochester *Rept.* Oct. (29, 48)

C. BÖHME, G.: Das Konzept von Primär—und Sekundärströmung bei der analytischen Berechnung der Bewegung nicht Newton'scher Fluide. *Methoden und Verfahren der Math. Phys.* **11**, 51-70. (44)

D. BÖHME, G.: On secondary flow phenomena in viscoelastic fluids near free boundaries. *Arch. Mech. Stos.* **26**, 729-743. (44)

E. BÖHME, G.: *Eine Theorie für sekundäre Strömungserscheinungen in nicht-Newtonschen Fluiden.* Deutsche Luft-und Raumfahrt Forschungsbericht 74-24. Tech. Hochschule Darmstadt. March. (44)

1975 A. HUILGOL, R. R.: *Upper and Lower Bounds for Die Swell.* School of Math. Sci., Flinders Univ. of South Australia, Bedford Park, April. (50)

WORK TOO RECENT TO BE CITED

1974 M. MCLEOD, J. B.: Over-determined systems and the rectilinear steady flow of simple fluids. In: *Proc. Confce. on Ordinary and Partial Differential Equations,* (Ed.) B. D. SLEEMAN and I. D. MICHAEL, pp. 193-204. Lecture Notes in Math. No. 415, Springer-Verlag. (30).

SUBJECT INDEX

SUBJECT INDEX

Stress flux 87
Stress power 97, 334
Stress relaxation 167, 168, 192
Stress tensor :
  body 103
  CAUCHY 78
  PIOLA-KIRCHHOFF 83
Stretch history, constant, *see* MWCSH
Subfluids 106
Superposed flow 208 *ff.*

T